Springer Handbook
of Electronic and Photonic Materials

Safa Kasap, Peter Capper (Eds.)

Springer Handbook of Electronic and Photonic Materials
Organization of the Handbook

Each chapter has a concise summary that provides a general overview of the subject in the chapter in a clear language. The chapters begin at fundamentals and build up towards advanced concepts and applications. Emphasis is on physical concepts rather than extensive mathematical derivations. Each chapter is full of clear color illustrations that convey the concepts and make the subject matter enjoyable to read and understand. Examples in the chapters have practical applications. Chapters also have numerous extremely useful tables that summarize equations, experimental techniques, and most importantly, properties of various materials. The chapters have been divided into five parts. Each part has chapters that form a coherent treatment of a given area. For example,

Part A contains chapters starting from basic concepts and build up to up-to-date knowledge in a logical easy to follow sequence. Part A would be equivalent to a graduate level treatise that starts from basic structural properties to go onto electrical, dielectric, optical, and magnetic properties. Each chapter starts by assuming someone who has completed a degree in physics, chemistry, engineering, or materials science.

Part A Fundamental Properties
2 Electrical Conduction in Metals and Semiconductors
3 Optical Properties of Electronic Materials: Fundamentals and Characterization
4 Magnetic Properties of Electronic Materials
5 Defects in Monocrystalline Silicon
6 Diffusion in Semiconductors
7 Photoconductivity in Materials Research
8 Electronic Properties of Semiconductor Interfaces
9 Charge Transport in Disordered Materials
10 Dielectric Response
11 Ionic Conduction and Applications

Part B provides a clear overview of bulk and single-crystal growth, growth techniques (epitaxial crystal growth: LPE, MOVPE, MBE), and the structural, chemical, electrical and thermal characterization of materials. Silicon and II–VI compounds and semiconductors are especially emphasized.

Part B Growth and Characterization
12 Bulk Crystal Growth-Methods and Materials
13 Single-Crystal Silicon: Growth and Properties
14 Epitaxial Crystal Growth: Methods and Materials
15 Narrow-Bandgap II–VI Semiconductors: Growth
16 Wide-Bandgap II–VI Semiconductors: Growth and Properties
17 Structural Characterization
18 Surface Chemical Analysis
19 Thermal Properties and Thermal Analysis: Fundamentals, Experimental Techniques and Applications
20 Electrical Characterization of Semiconductor Materials and Devices

Part C covers specific materials such as crystalline Si, microcrystalline Si, GaAs, high-temperature semiconductors, amorphous semiconductors, ferroelectric materials, and thin and thick films.

Part C Materials for Electronics
21 Single-Crystal Silicon: Electrical and Optical Properties
22 Silicon-Germanium: Properties, Growth and Applications
23 Gallium Arsenide
24 High-Temperature Electronic Materials: Silicon Carbide and Diamond
25 Amorphous Semiconductors: Structure, Optical, and Electrical Properties
26 Amorphous and Microcrystalline Silicon
27 Ferroelectric Materials
28 Dielectric Materials for Microelectronics
29 Thin Films
30 Thick Films

Part D examines materials that have applications in optoelectronics and photonics. It covers some of the state-of-the-art developments in optoelectronic materials, and covers III–V Ternaries, III–Nitrides, II–VI compounds, quantum wells, photonic crystals, glasses for photonics, nonlinear photonic glasses, nonlinear organic, and luminescent materials.

Part D Materials for Optoelectronics and Photonics
31 III–V Ternary and Quaternary Compounds
32 Group III Nitrides
33 Electron Transport within the III–V Nitride Semiconductors, GaN, AlN, and InN: A Monte Carlo Analysis
34 II–IV Semiconductors for Optoelectronics: CdS, CdSe, CdTe
35 Doping Aspects of Zn-Based Wide-Band-Gap Semiconductors
36 II–VI Narrow-Bandgap Semiconductors for Optoelectronics
37 Optoelectronic Devices and Materials
38 Liquid Crystals
39 Organic Photoconductors
40 Luminescent Materials
41 Nano-Engineered Tunable Photonic Crystals in the Near-IR and Visible Electromagnetic Spectrum
42 Quantum Wells, Superlattices, and Band-Gap Engineering
43 Glasses for Photonic Integration
44 Optical Nonlinearity in Photonic Glasses
45 Nonlinear Optoelectronic Materials

Part E provides a survey on novel materials and applications such as information recording devices (CD, video, DVD) as well as phase-change optical recording. The chapters also include applications such as solar cells, sensors, photoconductors, and carbon nanotubes. Both ends of the spectrum from research to applications are represented in chapters on molecular electronics and packaging materials.

Part E Novel Materials and Selected Applications
46 Solar Cells and Photovoltaics
47 Silicon on Mechanically Flexible Substrates for Large-Area Electronics
48 Photoconductors for X-Ray Image Detectors
49 Phase-Change Optical Recording
50 Carbon Nanotubes and Bucky Materials
51 Magnetic Information-Storage Materials
52 High-Temperature Superconductors
53 Molecular Electronics
54 Organic Materials for Chemical Sensing
55 Packaging Materials

Glossary of Defining Terms There is a glossary of *Defining Terms* at the end of the handbook that covers important terms that are used throughout the handbook. The terms have been defined to be clear and understandable by an average reader not directly working in the field.

使 用 说 明

1.《电子与光子材料手册》原版为一册，分为A、B、C、D、E五部分。考虑到各部分内容相对独立完整，为使用方便，影印版按部分分为5册。

2.各册在页脚重新编排页码，该页码对应中文目录。保留了原书页眉及页码，其页码对应原书目录及主题索引。

3.各册均有完整5册书的内容简介。

4.作者及其联系方式、缩略语表各册均完整呈现。

5.名词术语表、主题索引安排在第5册。

6.文前页基本采用中英文对照形式，方便读者快速浏览。

材料科学与工程图书工作室

联系电话　0451-86412421
　　　　　0451-86414559

邮　　箱　yh_bj@yahoo.com.cn
　　　　　xuyaying81823@gmail.com
　　　　　zhxh6414559@yahoo.com.cn

Springer 手册精选系列

电子与光子材料手册

电子材料

【第3册】

Springer
Handbook of
Electronic
and Photonic
Materials

﹝加拿大﹞Safa Kasap
﹝英　国﹞Peter Capper　主编

（影印版）

哈尔滨工业大学出版社
HARBIN INSTITUTE OF TECHNOLOGY PRESS

黑版贸审字08-2012-031号

Reprint from English language edition:
Springer Handbook of Electronic and Photonic Materials
by Safa Kasap and Peter Capper
Copyright © 2007 Springer US
Springer US is a part of Springer Science+Business Media
All Rights Reserved

This reprint has been authorized by Springer Science & Business Media for distribution in China Mainland only and not for export there from.

图书在版编目（CIP）数据

电子与光子材料手册. 第3册，电子材料=Handbook of Electronic and Photonic Materials Ⅲ Materials for Electronics：英文／（加）卡萨普（Kasap S.），（英）卡珀（Capper P.）主编. —影印本. —哈尔滨：哈尔滨工业大学出版社，2013.1

（Springer手册精选系列）
ISBN 978-7-5603-3762-3

Ⅰ. ①电… Ⅱ. ①卡…②卡… Ⅲ. ①电子材料-手册-英文②光学材料-手册-英文 Ⅳ. ①TN04-62②TB34-62

中国版本图书馆CIP数据核字（2012）第190658号

责任编辑	杨　桦　张秀华
出版发行	哈尔滨工业大学出版社
社　　址	哈尔滨市南岗区复华四道街10号 邮编150006
传　　真	0451-86414749
网　　址	http://hitpress.hit.edu.cn
印　　刷	哈尔滨市石桥印务有限公司
开　　本	787mm×960mm 1/16 印张20.5
版　　次	2013年1月第1版　2013年1月第1次印刷
书　　号	ISBN 978-7-5603-3762-3
定　　价	58.00元

（如因印刷质量问题影响阅读，我社负责调换）

序 言

本书的编辑、作者、出版人都将庆祝这本卓著书籍的出版,这对于电子与光子材料领域的工作者也将是无法衡量的好消息。从以往编辑的系列手册看,我认为本书的出版是值得的,坚持出版这样一本书也是必要的。本书之所以显得特别重要,是因为它在这个领域,内容覆盖范围广泛,涉及的方法也是当今的最新研究进展。在这样一个迅速发展的领域,这是一个相当大的挑战,它已经赢得了人们的敬意。

早期的手册和百科全书也都注重阐述半导体材料的发展趋势,而且必须覆盖半导体材料广泛的研究范围和所涉及的现象。这是可以理解的,原因在于半导体材料在电子领域中的主导地位。但没有多少人有足够的勇气预测未来的发展趋势。1992年,Mahajan和Kimerling在其《简明半导体材料百科全书和相关技术》一书的引言中做了尝试,并且预测未来的挑战将是纳米电子领域、低位错密度的III-V族衬底技术、半绝缘III-V族衬底技术、III-V族图形外延技术、替换电介质和硅接触技术、离子注入和扩散技术的发展。这些预测或多或少地成为了现实,但是这也同样说明做出这样的预测是多么的困难。

十年前没有多少人会想到III族氮化物在这本书中将成为重要的部分。与制备相关的问题是,作为高熔点材料,在受欢迎的能在光谱蓝端作光发射器的材料中它们的熔点并不高。这是一个很有意思的话题,至少与解决早期光谱红端的固体激光器工作寿命短的问题一样有趣。总地说来,光电子学和光子学在前十年中已经呈现出一些令人瞩目的研究进展,这些在本书中得到了体现,范围从可见光发光器件材料到红外线材料。书中Part D的内容范围很宽,包括III-V族和II-VI族光电子材料和能带隙工程,以及光子玻璃、液态晶体、有机光电导体和光子晶体的新领域。整个部分反映了材料的光产生、工艺、光传输和光探测,包括所有用光取代电子的必要内容。

在电子材料这一章(Part C)探讨了硅的进展。毋庸置疑地是,硅是占据了电子功能和电子电路整个范围主导地位的材料,包括新电介质和其他关于缩减电路和器件的几何尺寸以实现更高密度的封装方面的内容,以及其他书很少涉及的领域、薄膜、高温电子材料、非晶和微晶材料。增加硅使用寿命的新技术成果(包括硅/锗合金)在书中也有介绍,并且又一次提出了同样问题,即,预测硅过时时间是否过于超前!铁电体——一类与硅非常有效结合的材料同样也出现在书中。

Part E章节中（新型材料和选择性的应用）使用了一些极好的新方法开辟了新领地。我们大都知道且频繁使用信息记录器件，但是很少知道，涉及器件使用的材料或原理，比如说CD、视频、DVD等。本书介绍了磁信息存储材料，同样介绍了相变光记录材料，使我们充分与当前发展步伐保持同步更新。该章也同样介绍了太阳能电池、传感器、光导体和碳纳米管的应用，这样大量的工作也体现出编写内容汲取到了世界范围的广度。本章各节中的分子电子和封装材料从研究到应用都得到了呈现。

本书的突出优点在于它的内容覆盖了从基础科学（Part A）到材料的制备、特性（Part B）再到材料的应用（Part C～E）。实际上，书中介绍了涉及的所有材料的广泛应用，这就是本书为什么将会实用的原因之一。就像我之前提及的那样，我们之中没有多少人能够成功地预测未来的发展方向和趋势，在未来十年占领这个领域的主导地位。但是，本书教给我们关于材料的基本性能，可用它们去满足将来的需要。我热切地把这本书推荐给你们。

Prof. Arthur Willoughby
Materials Research Group,
University of Southampton,
UK

前　言

不同学科各种各样的手册，例如电子工程、电子学、生物医学工程、材料科学等手册被广大学生、教师、专业人员很好地使用着，大部分的图书馆也都藏有这些手册。这类手册一般包含许多章（至少50章）内容，在已确定的学科内覆盖广泛的课题；学科选材和论述水平吸引着本科生、研究生、研究员，乃至专业工程人员；最新课题提供广泛的信息，这对该领域所有初学者和研究人员是非常有帮助的；每隔几年，就会有增加新内容的新版本更新之前的版本。

电子和光子材料领域没有类似手册的出版，我们出版这本《电子与光子材料手册》的想法是源自于对手册的需求。它广泛覆盖当今材料领域内的课题，在工程学、材料科学、物理学和化学中都有需要。电子和光子材料真正是一门跨学科的学问，它包含了一些传统的学科，如材料科学、电子工程、化学工程、机械工程、物理学和化学。不难发现，机械工程人员对电子封装实施研究，而电子工程人员对半导体特性进行测量。只有很少的几所大学创建了电子材料或光子材料系。一般来说，电子材料作为一个"学科"是以研究组或跨学科的活动出现在"学院"中。有人可能会对此有异议，因为它事实上是一个跨学科领域，非常需要既包括基础学科又要有最新课题介绍的手册，这就是出版本手册的原因。

本手册是一部关于电子和光子材料的综合论述专著，每一章都是由该领域的专家编写的。本手册针对于大学四年级学生或研究生、研究人员和工作在电子、光电子、光子材料领域的专业人员。书中提供了必要的背景知识和内容广泛的更新知识。每一章都有对内容的一个介绍，并且有许多清晰的说明和大量参考文献。清晰的解释和说明使手册对所有层次的研究者有很大的帮助。所有的章节内容都尽可能独立。既有基础又有前沿的章节内容将吸引不同背景的读者。本手册特别重要的一个特点就是跨学科。例如，将会有这样一些读者，其背景（第一学历）是学化学工程的，工作在半导体工艺线上，而想要学习半导体物理的基础知识；第一学历是物理学的另外一些读者需要尽快更新材料科学的新概念，例如，液相外延等。只要可能，本手册尽量避免采用复杂的数学公式，论述将以半定量的形式给出。手册给出了名词术语表（Glossary of Defining Terms），可为读者提供术语定义的快速查找——这对跨学科工具书来说是必须的。

编者非常感激所有作者们卓越的贡献和相互合作，以及在不同阶段对撰写这本手册的奉献。真诚地感谢Springer Boston的Greg Franklin在文献整理以及手册出版的漫长的工作中给予的支持和帮助。Dr.Werner Skolaut在Springer Heidelberg非常熟练地处理了无数个出版问题，涉及审稿、绘图、书稿的编写和校样的修改，我们真诚地感谢他和他所做出的工作——使得手册能够吸引读者。他是我们见过的最有奉献精神和有效率的编者。

感谢Arthur Willoughby教授的诸多建设性意见使得本手册更加完善。他在材料科学杂志（Journal of Materials Science）积累了非常丰富的编辑经验：电子材料这一章在书中起着重要作用，不仅仅是选取章节，而且还要适应读者需要。

最后，编者感谢所有的成员（Marian, Samuel and Tomas; and Nicollette）在全部工作中的支持和付出的特别耐心。

Dr. Peter Capper
Materials Team Leader,
SELEX Sensors and Airborne Systems,
Southampton, UK

Prof. Safa Kasap
Professor and Canada Research Chair,
Electrical Engineering Department,
University of Saskatchewan,
Canada

Foreword

The Editors, Authors, and Publisher are to be congratulated on this distinguished volume, which will be an invaluable source of information to all workers in the area of electronic and photonic materials. Having made contributions to earlier handbooks, I am well aware of the considerable, and sustained work that is necessary to produce a volume of this kind. This particular handbook, however, is distinguished by its breadth of coverage in the field, and the way in which it discusses the very latest developments. In such a rapidly moving field, this is a considerable challenge, and it has been met admirably.

Previous handbooks and encyclopaedia have tended to concentrate on semiconducting materials, for the understandable reason of their dominance in the electronics field, and the wide range of semiconducting materials and phenomena that must be covered. Few have been courageous enough to predict future trends, but in 1992 Mahajan and Kimerling attempted this in the Introduction to their Concise Encyclopaedia of Semiconducting Materials and Related Technologies (Pergamon), and foresaw future challenges in the areas of nanoelectronics, low dislocation-density III-V substrates, semi-insulating III-V substrates, patterned epitaxy of III-Vs, alternative dielectrics and contacts for silicon technology, and developments in ion-implantation and diffusion. To a greater or lesser extent, all of these have been proved to be true, but it illustrates how difficult it is to make such a prediction.

Not many people would have thought, a decade ago, that the III-nitrides would occupy an important position in this book. As high melting point materials, with the associated growth problems, they were not high on the list of favourites for light emitters at the blue end of the spectrum! The story is a fascinating one – at least as interesting as the solution to the problem of the short working life of early solid-state lasers at the red end of the spectrum. Optoelectronics and photonics, in general, have seen one of the most spectacular advances over the last decade, and this is fully reflected in the book, ranging from visible light emitters, to infra-red materials. The book covers a wide range of work in Part D, including III-V and II-VI optoelectronic materials and band-gap engineering, as well as photonic glasses, liquid crystals, organic photoconductors, and the new area of photonic crystals. The whole Part reflects materials for light generation, processing, transmission and detection – all the essential elements for using light instead of electrons.

In the Materials for Electronics part (Part C) the book charts the progress in silicon – overwhelmingly the dominant material for a whole range of electronic functions and circuitry – including new dielectrics and other issues associated with shrinking geometry of circuits and devices to produce ever higher packing densities. It also includes areas rarely covered in other books – thick films, high-temperature electronic materials, amorphous and microcrystalline materials. The existing developments that extend the life of silicon technology, including silicon/germanium alloys, appear too, and raise the question again as to whether the predicted timetable for the demise of silicon has again been declared too early!! Ferroelectrics – a class of materials used so effectively in conjunction with silicon – certainly deserve to be here.

Prof. Arthur Willoughby
Materials Research Group,
University of Southampton,
UK

The chapters in Part E (Novel Materials and Selected Applications), break new ground in a number of admirable ways. Most of us are aware of, and frequently use, information recording devices such as CDs, videos, DVDs etc., but few are aware of the materials, or principles, involved. This book describes magnetic information storage materials, as well as phase-change optical recording, keeping us fully up-to-date with recent developments. The chapters also include applications such as solar cells, sensors, photoconductors, and carbon nanotubes, on which such a huge volume of work is presently being pursued worldwide. Both ends of the spectrum from research to applications are represented in chapters on molecular electronics and packaging materials.

A particular strength of this book is that it ranges from the fundamental science (Part A) through growth and characterisation of the materials (Part B) to applications (Parts C–E). Virtually all the materials covered here have a wide range of applications, which is one of the reasons why this book is going to be so useful. As I indicated before, few of us will be successful in predicting the future direction and trends, occupying the high-ground in this field in the coming decade, but this book teaches us the basic principles of materials, and leaves it to us to adapt these to the needs of tomorrow. I commend it to you most warmly.

Preface

Other handbooks in various disciplines such as electrical engineering, electronics, biomedical engineering, materials science, etc. are currently available and well used by numerous students, instructors and professionals. Most libraries have these handbook sets and each contains numerous (at least 50) chapters that cover a wide spectrum of topics within each well-defined discipline. The subject and the level of coverage appeal to both undergraduate and postgraduate students and researchers as well as to practicing professionals. The advanced topics follow introductory topics and provide ample information that is useful to all, beginners and researchers, in the field. Every few years, a new edition is brought out to update the coverage and include new topics.

There has been no similar handbook in electronic and photonic materials, and the present Springer Handbook of Electronic and Photonic Materials (SHEPM) idea grew out of a need for a handbook that covers a wide spectrum of topics in materials that today's engineers, material scientists, physicists, and chemists need. Electronic and photonic materials is a truly interdisciplinary subject that encompasses a number of traditional disciplines such as materials science, electrical engineering, chemical engineering, mechanical engineering, physics and chemistry. It is not unusual to find a mechanical engineering faculty carrying out research on electronic packaging and electrical engineers carrying out characterization measurements on semiconductors. There are only a few established university departments in electronic or photonic materials. In general, electronic materials as a "discipline" appears as a research group or as an interdisciplinary activity within a "college". One could argue that, because of the very fact that it is such an interdisciplinary field, there is a greater need to have a handbook that covers not only fundamental topics but also advanced topics; hence the present handbook.

This handbook is a comprehensive treatise on electronic and photonic materials with each chapter written by experts in the field. The handbook is aimed at senior undergraduate and graduate students, researchers and professionals working in the area of electronic, optoelectronic and photonic materials. The chapters provide the necessary background and up-to-date knowledge in a wide range of topics. Each chapter has an introduction to the topic, many clear illustrations and numerous references. Clear explanations and illustrations make the handbook useful to all levels of researchers. All chapters are as self-contained as possible. There are both fundamental and advanced chapters to appeal to readers with different backgrounds. This is particularly important for this handbook since the subject matter is highly interdisciplinary. For example, there will be readers with a background (first degree) in chemical engineering and working on semiconductor processing who need to learn the fundamentals of semiconductors physics. Someone with a first degree in physics would need to quickly update himself on materials science concepts such as liquid phase epitaxy and so on. Difficult mathematics has been avoided and, whenever possible, the explanations have been given semiquantitatively. There is a "*Glossary of Defining Terms*" at the end of the handbook, which can serve to quickly find the definition of a term – a very necessary feature in an interdisciplinary handbook.

The editors are very grateful to all the authors for their excellent contributions and for their cooperation in delivering their manuscripts and in the various stages of production of this handbook. Sincere thanks go to Greg Franklin at Springer Boston for all his support and help throughout the long period of commissioning, acquiring the contributions and the production of the handbook. Dr. Werner Skolaut at Springer Heidelberg has very skillfully handled the myriad production issues involved in copy-editing, figure redrawing and proof preparation and correction and our sincere thanks go to him also for all his hard

Dr. Peter Capper
Materials Team Leader,
SELEX Sensors and Airborne Systems,
Southampton, UK

Prof. Safa Kasap
Professor and Canada Research Chair,
Electrical Engineering Department,
University of Saskatchewan, Canada

work in making the handbook attractive to read. He is the most dedicated and efficient editor we have come across.

It is a pleasure to thank Professor Arthur Willoughby for his many helpful suggestions that made this a better handbook. His wealth of experience as editor of the Journal of Materials Science: Materials in Electronics played an important role not only in selecting chapters but also in finding the right authors.

Finally, the editors wish to thank all the members of our families (Marian, Samuel and Thomas; and Nicollette) for their support and particularly their endurance during the entire project.

Peter Capper and Safa Kasap
Editors

List of Authors

Martin Abkowitz
1198 Gatestone Circle
Webster, NY 14580, USA
e-mail: *mabkowitz@mailaps.org,
abkowitz@chem.chem.rochester.edu*

Sadao Adachi
Gunma University
Department of Electronic Engineering,
Faculty of Engineering
Kiryu-shi 376-8515
Gunma, Japan
e-mail: *adachi@el.gunma-u.ac.jp*

Alfred Adams
University of Surrey
Advanced Technology Institute
Guildford, Surrey, GU2 7XH,
Surrey, UK
e-mail: *alf.adams@surrey.ac.uk*

Guy J. Adriaenssens
University of Leuven
Laboratorium voor Halfgeleiderfysica
Celestijnenlaan 200D
B-3001 Leuven, Belgium
e-mail: *guy.adri@fys.kuleuven.ac.be*

Wilfried von Ammon
Siltronic AG
Research and Development
Johannes Hess Strasse 24
84489 Burghausen, Germany
e-mail: *wilfried.ammon@siltronic.com*

Peter Ashburn
University of Southampton
School of Electronics and Computer Science
Southampton, S017 1BJ, UK
e-mail: *pa@ecs.soton.ac.uk*

Mark Auslender
Ben-Gurion University of the Negev Beer Sheva
Department of Electrical
and Computer Engineering
P.O.Box 653
Beer Sheva 84105, Israel
e-mail: *marka@ee.bgu.ac.il*

Darren M. Bagnall
University of Southampton
School of Electronics and Computer Science
Southampton, S017 1BJ, UK
e-mail: *dmb@ecs.soton.ac.uk*

Ian M. Baker
SELEX Sensors and Airborne Systems Infrared Ltd.
Southampton, Hampshire S015 0EG, UK
e-mail: *ian.m.baker@selex-sas.com*

Sergei Baranovskii
Philipps University Marburg
Department of Physics
Renthof 5
35032 Marburg, Germany
e-mail: *baranovs@staff.uni-marburg.de*

Mark Baxendale
Queen Mary, University of London
Department of Physics
Mile End Road
London, E1 4NS, UK
e-mail: *m.baxendale@qmul.ac.uk*

Mohammed L. Benkhedir
University of Leuven
Laboratorium voor Halfgeleiderfysica
Celestijnenlaan 200D
B-3001 Leuven, Belgium
e-mail: *MohammedLoufti.Benkhedir
@fys.kuleuven.ac.be*

Monica Brinza
University of Leuven
Laboratorium voor Halfgeleiderfysica
Celestijnenlaan 200D
B-3001 Leuven, Belgium
e-mail: *monica.brinza@fys.kuleuven.ac.be*

Paul D. Brown
University of Nottingham
School of Mechanical, Materials and
Manufacturing Engineering
University Park
Nottingham, NG7 2RD, UK
e-mail: *paul.brown@nottingham.ac.uk*

Mike Brozel
University of Glasgow
Department of Physics and Astronomy
Kelvin Building
Glasgow, G12 8QQ, UK
e-mail: *mikebrozel@beeb.net*

Lukasz Brzozowski
University of Toronto
Sunnybrook and Women's Research Institute,
Imaging Research/
Department of Medical Biophysics
Research Building, 2075 Bayview Avenue
Toronto, ON, M4N 3M5, Canada
e-mail: *lukbroz@sten.sunnybrook.utoronto.ca*

Peter Capper
SELEX Sensors and Airborne Systems Infrared Ltd.
Materials Team Leader
Millbrook Industrial Estate, PO Box 217
Southampton, Hampshire SO15 0EG, UK
e-mail: *pete.capper@selex-sas.com*

Larry Comstock
San Jose State University
6574 Crystal Springs Drive
San Jose, CA 95120, USA
e-mail: *Comstock@email.sjsu.edu*

Ray DeCorby
University of Alberta
Department of Electrical
and Computer Engineering
7th Floor, 9107-116 Street N.W.
Edmonton, Alberta T6G 2V4, Canada
e-mail: *rdecorby@trlabs.ca*

M. Jamal Deen
McMaster University
Department of Electrical
and Computer Engineering (CRL 226)
1280 Main Street West
Hamilton, ON L8S 4K1, Canada
e-mail: *jamal@mcmaster.ca*

Leonard Dissado
The University of Leicester
Department of Engineering
University Road
Leicester, LE1 7RH, UK
e-mail: *lad4@le.ac.uk*

David Dunmur
University of Southampton
School of Chemistry
Southampton, SO17 1BJ, UK
e-mail: *d.a.dunmur@soton.ac.uk*

Lester F. Eastman
Cornell University
Department of Electrical
and Computer Engineering
425 Phillips Hall
Ithaca, NY 14853, USA
e-mail: *lfe2@cornell.edu*

Andy Edgar
Victoria University
School of Chemical and Physical Sciences SCPS
Kelburn Parade/PO Box 600
Wellington, New Zealand
e-mail: *Andy.Edgar@vuw.ac.nz*

Brian E. Foutz
Cadence Design Systems
1701 North Street, Bldg 257-3
Endicott, NY 13760, USA
e-mail: foutz@cadence.com

Mark Fox
University of Sheffield
Department of Physics and Astronomy
Hicks Building, Hounsefield Road
Sheffield, S3 7RH, UK
e-mail: mark.fox@shef.ac.uk

Darrel Frear
RF and Power Packaging Technology Development,
Freescale Semiconductor
2100 East Elliot Road
Tempe, AZ 85284, USA
e-mail: darrel.frear@freescale.com

Milan Friesel
Chalmers University of Technology
Department of Physics
Fysikgränd 3
41296 Göteborg, Sweden
e-mail: friesel@chalmers.se

Jacek Gieraltowski
Université de Bretagne Occidentale
6 Avenue Le Gorgeu, BP: 809
29285 Brest Cedex, France
e-mail: Jacek.Gieraltowski@univ-brest.fr

Yinyan Gong
Columbia University
Department of Applied Physics
and Applied Mathematics
500 W. 120th St.
New York, NY 10027, USA
e-mail: yg2002@columbia.edu

Robert D. Gould[†]
Keele University
Thin Films Laboratory, Department of Physics,
School of Chemistry and Physics
Keele, Staffordshire ST5 5BG, UK

Shlomo Hava
Ben-Gurion University of the Negev Beer Sheva
Department of Electrical
and Computer Engineering
P.O. Box 653
Beer Sheva 84105, Israel
e-mail: hava@ee.bgu.ac.il

Colin Humphreys
University of Cambridge
Department of Materials Science and Metallurgy
Pembroke Street
Cambridge, CB2 3!Z, UK
e-mail: colin.Humphreys@msm.cam.ac.uk

Stuart Irvine
University of Wales, Bangor
Department of Chemistry
Gwynedd, LL57 2UW, UK
e-mail: sjc.irvine@bangor.ac.uk

Minoru Isshiki
Tohoku University
Institute of Multidisciplinary Research
for Advanced Materials
1-1, Katahira, 2 chome, Aobaku
Sendai, 980-8577, Japan
e-mail: isshiki@tagen.tohoku.ac.jp

Robert Johanson
University of Saskatchewan
Department of Electrical Engineering
57 Campus Drive
Saskatoon, SK S7N 5A9, Canada
e-mail: johanson@engr.usask.ca

Tim Joyce
University of Liverpool
Functional Materials Research Centre,
Department of Engineering
Brownlow Hill
Liverpool, L69 3BX, UK
e-mail: t.joyce@liv.ac.uk

M. Zahangir Kabir
Concordia University
Department of Electrical and Computer
Engineering
Montreal, Quebec S7N5A9, Canada
e-mail: *kabir@encs.concordia.ca*

Safa Kasap
University of Saskatchewan
Department of Electrical Engineering
57 Campus Drive
Saskatoon, SK S7N 5A9, Canada
e-mail: *safa.kasap@usask.ca*

Alexander Kolobov
National Institute of Advanced
Industrial Science and Technology
Center for Applied Near-Field Optics Research
1-1-1 Higashi, Tsukuba
Ibaraki, 305-8562, Japan
e-mail: *a.kolobov@aist.go.jp*

Cyril Koughia
University of Saskatchewan
Department of Electrical Engineering
57 Campus Drive
Saskatoon, SK S7N 5A9, Canada
e-mail: *kik486@mail.usask.ca*

Igor L. Kuskovsky
Queens College, City University of New York (CUNY)
Department of Physics
65-30 Kissena Blvd.
Flushing, NY 11367, USA
e-mail: *igor_kuskovsky@qc.edu*

Geoffrey Luckhurst
University of Southampton
School of Chemistry
Southampton, SO17 1BJ, UK
e-mail: *g.r.luckhurst@soton.ac.uk*

Akihisa Matsuda
Tokyo University of Science
Research Institute for Science and Technology
2641 Yamazaki, Noda-shi
Chiba, 278-8510, Japan
e-mail: *amatsuda@rs.noda.tus.ac.jp,
a.matsuda@aist.go.jp*

Naomi Matsuura
Sunnybrook Health Sciences Centre
Department of Medical Biophysics,
Imaging Research
2075 Bayview Avenue
Toronto, ON M4N 3M5, Canada
e-mail: *matsuura@sri.utoronto.ca*

Kazuo Morigaki
University of Tokyo
C-305, Wakabadai 2-12, Inagi
Tokyo, 206-0824, Japan
e-mail: *k.morigaki@yacht.ocn.ne.jp*

Hadis Morkoç
Virginia Commonwealth University
Department of Electrical
and Computer Engineering
601 W. Main St., Box 843072
Richmond, VA 23284-3068, USA
e-mail: *hmorkoc@vcu.edu*

Winfried Mönch
Universität Duisburg-Essen
Lotharstraße 1
47048 Duisburg, Germany
e-mail: *w.moench@uni-duisburg.de*

Arokia Nathan
University of Waterloo
Department of Electrical
and Computer Engineering
200 University Avenue W.
Waterloo, Ontario N2L 3G1, Canada
e-mail: *anathan@uwaterloo.ca*

Gertrude F. Neumark
Columbia University
Department of Applied Physics
and Applied Mathematics
500W 120th St., MC 4701
New York, NY 10027, USA
e-mail: *gfn1@columbia.edu*

Stephen K. O'Leary
University of Regina
Faculty of Engineering
3737 Wascana Parkway
Regina, SK S4S 0A2, Canada
e-mail: *stephen.oleary@uregina.ca*

Chisato Ogihara
Yamaguchi University
Department of Applied Science
2-16-1 Tokiwadai
Ube, 755-8611, Japan
e-mail: *ogihara@yamaguchi-u.ac.jp*

Fabien Pascal
Université Montpellier 2/CEM2-cc084
Centre d'Electronique
et de Microoptoélectronique de Montpellier
Place E. Bataillon
34095 Montpellier, France
e-mail: *pascal@cem2.univ-montp2.fr*

Michael Petty
University of Durham
Department School of Engineering
South Road
Durham, DH1 3LE, UK
e-mail: *m.c.petty@durham.ac.uk*

Asim Kumar Ray
Queen Mary, University of London
Department of Materials
Mile End Road
London, E1 4NS, UK
e-mail: *a.k.ray@qmul.ac.uk*

John Rowlands
University of Toronto
Department of Medical Biophysics
Sunnybrook and Women's College
Health Sciences Centre
S656-2075 Bayview Avenue
Toronto, ON M4N 3M5, Canada
e-mail: *john.rowlands@sri.utoronto.ca*

Oleg Rubel
Philipps University Marburg
Department of Physics
and Material Sciences Center
Renthof 5
35032 Marburg, Germany
e-mail: *oleg.rubel@physik.uni-marburg.de*

Harry Ruda
University of Toronto
Materials Science and Engineering,
Electrical and Computer Engineering
170 College Street
Toronto, M5S 3E4, Canada
e-mail: *ruda@ecf.utoronto.ca*

Edward Sargent
University of Toronto
Department of Electrical
and Computer Engineering
ECE, 10 King's College Road
Toronto, M5S 3G4, Canada
e-mail: *ted.sargent@utoronto.ca*

Peyman Servati
Ignis Innovation Inc.
55 Culpepper Dr.
Waterloo, Ontario N2L 5K8, Canada
e-mail: *pservati@uwaterloo.ca*

Derek Shaw
Hull University
Hull, HU6 7RX, UK
e-mail: *DerekShaw1@compuserve.com*

Fumio Shimura
Shizuoka Institute of Science and Technology
Department of Materials and Life Science
2200-2 Toyosawa
Fukuroi, Shizuoka 437-8555, Japan
e-mail: shimura@ms.sist.ac.jp

Michael Shur
Renssellaer Polytechnic Institute
Department of Electrical, Computer,
and Systems Engineering
CII 9017, RPI, 110 8th Street
Troy, NY 12180, USA
e-mail: shurm@rpi.edu

Jai Singh
Charles Darwin University
School of Engineering and Logistics,
Faculty of Technology, B-41
Ellengowan Drive
Darwin, NT 0909, Australia
e-mail: jai.singh@cdu.edu.au

Tim Smeeton
Sharp Laboratories of Europe
Edmund Halley Road, Oxford Science Park
Oxford, OX4 4GB, UK
e-mail: tim.smeeton@sharp.co.uk

Boris Straumal
Russian Academy of Sciences
Institute of Sold State Physics
Institutskii prospect 15
Chernogolovka, 142432, Russia
e-mail: straumal@issp.ac.ru

Stephen Sweeney
University of Surrey
Advanced Technology Institute
Guildford, Surrey GU2 7XH, UK
e-mail: s.sweeney@surrey.ac.uk

David Sykes
Loughborough Surface Analysis Ltd.
PO Box 5016, Unit FC, Holywell Park, Ashby Road
Loughborough, LE11 3WS, UK
e-mail: d.e.sykes@lsaltd.co.uk

Keiji Tanaka
Hokkaido University
Department of Applied Physics,
Graduate School of Engineering
Kita-ku, N13 W8
Sapporo, 060-8628, Japan
e-mail: keiji@eng.hokudai.ac.jp

Charbel Tannous
Université de Bretagne Occidentale
LMB, CNRS FRE 2697
6 Avenue Le Gorgeu, BP: 809
29285 Brest Cedex, France
e-mail: tannous@univ-brest.fr

Ali Teke
Balikesir University
Department of Physics, Faculty of Art and Science
Balikesir, 10100, Turkey
e-mail: ateke@balikesir.edu.tr

Junji Tominaga
National Institute of Advanced Industrial
Science and Technology, AIST
Center for Applied Near-Field Optics Research,
CAN-FOR
Tsukuba Central 4 1-1-1 Higashi
Tsukuba, 3.5-8562, Japan
e-mail: j-tomonaga@aist.go.jp

Dan Tonchev
University of Saskatchewan
Department of Electrical Engineering
57 Campus Drive
Saskatoon, SK S7N 5A9, Canada
e-mail: dan.tonchev@usask.ca

Harry L. Tuller
Massachusetts Institute of Technology
Department of Materials Science and Engineering,
Crystal Physics and Electroceramics Laboratory
77 Massachusetts Avenue
Cambridge, MA 02139, USA
e-mail: tuller@mit.edu

Qamar-ul Wahab
Linköping University
Department of Physics,
Chemistry, and Biology (IFM)
SE-581 83 Linköping, Sweden
e-mail: *quw@ifm.liu.se*

Robert M. Wallace
University of Texas at Dallas
Department of Electrical Engineering
M.S. EC 33, P.O.Box 830688
Richardson, TX 75083, USA
e-mail: *rmwallace@utdallas.edu*

Jifeng Wang
Tohoku University
Institute of Multidisciplinary Research
for Advanced Materials
1-1, Katahira, 2 Chome, Aobaku
Sendai, 980-8577, Japan
e-mail: *wang@tagen.tohoku.ac.jp*

David S. Weiss
NexPress Solutions, Inc.
2600 Manitou Road
Rochester, NY 14653-4180, USA
e-mail: *David_Weiss@Nexpress.com*

Rainer Wesche
Swiss Federal Institute of Technology
Centre de Recherches en Physique des Plasmas
CRPP (c/o Paul Scherrer Institute), WMHA/C31,
Villigen PS
Lausanne, CH-5232, Switzerland
e-mail: *rainer.wesche@psi.ch*

Roger Whatmore
Tyndall National Institute
Lee Maltings, Cork , Ireland
e-mail: *roger.whatmore@tyndall.ie*

Neil White
University of Southampton
School of Electronics and Computer Science
Mountbatten Building
Highfield, Southampton SO17 1BJ, UK
e-mail: *nmw@ecs.soton.ac.uk*

Magnus Willander
University of Gothenburg
Department of Physics
SE-412 96 Göteborg, Sweden
e-mail: *mwi@fy.chalmers.se*

Jan Willekens
University of Leuven
Laboratorium voor Halfgeleiderfysica
Celestijnenlaan 200D
B-3001 Leuven, Belgium
e-mail: *jan.willekens@kc.kuleuven.ac.be*

Acknowledgements

C.23 Gallium Arsenide
by Mike Brozel

The author is delighted to acknowledge the help given to him over many years by his colleagues and friends both at UMIST and in industry. Specifically, he wishes to thank R. Blunt, I. R. Grant, and R. H. Wallis for their careful and critical reading of this manuscript.

C.24 High-Temperature Electronic Materials: Silicon Carbide and Diamond
by Magnus Willander, Milan Friesel, Qamar-ul Wahab, Boris Straumal

Magnus Willander and Milan Friesel would like to thank Dr. V. Narayan for checking the text, and Dr. A. Baranzahi for letting us use Figs. 24.2–24.3. Qamar-ul Wahab thanks Mr. Amir Karim for all his support.

C.25 Amorphous Semiconductors: Structure, Optical, and Electrical Properties
by Kazuo Morigaki, Chisato Ogihara

We wish to thank M. Ichihara, K. Suzuki and M. Yamaguchi, Institute for Solid State Physics, University of Tokyo, for providing us with their unpublished materials (Fig. 25.7a,b). Stimulating and helpful discussions were held with Prof. S. Kugler during the stay of one of us (K.M.) at the Budapest University of Technology and Economics, for which K.M. is grateful.

C.28 Dielectric Materials for Microelectronics
by Robert M. Wallace

RMW gratefully acknowledges the many discussions and hard work of his colleagues and students engaged in gate-stack research. This work is supported in part by the Texas Advanced Technology Program and the Semiconductor Research Corporation.

C.29 Thin Films
by Robert D. Gould[†]

The author wishes to acknowledge the general support and encouragement of Prof. C. A. Hogarth, Department of Physics, Brunel University, and of Prof. W. Fuller, Department of Physics, Keele University. Particular thanks are also due to Prof. E. W. Williams, Electronic Engineering Group, Keele University for collaborative work and permission to reproduce Figs. 29.6 and 29.7.

目 录

缩略语

Part C 电子材料

21 单晶硅：电学与光学特性 3
21.1 硅基 3
21.2 电学特性 13
21.3 光学特性 34
参考文献 40

22 硅-锗：特性、生长和应用 43
22.1 硅-锗物理特性 44
22.2 硅-锗光学特性 50
22.3 硅-锗生长 54
22.4 多晶硅-锗 56
参考文献 59

23 砷化镓 61
23.1 GaAs的体生长 64
23.2 GaAs的外延生长 69
23.3 GaAs的扩散 73
23.4 GaAs离子注入 75
23.5 GaAs晶格缺陷 76
23.6 GaAs的杂质与缺陷分析（化学）............ 79
23.7 GaAs的杂质与缺陷分析（电学）............ 80
23.8 GaAs的杂质与缺陷分析（光学）............ 83
23.9 复杂异质结的评估 84
23.10 GaAs的电接触 86

23.11 GaAs器件（微波） ...86
23.12 GaAs器件（电-光） ...89
23.13 GaAs的其他应用 ...94
23.14 结论 ...94
参考文献 ...95

24 高温电子材料：碳化硅与金刚石 ...99
24.1 材料的特性与制备 ...102
24.2 电子器件 ...109
24.3 总结 ...119
参考文献 ...120

25 非晶态半导体：结构、光学与电学特性 ...127
25.1 电子态 ...127
25.2 结构特性 ...130
25.3 光学特性 ...132
25.4 电学特性 ...135
25.5 光诱导现象 ...137
25.6 纳米非晶态结构 ...139
参考文献 ...140

26 非晶态与微晶硅 ...143
26.1 等离子体SiH_4与SiH_4/H_2的反应 ...143
26.2 表面薄生长 ...145
26.3 a-Si:H与μc-Si:H缺陷密度测定 ...151
26.4 器件应用 ...152
26.5 硅薄膜太阳能电池材料的相关问题研究进展 ...153
26.6 总结 ...156
参考文献 ...156

27 铁电体材料 ...159
27.1 铁电体材料 ...163
27.2 铁电体材料制备技术 ...170

27.3 铁电体应用..178

参考文献..184

28 微电子电介质材料..187

28.1 栅极电介质..192

28.2 隔离电介质..209

28.3 电容电介质..209

28.4 互连电介质..213

28.5 总结..215

参考文献..215

29 薄膜..221

29.1 淀积形成方法..223

29.2 结构..244

29.3 特性..254

29.4 结论..270

参考文献..273

30 厚膜..279

30.1 厚膜工艺..280

30.2 衬底..282

30.3 厚膜材料..283

30.4 组件与装配..286

30.5 传感器..290

参考文献..293

Contents

List of Abbreviations

Part C Materials for Electronics

21 Single-Crystal Silicon: Electrical and Optical Properties ... 441
21.1 Silicon Basics ... 441
21.2 Electrical Properties ... 451
21.3 Optical Properties ... 472
References ... 478

22 Silicon–Germanium: Properties, Growth and Applications ... 481
22.1 Physical Properties of Silicon–Germanium ... 482
22.2 Optical Properties of SiGe ... 488
22.3 Growth of Silicon–Germanium ... 492
22.4 Polycrystalline Silicon–Germanium ... 494
References ... 497

23 Gallium Arsenide ... 499
23.1 Bulk Growth of GaAs ... 502
23.2 Epitaxial Growth of GaAs ... 507
23.3 Diffusion in Gallium Arsenide ... 511
23.4 Ion Implantation into GaAs ... 513
23.5 Crystalline Defects in GaAs ... 514
23.6 Impurity and Defect Analysis of GaAs (Chemical) ... 517
23.7 Impurity and Defect Analysis of GaAs (Electrical) ... 518
23.8 Impurity and Defect Analysis of GaAs (Optical) ... 521
23.9 Assessment of Complex Heterostructures ... 522
23.10 Electrical Contacts to GaAs ... 524
23.11 Devices Based on GaAs (Microwave) ... 524
23.12 Devices based on GaAs (Electro-optical) ... 527
23.13 Other Uses for GaAs ... 532
23.14 Conclusions ... 532
References ... 533

24 High-Temperature Electronic Materials: Silicon Carbide and Diamond ... 537
24.1 Material Properties and Preparation ... 540
24.2 Electronic Devices ... 547
24.3 Summary ... 557
References ... 558

25 Amorphous Semiconductors: Structure, Optical, and Electrical Properties 565
- 25.1 Electronic States 565
- 25.2 Structural Properties 568
- 25.3 Optical Properties 570
- 25.4 Electrical Properties 573
- 25.5 Light-Induced Phenomena 575
- 25.6 Nanosized Amorphous Structure 577
- References 578

26 Amorphous and Microcrystalline Silicon 581
- 26.1 Reactions in SiH_4 and SiH_4/H_2 Plasmas 581
- 26.2 Film Growth on a Surface 583
- 26.3 Defect Density Determination for a-Si:H and μc-Si:H 589
- 26.4 Device Applications 590
- 26.5 Recent Progress in Material Issues Related to Thin-Film Silicon Solar Cells 591
- 26.6 Summary 594
- References 594

27 Ferroelectric Materials 597
- 27.1 Ferroelectric Materials 601
- 27.2 Ferroelectric Materials Fabrication Technology 608
- 27.3 Ferroelectric Applications 616
- References 622

28 Dielectric Materials for Microelectronics 625
- 28.1 Gate Dielectrics 630
- 28.2 Isolation Dielectrics 647
- 28.3 Capacitor Dielectrics 647
- 28.4 Interconnect Dielectrics 651
- 28.5 Summary 653
- References 653

29 Thin Films 659
- 29.1 Deposition Methods 661
- 29.2 Structure 682
- 29.3 Properties 692
- 29.4 Concluding Remarks 708
- References 711

30 Thick Films 717
- 30.1 Thick Film Processing 718
- 30.2 Substrates 720
- 30.3 Thick Film Materials 721
- 30.4 Components and Assembly 724
- 30.5 Sensors 728
- References 731

List of Abbreviations

2DEG	two-dimensional electron gas

A

AC	alternating current
ACCUFET	accumulation-mode MOSFET
ACRT	accelerated crucible rotation technique
AEM	analytical electron microscopes
AES	Auger electron spectroscopy
AFM	atomic force microscopy
ALD	atomic-layer deposition
ALE	atomic-layer epitaxy
AMA	active matrix array
AMFPI	active matrix flat-panel imaging
AMOLED	amorphous organic light-emitting diode
APD	avalanche photodiode

B

b.c.c.	body-centered cubic
BEEM	ballistic-electron-emission microscopy
BEP	beam effective pressure
BH	buried-heterostructure
BH	Brooks–Herring
BJT	bipolar junction transistor
BTEX	m-xylene
BZ	Brillouin zone

C

CAIBE	chemically assisted ion beam etching
CB	conduction band
CBE	chemical beam epitaxy
CBED	convergent beam electron diffraction
CC	constant current
CCD	charge-coupled device
CCZ	continuous-charging Czochralski
CFLPE	container-free liquid phase epitaxy
CKR	cross Kelvin resistor
CL	cathodoluminescence
CMOS	complementary metal-oxide-semiconductor
CNR	carrier-to-noise ratio
COP	crystal-originated particle
CP	charge pumping
CPM	constant-photocurrent method
CR	computed radiography
CR-DLTS	computed radiography deep level transient spectroscopy
CRA	cast recrystallize anneal
CTE	coefficient of thermal expansion
CTO	chromium(III) trioxalate
CuPc	copper phthalocyanine
CuTTBPc	tetra-tert-butyl phthalocyanine
CV	chemical vapor
CVD	chemical vapor deposition
CVT	chemical vapor transport
CZ	Czochralski
CZT	cadmium zinc telluride

D

DA	Drude approximation
DAG	direct alloy growth
DBP	dual-beam photoconductivity
DC	direct current
DCPBH	double-channel planar buried heterostructure
DET	diethyl telluride
DFB	distributed feedback
DH	double heterostructure
DIL	dual-in-line
DIPTe	diisopropyltellurium
DLC	diamond-like carbon
DLHJ	double-layer heterojunction
DLTS	deep level transient spectroscopy
DMCd	dimethyl cadmium
DMF	dimethylformamide
DMOSFET	double-diffused MOSFET
DMS	dilute magnetic semiconductors
DMSO	dimethylsulfoxide
DMZn	dimethylzinc
DOS	density of states
DQE	detective quantum efficiency
DSIMS	dynamic secondary ion mass spectrometry
DTBSe	ditertiarybutylselenide
DUT	device under test
DVD	digital versatile disk
DWDM	dense wavelength-division multiplexing
DXD	double-crystal X-ray diffraction

E

EBIC	electron beam induced conductivity
ED	electrodeposition
EDFA	erbium-doped fiber amplifier
EELS	electron energy loss spectroscopy
EFG	film-fed growth
EHP	electron–hole pairs
ELO	epitaxial lateral overgrowth
ELOG	epitaxial layer overgrowth
EM	electromagnetic
EMA	effective media approximation

ENDOR	electron–nuclear double resonance		IFIGS	interface-induced gap states
EPD	etch pit density		IFTOF	interrupted field time-of-flight
EPR	electron paramagnetic resonance		IGBT	insulated gate bipolar transistor
ESR	electron spin resonance spectroscopy		IMP	interdiffused multilayer process
EXAFS	extended X-ray absorption fine structure		IPEYS	internal photoemission yield spectroscopy
			IR	infrared
			ITO	indium-tin-oxide

F

FCA	free-carrier absorption
f.c.c.	face-centered cubic
FET	field effect transistor
FIB	focused ion beam
FM	Frank–van der Merwe
FPA	focal plane arrays
FPD	flow pattern defect
FTIR	Fourier transform infrared
FWHM	full-width at half-maximum
FZ	floating zone

J

JBS	junction barrier Schottky
JFET	junction field-effect transistors
JO	Judd–Ofelt

K

KCR	Kelvin contact resistance
KKR	Kramers–Kronig relation
KLN	$K_3Li_2Nb_5O_{12}$
KTPO	$KTiOPO_4$

G

GDA	generalized Drude approximation
GDMS	glow discharge mass spectrometry
GDOES	glow discharge optical emission spectroscopy
GF	gradient freeze
GMR	giant magnetoresistance
GOI	gate oxide integrity
GRIN	graded refractive index
GSMBE	gas-source molecular beam epitaxy
GTO	gate turn-off

L

LB	Langmuir–Blodgett
LD	laser diodes
LD	lucky drift
LDD	lightly doped drain
LEC	liquid-encapsulated Czochralski
LED	light-emitting diodes
LEIS	low-energy ion scattering
LEL	lower explosive limit
LF	low-frequency
LLS	laser light scattering
LMA	law of mass action
LO	longitudinal optical
LPE	liquid phase epitaxy
LSTD	laser light scattering tomography defect
LVM	localized vibrational mode

H

HAADF	high-angle annular dark field
HB	horizontal Bridgman
HBT	hetero-junction bipolar transistor
HDC	horizontal directional solidification crystallization
HEMT	high electron mobility transistor
HF	high-frequency
HOD	highly oriented diamond
HOLZ	high-order Laue zone
HPc	phthalocyanine
HPHT	high-pressure high-temperature
HRXRD	high-resolution X-ray diffraction
HTCVD	high-temperature CVD
HVDC	high-voltage DC
HWE	hot-wall epitaxy

M

MBE	molecular beam epitaxy
MCCZ	magnetic field applied continuous Czochralski
MCT	mercury cadmium telluride
MCZ	magnetic field applied Czochralski
MD	molecular dynamics
MEED	medium-energy electron diffraction
MEM	micro-electromechanical systems
MESFET	metal-semiconductor field-effect transistor
MFC	mass flow controllers
MIGS	metal-induced gap states
ML	monolayer
MLHJ	multilayer heterojunction
MOCVD	metal-organic chemical vapor deposition
MODFET	modulation-doped field effect transistor

I

IC	integrated circuit
ICTS	isothermal capacitance transient spectroscopy
IDE	interdigitated electrodes

MOMBE	metalorganic molecular beam epitaxy	PL	photoluminescence
MOS	metal/oxide/semiconductor	PM	particulate matter
MOSFET	metal/oxide/semiconductor field effect transistor	PMMA	poly(methyl-methacrylate)
MOVPE	metalorganic vapor phase epitaxy	POT	poly(n-octyl)thiophene
MPc	metallophthalocyanine	ppb	parts per billion
MPC	modulated photoconductivity	ppm	parts per million
MPCVD	microwave plasma chemical deposition	PPS	polyphenylsulfide
MQW	multiple quantum well	PPY	polypyrrole
MR	magnetoresistivity	PQT-12	poly[5,5'-bis(3-alkyl-2-thienyl)-2,2'-bithiophene]
MS	metal–semiconductor	PRT	platinum resistance thermometers
MSRD	mean-square relative displacement	PSt	polystyrene
MTF	modulation transfer function	PTC	positive temperature coefficient
MWIR	medium-wavelength infrared	PTIS	photothermal ionisation spectroscopy
		PTS	1,1-dioxo-2-(4-methylphenyl)-6-phenyl-4-(dicyanomethylidene)thiopyran

N

NDR	negative differential resistance
NEA	negative electron affinity
NeXT	nonthermal energy exploration telescope
NMOS	n-type-channel metal–oxide–semiconductor
NMP	N-methylpyrrolidone
NMR	nuclear magnetic resonance
NNH	nearest-neighbor hopping
NSA	naphthalene-1,5-disulfonic acid
NTC	negative temperature coefficient
NTD	neutron transmutation doping

PTV	polythienylene vinylene
PV	photovoltaic
PVD	physical vapor transport
PVDF	polyvinylidene fluoride
PVK	polyvinylcarbazole
PVT	physical vapor transport
PZT	lead zirconate titanate

Q

QA	quench anneal
QCL	quantum cascade laser
QCSE	quantum-confined Stark effect
QD	quantum dot
QHE	quantum Hall effect
QW	quantum well

O

OLED	organic light-emitting diode
OSF	oxidation-induced stacking fault
OSL	optically stimulated luminescence
OZM	overlap zone melting

R

RAIRS	reflection adsorption infrared spectroscopy
RBS	Rutherford backscattering
RCLED	resonant-cavity light-emitting diode
RDF	radial distribution function
RDS	reflection difference spectroscopy
RE	rare earth
RENS	resolution near-field structure
RF	radio frequency
RG	recombination–generation
RH	relative humidity
RHEED	reflection high-energy electron diffraction
RIE	reactive-ion etching
RIU	refractive index units
RTA	rapid thermal annealing
RTD	resistance temperature devices
RTS	random telegraph signal

P

PAE	power added efficiency
PAni	polyaniline
pBN	pyrolytic boron nitride
Pc	phthalocyanine
PC	photoconductive
PCA	principal component analysis
PCB	printed circuit board
PDMA	poly(methylmethacrylate)/poly(decyl methacrylate)
PDP	plasma display panels
PDS	photothermal deflection spectroscopy
PE	polysilicon emitter
PE BJT	polysilicon emitter bipolar junction transistor
PECVD	plasma-enhanced chemical vapor deposition
PEN	polyethylene naphthalate
PES	photoemission spectroscopy
PET	positron emission tomography
pHEMT	pseudomorphic HEMT

S

SA	self-assembly
SAM	self-assembled monolayers

SAW	surface acoustic wave	TMA	trimethyl-aluminum
SAXS	small-angle X-ray scattering	TMG	trimethyl-gallium
SCH	separate confinement heterojunction	TMI	trimethyl-indium
SCVT	seeded chemical vapor transport	TMSb	trimethylantimony
SE	spontaneous emission	TO	transverse optical
SEM	scanning electron microscope	TOF	time of flight
SIMS	secondary ion mass spectrometry	ToFSIMS	time of flight SIMS
SIPBH	semi-insulating planar buried heterostructure	TPC	transient photoconductivity
		TPV	thermophotovoltaic
SIT	static induction transistors	TSC	thermally stimulated current
SK	Stranski–Krastanov	TSL	thermally stimulated luminescence
SNR	signal-to-noise ratio		

U

SO	small outline
SOA	semiconductor optical amplifier
SOC	system-on-a-chip
SOFC	solid oxide fuel cells
SOI	silicon-on-insulator
SP	screen printing
SPECT	single-photon emission computed tomography

ULSI	ultra-large-scale integration
UMOSFET	U-shaped-trench MOSFET
UPS	uninterrupted power systems
UV	ultraviolet

V

SPR	surface plasmon resonance	VAP	valence-alternation pairs
SPVT	seeded physical vapor transport	VB	valence band
SQW	single quantum wells	VCSEL	vertical-cavity surface-emitting laser
SSIMS	static secondary ion mass spectrometry	VCZ	vapor-pressure-controlled Czochralski
SSPC	steady-state photoconductivity	VD	vapor deposition
SSR	solid-state recrystallisation	VFE	vector flow epitaxy
SSRM	scanning spreading resistance microscopy	VFET	vacuum field-effect transistor
STHM	sublimation traveling heater method	VGF	vertical gradient freeze
SVP	saturated vapor pressure	VIS	visible
SWIR	short-wavelength infrared	VOC	volatile organic compounds
		VPE	vapor phase epitaxy
		VRH	variable-range hopping
		VUVG	vertical unseeded vapor growth
		VW	Volmer–Weber

T

W

TAB	tab automated bonding	WDX	wavelength dispersive X-ray
TBA	tertiarybutylarsine	WXI	wide-band X-ray imager
TBP	tertiarybutylphosphine		

X

TCE	thermal coefficient of expansion		
TCNQ	tetracyanoquinodimethane		
TCR	temperature coefficient of resistance	XAFS	X-ray absorption fine-structure
TCRI	temperature coefficient of refractive index	XANES	X-ray absorption near-edge structure
TDCM	time-domain charge measurement	XEBIT	X-ray-sensitive electron-beam image tube
TE	transverse electric	XPS	X-ray photon spectroscopy
TED	transient enhanced diffusion	XRD	X-ray diffraction
TED	transmission electron diffraction	XRSP	X-ray storage phosphor
TEGa	triethylgallium		

Y

TEM	transmission electron microscope		
TEN	triethylamine	YSZ	yttrium-stabilized zirconia
TFT	thin-film transistors		
THM	traveling heater method		
TL	thermoluminescence		
TLHJ	triple-layer graded heterojunction		
TLM	transmission line measurement		
TM	transverse magnetic		

Part C Materials for Electronics

21 Single-Crystal Silicon: Electrical and Optical Properties
Shlomo Hava, Beer Sheva, Israel
Mark Auslender, Beer Sheva, Israel

22 Silicon–Germanium: Properties, Growth and Applications
Peter Ashburn, Southampton, UK
Darren M. Bagnall, Southampton, UK

23 Gallium Arsenide
Mike Brozel, Glasgow, UK

24 High-Temperature Electronic Materials: Silicon Carbide and Diamond
Magnus Willander, Göteborg, Sweden
Milan Friesel, Göteborg, Sweden
Qamar-ul Wahab, Linköping, Sweden
Boris Straumal, Chernogolovka, Russia

25 Amorphous Semiconductors: Structure, Optical, and Electrical Properties
Kazuo Morigaki, Tokyo, Japan
Chisato Ogihara, Ube, Japan

26 Amorphous and Microcrystalline Silicon
Akihisa Matsuda, Chiba, Japan

27 Ferroelectric Materials
Roger Whatmore, Lee Maltings, Ireland

28 Dielectric Materials for Microelectronics
Robert M. Wallace, Richardson, USA

29 Thin Films
Robert D. Gould[†], Keele, UK

30 Thick Films
Neil White, Highfield, UK

21. Single-Crystal Silicon: Electrical and Optical Properties

Electrical and optical properties of crystalline semiconductors are important parts of pure physics and material science research. In addition, knowledge of parameters related to these properties, primarily for silicon and III–V semiconductors, has received a high priority in microelectronics and optoelectronics since the establishment of these industries. For control protocols, emphasis has recently been placed on novel optical measurement techniques, which have proved very promising as nondestructive and even non-contact methods. Earlier they required knowledge of the free-carrier-derived optical constants, related to the electrical conductivity at infrared frequencies, but interest in the optical constants of silicon in the visible, ultraviolet (UV) and soft-X-ray ranges has been revived since the critical dimensions in devices have become smaller.

This chapter surveys the electrical (Sect. 21.2) and optical (Sect. 21.3) properties of crystalline silicon. Section 21.2 overviews the basic concepts. Though this section is bulky and its material is documented in textbooks, it seems worth including since the consideration here focuses primarily on silicon and is not spread over other semiconductors – this makes the present review self-contained. To avoid repeated citations we, in advance, refer the reader to stable courses on solid-state physics (e.g. [21.1, 2]), semiconductor physics (e.g. [21.3]), semiconductor optics (e.g. [21.4]) and electronic devices (e.g. [21.5]); seminal papers are cited throughout Sect. 21.2.

21.1	**Silicon Basics**..	441
	21.1.1 Structure and Energy Bands........	441
	21.1.2 Impurity Levels and Charge-Carrier Population ...	443
	21.1.3 Carrier Concentration, Electrical and Optical Properties	446
	21.1.4 Theory of Electrical and Optical Properties	447
21.2	**Electrical Properties**	451
	21.2.1 Ohm's Law Regime	451
	21.2.2 High-Electric-Field Effects..........	465
	21.2.3 Review Material	471
21.3	**Optical Properties**..............................	472
	21.3.1 Diversity of Silicon as an Optical Material................	472
	21.3.2 Measurements of Optical Constants.....................	472
	21.3.3 Modeling of Optical Constants.....	474
	21.3.4 Electric-Field and Temperature Effects on Optical Constants.........	477
References		478

We realize how formidable our task is – publications on electrical and optical properties of silicon amount to a huge number of titles, most dating back to the 1980s and 1990s – so any review of this subject will inevitably be incomplete. Nevertheless, we hope that our work will serve as a useful shortcut into the silicon world for a wide audience of applied physics, electrical and optical engineering students.

21.1 Silicon Basics

21.1.1 Structure and Energy Bands

Normally silicon (Si) crystallizes in a diamond structure on a face-centered cubic (f.c.c.) lattice, with a lattice constant of $a_0 = 5.43$ Å. The basis of the diamond structure consists of two atoms with coordinates $(0, 0, 0)$ and $a_0/4(1, 1, 1)$, as seen in Fig. 21.1. Other solids that can crystallize in the diamond structure are C, Ge and Sn. The important notion for the electronic band structure is the Brillouin zone (BZ). The BZ is a primitive cell in the reciprocal-space lattice, which proves to be a body-centered cubic (b.c.c.) lattice for an f.c.c. real-space lattice. For this case, the BZ with important reference points and directions within it is shown in Fig. 21.2.

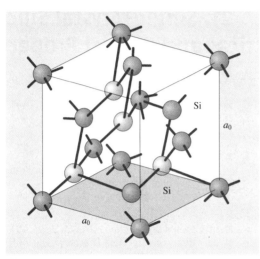

Fig. 21.1 Diamond crystal structure of Si

The states of electrons in solids are described by wave functions of the Bloch type

$$\psi(r) = e^{ik \cdot r} u_{sk}(r), \qquad (21.1)$$

where k is the wave vector that runs over reciprocal space, s is a band index and $u_{sk}(r)$ is the periodic function of the direct lattice (Bloch amplitude). Both $u_{sk}(r)$ and the corresponding energy-band spectrum $E_s(k)$ are periodic in k, which allows one to restrict consideration

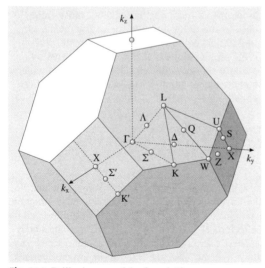

Fig. 21.2 Brillouin zone of the f.c.c. lattice

to within the BZ. The bands are arranged so that there are energy regions for which no states given by (21.1) exist. Such forbidden regions are called energy gaps or band gaps and result from the interaction of valence electron with the ion cores of crystal. In semiconductor science the term *band gap* is accepted for the energy distance between the maximum of $E_s(k)$ for the highest filled (valence) band and the minimum of $E_s(k)$ for the lowest empty (conduction) band (denoted by E_g). The band gap is called *direct* if the aforementioned maximum and minimum occur at the same point of the BZ, e.g. Γ (Fig. 21.2), and *indirect* if they occur at different points of the BZ, e.g. Γ and X (Fig. 21.2).

Si is an indirect-band-gap semiconductor with $E_g = 1.1700$ eV at 4.2 K. The calculated energy-band structure, that is the curves of $E_v(k)$ for selected directions in the BZ, is shown in Fig. 21.3a. The conduction-band minimum lies at six equivalent points Δ on the Γ–X lines (Fig. 21.2). In some vicinity (called the *valley*) of every such point the band spectrum is quadratic in k, e.g. for the valley $\langle 100 \rangle$

$$E_c(k) = E_{c0} + \frac{\hbar^2 (k_x - k_0)^2}{2m_l} + \frac{\hbar^2 \left(k_y^2 + k_z^2\right)}{2m_t}, \qquad (21.2)$$

where $k_0 \approx 1.72\pi/a_0$, m_l and m_t are the longitudinal and transverse effective masses. The spectra for other five valleys are obtained from (21.2) by 90° rotations and inversions $k_0 \to -k_0$. Though the constant-energy surface for (21.2) is an ellipsoid (Fig. 21.3b), the density of states (DOS) proves to be the same as for an isotropic parabolic spectrum with an effective mass

$$m_{de} = 6^{2/3} m_l^{1/3} m_t^{2/3}, \qquad (21.3)$$

which is called the *DOS effective mass*. Another mass, m_{ce}, which appears in the direct-current (DC) and optical conductivity formulas, is defined via the harmonic mean

$$\frac{1}{m_{ce}} = \frac{1}{3}\left(\frac{1}{m_l} + \frac{2}{m_t}\right). \qquad (21.4)$$

For this reason, m_c is called the *conductivity/optical effective mass*. Equation (21.2) holds at $E_c - E_{c0} < 0.15$ eV, but at larger energies the ellipsoids warp strongly, especially near the X point; the change of the spectrum with energy is mostly due to the increasing m_t, while m_l increases weakly [21.6].

The valence-band maximum is at the Γ point ($k = 0$), where the Bloch-wave state $u_{n0}(r)$ has the full symmetry of an atomic p-orbital, being six-fold degenerate in the nonrelativistic limit. The spin–orbit interaction splits off from the bare band top a four-fold degenerate $p_{3/2}$ level up by $1/3\Delta_{so}$, and a two-fold degenerate $p_{1/2}$ level down by $-2/3\Delta_{so}$. The spectrum at $\boldsymbol{k} \neq 0$, even at energies near the top E_{v0}, is very complex. It consists of three branches, which are in general nonparabolic and nonisotropic [21.7]. At $E_{v0} - E_v \ll \Delta_{so}$ there are two nonparabolic anisotropic $p_{3/2}$-derived sub-bands with the energy spectra

$$E_{v1,2}(\boldsymbol{k}) = E_{v0} + \frac{\hbar^2}{2m_0}\left[Ak^2 \right.$$
$$\left. \pm \sqrt{B^2k^4 + C^2\left(k_x^2k_y^2 + k_x^2k_z^2 + k_y^2k_z^2\right)}\right],$$
$$A < 0 \qquad (21.5)$$

and at $E_{v0} - E_v \ll 2\Delta_{so}$ these is an isotropic parabolic $p_{1/2}$-derived split-off band with the energy spectrum

$$E_{v3}(\boldsymbol{k}) = E_{v0} - \Delta_{so} + \frac{\hbar^2 k^2}{2m_0} A',$$
$$A' < 0. \qquad (21.6)$$

Here m_0 is the free electron mass, A, B, C and A' are inverse hole-mass parameters (for small Δ_{so}, $A' \approx A$). In (21.5) the $+$ sign corresponds to heavy holes and the $-$ sign to light holes. The constant-energy surfaces for (21.5) are warped spheres (Fig. 21.3c), the DOS is nevertheless parabolic and is described by the effective mass

$$m_{dh} = \left(m_{d1}^{3/2} + m_{d2}^{3/2}\right)^{2/3}, \qquad (21.7)$$

where $m_{d1,2}$ are the partial DOS effective masses. Quite different masses enter various physical quantities for heavy (light) holes, and are complicated functions of A, B and C ([21.9]); however, the split-off band is characterized by only one effective mass: $m_3 = m_0/|A'|$.

These band-structure parameters, obtained from cyclotron-resonance data and calculations, are presented in Tables 21.2 and 21.3. The unreferenced values of m_{de}, m_{ce} and m_3 were calculated from the referenced data using (21.3, 4) and the assumption $A' = A$. Optical data for m_{ce} are discussed in Sect. 21.3, while the specific-heat data for m_{de} are not considered here.

Considerable uncertainty and errors in the values of B and C have a small effect on the light-hole effective-mass values (m_{d2} and optical m_{c2}) but lead to an ambiguity in the heavy-hole effective-mass (m_{d1} and optical m_{c1}) values (Table 21.3). In each data set $m_{d2} \approx m_{c2}$, which means that an isotropic approximation is reasonable for the light-hole spectrum.

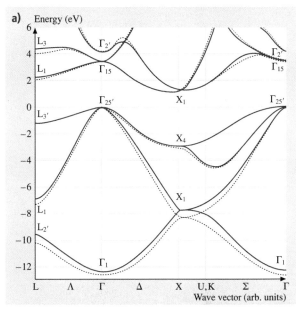

Fig. 21.3 Electronic band structure of Si:(a) Energy dispersion curves near the fundamental gap. After [21.8] with permission;

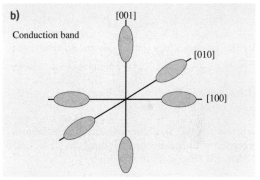

Fig. 21.3 (b) The constant-energy ellipsoids of the conduction band;

The experimental data [21.10, 11] and band-structure calculations [21.12] are in good agreement; the former is used in theoretical papers on hole transport in Si [21.13, 14].

Nonparabolic parts in the electron [21.6] and hole spectra [21.7] lead to apparent dependence of effective-mass measurements on the temperature and carrier concentration [21.15]. Recently the DOS mass issue was revisited [21.16] in connection with the intrinsic carrier concentration.

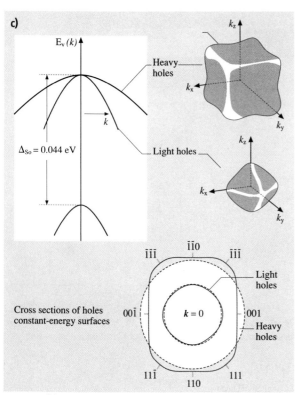

Fig. 21.3 (c) The constant-energy warped spheres and schematic of the valence band

21.1.2 Impurity Levels and Charge-Carrier Population

In highly purified Si electrons and holes are thermally generated with equal concentrations ($n = p$), equal to the intrinsic concentration n_i given by

$$n_i = 2\left(\frac{2\pi\sqrt{m_{de}m_{dh}}k_B T}{h^2}\right)^{3/2} \exp\left(-\frac{E_g}{2k_B T}\right), \quad (21.8)$$

where k_B and h are the Boltzmann and Planck constants, respectively, and T is the absolute temperature. To assess $n_i(T)$ theoretically one should take into account possible changes of $m_{de,h}$ and E_g with T. An early experimental plot of $n_i(T)$ for Si is shown in Fig. 21.4, giving $n_i(300\,\text{K}) = 1.38 \times 10^{10}\,\text{cm}^{-3}$. This and the early textbook value ($1.45 \times 10^{10}\,\text{cm}^{-3}$) do not fit (21.8) with the established low-temperature values of $m_{de,h}$ and $E_g(300\,\text{K}) = 1.1242\,\text{eV}$, giving $(0.662\text{–}0.694) \times 10^{10}\,\text{cm}^{-3}$ for the best available data.

This discrepancy was eliminated [21.16] by critically evaluating measurements of $n_i(T)$ and incorporating the latest data for the hole DOS effective mass. The reassessed value of $n_i(300\,\text{K})$, verified by measurements on solar-cell devices [21.17], is $1.08(8) \times 10^{10}\,\text{cm}^{-3}$ (in a very recent textbook [21.18] the unreferenced value $n_i(300\,\text{K}) = 1.00 \times 10^{10}\,\text{cm}^{-3}$ is adopted). This is obtained from (21.8) using $m_{de}(300\,\text{K}) = 1.08 m_0$, which is quite close to the tabular values presented, and $m_{dh}(300\,\text{K}) = 1.15 m_0$, which is 1.8–2.1 times larger than the tabular values presented. Such a strong increase of m_{dh} when increasing T from 4.2 to 300 K is anticipated from the nonparabolic part of the hole spectra [21.7].

In impure samples $n \neq p$; the charge-neutrality condition is then

$$n - N_d^+ = p - N_a^- \quad (21.9)$$

together with the known expressions for n, p (with $np = n_i^2$ holding until the carriers become degenerate) and the concentrations of charged donors N_d^+ and acceptors N_a^- via the Fermi energy E_F and impurity ionization energies

Table 21.1 Energy levels of impurities in Si

Impurity	Donor/acceptor	Below conduction band (eV)	Above conduction band (eV)
Ag	D	0.310	
	A		0.210
Al	A		0.057
As	D	0.049	
Au	D		0.330
	A	0.540	
B	A		0.045
Cu	A		0.490
	D		0.240
Fe	D	0.550	
	D		0.400
Ga	A		0.065
In	A		0.160
Li	D	0.033	
Mn	D	0.530	
P	D	0.044	
Sb	D	0.039	
S	D	0.180	
	D		0.370
Zn	A	0.550	
	A		0.300

Table 21.2 The conduction-band effective masses and valence-band parameters for Si

Conduction band Mass	m_l/m_0	m_t/m_0	m_{de}/m_0	m_{ce}/m_0
Experiment	0.97 ± 0.04[a]	0.19 ± 0.01[a]	1.08 ± 0.05	0.26 ± 0.01
	0.9163 ± 0.04[b]	0.1905 ± 0.0001[b]	1.0618 ± 0.0005	0.2588 ± 0.0001
Theory	0.971[c]	0.205[c]	1.137	0.2652
	0.9716[d]	0.1945[d]	1.0978	0.2780

Valence band Parameter	Δ_{so}	A	$|B|$ or B	$|C|$	m_3/m_0
Experiment	0.0441 ± 0.004[g]	-4.1 ± 0.2[a]	1.6 ± 0.2[a]	3.3 ± 0.2[a]	0.24 ± 0.01
		-4.28 ± 0.02[e]	-0.75 ± 0.04[e]	5.25 ± 0.05[e]	0.234 ± 0.001
		-4.27 ± 0.02[h]	-0.63 ± 0.08[h]	4.93 ± 0.15[h]	0.234 ± 0.001
Theory		-4.38[c]	0.84[c]	4.11[c]	0.23
	0.04[d]	-4.38[d]	-1.00[d]	4.80[d]	0.23
	0.044[f]	-4.22[f]	-0.78[f]	4.80[f]	0.24

[a] [21.19], [b] [21.20], [c] [21.21], [d] [21.22], [e] [21.10], [f] [21.12], [g] [21.23], [h] [21.11]

Table 21.3 The valence-band effective masses calculated using experimental data

Mass	m_{d1}/m_0	m_{d2}/m_0	m_d/m_0	m_{c1}/m_0	m_{c2}/m_0
Exp.[a]	0.55 ± 0.12	0.16 ± 0.01	0.61 ± 0.12	0.51 ± 0.10	0.16 ± 0.01
Exp.[b]	0.58 ± 0.02	0.151 ± 0.001	0.63 ± 0.02	0.43 ± 0.01	0.145 ± 0.001
Exp.[c]	$0.49 - 0.56$	$0.153 - 0.158$	$0.54 - 0.62$	$0.40 - 0.43$	$0.147 - 0.152$

[a] [21.19], [b] [21.10], [c] [21.11]

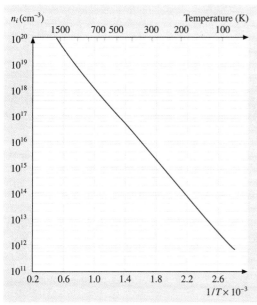

Fig. 21.4 Intrinsic concentration in Si versus temperature. After [21.24] with permission

E_d and E_a allows one to calculate E_F (and hence all the concentrations) as a function of T and doping.

An impurity is called *shallow* if $|E_{c(v)0} - E_{d(a)}| \ll E_g$, and *deep* if $|E_{c(v)0} - E_{d(a)}| \approx 0.5 E_g$. Shallow group V donors (Sb, P, As) and group III acceptors (B, Al, Ga, In) are well soluble in Si [21.25]. The ionization energy was calculated using the effective-mass approximation, and the Bohr model for donors [21.26–28] (Kohn and Luttinger, 1955; Kohn, 1955) and acceptors (Luttinger and Kohn, 1955). The value of $E_{d(a)}$ calculated in this way is insensitive to the specific shallow donor (acceptor). Actual thermal (i.e. retrieved from electrical measurements) ionization energies are given in Table 21.1. It is seen that shallow impurities have different ionization energies. This difference is small for donors and larger for acceptors. The same trend was observed for optical ionization energies, although they are different from the thermal values [21.29, 30]. For shallow impurities, $E_{d(a)}$ decreases as doping becomes heavier, and becomes zero as $N_{d(a)}$ approaches the corresponding insulator–metal transition concentration [21.31].

Deep impurities (except for Mn, Fe and Zn) are amphoteric, i.e. they act simultaneously as donors and acceptors. For such a donor (acceptor) state, the level may lie closer to the valence (conduction) than the conduction (valence) band. The deep impurities are mostly unionized at room temperature due to their large $E_{d(a)}$, so their direct contribution to n or p is negligible. The unionized deep impurities may, however, trap the carriers available from the shallow impurities or injection, thus decreasing the conductivity or the minority-carrier lifetime. Atoms that behave in Si in this manner, for example Au, Ag and Cu, are added for lifetime control. The properties of these impurities in Si have been studied in detail (see, e.g., [21.32, 33]).

21.1.3 Carrier Concentration, Electrical and Optical Properties

Concentration and Electrical Measurements
Measurements of carrier concentrations, as well as electrical and optical characteristics are most tractable if either $n \gg p$ (strongly n-type conduction) or $p \gg n$ (strongly p-type conduction). Since the np product is constant versus doping, the contribution of minority carriers to the conductivity becomes unimportant when $N_{d(a)}$ increases significantly over n_i. A standard route for determining $n(p)$ is Hall-effect measurements. The Hall coefficient R^H, measured directly on a long thin slab in a standard crossed electric and magnetic field configuration, is retrieved by

$$R^H = 10^{-8} \frac{V_H d}{IB} , \qquad (21.10)$$

where V_H is the Hall voltage (Volts), I is the current (Amps), d is the sample thickness (cm) in the z-direction, and B is the magnetic field strength (Gauss) applied in this direction. There are two limiting cases. One, the high-field regime is defined by $qB\tau/mc \gg 1$, where m and τ are the appropriate mass and relaxation-time parameters, respectively. In this case

$$R_e^H(\infty) = -\frac{1}{qn} ,$$
$$R_h^H(\infty) = \frac{1}{qp} . \qquad (21.11)$$

The other, low-field, regime holds with the opposite inequality; in this regime

$$R_e^H(0) = -\frac{r_e}{qn} ,$$
$$R_h^H(0) = \frac{r_h}{qp} , \qquad (21.12)$$

where the constant of proportionality $r_{e(h)}$, called the electron (hole) Hall factor, depends on the details of the scattering process and band structure. Thus the majority-carrier concentration is determined directly from $R_{e(h)}^H$ using a high-field Hall measurement. For typical laboratory magnetic fields, this regime is attainable only with extremely high mobility and low effective mass, which excludes moderately and heavily doped Si, for which very high magnetic fields are required. In some cases the Hall factor is quite close to unity, e.g. $r_e = 3\pi/8$ for the phonon scattering in the isotropic and parabolic (standard) band.

The electrical properties are fully described by the drift-diffusion relation for the electron (hole) current density $\boldsymbol{j}_{e(h)}$

$$\boldsymbol{j}_e = -qn\boldsymbol{v}_{de} + qD_e \nabla n ,$$
$$\boldsymbol{j}_h = qp\boldsymbol{v}_{dh} - qD_h \nabla p , \qquad (21.13)$$

where $\boldsymbol{v}_{de} = -\mu_e E$ ($\boldsymbol{v}_{dh} = \mu_h E$) is the drift velocity, $\mu_{e(h)}$ is the drift mobility, E is the electric field strength, and $D_{e(h)}$ is the diffusion coefficient; in general, $\mu_{e(h)}$ and $D_{e(h)}$ depend on E. In the homogeneous case (21.13) converts into the material equation $\boldsymbol{j}_{e(h)} = \sigma_{e(h)} E$, where $\sigma_{e(h)} = qn\mu_{e(h)}$ is the electron (hole) conductivity; the total conductivity equals $\sigma = \sigma_e + \sigma_h$. In the weak-field DC (Ohm) and alternating current (AC: microwave or light, except for intense laser, irradiation) regimes, $D_{e(h)}$ is proportional to $\mu_{e(h)}$ being both constant versus E, depending on the radiation frequency ω.

Combining the high-induction Hall and Ohm resistivity ($\rho = \sigma^{-1}$) measurements one obtains the drift mobility

$$\mu_{e(h)} = R_{e(h)}^H(\infty)\sigma . \qquad (21.14)$$

Replacing $R_e^H(\infty)$ by $R_e^H(0)$ in the right-hand side of (21.14), one arrives at the so-called Hall mobility

$$\mu_{e(h)}^H = R_{e(h)}^H(0)\sigma = r_{e(h)}\mu_{e(h)} , \qquad (21.15)$$

which never equals the drift mobility, although it may be fairly close to it in the cases mentioned above. In general, to extract $n(p)$ from $R_e^H(0)$, the calculation of the $r_{e(h)}$ factor is completed. Magnetoresistance (MR), i.e. ρ versus B measurement, is an important experimental tool as well. Another established method is the Haynes–Shockley experiment, which allows one to measure the minority-carrier drift mobility. In high-electric-field conditions, a noise-measurement technique is used. A relatively novel, time-of-flight technique was used in the latest (to our knowledge) mobility and diffusion-coefficient measurements on lightly doped crystalline Si samples, both in the low- and high-field regimes [21.14].

Basic Optical Parameters

The electromagnetic response of homogeneous nonmagnetic material is governed by the dielectric constant tensor ε, which connects the electric displacement vector $\boldsymbol{D}$ inside the material to $\boldsymbol{E}$ through the material equation $\boldsymbol{D} = \varepsilon \boldsymbol{E}$. For cubic crystals, such as Si, ε is a scalar. An effective-medium homogeneous dielectric constant may be attributed to inhomogeneous and composite materials if the nonhomogeneity feature size is smaller than the radiation wavelength $\lambda = 2\pi c/\omega$. Actually, ε characterizes the material's bulk and therefore loses its sense in nanoscale structures (superlattices, quantum wells etc). The dependence $\varepsilon(\omega)$ expresses the optical dispersion in the material. The dielectric constant is usually represented via its real and imaginary parts: $\varepsilon = \varepsilon_1 + \mathrm{i}\varepsilon_2$ ($\varepsilon_2 \geq 0$), connected to each other by the Kramers–Kronig relation (KKR)

$$\varepsilon_1(\omega) = 1 + \frac{2}{\pi} \int_0^\infty \frac{\Omega}{\Omega^2 - \omega^2} \varepsilon_2(\Omega) \mathrm{d}\Omega \ . \tag{21.16}$$

At low frequency (radio, microwave), in the absence of magnetic fields, $\varepsilon_1 \approx \varepsilon(0)$ and ε_2, which is responsible for dielectric loss, is small. At optical wavelengths, from far-IR to soft X-rays, the basic quantity is the complex refractive index $\sqrt{\varepsilon} = n + \mathrm{i}k$. The real refractive index n, which is responsible for wave propagation properties, and the extinction index k, responsible for the field attenuation, are referred to as *optical constants*. They are related to the dielectric constant via:

$$\varepsilon_1 = n^2 - k^2 \ ,$$
$$\varepsilon_2 = 2nk \ ,$$
$$n = \sqrt{\frac{\left(\varepsilon_1^2 + \varepsilon_2^2\right)^{1/2} + \varepsilon_1}{2}} \ ,$$
$$k = \sqrt{\frac{\left(\varepsilon_1^2 + \varepsilon_2^2\right)^{1/2} - \varepsilon_1}{2}} \ . \tag{21.17}$$

21.1.4 Theory of Electrical and Optical Properties

Boltzmann-Equation Approach

The response of carriers in a band to perturbations away from the thermal–equilibrium state, such as applied electric and magnetic fields or impinging electromagnetic radiation, is described by the deviation of the carrier distribution function $f_s(\boldsymbol{k}, \boldsymbol{r}, t)$ from the equilibrium Fermi–Dirac distribution $f_0[E_s(\boldsymbol{k})]$. The current density equals $\boldsymbol{j}_s = q \int \boldsymbol{v}_s(\boldsymbol{k}) f_s(\boldsymbol{k}, \boldsymbol{r}, t) \mathrm{d}\boldsymbol{k}$, where $\boldsymbol{v}_s(\boldsymbol{k}) = \frac{\partial E_s(\boldsymbol{k})}{\hbar \partial \boldsymbol{k}}$ is the microscopic carrier velocity and the integration is performed over the BZ. The process that balances the external perturbations is scattering of carriers by lattice vibrations (phonons), impurities and other carriers. Impurity scattering dominates transport at low temperatures and remains important at room temperature for moderate and high doping levels, although carrier–carrier scattering also becomes appreciable. Under appropriate conditions, one being that $\hbar\omega \ll \bar{E}$ (where $\bar{E}$ is the average carrier kinetic energy), $f_s(\boldsymbol{k}, \boldsymbol{r}, t)$ satisfies the quasi-classical Boltzmann kinetic equation. In the opposite, quantum, range, radiation influences the scattering process. Generalized kinetic equations, which interpolate between the quasi-classical and quantum regimes, have also been derived [21.34].

There exist various methods of solving the quasi-classical Boltzmann equation. The relaxation-time method, variational method [21.35–37] for low electric fields, and displaced–Maxwell–distribution approximation [21.38] for high electric fields, were used in early studies. In the last three decades the Monte Carlo technique [21.39], which overcomes limitations inherent to these theories and allows one to calculate subtle details of the carrier distribution, has been applied to various semiconductors, including crystalline Si [21.14]. If $\hbar\omega \ll \bar{E}$, the kinetic and optical characteristics are calculated well using transition probabilities between carrier states, with the radiation quantum absorbed or emitted [21.40]. The most problematic is the interme-

Table 21.4 Parameters of the phonon modes in crystalline Si

Mode	Energy (K)				Sound velocity
LO	760	700–735[a]	560	580	
TO	760	–	630–690[a]	680	
LA	0	240–260	500–510[a]	580	8.99×10^5 cm/s
TA	0	140–160	210–260	220	5.39×10^5 cm/s
q	Γ	Δ	S	X	

[a] [21.19]

diate range – generalized kinetic equations have been shown to recover the extreme ranges, but no working methods have been developed for solutions at $\hbar\omega \approx \bar{E}$, to the best of our knowledge. The relaxation-time approximation has proved to work well in many cases of scattering in Si. In this framework, the basic quantity is the relaxation-time tensor $\tau_s(E)$, where s is the index indicating the conduction or valence band, and E is the carrier kinetic energy. For electron valleys, $\tau_c(E)$ has the same symmetry as the respective effective-mass tensor – it is diagonal with principal values of, e.g., τ_l, τ_t, τ_t for $\langle 100 \rangle$ etc. For scalar holes $\tau_{1,2}(E)$ can only be introduced in the isotropic-bands approximation. The mobility and the Hall factor for weak electric fields in the relaxation-time approximation are given by

$$\mu_e(\omega) = \frac{q}{3}\left\langle \frac{\tau_l/m_l}{1-i\omega\tau_l} + \frac{2\tau_t/m_t}{1-i\omega\tau_t} \right\rangle,$$

$$r_e = \frac{3\langle 2\tau_l\tau_t/m_l m_t + (\tau_t/m_t)^2 \rangle}{\langle \tau_l/m_l + 2\tau_t/m_t \rangle^2};$$

$$\mu_h(\omega) = \frac{q}{1+\beta}\left\langle \frac{\tau_1/m_1}{1-i\omega\tau_1} + \frac{\beta\tau_2/m_2}{1-i\omega\tau_2} \right\rangle,$$

$$r_h = \frac{(1+\beta)\langle (\tau_1/m_1)^2 + \beta(\tau_2/m_2)^2 \rangle}{\langle \tau_1/m_1 + \beta\tau_2/m_2 \rangle^2}, \quad (21.19)$$

where the angular brackets indicate averaging with the weight $-E^{3/2} f_0'(E)$, $\beta = (m_{d2}/m_{d1})^{3/2}$ is the density ratio of light holes to heavy holes, and the option for a nonparabolic band is retained.

Lattice Scattering

Deformational phonons – longitudinal, transverse acoustical (LA, TA) and optical (LO, TO) – mediate carrier–lattice scattering in Si. The phonon modes are presented in Table 21.4, where q is a point in the phonon BZ. The points Δ and S correspond to the scattering process, where the electron transits between the bottoms of two perpendicularly (f) and parallel (g) oriented valleys, respectively. The phonon energies are precise at the points Γ and X, as determined by neutron-scattering techniques [21.42], but uncertain at Δ and S, since in this case only estimation and fitting methods were available.

The rigid- and deformable-ion lattice models have been used to obtain the carrier–phonon interaction for electrons [21.43] and holes [21.44]. A deformation-potential theory of the interaction with long-wavelength phonons, which takes the crystal symmetry and band structure fully into account in a phenomenological manner, has been developed. This theory deduces two constants, Ξ_u, Ξ_d, for the conduction band [21.45] and four, a, b, d, d_{opt}, for the valence band [21.46,47], which are presented in Table 21.5. The deformation-potential theory has been used to calculate the acoustic scattering-limited mobility in n-Si [21.48] and p-Si [21.49]. Later, optical-phonon scattering, along with an approximate valence-band spectrum instead of (21.5), were taken into account [21.13, 14].

The matrix elements of electron–phonon interaction between wave functions of different valleys are not taken into account by the deformation-potential theory. For inter-valley transitions, other than the three marked in Table 21.4 by the superscript 'a', the matrix elements calculated at the valley-bottom wave vectors are zero [21.43, 50]. The actual scattering probabilities are never zero; for those forbidden by selection rules [21.43, 50] one should take into account the wave-vector offset at the final scattering state, which gives nominally small, but unknown values. Several inter-valley scattering models have been tried to fit the theoretical formulas to the mobility data in lightly doped n-Si: with one allowed TO and one forbidden TA phonon [21.51], one allowed TO phonon [21.52] and more involved combinations of the transitions [21.53, 54]. The scattering of electrons by long-wavelength optical phonons, regarded as a cause of drift-velocity saturation at high electric fields [21.38], is forbidden in Si [21.43]. In n-Si, by all accounts, the cause may be the allowed g-phonon (Table 21.4) scattering [21.54].

Impurity Scattering

There are two types of impurity scattering – by ionized and neutral impurities. The latter is the dominant impurity scattering for uncompensated, light or moderate shallow-impurity doping, at low T. In samples doped with deep impurities, neutral-impurity scattering may also show up. At elevated T, when shallow impurities are increasingly ionized, ionized-impurity scattering is the dominant impurity scattering and may compete with

Table 21.5 Deformation-potential parameters [21.41]

T(K)	Ξ_u(eV)	$\Xi_d + 1/3\Xi_u - a$(eV)	$|b|$(eV)	$|d|$(eV)
80	8.6 ± 0.2	3.8 ± 0.5	2.4 ± 0.2	5.3 ± 0.4
295	9.2 ± 0.3	3.1 ± 0.5	2.2 ± 0.3	–

phonon scattering, depending on $N_{d(a)}$. Lastly, in heavily doped samples, where the impurities are ionized for all T, ionized-impurity scattering is dominant up to 300 K.

Early theories of impurity scattering were developed for carriers in parabolic bands, scattered by hydrogen-like centers. For neutral impurities the s-scattering cross section [21.55] and a cross section that takes allowance of the scattered carrier's bound state [21.56] were adopted. For ionized centers, use was made of the Coulomb scattering cross section, cut off at a small angle depending on $N_{d(a)}$ [21.57], and the screened Coulomb potential cross section, calculated in the Born approximation [21.58–60]. These theories consider scattering by the donors and acceptors on an equal footing. The use of the Conwell–Weisskopf formula for $\tau_s(\varepsilon)$ declined towards the end of the 1950s, while the corresponding Brooks–Herring (BH) formula became widespread, mostly due to the consistency of its derivation, even though none of the assumptions for its validity are completely satisfied. This formula was corrected [21.61], on account of the band carrier's degeneracy, compensation and screening by carriers on impurity centers. The discrepancy between experiment and the BH formula, revealed during three decades of studies, have been thoroughly analyzed [21.62].

Modifications and developments made to overcome the drawbacks of the BH formula are worth mentioning. Taking the multi-valley band structure into account did not invalidate the relaxation-time method as such, but resulted in essentially different $\tau_l(E)$ and $\tau_t(E)$ [21.63, 64]. These formulas have also been discussed [21.65] in light of the scattering anisotropies measured in n-Si [21.66]. To overcome limitations of the Born approximation exact, although limited only to the standard band, phase-shift analysis was employed [21.67]. Including a non-Coulomb part of the impurity potential [21.68] made it possible to explain in part the difference in mobility of n-Si samples doped with different donors [21.25]. Lastly, Monte Carlo simulations of the impurity scattering, improving the agreement between theory and experiment for n-Si, have recently been reported [21.69]. The ionized-impurity scattering in p-Si was considered in the approximation of isotropic hole bands [21.70]. We are not aware of any theoretical papers on the subject, which used anisotropic energy spectra given by (21.26) in the case of p-Si.

Carrier–Carrier Scattering

Carrier–carrier scattering becomes important as n or p increases, along with increasing N_d^+ or N_a^-. The relaxation-time concept does not apply for this mechanism. Carrier–carrier collisions redistribute the carrier's energy in a chaotic manner that was presumed to cause a decrease in the net mobility due to other mechanisms [21.71]. For the standard band, the effects of electron–electron scattering were modeled using the variational method, which predicted a $\approx 30\%$ reduction in the ionized-impurity scattering-limited mobility [21.72]; close results were obtained using another, quite different, method [21.73]. Hole–hole scattering and electron–hole scattering were also considered [21.74] in the standard band. Due to ignorance of the specific band-structure features, the results of these papers had limited relevance to Si. The effect of electron–electron scattering has been recast [21.75] for the multi-valley band structure using the generalized Drude approximation (GDA). At DC, the GDA corresponds to the zeroth-order approximation of the variational method, which highly overestimates [21.72] the effect considered.

Dielectric Constant

In Si the current carriers are well decoupled from the host electrons, so the Maxwell equations result in a unique decomposition of the dielectric constant

$$\varepsilon(\omega) = \varepsilon_L(\omega) + \varepsilon_C(\omega), \; \varepsilon_C(\omega) = i\frac{4\pi\sigma(\omega)}{\omega}. \quad (21.20)$$

Here $\varepsilon_L(\omega)$ is the host contribution, which is indirectly influenced by the carriers. The direct effect of the carriers on the dielectric constant is the conductivity contribution $\varepsilon_C(\omega)$. As seen from (21.20), doped Si behaves at DC as a metal. The asymptote at high frequencies (IR for Si) is $\varepsilon_{C,s}(\omega) \approx -(\Omega_{\text{pl},s}/\omega)^2$, where $s = $ e or h, and $\Omega_{\text{pl},s}$ are the bare plasma frequencies given by

$$\Omega_{\text{pl,e}}^2 = \frac{4\pi q^2 n}{m_{ce}}, \; \frac{1}{m_{ce}} = \frac{1}{3}\left\langle\frac{1}{m_l} + \frac{2}{m_t}\right\rangle,$$

$$\Omega_{\text{pl,h}}^2 = \frac{4\pi q^2 p}{m_{ch}}, \; \frac{1}{m_{ch}} = \frac{1}{1+\beta}\left\langle\frac{1}{m_1} + \frac{\beta}{m_2}\right\rangle,$$

$$(21.21)$$

resulting from (21.18, 19, 20), irrespective of the scattering model. Due to this asymptote and (21.20) $\varepsilon_1(\omega)$ should become zero at some frequency $\omega_{\text{pl},s}$ and doped Si should behave optically as a dielectric at $\omega < \omega_{\text{pl},s}$ and as a metal at $\omega > \omega_{\text{pl},s}$. True plasma frequencies are estimated roughly as $\omega_{\text{pl},s} \approx \Omega_{\text{pl},s}/n_L$, where $n_L = valrange3.423.44$ is the Si host refractive index in the IR. For the parabolic bands, m_{ce} has already been presented in Table 21.2, and $m_{ch} = (0.33–0.39)m_0$ using the data of Table 21.2, with the assumption that $m_{1,2} = m_{c1,2}$.

General formulas for $\varepsilon_C(\omega)$ are rather involved because of the averaging over E they contain, and so are rarely used. The Drude formula

$$\varepsilon_{C,s}(\omega)\big|_{\text{Drude}} = \frac{i}{\omega} \cdot \frac{\Omega_{\text{pl},s}^2}{\gamma_s - i\omega}, \quad (21.22)$$

where γ_s are adjustable phenomenological damping parameters, is often employed instead [21.76]. To match the behavior of $\varepsilon_C(\omega)$ at $\omega \to 0$ one should put $\gamma_s = 1/\tau_{0,s}$ in (21.22), where

$$\tau_{0,e} = \frac{m_{ce}}{3}\left\langle \frac{\tau_l}{m_l} + \frac{2\tau_t}{m_t} \right\rangle;$$

$$\tau_{0,h} = \frac{m_{ch}}{1+\beta}\left\langle \frac{\tau_1}{m_1} + \frac{\beta\tau_2}{m_2} \right\rangle \quad (21.23)$$

are the DC mobility relaxation times. Such an adjustment was shown to work poorly in n-Si [21.77]. On the other hand, putting $\gamma_s = \gamma_{\infty,s}$ in (21.22), where

$$\gamma_{\infty,e} = \frac{m_{ce}}{3}\left\langle \frac{1}{m_l\tau_l} + \frac{2}{m_t\tau_t} \right\rangle;$$

$$\gamma_{\infty,h} = \frac{m_{ch}}{1+\beta}\left\langle \frac{1}{m_1\tau_1} + \frac{\beta}{m_2\tau_2} \right\rangle, \quad (21.24)$$

allows one using the Drude formula to match *two* leading at $\omega \to \infty$ terms in the power series expansion of $\varepsilon_C(\omega)$ with respect to ω^{-1}. Thus (21.22) with the above adjustments may serve as an overall interpolation if the *high-frequency* relaxation time, $\tau_{\infty,s} = 1/\gamma_{\infty,s}$ turns out to be close to $\tau_{0,s}$. A Drude formula, empirically adjusted in IR has been devised for n-Si [21.78].

Using $\tau_{\infty,s}$ instead of $\tau_{0,s}$ in the mobility is a prerequisite for GDA at DC. As discussed above, the Boltzmann-equation-based formulas are valid in the range $\lambda \gg \lambda_q = hc/\bar{E}$. For nondegenerate carriers $\lambda_q(\text{cm}) \approx 1.4388/T$, while in Si for n or p up to $\approx 10^{20}$ cm^{-3} the carrier degeneracy (if present) stops much below 300 K, so the validity of the $\varepsilon_C(\omega)$ formulas considered is restricted to the far-IR and longer wavelengths. To properly describe $\varepsilon_C(\omega)$ in the near- and mid-IR range, ω-dependent GDA has been suggested [21.79]. In this approximation one replaces γ_s in (21.22) by an ω-dependent damping $\gamma_s(\omega)$, which is then determined by comparison of the first imaginary term in the expansion of the thus-generalized Drude formula, i.e. $(\Omega_{\text{pl},s}/\omega)^2\gamma_s(\omega)/\omega$, with $\varepsilon_2(\omega)$ calculated using the methods of transition probabilities or perturbations for correlation functions.

In Si, unlike semiconductors with ionic bonds (e.g. $A^{III}B^V$), the elementary cell has no dipole moment and hence no quasi-classical optical-phonon contribution in $\varepsilon_L(\omega)$ is present. However, several weak IR absorption bands, attributed to two-phonon interaction of light with the Si lattice, are observed. Of these bands the most prominent is that peaked at 16.39 μm, which undergoes about a twofold increase in absorption upon increasing the temperature from 77 to 290 K. Comparable to that, the 9.03 − μm absorption band, observed in pulled Si crystals, was attributed to Si−O bond stretching vibrations [21.80]. At $\lambda < 1.2$ μm, where $\varepsilon_L(\omega)$ dominates the dielectric constant irrespective of the doping, $\varepsilon_{2L}(\omega)$ is accounted for by inter-band electronic transitions. The spectral bands of $\varepsilon_{2L}(\omega)$ correspond to the absorption of photons with energies close to the band gaps. Bare indirect-band-gap transitions, that necessitates lowest energy, is forbidden, but perturbation correction in the electron–phonon interaction to $\varepsilon_{2L}(\omega)$ suffices to describe the observed indirect absorption band. Due to the phonons the lowest fundamental absorption, although smaller than in direct bands, increases with increasing T. In principle, $\varepsilon_{2L}(\omega)$ may be calculated by the band-structure simulation route (e.g. Kleinman and Phillips [21.6, 8, 21]), especially with the present state-of-the-art theory − $\varepsilon_{1L}(\omega)$ is then calculated using the KKR (21.16). However, applications need fast modeling, and such formulas have been developed for Si [21.81–83]. There are a few distinct doping effects on $\varepsilon_L(\omega)$:

1. Effects on the lowest band edges, both direct and indirect: the Burstein–Moss shift with increasing $n(p)$ in heavily doped samples at the degeneracy due to filling of the conduction (valence) band below (above) E_F; band-gap shrinkage due to carrier–carrier and carrier–impurity interactions [21.84, 85] that work against the Burstein–Moss shift; and the formation of band tails because of the random potential of impurities (e.g. [21.31]).

2. Effects on higher edges, such as: E_1 (3.4 eV) – due to transitions between the highest valence band and the lowest conduction band along the Λ line in a region from $\pi/4a_0(1, 1, 1)$ to the L point on the BZ edge; and E_2(4.25 eV) – due to transitions between the valence band at the X point and the conduction band at $2\pi/a_0(0.9, 0.1, 0.1)$ [21.8]. In this case [21.86], the electron–electron interaction plays a small role because carriers are located in a small region of the BZ, different from that where the transitions take place,

and the effect of the electron–impurity interaction is calculated using standard perturbation theory.
3. Absorption due to direct inter-conduction-band (inter-valence-band) transitions specific to the type of doping. In n-Si this is a transition from the lowest conduction band to the band that lies higher at the Δ point but crosses the former at the X point, and which gives rise to a broad absorption band peaked around 0.54 eV [21.87] and tailing of the indirect gap at heavy doping [21.88]; the theory of this contribution to $\varepsilon_{2L}(\omega)$ has been developed [21.85]. In p-Si these transitions are those between the three highest valence bands [21.7]. Absorption due to the 1→2 transition has no energy threshold, and so resembles the usual free-carrier absorption; that due to the 1→3 and 2→3 transitions appears at $\hbar\omega = \Delta_{so}$. A high-energy threshold, above which the inter-valence-band absorption becomes negligible, exists due to the near congruency of all three valence sub-bands [21.7] at large well k (this is not accounted for by (21.5, 6), which are valid at small k). In contrast, with p-Ge, the manifestation of the inter-valence-band transitions in the reflection, was proposed for p-Si [21.89], but fully reconciled later [21.85, 90]. Kane's theory of the inter-valence-band absorption [21.7] has also been revisited [21.90].

21.2 Electrical Properties

An extensive investigation of basic electrical properties was started 55 years ago, when polycrystalline Si containing B and P, was reported [21.91]. This seminal work was necessarily limited because neither single crystals nor the means for measuring below 77 K were then available. The first papers on single crystals were published five years afterwards. Since then, as techniques for fabricating quality single-crystalline silicon, such as the pulling, e.g. Czochralski (CZ), Teal–Little (TL), and floating-zone (FZ) techniques, became highly developed, many experiments on electrical properties have been published. A number of papers are considered below in historical retrospect. In the accompanying graphs additional, less cited, papers are referenced.

Though the physical mechanisms behind the electrical properties of crystalline Si have been studied and partially understood for a long time, the resulting formulas and procedures are too complicated and time-consuming to be used in electronics device modeling. In this connection, several useful, simplified but accurate, procedures for modeling mobility versus temperature, doping, injection level and electric field strength have been developed. For this issue we refer to points 2–4 in Sect. 21.2.3.

21.2.1 Ohm's Law Regime

Drift Measurements
1. The minority-carrier mobility as a function of N_d, N_a, T and ρ in n- and p-type samples is in the range 0.3–30 Ωcm [21.92]. This cited paper revealed for the first time the inapplicability, at least for holes, of the simple $T^{-1.5}$ lattice mobility law, and presented curves of ρ versus exhaustion concentration $N = |N_d - N_a|$ in the range 10^{14} cm^{-3} ≤ N ≤ 10^{17} cm^{-3}.
2. Measurements of μ_e in p-type and μ_h in n-type Si on 11 single crystals ranging in ρ from 19 to 180 Ωcm [21.93]. In the purest crystals, in the range 160–400 K, μ_e and μ_h obeyed the dependencies $T^{-2.5\pm0.1}$ and $T^{-2.7\pm0.1}$, respectively. The conductivity of some of these crystals, measured from 78 to 400 K, provided independent evidence for the temperature dependencies of the mobility quoted above.
3. Room-temperature drift and conductivity plus Hall-effect measurements [21.94] of μ_e and μ_h versus resistivity on an unprecedentedly large number of samples cut from CZ crystals. The largest ρ was above 200 Ωcm. The values of μ_e and μ_h obtained from both experiments in the purest crystals were reported and compared with those obtained by other authors.

Resistivity and Galvanomagnetic Measurements
1. Room-temperature $\mu_{e(h)}^H$ as a function of $\rho_{e(h)}$ [21.95]. The crystals used were grown from Dupont hyperpure material, with ρ of 0.01–94 Ωcm for n-type and 0.025–110 Ωcm for p-type samples. Curves of $\mu_{e(h)}^H$ versus $\rho_{e(h)}$ were calculated using the BH and combined-mobility [21.71] formulas for $m_{e(h)} = m_0$ and compared with experimental curves.
2. The first extensive experimental study of electrical conductivity and the Hall effect in TL silicon [21.96]. The properties were measured at

temperatures of 10–1100 K on six arsenic-doped n-type samples, and one undoped, plus five boron-doped, p-type samples, covering the range from light ($N = 1.75 \times 10^{14}$ and 3.1×10^{14} cm^{-3}) to heavy ($N = 2.7 \times 10^{19}$ and 1.5×10^{19} cm^{-3}) doping. Compensation by unknown acceptors (donors) occurred in four lightly and moderately doped n(p)-type samples. A deviation of the lattice mobility from the $T^{-1.5}$ dependence was reported for both electrons and holes. Curves of $\mu_{e(h)}^H$ against $\rho_{e(h)}$ at 300 K were computed in the same way as by Debye and Kohane, but incomplete ionization of impurity centers was additionally taken into account.

3. First systematic study of μ_e and μ_h versus N_d and N_a, respectively, at $T = 300$ K [21.97]. Measurements were taken with several group V and group III impurities up to 6×10^{19} and 6×10^{18} cm^{-3} for n- and p-Si, respectively. Impurity concentrations were obtained by radioactive tracers or from thermal neutron activation analysis; μ_e and μ_e were calculated from these data by considering the N_d^+ and N_a^- percentages. The combination with measured $\mu_{e(h)}^H$ resulted in $r_{e(h)}$ values in agreement with theory. A comparison with a BH-formula-based theory yielded semiquantitative agreement for μ_e^H, while measured values of μ_h^H proved to be much smaller than the theoretical values.

4. Galvanomagnetic effects in p-Si: ρ and R^H versus T and B [21.98] and MR [21.99]. Boron-doped samples cut from CZ crystals were used. In the first paper four samples, two with $\rho(300 \text{ K}) = 35\, \Omega\text{cm}$ and two with $\rho(300 \text{ K}) = 85\, \Omega\text{cm}$, were measured in the range 77–320 K. The dependence $\mu_h \propto T^{-2.7 \pm 0.1}$ at $B = 0$, as observed by Ludwig and Watters, was typical of the results obtained on all the samples; r_h was observed to exhibit a weak linear decrease with T in the range 200–320 K, and to be almost entirely independent of B up to $B = 1.3$ T in the temperature interval studied. The dependence of MR on the relative directions of current, fields and crystallographic axes was studied at 77 K and 300 K as a function of B. Large values of longitudinal MR, as large as the transverse effects in some cases, were observed, contradicting the only calculations available at that time [21.9]. To obtain data sufficient for constructing a more satisfactory model, the above study was continued in the second paper on 10 more samples, with ρ of 0.15–115 Ωcm. Measurements of three MR coefficients were carried out at a number of temperatures in the range 77–350 K. The results showed a marked dependence of the band structure and scattering anisotropies on the temperature, yet no definite model of these effects was arrived at.

5. A comparative study of mobility in pulled and FZ crystals [21.100]. The question of the dependence of the intrinsic mobility on temperature was recast. The authors found that in FZ, contrary to pulled crystals, μ_e^H followed the $T^{-1.5}$ law in the range 20–100 K, although μ_h^H still displayed a different, viz. T^{-2}, variation with temperature. It was argued that such a disagreement with the work of Morin and Maita was due to the large content, up to 10^{18} cm^{-3}, of oxygen impurities in the pulled crystals they used, which resulted in scattering that obscured the phonon scattering.

6. Solid analysis of electrical properties of n-Si with respect to: the ionized-impurity scattering in isotropic approximation [21.101], scattering anisotropies [21.66] and lattice scattering [21.51]. The measurements were made from 30 to 100–350 K using a set of P-doped, B-compensated, n-type samples of rather wide impurity content, yet in the range from light to moderate doping ($N_d = 4.5 \times 10^{15}$ cm^{-3} at most). The purest samples were cut from FZ crystals while others were from CZ crystals. These authors developed a sophisticated, but robust, method of determining N_d and N_a by analysis of the R^H versus T data. With this method, in the first paper they obtained curves of μ_e versus T in the range 30–100 K, which were used to test the BH formula. In comparing the formula with the data, correction to the observed μ_e because of the phonon-scattering contribution was necessary. The BH formula was shown to provide a good quantitative description of the data when $m_e = 0.3 m_0$ was used, provided that ion scattering was not too strong. When ion scattering was dominant, viz. in moderately doped samples at low T, they observed a discrepancy between the theory and data, which was attributed to electron–electron interaction. For the purpose of detecting scattering anisotropy, the MR coefficients were measured in the second paper on several relatively pure ($N_d = 8.0 \times 10^{14}$ cm^{-3} at most) samples. The results indicated that $\tau_l/\tau_t \approx 0.67$ and $\tau_l/\tau_t > 1$ for acoustic-phonon and ionized-impurity scattering, respectively. The inter-valley phonon scattering, important at higher T, proved to be isotropic. In the third paper lattice scattering was treated. A model assuming inter-valley scattering by two-phonon modes, in addition to the intra-valley acoustic-phonon scattering, was applied to the re-

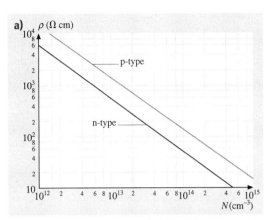

Fig. 21.5 Room-temperature resistivity of Si versus N (defined in the text) for overlapping ranges of doping: (**a**) light to moderate;

sults of the electrical and MR measurements in the purest samples. By using the coupling constants with the phonon modes as fitting parameters, fairly good agreement was achieved between the model and experimental data on curves of μ_e versus T in the range 30–350 K, though one of the inter-valley phonon energies corresponded to a forbidden transition (Table 21.4).

7. Experimental verification of the anisotropic ion-scattering theory [21.102]. Measurements of saturated longitudinal MR and μ_e were made in fields up to 9 T at 78 K on a series of [111]-oriented phosphorus-doped samples with N_d in the range 2×10^{13}–6×10^{16} cm^{-3}. A quantitative analysis was made that involved combined re-

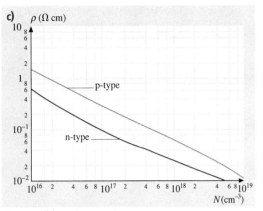

Fig. 21.5 (**c**) moderate to heavy;

laxation times from impurity, intra-valley, and inter-valley lattice scattering. The data agreed very well with the theory [21.103]. In samples with $N_d > 10^{16}$ cm^{-3} the neutral-impurity scattering effect was observed.

8. Refined analysis of the Hall effect and mobility in n-Si [21.53]. μ_e and n, as determined from the high-field Hall effect, were numerically analyzed for a series of n-type samples doped with Sb, P, and As. The calculations were based on the general treatment for an anisotropic parabolic band [21.48]. Lattice-scattering parameters for the inter-valley and acoustic modes were determined from a comparison of the results between theory and experiment, using as many as four inter-valley phonons. The conclusions supported the two-phonon Long model. Ionized-impurity scatter-

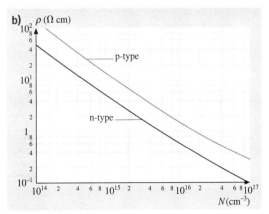

Fig. 21.5 (**b**) moderate;

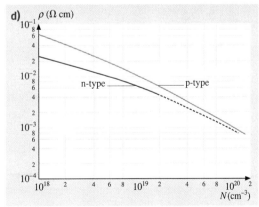

Fig. 21.5 (**d**) heavy. After [21.24] with permission

ing was calculated from the BH formula and a newer theory [21.103] and compared with experiment in favor of the latter. Neutral-impurity scattering proved to be temperature-dependent, unlike the available model [21.55].

9. The Hall factor in n-Si versus $N_I = N_d^+ + N_a^-$ [21.106]. The experimental value of r_e was obtained from the ratio of $R_e^H(0)$ to $R_e^H(\infty)$, determined by independent methods at 77 and 350 K over a wide range of N_I. Samples were cut from non-oxygenated ingots grown in vacuum, in which $R_e^H(0)$ was measured by the standard method (Sect. 21.2). Discs were then cut from the same samples, in proximity to Hall probes, and their total resistances (given by $BR_e^H(\infty)/d$ at $\mu_e B/c \gg 1$ (Sect. 21.2), where d is the disc width) were measured in fields up to $B = 3.5$ T. To check these results, standard measurements of R_e^H in pulsed fields (Sect. 21.2) up to $B = 15$ T were made. The values of r_e were calculated from theory [21.103]. Good agreement of experiment with theoretical calculations was obtained.

Minority Carriers – Miscellaneous

Minority carriers play a crucial role in silicon-based electronic devices such as bipolar junction transistors (BJT) and solar cells, so their charge-transport parameters – concentration, recombination time, mobility and diffusion coefficient – are importany for device modeling. We do not present a literature overview nor graphical and tabular material here, as we do for the case of the majority carriers, and refer the reader directly to the review papers outlined in points 5 and 6 of Sect. 21.2.3.

Irvin Curves

The curves of $\rho(300\,\text{K})$ versus N, over the entire range up to the solubility limit, compiled using published data [21.104, 107] are shown in Fig. 21.5a–d, with the high-doping range shown Fig. 21.6a–d. The resistivity decreases with increasing N since the conductivity is proportional to $n(p)$ while $n(p) = N$ at $T = 300$ K. However, at large N the decrease is slower than $\propto N^{-1}$ due to the notable dependence of $\mu_{e(h)}$ on doping (see below).

Mobility and Hall Effect
Versus Resistivity and Doping

Many studies have reported μ versus ρ graphically. Figure 21.7a–d summarize this information obtained before 1965. For the Hall measurements the drift mobility was recalculated from the Hall value using the Hall factor for the standard band (Sect. 21.1.3). The dependencies of $\mu_{e(h)}$ on N, $N_{d(a)}$ and on N_I at low T are more physical and allow comparison with available theories. Exam-

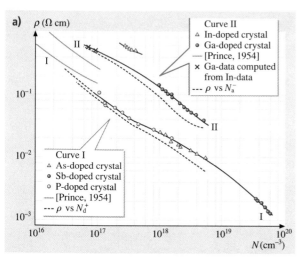

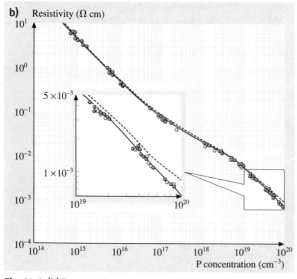

Fig. 21.6 Room-temperature resistivity of highly doped Si versus: (a) N in compensated n- and p-type crystals; the curves corrected for ionized impurities content are also presented [21.97] with permission;

Fig. 21.6 (b) P content in uncompensated P-doped Si. *Dashed curve* is due to *Irvin* [21.104]. P concentration by: *solid circle* – neutron activation analysis, *open circle* – Hall-effect measurement. *Inset* – heavily doped range enlarged. After [21.105] with permission

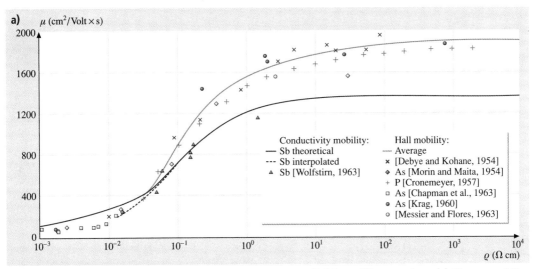

Fig. 21.7 Mobility versus resistivity in Si at room temperature compiled from different authors: (**a**) Electron mobility. After [21.24] with permission;

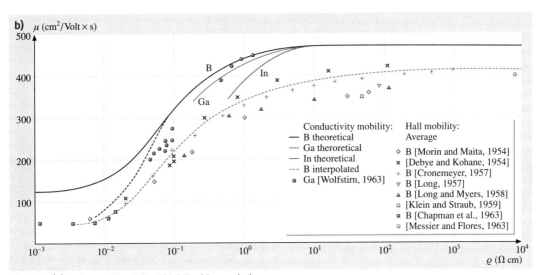

Fig. 21.7 (**b**) Hole mobility. After [21.24] with permission;

ples of such dependencies obtained by different authors are presented in Fig. 21.8a–e. Several phenomenological expressions for $\mu_{e(h)}(N, 300\,\mathrm{K})$ have been given in the literature (Sect. 21.2.3) in agreement with experimental data on charge transport in Si. We consider here two of them, for obvious reasons of space. The first [21.108] reads

$$\mu(N_i) = \mu_{\min} + \frac{\mu_{\max} - \mu_{\min}}{1 + (N_i/N_{\mathrm{ref}})^\alpha}, \qquad (21.25)$$

where $i = \mathrm{a, d}$, and the other fitting parameters are given in Table 21.6, the electron column of which corre-

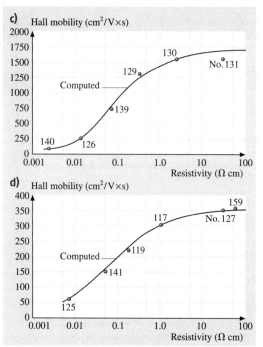
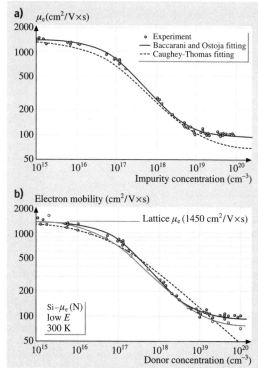

Fig. 21.7 (c) Hall mobility of electrons in Si:As. After [21.96] with permission; **(d)** Hall mobility of holes in Si:B. After [21.96] with permission. Numbers on graphs (c) and (d) denote samples, details of which are given in Table 21.10

sponds to a later best fit for Si:P [21.109], while the hole column is due to *Caughey* and *Thomas* [21.108]. The particular case of (21.25) with $\mu_{\min} = 0$, $\alpha = 0.5$, $N_{\text{ref}} = 10^{17}$ cm^{-3} [21.110] has also proved useful. The second expression for Ohm's mobility versus impurity concentration,

$$\mu(N_i) = \frac{\mu_0}{\sqrt{1 + \frac{(N_i/N_{\text{eff}})S}{N_i/N_{\text{eff}}+S}}} \, , \qquad (21.26)$$

where N_{eff} and S are fitting parameters, follows from the weak electric field limit of the more general phenomenological expression (21.31) for v_d versus E and N_i. The best-fit parameters for (21.26) can be found in Table 21.7. Fits to (21.25) and (21.26) are used in Fig. 21.8a–c. Physically, $\mu_{\max}$ in (21.25) and μ_0 in (21.26) are $\mu_{e,\text{hL}}(300\,\text{K})$, the intrinsic mobility (see below) at room temperature. Experimental data on this parameter due to different authors are presented in Table 21.8. Figure 21.8d shows a com-

Fig. 21.8 Mobility versus doping in Si: **(a)** Room-temperature electron mobility in n-type Si: P versus P concentration, *solid square* – experimental data [21.105], *full line* – the best fit due to Baccarani and Ostoja ((21.25), Table 21.6) and *dashed line* – fitting of Irvin's data by (21.25) due to Caughey and Thomas [21.109], with permission; **(b)** Summary of data on and fits to room-temperature electron mobility versus donor concentration in n-Si: *open circle* – compiled experimental data [21.104], *solid circle* – experimental data [21.105], *full line* – the best fit due to Baccarani and Ostoja ((21.25), Table 21.6), *broken line* – the best fit by (21.25) due to Hilsum and *dash-dotted line* – the best fit due to Scharfetter and Gummel ((21.26), Table 21.9) [21.14], with permission;

parison between the mobility of holes, as minority and majority carriers, plotted against the concentration of majority impurities. At the same concentration the minority-hole mobility is strikingly larger than the majority-hole mobility. This effect may be due to the difference between scattering by attractive and repulsive centers as well as to hole–hole interaction. The variation of $\mu(N, 300\,\text{K})$ with shallow donor and accep-

Fig. 21.8 (c) Summary of data on and fits to room-temperature hole mobility versus acceptor concentration in p-Si, *open circle* – compiled experimental data [21.104], *solid circle* – lattice mobility, *full line* – fitting of Irvin's data by (21.25) due to Caughey and Thomas, and *dash-dotted line* – the best fit due to Scharfetter and Gummel ((21.26), Table 21.9) [21.14], with permission; **(d)** Mobility of holes as being minority carrier (*symbols* and *dashed line*), measured on devices by different authors indicated in captions, and majority carrier (*full line*) due to Thurber et al. After [21.111] with permission, the references are found in this review;

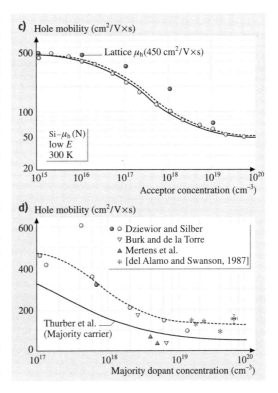

tor species, noted in Sect. 21.1.4, is shown in Table 21.9. As can be seen, the effect – the larger $E_{d(a)}$ the lower $\mu_{e(h)}$ – is not drastic but quite noticeable. At low temperatures the impurity-scattering-limited mobility depends on the density of ionized centers, as predicted by the BH formula (or its modifications), e.g. for electrons

$$\mu_{eI} = \frac{\eta T^{3/2}}{N_I (\ln b - 1)},$$

$$\eta = \frac{2^{7/2} k_B^{3/2} \varepsilon'^2}{\pi^{3/2} q^3 m^{*1/2}}, \quad (21.27)$$

where m^* is an adjusted effective mass, $\varepsilon' = \varepsilon(0)$ is the static dielectric constant and

$$b = \frac{24\pi m \varepsilon' (k_B T)^2}{n' q^2 \hbar^2},$$

$$n' = n + (n + N_a)\left(1 - \frac{n + N_a}{N_d}\right). \quad (21.28)$$

For compensated n-Si $n \ll N_a$ at sufficiently low T, so $N_I = 2N_a$ and $n' = N_a(N/N_d) = N_I(1-K)/2$, where K is the compensation ratio. Thus, disregarding the weak logarithmic dependence, μ_{eI}^{-1} proves to be linear with N_I. Such dependence is seen in the experimental data shown in Fig. 21.8e.

The Hall coefficient for lightly compensated n-Si as a function of B is shown in Fig. 21.9a. The curve for

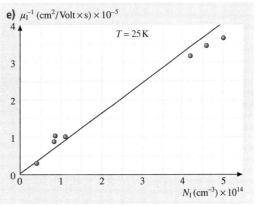

Fig. 21.8 (e) Inverse ionized-scattering mobility in n-Si at $T = 25$ K versus N_I. After [21.100] with permission

$T = 300$ K shows no visible variation when sweeping from low to high fields, which means that either the high-field regime is not attained or $r_e = 1$. The transition between the two regimes is clearly seen with $r_e < 1$, at the lower indicated values of T. The data below 20 K

Table 21.6 Best fit parameters for the Caughey–Thomas formula (21.25)

Parameters	Electrons	Holes	Units
μ_{min}	92.0	47.7	cm^{2}V^{-1}s^{-1}
μ_{max}	1360	495	cm^{2}V^{-1}s^{-1}
N_{ref}	1.3×10^{17}	6.3×10^{16}	cm^{-3}
α	0.91	0.76	–

Table 21.7 Best fit parameters for (21.26, 31, 32)

Parameters	Electrons	Holes	Units
μ_0	1400	480	cm^2/Vs
N_{ref}	3×10^{16}	4×10^{16}	cm^{-3}
S	350	81	...
E_1	3.5×10^3	6.1×10^3	V/cm
E_2	7.4×10^3	2.5×10^3	V/cm
F	8.8	1.6	...
v_m	$1.53 \times 10^9 \times T^{-0.87}$	$1.62 \times 10^8 \times T^{-0.87}$	cm/s
E_c	$1.01 \times T^{1.55}$	$1.01 \times T^{1.68}$	V/cm
β	$2.57 \times 10^{-2} \times T^{0.66}$	$0.46 \times T^{0.17}$	...

Table 21.8 Intrinsic mobility in crystalline Si at room temperature

Carriers	Hall mobility (cm^{2}V^{-1}s^{-1})			Drift mobility (cm^{2}V^{-1}s^{-1})			
Electrons	1610[a]	1450[b]	1560[c]	1500[d]	1610[a]	1350[e]	1360[c]
Holes	365	298	345	500	360	480	510

[a] [21.95], [b] [21.96], [c] [21.94], [d] [21.92], [e] [21.93]

Table 21.9 Room-temperature mobility of Si at $n(p) = 2 \times 10^{18}$ cm^{-3} [21.112]

Impurity	Donor		Acceptor	
	Sb	As	B	Ga
Ionization energy (eV)	0.039	0.049	0.045	0.065
Mobility (cm^{2}V^{-1}s^{-1})	235	220	110	100

Table 21.10 Si samples [21.96]

Sample number	Impurity	E_d or E_a (eV)	N (cm^{-3})	N_a or N_d (cm^{-3})	m^*/m_0
n-type					
131	As	0.056	1.75×10^{14}	1.00×10^{14}	0.5
130	As	0.049	2.10×10^{15}	5.25×10^{14}	1.0
129	As	0.048	1.75×10^{16}	1.48×10^{15}	1.2
139	As	0.046	1.30×10^{17}	2.20×10^{15}	1.0
126	As	?	2.00×10^{18}	...	...
140	As	Degenerate	2.70×10^{19}	...	...
p-type					
159	B	0.045	3.10×10^{14}	4.10×10^{14}	0.4
127	B	0.045	7.00×10^{14}	2.20×10^{14}	0.4
117	B	0.043	2.40×10^{16}	2.30×10^{15}	0.6
119	B	0.043	2.00×10^{17}	4.90×10^{15}	0.7
141	B	?	1.00×10^{18}	...	...
125	B	Degenerate	1.50×10^{19}	...	...

Table 21.11 Si samples [21.100]

Sample	Impurity	N_d (cm^{-3})	N_a (cm^{-3})	N_{Oxygen} (cm^{-3})
A (FZ)	P	1.14×10^{14}	4.00×10^{12}	$\approx 10^{16}$
B (FZ)	P	9.00×10^{13}	2.00×10^{13}	$\approx 10^{16}$
C (CZ)	P	2.50×10^{14}	5.50×10^{13}	$\approx 10^{18}$
D (CZ)	P	4.90×10^{14}	2.10×10^{14}	5.0×10^{17}
E (CZ)	P	3.30×10^{14}	2.30×10^{14}	7.7×10^{17}
F (FZ)	B	2.00×10^{12}	3.40×10^{14}	$\approx 10^{16}$

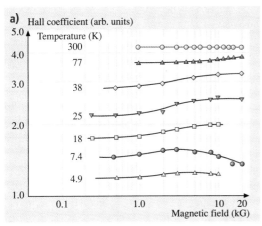

Fig. 21.9 Hall-effect parameters in Si: (**a**) Hall coefficient in n-Si versus magnetic field strength as measured for Si:As1 (Table 21.13) at several indicated temperatures. After [21.53] with permission;

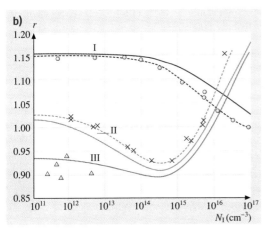

Fig. 21.9 (**b**) Hall factor in n-Si versus N_I, *cross* and *open circle* – experimental data at $T = 77$ K and $T = 300$ K, respectively, *open triangle* – experimental data at $T = 77$ K for samples with lowest μ_e (supposedly compensated), *dashed lines* – guides for the eye, *full lines* are the dependencies calculated using theory [21.103] at $T = 77$ K (I) and 300 K (II) without account of neutral-impurity scattering, and at $T = 77$ K on account of neutral impurities with the density 6×10^{16} cm^{-3} (III). After [21.106] with permission;

were obtained by sweeps of the photo-Hall coefficient, so the decrease of R_e^H at high B is fictitious, due to the field dependence of the photo-carriers lifetime [21.53].

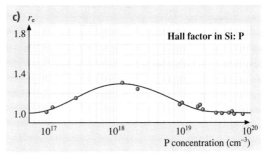

Fig. 21.9 (**c**) Hall factor in highly doped n-Si:P at $T = 300$ K versus phosphorous concentration [21.105];

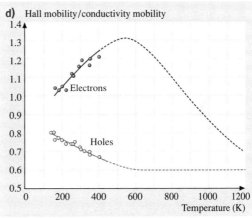

Fig. 21.9 (**d**) Hall factors for electrons and holes versus T, *solid circle* – measured, *dashed line* – computed dependencies. After [21.96] with permission;

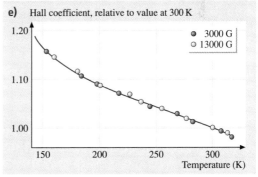

Fig. 21.9 (**e**) Hall coefficient, relative to its value at 300 K, versus T at two indicated strengths of magnetic field in a sample of p-Si with room-temperature $\rho = 35$ Ωcm. After [21.98] with permission;

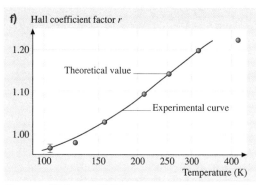

Fig. 21.9 (**f**) Electron Hall factor versus T in n-type (SP6A, Table 21.12) sample, *full line* – experimental data, *open circle* – theory. After [21.51] with permission

Figures 21.9b and 21.9c display the dependence of r_e on the impurity concentration for light to moderate and high doping, respectively. Figures 21.9a and 21.9b are in qualitative agreement with the inequality $r_e < 1$ at $T = 77\,\mathrm{K}$ for comparable concentrations. The overall observed behavior of $r_e(N)$ at $T = 300\,\mathrm{K}$, as shown in Fig. 21.9c,

Table 21.12 Compensated Si:P samples [21.51, 66]

Sample	N_d (cm^{-3})	N_a (cm^{-3})
SP6X	1.7×10^{13}	8.0×10^{12}
SP6S	2.1×10^{13}	8.0×10^{12}
SP6A	2.8×10^{13}	7.5×10^{12}
SP8A	5.4×10^{13}	1.0×10^{13}
SP4M	1.3×10^{14}	1.0×10^{13}
SP4A	2.4×10^{14}	1.0×10^{13}
SP1A	6.7×10^{14}	3.8×10^{14}
SP1D	8.0×10^{14}	4.0×10^{14}
SP2A	2.3×10^{15}	8.3×10^{14}
SM 2	3.9×10^{15}	3.3×10^{15}
SM 3	4.5×10^{15}	3.3×10^{15}

viz. $r_e \approx 1$ on both the moderate and heavy doping sides and a maximum in between at which $r_e > 1$, is deduced from (21.23) using knowledge of the anisotropy of the mass and relaxation-time tensors. Compared to these, results for the Hall effect for holes are not available, to our knowledge.

Hall Effect and Mobility Versus Temperature

The dependencies of $r_e(T)$ and $r_h(T)$ for samples 131, 130 and 159, 127 from Table 21.10 are shown in Fig. 21.9d. The temperature dependence of R_h^H relative to its value at $T = 300\,\mathrm{K}$ for another p-type sample is shown in Fig. 21.9e. Since, in the considered temperature range, p is constant with T, the dependence is fully congruent to $r_h(T)$ – in this regard the lower curve in Fig. 21.9d and the curve in Fig. 21.9e are in agreement. There is good quantitative agreement between experiment and theory for $r_e(T)$ using an exact effective-mass tensor, as shown in Fig. 21.9f. On the contrary, the Hall effect and MR in p-Si are poorly understood [21.98, 99] as the energy spectra of holes is much more complicated than (21.5, 6), even for energies well above 200 K [21.7].

The $\mu_{e,h}(T)$ curves have been the subject of extensive, but to our opinion not exhaustive, studies. As discussed in Sect. 21.2, these dependencies reflect the scattering mechanisms. The basic question is that of temperature dependence due to lattice scattering, which can be measured on technologically pure, very lightly doped and negligibly compensated samples. Since such samples were not available in early studies, researchers tried to apportion the intrinsic $\mu_{e,h}(T)$ from the drift measurements, as shown in Fig. 21.10a, or from conductivity and Hall measurements combined with calculations, as shown in Fig. 21.10b. In practice the intrinsic $\mu_{e,h}(T)$ is described by the phenomenological best-fit equation

$$\mu_{e,hL}(T) = AT^{-\gamma}, \tag{21.29}$$

Table 21.13 Weakly compensated n-Si samples [21.53]

Sample	N_d (cm^{-3})	N_a (cm^{-3})	E_d (meV)	m^*/m_0	ρ_{300} ($\Omega \cdot$cm)
Si:P 1	9.5×10^{15}	4.2×10^{12}	45.64	0.3218	0.66
Si:P 3	2.5×10^{14}	2.3×10^{13}	45.14	0.3218	17.1
Si:P 4	9.6×10^{14}	2.0×10^{14}	43.39	0.3218	4.76
Si:P 5	3.3×10^{14}	8.6×10^{12}	45.38	0.3218	13.6
Si:P 6	4.3×10^{13}	7.7×10^{12}	45.42	0.3218	123
Si:As 1	1.4×10^{16}	6.0×10^{12}	53.64	0.3218	0.43
Si:As 2	7.9×10^{15}	4.3×10^{13}	52.54	0.3218	0.66
Si:As 3	7.5×10^{16}	1.8×10^{13}	52.52	0.3218	0.13
Si:Sb 1	7.4×10^{15}	5.3×10^{12}	42.59	0.3218	0.68

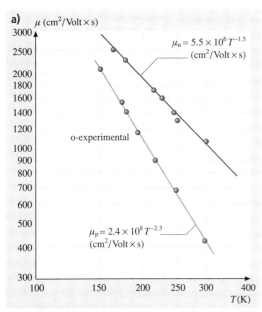

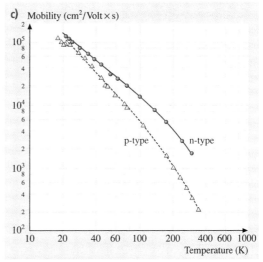

Fig. 21.10 Apportionment of dependence of lattice mobility on T in Si: (**a**) Minority electron and hole drift mobility in high-resistivity crystals. After [21.92] with permission;

Fig. 21.10 (**c**) Conductivity mobility in FZ crystals of n-type (A, Table 21.11) and of p-type (F, Table 21.11). After [21.100] with permission;

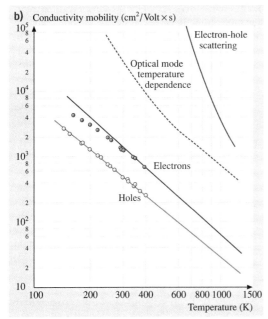

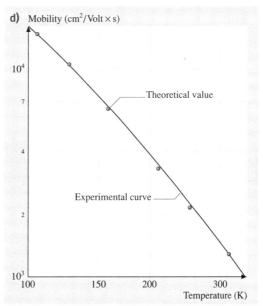

Fig. 21.10 (**b**) Conductivity mobility in TL crystals of n-type (130, 131, Table 21.10) and p-type (127, 159, Table 21.10). After [21.96] with permission;

Fig. 21.10 (**d**) Conductivity mobility in FZ crystal of n-type (SP6A, Table 21.12), *full line* – experimental data, *open circle* – theory. After [21.51] with permission;

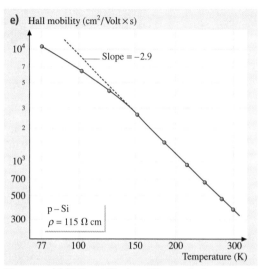

Fig. 21.10 (e) Hall mobility in a CZ crystal of p-type with the indicated resistivity. After [21.99] with permission

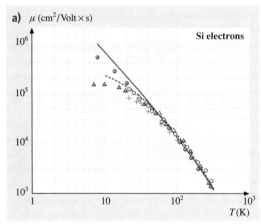

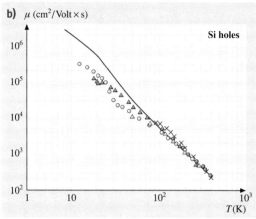

Fig. 21.11 Summary of data on μ versus T in nearly intrinsic to lightly doped Si; *symbols* and *lines* indicate experimental data and simulations, respectively: (**a**) n-type. After [21.114] with permission. *Solid circle* – time-of-flight technique on high-purity n-Si (Ohm's regime achieved only at $T \geq 45$ K) by Canali et al., *plus* – [21.115], *open triangle* – [21.101], *open circle* – [21.100], *solid triangle* – [21.53], *dashed* and *full lines* – Monte Carlo method by Canali et al. using three allowed and six lowest inter-valley phonon modes (Table 21.4), respectively, and no impurities, *dash-dotted line* – the same as full, but with inclusion of 10^{13} cm^{-3} ionized donors;

Fig. 21.11 (**b**) p-type. After [21.13] with permission. *Open circle* – time-of-flight technique on high-purity p-Si (Ohm's regime achieved only at $T \geq 100$ K) by Ottaviani et al., *solid triangle* – [21.100], *cross* – minority hole drift mobility [21.93], *open triangle* – [21.96], *solid circle* – [21.96], *full line* – relaxation time method by Ottaviani et al. with the use of simplified models of the hole energy spectrum and hole–phonon interactions

where A and γ may vary with the carrier and measurement type, fitting interval and sample quality. Different drift measurements, in the ranges 150–300 K [21.92], and 160–400 K [21.93], resulted in very different values of A and γ for both electrons and holes. The first measurements of the Hall effect and conductivity against temperature [21.96] gave the same γ for holes as that obtained by Prince, but a drastically larger γ for electrons, as the n-type samples with the lowest N_d used were inappropriate for that purpose – one sample (131, Table 21.10), though lightly doped, was compensated ($K \approx 0.57$) and the other (130, Table 21.10) was doped in excess of 10^{15} cm^{-3}. Therefore, Morin and Maita numerically extracted the ionized-impurity scattering mobility portion (21.27, 28) and then determined γ. This procedure seems unreliable in view of the inapplicability of the BH formula with strong compensation, neglect of scattering by neutral impurities and of the spectrum and scattering anisotropies. Analogously, using (21.29) for a high-resistivity p-type Si sample [21.99], fitting to $\mu_h(T)$ resulted in the same γ as that obtained by Ludwig and Waters while fitting to $\mu_h^H(T)$ (Fig. 21.10e) resulted in an even larger γ. Using high-purity samples, for which the $\mu_e(T)$ and $\mu_h(T)$ values are presented in Fig. 21.10c, made it possible to determine a γ for $\mu_e(T)$ in agreement with that obtained by Prince, but failed to fit $\mu_h(T)$

Table 21.14 Parameters of the best fit of the intrinsic mobility to (21.29)

Carriers	γ					$A(10^8\,\mathrm{K}^\gamma\,\mathrm{cm}^2\,\mathrm{V}^{-1}\,\mathrm{s}^{-1})$				
Electrons	1.5[a]	2.6[b]	2.5[c]	...	1.5[e]	2.42[f]	5.5[a]	40[b]	21[c]	14.3[f]
Holes	2.3[a]	2.3[b]	2.7[c]	2.7–2.9[d]	2.0[e]	2.20[f]	2.4[a]	2.5[b]	23[c]	1.35[f]

[a] [21.92], [b] [21.96], [c] [21.93], [d] [21.99], [e] [21.100] at $T \leq 100$ K, [f] [21.14]

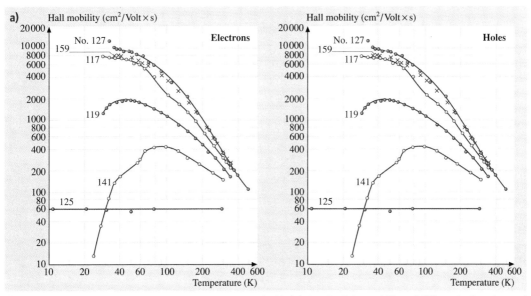

Fig. 21.12 Effects of doping on mobility versus T curves in Si: (a) Conductivity mobility of differently doped n- and p-type TL crystals. Numbers indicate samples from Table 21.10. After [21.96] with permission;

to (21.29) [21.100]. *Long* did not fit using (21.29) and calculated $\mu_{\mathrm{eL}}(T)$ using a realistic electron spectrum and a specific inter-valley electron–phonon scattering model (Sect. 21.1.4TSnotePlease check this link.), and compared the calculation to experiment [21.51], with excellent agreement, as can be seen in Fig. 21.10d. The validity of the model [21.51] was confirmed by an independent study [21.53]. The calculation of $\mu_{\mathrm{eL}}(T)$ was revisited on the basis of the Monte Carlo method and extended inter-valley scattering models [21.113]; the results are displayed in Fig. 21.11a along with the experimental data and fit line. We are unaware of analysis for $\mu_{\mathrm{hL}}(T)$ comparable to that done for $\mu_{\mathrm{eL}}(T)$, taking both the realistic hole spectrum [21.7] and full hole–phonon scattering into consideration. Another analysis [21.70] greatly simplified the problem by using relaxation-time equalization ($\tau_1(E) = \tau_2(E)$), resulting from the isotropic approximation adopted for the hole spectra (21.5), but proved insufficient to account for both Ohm's mobility and hot-holes phenomena [21.13]. The results

of calculations based on the relaxation-time method, using only one isotropic but nonparabolic heavy-hole band [21.13], are displayed in Fig. 21.11b along with the experimental data and fit line. The best-fit parameters of (21.29) obtained by different authors are given in Table 21.14.

The effect of doping on mobility is shown in Fig. 21.12. In doped samples $\mu_{\mathrm{e,h}}(T)$ increases with decreasing T up a maximum at some $T_{\max}$ (the lower the sample's purity the larger $T_{\max}$) and then starts to decrease. This behavior is due to competition between lattice scattering, which weakens with decreasing T, and impurity scattering, which strengthens with decreasing T, (21.27, 28, 29). In heavily doped samples scattering by impurities, while dominating the mobility, ceases to depend notably on T [(21.27, 28, 29) do not hold in this case], which results in an overall weak dependence of the observed $\mu_{\mathrm{e,h}}(T)$ on T. These types of behavior are clearly seen in Fig. 21.12a–b. The evolution of $T_{\max}$ with the impurity content for lightly doped, differ-

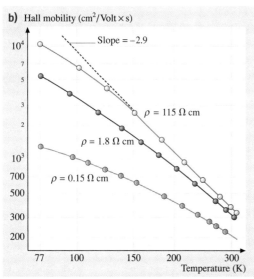

Fig. 21.12 (b) Hall mobility of p-type CZ crystals with different ρ_{300}. After [21.99] with permission;

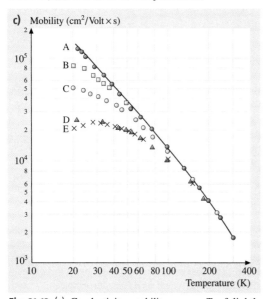

Fig. 21.12 (c) Conductivity mobility versus T of lightly doped, differently compensated, n-Si samples from Table 21.11. After [21.100] with permission;

ently compensated, n-Si is readily traced in Fig. 21.12c. The compensation is crucial for the rate of downturn of the mobility, as follows from (21.27, 28) and is confirmed by the curves for samples C and D (Table 21.11)

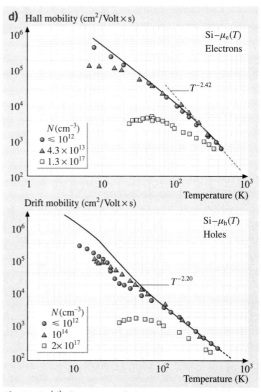

Fig. 21.12 (d) Summary of data on μ versus T; symbols and lines indicate experimental data and simulations, respectively. After [21.14] with permission. Electrons: solid circle – high purity Si ($N \leq 10^{12}\,\text{cm}^{-3}$, time-of-flight) [21.114], solid triangle – lightly doped compensated (Si: P 6, Table 21.13, photo–Hall effect) [21.53], open square – moderately doped with $K \approx 0.01$ (139, Table 21.10, Hall effect) [21.96], full line indicates the theoretical results for the lattice mobility [21.114], dot-dashed line gives the best fit of that mobility by an inverse power of T around room temperature ((21.29), Table 21.9); Holes: solid circle – high purity Si ($N \leq 10^{12}\,\text{cm}^{-3}$, time-of-flight) [21.13], open triangle – lightly doped with $K \approx 0.01$ (F, Table 21.11, Hall effect) [21.100], open square – moderately doped with $K \approx 0.1$ (119, Table 21.10, Hall effect) [21.96], full line indicates the theoretical results for the lattice mobility, dot-dashed line gives the best fit of that mobility by an inverse power of T around room temperature ((21.29), Table 21.9) [21.13]

in Fig. 21.12c. Figure 21.12d summarizes the information on $\mu_{e,h}(T)$ for doped Si. For estimations of the

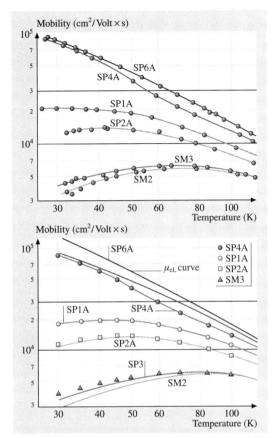

Fig. 21.13 Comparison of the theoretical results on $\mu_e(T)$ and $\mu_e^H(T)$ (*symbols*) with the experimental data (*full lines*). The indicated samples are those from Table 21.12. After [21.101] with permission

combined-scattering mobility e.g. for electrons, the Mattisen rule $\mu_e^{-1}(T) = \mu_{eL}^{-1} + \mu_{eI}^{-1}$, where μ_{eL} and μ_{eI} are given by (21.29) and (21.27, 28), respectively, is used (e.g. [21.100, 101]). The formula, consistently derived for nondegenerate carriers in the standard band, is

$$\mu_e(T) = \mu_{eL} + \mu_{eL} x^2$$
$$\times \left\{ \mathrm{Ci}(x) \cos x + \left[\mathrm{Si}(x) - \frac{\pi}{2} \right] \sin x \right\},$$
$$x = \sqrt{6 \frac{\mu_{eL}}{\mu_{eI}}}, \qquad (21.30)$$

where $\mathrm{Ci}(x)$ and $\mathrm{Si}(x)$ are integral cosine and sine, respectively, and μ_{eL} is due to acoustic-phonon scattering (with $\gamma = 1.5$) [21.71], which proved more appropriate for quantitative analysis. The comparison of (21.30) and the expression for $\mu_e^H(T)$, derived under the same conditions, with the corresponding experimental curves, using $\eta = 8.6 \times 10^{17} \, \mathrm{K}^{-1.5} \, \mathrm{cm}^2 \, \mathrm{V}^{-2} \, \mathrm{s}^{-1}$ (21.27) for an effective isotropic band mass of $m^* = 0.3 m_0$, is shown in Fig. 21.13. Good quantitative agreement between the theory and experiment is observed except for the samples SM3 and SM2, in which the ionized-impurity scattering is strongest (to avoid confusion the data points for SM2 are not shown but they are as discrepant with respect to the observed mobility curve as those shown for SM3). The generalization of (21.30) to the case with spectrum and scattering anisotropies [21.103] was employed to calculate $r_e(T)$ [21.65] and to extract N_a from the mobility data [21.53]. Dakhoskii numerically explained why the above isotropic approximation [21.101] worked well, but presented no graphs of mobility versus T. Norton et al. presented graphically calculated $\mu_e(T)$ only for Si:P6, the purest of their samples. No analysis for $\mu_h(N, T)$ comparable to that done for $\mu_e(N, T)$, i.e. taking both the realistic hole spectrum and hole–phonon and hole–impurity scatterings into account, exists to our knowledge in the literature. Success of modeling based on empirically adjusted phenomenological expressions (point 2–4 in Sect. 21.2.3), though valuable for the device community, cannot be considered as such from the viewpoint of solid-state and semiconductors theories.

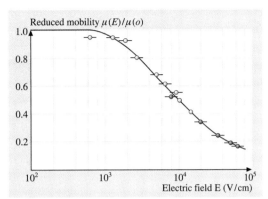

Fig. 21.14 Variation of reduced mobility with electric field applied to a 70 Ωcm n-Si sample along the direction $\langle 111 \rangle$. The electron concentration is assumed to be constant. After [21.24] with permission

21.2.2 High-Electric-Field Effects

The group of effects associated with the carrier transport in high electric fields is traditionally called

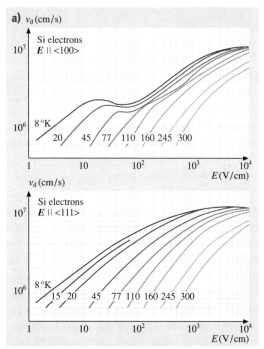
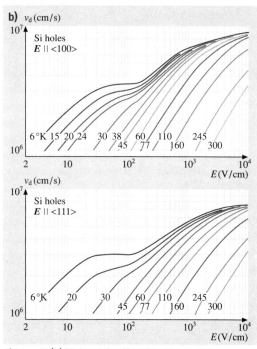

Fig. 21.15 Carriers' drift velocity for E, applied along two indicated directions, measured at different indicated T by the time-of-flight method: (**a**) Electrons. After [21.116] with permission;

Fig. 21.15 (**b**) Holes. After [21.13] with permission

hot-carrier phenomena. This name was borrowed from the displaced-Maxwell-distribution method, where an effective parameter is the carrier temperature, which becomes much higher than that of the lattice, as if the field heats the carriers [21.38]. Not too much work has been published on hot-carrier phenomena in Si, but most of the existing work was overviewed in the two references cited below, and in point 3 in Sect. 21.2.3.

1. The electron drift velocity ([21.113]). Experimental results for electrons in Si obtained using a time-of-flight technique were presented for $8\,\mathrm{K} \leq T \leq 300\,\mathrm{K}$ and E in the magnitude range $1.5 \times 10^4 – 5 \times 10^4$ V/cm, oriented along the $\langle 111 \rangle$, $\langle 110 \rangle$ and $\langle 100 \rangle$ crystallographic directions. At 8 K the dependence of the transit time upon sample thickness allowed a measurement of the valley repopulation time when the electric field is $\langle 100 \rangle$ oriented. These experimental results were interpreted with Monte Carlo calculations in the same ranges of T and E. The model included the many-valley structure of the conduction band of Si, acoustic intra-valley scattering with the correct momentum and energy relaxation and correct equilibrium phonon population, several inter-valley scatterings, and ionized-impurity scattering.

2. Hole drift velocity [21.13]. Drift velocities for holes in high-purity Si, were measured by a time-of-flight technique with E in the amplitude range from $3 \times 10^4 – 5 \times 10^4$ V/cm along the $\langle 100 \rangle$, $\langle 110 \rangle$, and $\langle 111 \rangle$ directions, and at $8\,\mathrm{K} \leq T \leq 300\,\mathrm{K}$. The Ohm's mobility is theoretically interpreted on the basis of a two-band model consisting of a spherical parabolic and a spherical nonparabolic band, and relaxation-time approximation. The low-T Ohm's mobility proved to be strongly influenced by the nonparabolic structure of the heavy-hole band. The high-field region, $E = |E| \geq 10^3$ V/cm, was analyzed using a single, warped heavy-hole band model and a Monte Carlo technique. Anisotropy of the hot-hole drift velocity due to warping of the valence band was measured. Optical- and acoustic-phonon scattering were found to have comparable rates.

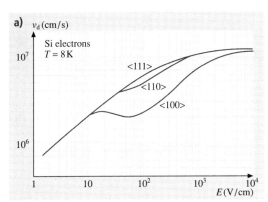

Fig. 21.16 Carriers' drift velocity anisotropy in details at different indicated T: **(a)** Electrons. After [21.116] with permission;

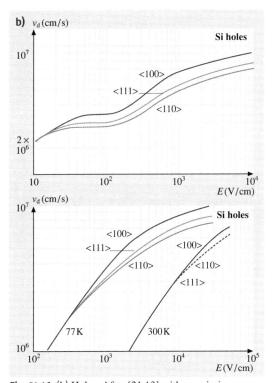

Fig. 21.16 (b) Holes. After [21.13] with permission;

Impact ionization phenomenon are different in appearance and are not considered here in detail; see point 7 in Sect. 21.2.3.

Drift Velocity – Electric-Field Relationship

As the field strength E overcomes some limit (which depends on the lattice temperature, doping and type of carrier) the drift velocity (Sect. 21.2.3) becomes a sublinear function of E, i.e. the mobility starts to decrease with increasing E. As revealed by the early study for n-Si [21.117], in the range $10^3 – 10^5$ V/cm, n does remain constant, but the mobility decreases appreciably, as seen Fig. 21.14. Further studies obtained, in addition, the following patterns: (e.g. [21.13, 14, 113]).

1. Drift velocity $v_d(E)$ is anisotropic in the orientation of E relative to the crystallographic axes: different values of v_d are obtained for the same E for E along different high-symmetry directions $\langle 111 \rangle$, $\langle 110 \rangle$ and $\langle 100 \rangle$. For less-symmetric directions v_d is not even parallel to E, which is known as the Sasaki–Shibuya effect [21.38]. The v_d–E relation depends strongly on T, as seen from Fig. 21.15a–b for the $\langle 111 \rangle$ and $\langle 100 \rangle$ directions. The anisotropy becomes stronger as T decreases, as seen from Fig. 21.16a–b. For electrons, the curves of v_d versus E for different T (Fig. 21.15a) tend to join together at some value of E, which increases with T, even though such a rejoining is not reached at all considered T. For holes (Fig. 21.15b), this tendency is much weaker and the rejoining field is much higher than for electrons.

2. For electrons at the highest $E \| \langle 111 \rangle$, a region where v_d is independent of E (saturation) is obtained (Fig. 21.15a). For holes, this type of saturation is not approached for all available E (Fig. 21.15b).

3. Ohm's law is reached at $T > 45$ K and 100 K for electrons and holes, respectively, and the results for mobility are in agreement with those obtained by standard techniques for the linear regime. At $T < 45$ K for electrons, and $T < 100$ K for holes, even at the lowest E, v_d is not linear with E. At low temperatures the mobility is as high as 5×10^5 cm^{2}V^{-1}s^{-1} at $E = 1.5$ V/cm for electrons ($T = 8$ K), and 3.5×10^5 cm^{2}V^{-1}s^{-1} at $E = 3$ V/cm for holes ($T = 6$ K). A negative differential mobility (NDM) was found with $E \| \langle 100 \rangle$ at $T < 40$ K for electrons (Fig. 21.16a). The field at which v_d reaches the maximum before the NDM region (NDM threshold), decreases with decreasing T. For holes, a net low-field saturation region of v_d (50 V/cm $< E <$ 150 V/cm at $T = 6$ K) shows up for $E \| \langle 111 \rangle$, $\langle 110 \rangle$ and $\langle 100 \rangle$ (see top of Fig. 21.16b). This effect gradually disappears as T increases, and at $T > 38$ K it is no longer visible.

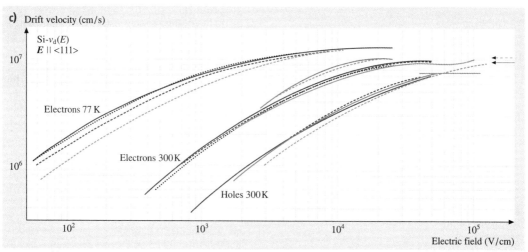

Fig. 21.16 (c) Summary of the data on the drift velocity obtained by different techniques. After [21.14] with permission. Holes, $T = 300$ K: *full* – time-of-flight [21.118], *dashed* – time-of-flight [21.119], *double-dot-dashed* – time-of-flight [21.120], *cross-dashed* – $I(V)$ in space-charge-limited current (SCLC) regime [21.121], *dot-dashed* – $I(V)$ [21.122], *arrow* – extrapolated value of v_{sh} [21.122]. Electrons, $T = 300$ K: *full* – time-of-flight [21.118], *long-dash-dot* – time-of-flight [21.119], *dashed* – time-of-flight [21.120], *double-dot-dash* – $I(V)$ in the SCLC regime [21.123], *dots* – $I(V)$ [21.124], *dash-crossed* – $I(V)$ [21.125], *dashed arrow* – v_{se} from $I(V)$ in avalanche diodes [21.126, 127]. Electrons, $T = 77$ K: *full* – time-of-flight [21.118], *dashed* – $I(V)$ [21.128], *dash-dot* – $I(V)$ [21.129], *dotted* – $I(V)$ [21.130, 131]

For electrons, the anisotropy is due to a repopulation of the valleys: when $E \| \langle 111 \rangle$ the six valleys (Fig. 21.3b) are equally oriented with respect to E and all of them give the same contribution to v_d. When, for example, $E \| \langle 100 \rangle$, two valleys exhibit the effective mass m_l in the direction of the field, while the remaining four exhibit the transverse mass $m_t < m_l$. Electrons in the transverse valleys respond with a higher mobility, are heated to a greater extent by the field and transfer electrons to the two longitudinal, *colder* and slower valleys, which results in a lower v_d than for $E \| \langle 111 \rangle$, as seen e.g. in Fig. 21.16a. For holes, the anisotropy is due to the two warped and degenerate valence sub-bands (Fig. 21.3c) resulting in different effective masses for holes with different k, and in the lowest and highest v_d for $E \| \langle 110 \rangle$ and $E \| \langle 100 \rangle$, as seen in Fig. 21.16b. A summary of the curves of v_d versus E is presented on Fig. 21.16c.

Anisotropy becomes stronger as T decreases, since a lower T leads to less-effective relaxation effects. In particular, at $T \approx 45$ K the repopulation of electron valleys may be so rapid with increasing $E \| \langle 100 \rangle$ that NDM occurs with E in the range 20–60 V/cm. NDM was observed via oscillations of the current [21.132], and in the $I–V$ [21.128, 132, 133] and v_d–E ([21.113, 134, 135]) characteristics. Whether the curves of electron v_d versus E for $E \| \langle 100 \rangle$ and $E \| \langle 111 \rangle$ join together at the high-field limit is still an open point. Theoretical considerations [21.136] seem to indicate that a small difference ($\approx 5\%$) between the two curves should remain. However, this difference is comparable to experimental error and merging of the two curves has been claimed at several temperatures in the experimental results [21.113, 118]. In the case of holes, both theoretical and experimental results [21.13] indicate that, up to the highest fields attained (5×10^4 V/cm), no merging occurs. The aforementioned anomalous behavior of the curves of v_d versus E for holes at low temperatures ($T < 38 K$) – that these curves tend to *saturate* at intermediate field strengths and then rise again – has been explained [21.13, 116] by a nonparabolic distortion of the heavy-hole spectrum.

A general tendency of $v_d(E)$ to saturate at the highest fields is an important phenomenon which was considered from the very beginning of the hot-electron investigations [21.137–139]. The original explanation [21.138] for this behavior was based on a rough physical model, in which it was assumed that the carriers emit an optical phonon as soon as they reach the energy $\hbar\omega_0$ of the phonon that limits the drift veloc-

Single-Crystal Silicon: Electrical and Optical Properties | 21.2 Electrical Properties

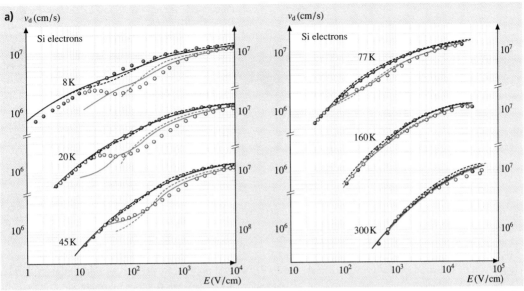

Fig. 21.17 Measured versus simulation results on drift velocity in Si: **(a)** $v_{de}(E)$ at different indicated temperatures. *Circles* indicate experimental data by time-of-flight method, *solid circle* – $E\|\langle 111\rangle$ and $E\|$ *open circle* – $\langle 100\rangle$, and lines Monte Carlo simulations results neglecting impurity scattering, *full* and *broken* – using six lowest and three allowed inter-valley phonons (Table 21.4), respectively. For $T \leq 45$ K, the theoretical curves which refer to the $\langle 100\rangle$ direction are interrupted, as it was not possible to reach a sufficient precision in the simulated v_d. After [21.113, 114] with permission;

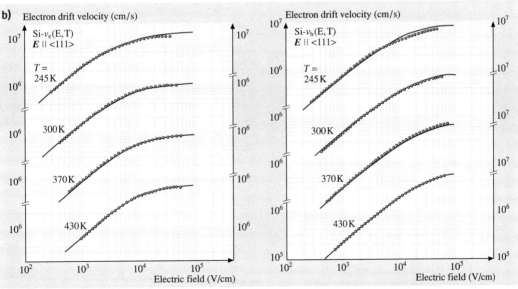

Fig. 21.17 (b) The $v_{de}(E)$ and $v_{de}(E)$, for $E\|\langle 111\rangle$, in high-purity Si at four different T. After [21.14] with permission, *solid circle* – the experimental data [21.118] and *lines* are the best-fit curves obtained with (21.32) using the parameters listed in Table 21.7;

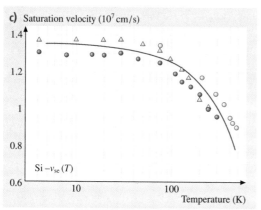

Fig. 21.17 (c) $v_{se}(T)$ obtained experimentally with different techniques and numerically fitted. After [21.14] with permission, *solid circle* – time-of-flight [21.113, 114, 118, 142], *open circle* – $I(V)$ in avalanche diodes [21.126, 127], *open triangle* – $I(V)$ in the SCLC regime [21.123]; the latter data have been normalized to 9.6×10^6 cms^{-1} at 300 K; *line* represents the best fitting curve of (21.33)

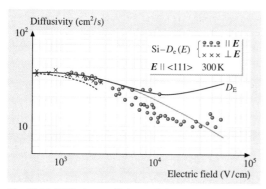

Fig. 21.18 Diffusion coefficients of electrons versus E in Si at $T = 300$ K. After [21.14] with permission: *solid circle* – $D_{\parallel}$, time-of-flight data [21.113, 114]; *dashed* – $D_{\parallel}$, noise data [21.143]; *crosses* – $D_{\perp}$, the geometrical technique data [21.144]; *full lines* – the Monte Carlo computed $D_{\parallel}$ and D_E [21.113, 114]

ity by the saturation value $v_s \approx \sqrt{\hbar\omega_0/m^*}$ [21.138]. For electrons, inter-valley instead of optical phonons should be considered (Sect. 21.2.3). Subsequent calculations [21.140, 141] showed, that the Shockley formula should be considered only as a rough evaluation, since at very high electric field the transport process depends on many factors including several phonon dispersion curves, phonon emission and absorption, and non-parabolic band corrections. Good agreement between the Monte Carlo calculations and the experimental data for high-purity n-Si is seen in Fig. 21.17a. In the case of holes the saturation was neither predicted theoretically, nor found experimentally. A nearly saturated drift velocity was observed [21.122] only at room temperature, and the saturation should occur, by extrapolation, at $E \geq 2 \times 10^5$ V/cm with v_{sh} around 10^7 cm/s (Fig. 21.16c).

Empirically adjusted formulas for $v_d(E)$ have been proposed for varying N_i and $T = 300$ K

$$v_d = \frac{\mu_0 E}{\sqrt{1 + \frac{S(N_i/N_{\text{ref}})}{N_i/N_{\text{ref}} + S} + \frac{(E/E_1)^2}{E/E_1 + F}} + (E/E_2)^2} \quad (21.31)$$

[21.145], and for $N_i = 0$ and varying T

$$v_d = v_m \frac{E/E_c}{[1 + (E/E_c)^\beta]^{1/\beta}} \quad (21.32)$$

[21.108]; the best-fit parameters are presented in Table 21.7. Also a formula was suggested [21.14] for the electron saturated drift velocity as a function of temperature

$$v_{se}(T) = \frac{2.4 \times 10^7 \text{cm/sec}}{1 + 0.8 \times \exp\left(\frac{600\,\text{K}}{T}\right)} \quad (21.33)$$

The applicability of (21.32) and (21.33) is clearly demonstrated by Figs. 21.17b and 21.17c, respectively.

Hot-Carrier Diffusion

Knowledge of diffusion processes is useful for a better understanding of charge-transport phenomena and correct simulation of high-frequency devices. At low fields the diffusion coefficient D is related to the mobility by the Einstein relation (Sect. 21.1.3). At high fields it becomes a field-dependent tensor, which describes the diffusion with respect to E depending on its direction in the crystallographic frame. The difference between diffusion parallel to and perpendicular to E is greater than the variations caused by changes in the crystallographic direction. Therefore, investigations set out to determine the field dependence of both the longitudinal ($D_{\parallel}$) and transverse ($D_{\perp}$) components of the diffusivity tensor with respect to E irrespective of its orientation in the crystal.

The longitudinal diffusion coefficient can be measured by the Haynes–Shockley or time-of-flight [21.120] technique by observing the difference between the fall and rise times of the current pulse. This difference is caused by the spread of the carriers traveling across the sample and is simply related to $D_\parallel$ (e.g. [21.146]). Analogously, $D_\perp$ can be obtained by observing the spread of the current perpendicular to the direction of the field. The current is originated by a point excitation on one surface of a Si wafer and is collected on the opposite surface by several electrodes of appropriate geometry [21.144]; this technique is sometimes called *geometrical* [21.14]. Finally, both $D_\parallel$ and $D_\perp$ have been related to noise measurements, parallel and perpendicular respectively to the current direction [21.143].

Figure 21.18 shows some experimental results on the field dependence of $D_\parallel$ and $D_\perp$ for electrons in Si at room temperature with $\boldsymbol{E}\|\langle 111\rangle$. The data obtained by the noise measurements are in a reasonable agreement with the time-of-flight results, although the former cover a narrower range of E, just outside Ohm's region. As E increases $D_\parallel$ decreases to about one third of its low-field value ($\approx 36\,\text{cm}^2/\text{s}$), which is in substantial agreement with theoretical Monte Carlo computations for the nonparabolic band [21.113]. The results for transverse diffusion showed that, as E increases, $D_\perp$ also decreases, but to a lesser extent than $D_\parallel$. There exists a hypothesis of validity to extrapolate the Einstein relation outside the linear region by $D_E = \frac{2\bar{E}}{3q}\mu(E)$. As seen from Fig. 21.18, this yields a qualitative interpretation of $D_\perp$ for not too high fields. As far as $D_\parallel$ is concerned, the diffusion process seems much more complex than pictured by the Einstein relation. For holes, the dependence of $D_\parallel$ on E was found to be similar to that for electrons [21.14].

Impact Ionization

Impact ionization is an important charge-generation mechanism. It occurs in many silicon-based devices, either determining the useful characteristic of the device or causing an unwanted parasitic effect. The breakdown of a silicon p–n diode is caused by impact ionization if its breakdown voltage is larger than about 8 V. The operation of such devices as thyristors, impact avalanche transit time (IMPATT) diodes and trapped plasma avalanche-triggered transit (TRAPATT) diodes is based on avalanche generation, the phenomenon that results from impact ionization. The avalanche generation also plays an increasing role in degradation due to hot-carrier effects and bipolar parasitic breakdown of metal–oxide–semiconductor (MOS) devices, the geometrical dimensions of which have been scaled down recently.

The ionization rate is defined as the number of electron–hole pairs generated by a carrier per unit distance traveled in a high electric field, and is different for electrons and for holes. Impact ionization can only occur when the particle gains at least the threshold energy for ionization from the electrical field. This can be derived from the application of the energy and momentum conservation laws to the amount $E_i \approx 1.5 E_g$ (assuming that the effective masses of electron and hole are equal). A large spread of experimental values for E_i exists, with a breakdown field of order of 3×10^5 V/cm. For more detailed consideration we refer the reader to the review article [21.147] noted in Sect. 21.2.3.

21.2.3 Review Material

The following materials may be recommended for further reading.

1. Electrical properties of Si [21.24]. Summary of papers on the subject that were published over a decade until 1965 are overviewed. Miscellaneous properties, such as piezoresistance and high-electric-field mobility, were also presented.
2. Electron mobility and resistivity in n-Si versus dopant density and temperature [21.148]. An improved model for computing μ_e as a function of N_d and T in uncompensated n-Si was formulated. The effects of electron–electron interaction on conventional scattering processes, as well as their anisotropies were incorporated empirically. The model was verified to $\pm 5\%$ of the mobility measured on wafers doped by phosphorous in the range $10^{13} - 10^{19}$ cm^{-3}.
3. Bulk charge-transport properties of Si [21.14]. Review of knowledge on the subject with special emphasis on application to solid-state devices. Most attention was devoted to experimental findings at room temperatures and to high-field properties. The techniques for drift-velocity measurements and the principles of Monte Carlo simulation were overviewed. Empirical expressions were given, when possible, for the most important transport quantities as functions of T, $N_{d(a)}$ and E.
4. Semi-empirical relations for the carrier mobilities [21.149]. From a review of different publications on $\mu_{e,h}$ in Si, the authors proposed an approximated calculation procedure, analogous to that of Li and Thurber, which permits a quick and accurate evaluation of $\mu_{e(h)}$ over a wide range of T, $N_{d(a)}$ and $n(p)$. The proposed relations are well adapted to device simulation since they allow short computation times.
5. Minority-carrier recombination in heavily doped silicon [21.150]. A review of understanding of

the recombination of minority carriers in heavily doped Si. A short phenomenological description of the carrier recombination process and lifetime was provided and the main theories of these were briefly reviewed with indications for their expected contributions in heavily doped Si. The various methods used for measuring the minority-carrier lifetime in heavily doped Si were described and critically examined. The insufficiency of existing theories to explain the patterns of lifetime versus doping was clearly demonstrated.
6. Minority-carrier transport modeling in heavily doped silicon emitters [21.111]. The experimental and theoretical efforts that addressed such important issues as: (i) the incomplete understanding of the minority-carrier physics in heavily doped Si, (ii) the lack of precise measurements for the minority-carrier parameters, (iii) the difficulties encountered with the modeling of transport and recombination in nonhomogeneously doped regions, and (iv) problems with the characterization of real emitters in bipolar devices, were reviewed with the goal of being able to achieve accurate modeling of the current injected into an arbitrarily heavily doped region in a silicon device.
7. Impact ionization in silicon: a review and update [21.147]. The multiplication factor and the ionization rate were revisited. The interrelationship between these parameters together with the multiplication and breakdown models for diodes and MOS transistors were discussed. Different models were compared and test structures were discussed to measure the multiplication factor accurately enough for reliable extraction of the ionization rates. Multiplication measurements at different T were performed on a BJT, and yielded new electron ionization rates at relatively low electric fields. An explanation for the spread of existed experimental data on ionization rate was given. A new implementation method for a local avalanche model into a device simulator was presented.

21.3 Optical Properties

21.3.1 Diversity of Silicon as an Optical Material

In *dielectric*-like material $n > k$ ($\varepsilon_1 > 0$), where for transparency and opacity it is necessary that $n \gg k$ ($\varepsilon_1 \gg \varepsilon_2$) and $n \approx k$ ($\varepsilon_1 \ll \varepsilon_2$), respectively (21.16). In *metallic*-like material $n < k$ ($\varepsilon_1 < 0$), where for good reflectivity and bad reflectivity it is necessary that $n \ll k(|\varepsilon_1| \gg \varepsilon_2)$ and $n \approx k(|\varepsilon_1| \ll \varepsilon_2)$, respectively. Since in Si ε depends on the wavelength and carrier concentration, it may exhibit all these types of optical behavior ranging from dielectric-like to metallic-like. For example, $\langle 111 \rangle$ undoped Si ($n = 2.3 \times 10^{14}$ cm^{-3}) at $\lambda = 0.62\,\mu$m behaves as a transparent dielectric, as $\varepsilon_1 = 15.254$ and $\varepsilon_2 = 0.172$ at that wavelength, while it is an opaque dielectric at $\lambda = 0.295\,\mu$m, where $\varepsilon_1 = 2.371$ and $\varepsilon_2 = 45.348$. For heavily doped Si ($n = 10^{20}$ cm^{-3}), $n = 1.911$ and $k = 8.63$ at $\lambda = 16.67\,\mu$m, so it behaves as a good metallic reflector while at $\lambda = 2\,\mu$m, where $n = 3.47$ and $k = 6.131 \times 10^{-3}$ it is a transparent dielectric.

21.3.2 Measurements of Optical Constants

Various methods are used to measure the dielectric constant of single-crystalline Si including transmission, reflection, and ellipsometric methods. For a smooth opaque sample, the quantity of interest is the complex reflection amplitude ρ_r (at normal incidence) defined by

$$\rho_r = -\frac{\sqrt{\varepsilon}-1}{\sqrt{\varepsilon}+1},$$

$$|\rho_r|^2 = \frac{(n-1)^2+k^2}{(n+1)^2+k^2} = R_0,$$

$$\phi = \arctan\frac{2k}{n^2+k^2-1} = \arg(\rho_r),$$

$$\mod(\pi). \qquad (21.34)$$

The Fresnel reflectance spectrum $R_0(\omega)$ is measured using reflectometry. The KKR analysis also applies to the causal function $\ln(\rho_r) = \frac{1}{2}\ln R_0 + i\phi$ that gives

$$\phi(\omega) = \frac{1}{2\pi}\int_0^\infty \ln\left|\frac{\omega+\Omega}{\omega-\Omega}\right|\frac{d}{d\Omega}\ln R_0(\Omega)d\Omega. \quad (21.35)$$

With the $\psi(\omega)$ retrieved in this way, the last two relations in (21.34) are simultaneously solved to yield the n and k spectra. The KKR method requires, in principle, data over an infinite ω range, which are supplied by measurements over a confined range and an appeal to simple models for high and low frequencies. This limi-

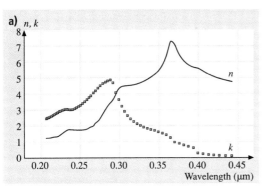

Fig. 21.19 Refractive index n and the extinction coefficient k versus wavelength, using Adachi and Geist models in the wavelength ranges: (**a**) 0.2–0.45 μm,

tation is avoided using optical measurements on smooth slabs.

For a slab of thickness d, the reflectance and transmittance at normal incidence are given by

$$R = R_0 \frac{1 - 2T_0 \cos 2\varphi + T_0^2}{1 - 2R_0 T_0 \cos 2\chi + R_0^2 T_0^2},$$

$$T = T_0 \frac{1 - 2R_0 \cos 2\psi + R_0^2}{1 - 2R_0 T_0 \cos 2\chi + R_0^2 T_0^2},$$

$$T_0 = e^{-\alpha d}, \qquad (21.36)$$

where $\varphi = 2\pi n d / \lambda$ is the optical-path phase, $\chi = \varphi + \psi$ and $\alpha = 4\pi k / \lambda$ is the power attenuation coefficient. The parameter αd characterizes the slab's opacity. At $n > k$ ($\varepsilon_1 > 0$, (21.17)), i.e. in the optically dielectric range, α is referred to as *absorption coefficient*. In the optically metallic range that exists for doped Si in the IR, where $k > n$ ($\varepsilon_1 < 0$, (21.17)), α^{-1} is the skin depth multiplied by a factor of ≈ 1. If the dispersion of the optical constants is disregarded, R and T are periodic in $1/\lambda$ (21.20). For nonopaque slabs, these interference effects show up at $nd/\lambda > 1$. For thick slabs ($nd/\lambda \gg 1$), even small fluctuation of n or/and λ causes changes in φ of the order 2π. Therefore, if such factors as bulk defects, source incoherence and low spectral resolution are present in an experiment, then the averages of R and T over φ, given by

$$\langle R \rangle = R_0 (1 + T_0 \langle T \rangle),$$

$$\langle T \rangle = T_0 \frac{1 - 2R_0 \cos 2\psi + R_0^2}{1 - R_0^2 T_0^2}, \qquad (21.37)$$

are more appropriate than (21.36). This *multiple-reflection* approximation [21.151] has proved effective

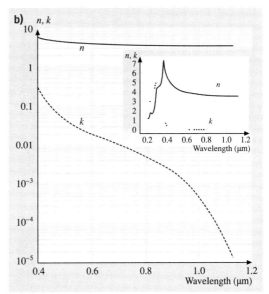

Fig. 21.19 (**b**) 0.4–1.127 μm. *Inset* – the complete range from 0.2 to 1.127 μm

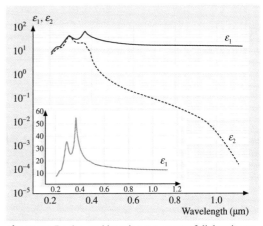

Fig. 21.20 Real ε_1 and imaginary ε_2 parts of dielectric constant versus wavelength, using Adachi and Geist models. *Inset*: real dielectric constant ε_1 versus wavelength

in countless studies. The apparent absorbance of the slab in the multiple-reflection approximation is calculated by

$$\langle A \rangle = 1 - \langle R \rangle - \langle T \rangle$$
$$= \frac{(1 - R_0)(1 - T_0) - 4R_0 T_0 \sin^2 \psi}{1 - R_0 T_0}. \qquad (21.38)$$

Table 21.15 The refractive index n and the extinction coefficient k of n-Si with electron concentration $N = 10^{16}$ cm^{-3} at various wavelength

Energy (eV)	Wavenumber (cm^{-1})	Wavelength (μm)	n HW[a]	n GDA[b]	k HW[a]	k GDA[b]
0.6199	5000	2.000	3.453	3.453	1.160×10^{-7}	1.514×10^{-7}
0.5579	4500	2.222	3.447	3.447	1.594×10^{-7}	2.006×10^{-7}
0.4959	4000	2.500	3.441	3.441	2.273×10^{-7}	2.751×10^{-7}
0.4339	3500	2.857	3.435	3.435	3.398×10^{-7}	3.945×10^{-7}
0.3720	3000	3.333	3.431	3.431	5.403×10^{-7}	6.001×10^{-7}
0.3100	2500	4.000	3.427	3.427	9.347×10^{-7}	9.912×10^{-7}
0.2480	2000	5.000	3.424	3.424	1.827×10^{-6}	1.850×10^{-6}
0.1860	1500	6.667	3.421	3.421	4.334×10^{-6}	4.210×10^{-6}
0.1240	1000	10.00	3.419	3.419	1.463×10^{-5}	1.392×10^{-5}
0.1116	900	11.11	3.419	3.419	2.006×10^{-5}	1.910×10^{-5}
0.09919	800	12.50	3.419	3.419	2.856×10^{-5}	2.727×10^{-5}
0.08679	700	14.29	3.418	3.418	4.262×10^{-5}	4.092×10^{-5}
0.07439	600	16.67	3.418	3.418	6.765×10^{-5}	6.549×10^{-5}
0.06199	500	20.00	3.417	3.417	1.168×10^{-4}	1.143×10^{-4}
0.04959	400	25.00	3.416	3.416	2.278×10^{-4}	2.259×10^{-4}
0.03720	300	33.33	3.413	3.413	5.382×10^{-4}	5.413×10^{-4}
0.03472	280	35.71	3.412	3.412	6.612×10^{-4}	6.669×10^{-4}
0.03224	260	38.46	3.411	3.411	8.247×10^{-4}	8.341×10^{-4}
0.02976	240	41.67	3.410	3.410	1.047×10^{-3}	1.061×10^{-3}
0.02728	220	45.45	3.408	3.408	1.356×10^{-3}	1.379×10^{-3}
0.02480	200	50.00	3.406	3.406	1.799×10^{-3}	1.834×10^{-3}
0.02232	180	55.56	3.399	3.399	3.481×10^{-3}	3.563×10^{-3}
0.01984	160	62.50	3.403	3.403	2.458×10^{-3}	2.511×10^{-3}
0.01736	140	71.43	3.394	3.394	5.154×10^{-3}	5.284×10^{-3}
0.01488	120	83.33	3.385	3.385	8.085×10^{-3}	8.297×10^{-3}
0.01240	100	100.0	3.372	3.372	1.370×10^{-2}	1.405×10^{-2}

[a] Values calculated using an empiricial fit [21.78], [b] values calculated using GDA [21.79]

Equation (21.38) at $k \ll n$ is extensively used in silicon-wafer thermometry [21.152, 153]. Given $\langle R \rangle$, $\langle T \rangle$ and d, (21.37) builds up a system of two equations for the two unknowns n and k. Thus, measurement of the reflection and transmission on the same slab of known thickness allows one to retrieve the optical constants. This R–T measurement method [21.154] is greatly simplified at $2k \ll n^2 + k^2 - 1$. Under this low-loss condition the above system can be solved analytically for T_0 and R_0. The calculated T_0 directly yields k, and n is then found using the calculated R_0. The R–T technique is the best method at $\alpha d \leq 1$, while at $\alpha d \gg 1$, where solving the aforementioned system becomes an ill-conditioned problem, the KKR analysis is more reliable. In two last decades, spectroscopic ellipsometry has gained wide recognition for being more precise than photometric methods. In ellipsometry, the ratio of reflectance for s- and p-polarized radiation, and the relative phase shift between the two, are both measured at large angles of incidence [21.155].

The measured results are affected by the structural atomic-scale properties of the samples. These properties are defined by polishing processes – mechanical or chemical – that affect the surface damage and roughness, the properties of the surface native oxide, the growth mechanism of the measured layer, grain boundaries, and the quality of the cleaved surface. Since Si samples may be optically inhomogeneous, retrieving the optical constants from measurements may become a complicated inverse electromagnetic problem [21.156, 157], which is why some of the reported data for ε disagree by up to 30%. A detailed list of publications on the subject can be found in [21.158, 159]. Emphasis on these effects should be especially considered when transmission measurement is done for a wavelength range in which the absorption coefficient is large and thin samples are therefore required.

Table 21.16 The refractive index n and the extinction coefficient k of n-Si with electron concentration $N = 10^{20}$ cm^{-3} at various wavelength

Energy (eV)	Wavenumber (cm^{-1})	Wavelength (μm)	n HW[a]	GDA[b]	k HW[a]	GDA[b]
0.6199	5000	2.000	3.270 P		1.834×10^{-2}	
			3.257 As	3.247	2.403×10^{-2}	6.131×10^{-3}
0.5579	4500	2.222	3.219		2.549×10^{-2}	
			3.203	3.190	3.341×10^{-2}	9.701×10^{-3}
0.4959	4000	2.500	3.151		3.698×10^{-2}	
			3.130	3.112	4.845×10^{-2}	1.664×10^{-2}
0.4339	3500	2.857	3.053		5.671×10^{-2}	
			3.027	3.000	7.432×10^{-2}	3.232×10^{-2}
0.3720	3000	3.333	2.902		9.410×10^{-2}	
			2.867	2.828	1.233×10^{-1}	8.053×10^{-2}
0.3100	2500	4.000	2.644		1.765×10^{-1}	
			2.597	2.577	2.314×10^{-1}	2.549×10^{-1}
0.2480	2000	5.000	2.138		4.176×10^{-1}	
			2.087	1.990	5.461×10^{-1}	5.336×10^{-1}
0.2356	1900	5.263	1.981		5.226×10^{-1}	
			1.939		6.796×10^{-1}	
0.2232	1800	5.556	1.800		6.717×10^{-1}	
			1.780		8.617×10^{-1}	
0.2108	1700	5.882	1.604		8.873×10^{-1}	
			1.626		1.107×10^{0}	
0.1984	1600	6.250	1.423		1.188×10^{0}	
			1.503		1.417×10^{0}	
0.1860	1500	6.667	1.295		1.566×10^{0}	
			1.429	0.971	1.777×10^{0}	1.902×10^{0}
0.1736	1400	7.143	1.237		1.989×10^{0}	
			1.411	0.893	2.172×10^{0}	2.369×10^{0}
0.1612	1300	7.692	1.239		2.437×10^{0}	
			1.442	0.873	2.591×10^{0}	2.859×10^{0}
0.1488	1200	8.333	1.291		2.911×10^{0}	
			1.518	0.872	3.037×10^{0}	3.387×10^{0}
0.1364	1100	9.091	1.390		3.418×10^{0}	
			1.642	0.913	3.514×10^{0}	3.960×10^{0}
0.1240	1000	10.00	1.540		3.971×10^{0}	
			1.817	0.985	4.034×10^{0}	4.603×10^{0}

[a] Values calculated using an empiricial fit [21.78], [b] values calculated using GDA [21.79]

21.3.3 Modeling of Optical Constants

A method for calculating ε_1 and ε_2 and then n, k and α for photon energies of 0–6 eV ($\lambda > 0.2$ μm) has been reported [21.82]. The calculated data are in excellent agreement with experimental data for the wavelength range 0.2–4 μm [21.156, 160]. The model is based on the KKR (Sect. 21.1.3) and takes into account the dependence of ε on the energy-band structure. It considers the effect of indirect-band-gap and inter-band transitions as well as the electron (conduction bands) and hole (valence bands) density of states. The fundamental absorption (generation of electron–hole pair) edge energy of 1.12 eV corresponds to the indirect transition from the highest valence band to the lowest conduction band. Sharp changes in the

Table 21.17 Table 21.16 cont.

Energy (eV)	Wavenumber (cm^{-1})	Wavelength (μm)	n HW[a]	GDA[b]	k HW[a]	GDA[b]
0.1116	900	11.11	1.751 [P]		4.584×10^0	
			2.057 [As]	1.105	4.608×10^0	5.341×10^0
0.09919	800	12.50	2.042		5.278×10^0	
			2.378	1.278	5.251×10^0	6.216×10^0
0.08679	700	14.29	2.444		6.079×10^0	
			2.810	1.530	5.982×10^0	7.282×10^0
0.08431	680	14.71	2.542		6.255×10^0	
			2.913		6.141×10^0	
0.08183	660	15.15	2.646		6.437×10^0	
			3.022		6.305×10^0	
0.07935	640	15.63	2.758		6.625×10^0	
			3.139		6.474×10^0	
0.07690	620	16.13	2.877		6.820×10^0	
			3.262		6.649×10^0	
0.07439	600	16.67	3.006		7.022×10^0	
			3.395	1.911	6.829×10^0	8.630×10^0
0.07180	580	17.24	3.143		7.231×10^0	
			3.536		7.015×10^0	
0.06933	560	17.86	3.291		7.448×10^0	
			3.686		7.208×10^0	
0.06685	540	18.52	3.451		7.674×10^0	
			3.846		7.407×10^0	
0.06438	520	19.23	3.622		7.909×10^0	
			4.018		7.614×10^0	
0.06199	500	20.00	3.807		8.154×10^0	
			4.201	2.517	7.829×10^0	1.040×10^1
0.05951	480	20.83	4.006		8.409×10^0	
			4.398		8.052×10^0	
0.05703	460	21.74	4.222		8.675×10^0	
			4.608		8.285×10^0	
0.05455	440	22.73	4.455		8.954×10^0	
			4.835		8.528×10^0	
0.05207	420	23.81	4.707		9.245×10^0	
			5.078		8.782×10^0	
0.04959	400	25.00	4.982		9.551×10^0	
			5.340	3.551	9.048×10^0	1.285×10^1

[a] Values calculated using an empiricial fit [21.78], [b] values calculated using GDA [21.79]

optical constants are obtained at wavelengths around 0.367, 0.29 and 0.233 μm, which correspond to the energy-band critical points of 3.38, 4.27 and 5.317 eV, respectively. An additional analytical model for calculating the n and k values for the wavelength range 0.4–1.127 μm has been developed [21.158]. The model is based on measured k and n data [21.161–163], where the calculated values are within ±10% of the measured values. Using Adachi and Geist models we have calculated and plotted n, k, ε_1 and ε_2 for the wavelength range 0.2–1.127 μm, as seen in Figs. 21.19 and 21.20.

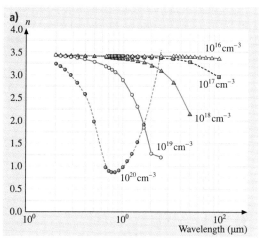

Fig. 21.21 Optical constants versus wavelength at various electron concentrations using the GDA model. (a) Refractive index n.

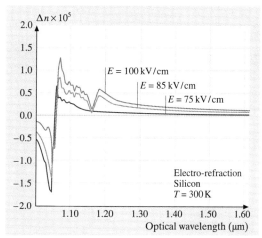

Fig. 21.22 Refractive index change versus wavelength at various electric fields [21.164]

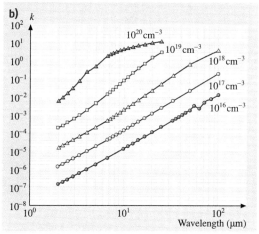

Fig. 21.21 (b) Extinction coefficient k

In the IR range the lattice and free carriers, which arise from doping, give additive contributions. The free-carrier contribution to the dielectric function when the radiation quantum is much smaller than the mean electron energy, $\hbar\omega \ll \bar{E}$, is considered using the semiclassical Drude approximation (DA, Sect. 21.1.4), which leads to the simple formula

$$\varepsilon = \varepsilon_L - \frac{n_L^2 \omega_p^2}{\omega(\omega + i\gamma)}, \quad \gamma = \frac{1}{\tau}, \quad (21.39)$$

where τ is the phenomenological relaxation time, ω_p is the electron plasma frequency given by

$$\omega_p = \sqrt{\frac{4\pi q^2 N}{\varepsilon_L m_c}}. \quad (21.40)$$

Here N denotes the carriers' concentration, $m_c \approx 0.26 m_0$ for electrons (Table 21.2) and $m_c = (0.33$–$0.39) m_0$ for holes (Sect. 21.1.4).

Advanced GDA, which extrapolates the free-carrier contribution to the range of $\hbar\omega > \bar{E}$ was considered [21.79]. In Table 21.15 we tabulated the values of n and k for practical use at two extreme electron concentrations: 10^{16} cm^{-3} and 10^{20} cm^{-3}. The results placed in the columns headed by the abbreviation 'HW' were obtained using DA (21.21, 40, 41) and the empirically adjusted τ versus N fit [21.78]. The columns of theoretical values calculated by us using GDA are headed by the corresponding abbreviation. A disagreement between the theory and the empirical fit notable at heavy doping has been discussed in detail [21.79]. The dependencies of n and k on wavelength in the range at carrier concentration of 10^{16}–10^{20} cm^{-3} in steps of 10 cm^{-3} are shown in Fig. 21.21.

21.3.4 Electric-Field and Temperature Effects on Optical Constants

It has been known for many years that the optical absorption spectrum of Si can also be affected by external electric fields (the Franz–Keldysh effect). The Franz–

Keldysh effect, which alters the α spectrum of crystalline Si, is field-induced tunneling between valence- and conduction-band states. In recent years, the generic term *electroabsorption* has been adopted for $\Delta\alpha$ versus E effects. The effect of electric field on the refractive index is shown in Fig. 21.22 [21.164]. Sharp changes occur around the wavelengths correspond to the band-gap transition.

The temperature dependence of the refractive index of high-purity damage-free Si, for photon energies less than 3 eV in the temperature range 300–500 K is given by [21.162]

$$\frac{\Delta n}{\Delta T} \approx 1.3 \times 10^{-4} n \quad (\text{K}^{-1}) \, . \tag{21.41}$$

References

21.1 C. Kittel: *Introduction to Solid State Physics*, 6th edn. (Wiley, New York 1986)
21.2 C. Kittel: *Quantum Theory of Solids*, 2nd edn. (Wiley, New York 1987)
21.3 K. Seeger: *Semiconductor Physics* (Springer, New York 1982)
21.4 T. S. Moss: *Optical Properties of Semiconductors* (Butterworths, London 1959)
21.5 S. M. Sze: *Physics of Semiconductor Devices* (Wiley, New York 1981)
21.6 H. M. van Driel: Appl. Phys. Lett. **44**, 617 (1984)
21.7 E. O. Kane: J. Phys. Chem. Solids **1**, 82 (1956)
21.8 J. R. Chelikowsky, M. L. Cohen: Phys. Rev. B **14**, 556 (1976)
21.9 B. Lax, J. G. Mavroides: Phys. Rev. **100**, 1650 (1955)
21.10 J. C. Hensel, G. Feher: Phys. Rev. **129**, 1041 (1963)
21.11 I. Balslev, P. Lawaetz: Phys. Lett. **19**, 3460 (1965)
21.12 P. Lawaetz: Phys. Rev. B **4**, 3460 (1971)
21.13 G. Ottaviani, L. Reggiani, C. Canali, F. Nava, A. A-Quranta: Phys. Rev. B **12**, 3318 (1975)
21.14 C. Jacoboni, C. Canali, G. Ottaviani, A. A-Quranta: Solid State Electron. **20**, 77 (1977)
21.15 H. D. Barber: Solid State Electron. **10**, 1039 (1967)
21.16 M. A. Green: J. Appl. Phys. **67**, 2944 (1990)
21.17 A. B. Sproul, M. A. Green: J. Appl. Phys. **70**, 846 (1991)
21.18 R. F. Pierret: *Advanced Semiconductor Fundamentals*, Modular Series on Solid State Devices, ed. by G. W. Neudeck, R. F. Pierret (Pearson Education, New York 2003)
21.19 G. Dresselhaus, A. F. Kip, C. Kittel: Phys. Rev. **98**, 368 (1955)
21.20 J. C. Hensel, H. Hasegawa, M. Nakayama: Phys. Rev. **138**, 225 (1965)
21.21 L. Kleinmann, J. C. Phillips: Phys. Rev. **118**, 1153 (1960)
21.22 M. Cardona, F. H. Pollak: Phys. Rev. **142**, 530 (1966)
21.23 S. Zwerdling, K. J. Button, B. Lax, L. M. Roth: Phys. Rev. Lett. **4**, 173 (1960)
21.24 W. R. Runyan: *Silicon Semiconductor Technology* (McGraw–Hill, New York 1965) Chap. 8
21.25 V. I. Fistul: *Heavily Doped Semiconductors* (Plenum, New York 1969)
21.26 W. Kohn, J. M. Luttinger: Phys. Rev. **97**, 1721 (1955)
21.27 W. Kohn, J. M. Luttinger: Phys. Rev. **98**, 915 (1955)
21.28 W. Kohn, D. Schechter: Phys. Rev. **99**, 1903 (1955)
21.29 E. Burstein, G. Picus, R. Henvis, R. Wallis: J. Phys. Chem. Solids **1**, 65 (1956)
21.30 G. Picus, E. Burstein, B. Henvis: J. Phys. Chem. Solids **1**, 75 (1956)
21.31 N. F. Mott: *Metal–Insulator Transitions*, 2nd edn. (Taylor & Francais, London 1990) p. 2
21.32 R. H. Hall, J. H. Racette: J. Appl. Phys. **35**, 379 (1964)
21.33 W. M. Bullis: Solid State Electron. **9**, 143 (1966)
21.34 B. Jensen: *Handbook of Optical Constants of Solids*, Vol. 2 (Academic, Orlando 1985) p. 169
21.35 M. Kohler: Z. Physik **124**, 777 (1948)
21.36 M. Kohler: Z. Physik **125**, 679 (1949)
21.37 B. R. Nag: *Theory of Electrical Transport in Semiconductors* (Pergamon, Oxford 1972)
21.38 E. M. Conwell: *High Field Transport in Semiconductors* (Academic, New York 1967)
21.39 W. Fawcett, A. D. Boardman, S. Swain: J. Phys. Chem. Solids **31**, 1963 (1970)
21.40 W. Dumke: Phys. Rev. **124**, 1813 (1961)
21.41 I. Balslev: Phys. Rev. **143**, 636 (1966)
21.42 B. N. Brockhouse: Phys. Rev. Lett. **2**, 256 (1959)
21.43 W. A. Harrison: Phys. Rev. **104**, 1281 (1956)
21.44 H. Ehrenreich, A. W. Overhauser: Phys. Rev. **104**, 331 (1956)
21.45 J. Bardeen, W. Shockley: Phys. Rev. **80**, 72 (1950)
21.46 G. L. Bir, G. E. Pikus: Fiz. Tverd. Tela **22**, 2039 (1960) *Soviet Phys. – Solid State* **2** (1961) 2039
21.47 M. Tiersten: IBM J. Res. Devel. **5**, 122 (1961)
21.48 C. Herring, E. Vogt: Phys. Rev. **101**, 944 (1956)
21.49 M. Tiersten: J. Phys. Chem. Solids **25**, 1151 (1964)
21.50 H. W. Streitwolf: Phys. Stat. Sol. **37**, K47 (1970)
21.51 D. Long: Phys. Rev. **120**, 2024 (1960)
21.52 D. L. Rode: Phys. Stat. Sol. (b) **53**, 245 (1972)
21.53 P. Norton, T. Braggins, H. Levinstein: Phys. Rev. B **8**, 5632 (1973)
21.54 C. Canali, C. Jacobini, F. Nava, G. Ottaviani, A. Alberigi: Phys. Rev. B **12**, 2265 (1975)
21.55 C. Erginsoy: Phys. Rev. **79**, 1013 (1950)
21.56 N. Sclar: Phys. Rev. **104**, 1559 (1956)
21.57 E. M. Conwell, V. F. Weisskopf: Phys. Rev. **77**, 338 (1950)
21.58 H. Brooks: Phys. Rev. **83**, 388 (1951)
21.59 C. Herring: Bell Syst. Tech. J. **36**, 237 (1955)
21.60 R. Dingle: Phil. Mag. **46**, 831 (1955)

21.61 H. Brooks: *Advances in Electronics and Electron Physics*, Vol. 7 (Academic, New York 1955) p. 85
21.62 D. Chattopadhyay, H. J. Queisser: Rev. Mod. Phys. **53**, 745 (1981)
21.63 A. G. Samoilovich, I. Ya. Korenblit, I. V. Dakhovskii, V. D. Iskra: Fiz. Tverd. Tela **3**, 3285 (1961) *Soviet Phys. – Solid State* (1962) 2385
21.64 P. M. Eagles, D. M. Edwards: Phys. Rev. **138**, A1706 (1965)
21.65 I. V. Dakhovskii: Fiz. Tverd. Tela **55**, 2332 (1963) *Soviet Phys. – Solid State* **5** (1964) 1695
21.66 D. Long, J. Myers: Phys. Rev. **120**, 39 (1960)
21.67 J. R. Meyer, F. J. Bartoli: Phys. Rev. B **23**, 5413 (1981)
21.68 H. I. Ralph, G. Simpson, R. J. Elliot: Phys. Rev. B **11**, 2948 (1975)
21.69 H. K. Jung, H. Ohtsuka, K. Taniguchi, C. Hamaguchi: J. Appl. Phys. **79**, 2559 (1996)
21.70 G. L. Bir, E. Normantas, G. E. Pikus: Fiz. Tverd. Tela **4**, 1180 (1962)
21.71 P. P. Debye, E. M. Conwell: Phys. Rev. **93**, 693 (1954)
21.72 J. Appel: Phys. Rev. **122**, 1760 (1961)
21.73 M. Luong, A. W. Shaw: Phys. Rev. B **4**, 30 (1971)
21.74 J. Appel: Phys. Rev. **125**, 1815 (1962)
21.75 B. E. Sernelius: Phys. Rev. B **41**, 2436 (1990)
21.76 P. A. Shumann, R. P. Phillips: Solid State Electron. **10**, 943 (1967)
21.77 M. A. Saifi, R. H. Stolen: J. Appl. Phys. **43**, 1171 (1972)
21.78 J. Humlicek, K. Wojtechovsky: Czech. J. Phys. B **38**, 1033 (1988)
21.79 M. Auslender, S. Hava: *Handbook of Optical Constants of Solids*, Vol. 3, ed. by D. Palik E. (Academic, New York 1998) p. 155
21.80 W. Kaiser, P. H. Keck, C. F. Lange: Phys. Rev. **101**, 1264 (1956)
21.81 S. Adachi: Phys. Rev. B **38**, 12966 (1988)
21.82 S. Adachi: J. Appl. Phys. **66**, 3224 (1989)
21.83 T. Aoki, S. Adachi: J. Appl. Phys. **69**, 1574 (1991)
21.84 K.-F. Berggren, B. E. Sernelius: Phys. Rev. B **24**, 1971 (1981)
21.85 P. E. Schmid: Phys. Rev. B **23**, 5531 (1981)
21.86 L. Viña, M. Cardona: Phys. Rev. B **29**, 6739 (1984)
21.87 W. G. Spitzer, H. Y. Fan: Phys. Rev. **106**, 882 (1957)
21.88 M. Balkanski, A. Aziza, E. Amzallag: Phys. Stat. Sol. **31**, 323 (1969)
21.89 M. Cardona, W. Paul, H. Brooks: Zeitschr. Naturforsch **101**, 329 (1960)
21.90 L. M. Lambert: Phys. Stat. Sol. **11**, 461 (1972)
21.91 L. Pearson, J. Bardeen: Phys. Rev. **75**, 865 (1961)
21.92 M. B. Prince: Phys. Rev. **93**, 1204 (1954)
21.93 G. W. Ludwig, R. L. Watters: Phys. Rev. **101**, 1699 (1956)
21.94 D. C. Cronemeyer: Phys. Rev. **105**, 522 (1957)
21.95 P. P. Debye, T. Kohane: Phys. Rev. **94**, 724 (1954)
21.96 F. J. Morin, J. P. Maita: Phys. Res. **96**, 28 (1954)
21.97 G. Backenstoss: Phys. Rev. **108**, 579 (1957)
21.98 D. Long: Phys. Rev. **107**, 672 (1957)
21.99 D. Long, J. Myers: Phys. Rev. **109**, 1098 (1958)
21.100 R. A. Logan, A. J. Peters: J. Appl. Phys. **31**, 122 (1960)
21.101 D. Long, J. Myers: Phys. Rev. **115**, 1107 (1959)
21.102 L. J. Neuringer, D. Long: Phys. Rev. **135**, A788 (1964)
21.103 A. G. Samoilovich, I. Ya. Korenblit, I. V. Dakhovskii, V. D. Iskra: Fiz. Tverd. Tela **3**, 2939 (1961) *Soviet Phys. – Solid State* **3** (1962) 2148
21.104 J. C. Irvin: Bell Syst. Tech. J. **41**, 387 (1962)
21.105 F. Mousty, P. Ostoja, L. Passari: J. Appl. Phys. **45**, 4576 (1974)
21.106 I. G. Kirnas, P. M. Kurilo, P. G. Litovchenko, V. S. Lutsyak, V. M. Nitsovich: Phys. Stat. Sol. (a) **23**, K123 (1974)
21.107 S. M. Sze, J. C. Irvin: Solid State Electron. **11**, 559 (1968)
21.108 D. M. Caughey, R. F. Thomas: Proc. IEEE **55**, 2192 (1967)
21.109 G. Baccarani, P. Ostoja: Solid State Electron. **18**, 1039 (1975)
21.110 C. Hilsum: Electron. Lett. **10**, 259 (1074)
21.111 J. A del Alamo, R. M. Swanson: Solid State Electron. **30**, 1127 (1987)
21.112 Y. Furukawa: J. Phys. Soc. Japan **16**, 577 (1961)
21.113 C. Canali, C. Jacoboni., G. Ottaviani, A. Alberigi Quaranta: Appl. Phys. Lett. **27**, 278 (1975)
21.114 C. Canali, C. Jacoboni, F. Nava, G. Ottaviani, A. Alberigi Quaranta: Phys. Rev. B **12**, 2265 (1975)
21.115 E. H. Putley, W. H. Mitchell: Proc. Phys. Soc. (London) A **72**, 193 (1958)
21.116 C. Canali, M. Costato, G. Ottaviani, L. Reggiani: Phys. Rev. Lett. **31**, 536 (1973)
21.117 E. A. Davies, D. S. Gosling: J. Phys. Chem. Solids **23**, 413 (1962)
21.118 C. Canali, G. Ottaviani, A. Alberigi Quaranta: J. Phys. Chem. Solids **32**, 1707 (1971)
21.119 C. B. Norris, J. F. Gibbons: IEEE Trans. Electron. Dev. **14**, 30 (1967)
21.120 T. W. Sigmon, J. F. Gibbons: Appl. Phys. Lett. **15**, 320 (1969)
21.121 V. Rodriguez, H. Ruegg, M.-A. Nicolet: IEEE Trans. Electron. Dev. **14**, 44 (1967)
21.122 T. E. Seidel, D. L. Scharfetter: J. Phys. Chem. Solids **28**, 2563 (1967)
21.123 V. Rodriguez, M.-A. Nicolet: J. Appl. Phys. **40**, 496 (1969)
21.124 B. L. Boichenko, V. M. Vasetskii: Soviet Phys. Solid State **7**, 1631 (1966)
21.125 A. C. Prior: J. Phys. Chem. Solids **12**, 175 (1959)
21.126 C. Y. Duh, J. L. Moll: IEEE Trans. Electron. Dev. **14**, 46 (1967)
21.127 C. Y. Duh, J. L. Moll: Solid State Electron. **11**, 917 (1968)
21.128 M. H. Jorgensen, N. O. Gram, N. I. Meyer: Solid-State Comm. **10**, 337 (1972)
21.129 M. Asche, B. L. Boichenko, O. G. Sarbej: Phys. Stat. Sol. **9**, 323 (1965)
21.130 J. G. Nash, J. W. Holm-Kennedy: Appl. Phys. Lett. **24**, 139 (1974)
21.131 J. G. Nash, J. W. Holm-Kennedy: Appl. Phys. Lett. **25**, 507 (1974)

21.132 M. Asche, O. G. Sarbej: Phys. Stat. Sol. (a) **38**, K61 (1971)
21.133 N. O. Gram: Phys. Lett. A **38**, 235 (1972)
21.134 C. Canali, A. Loria, F. Nava, G. Ottaviani: Solid-State Comm. **12**, 1017 (1973)
21.135 M. Asche, O. G. Sarbej: Phys. Stat. Sol. (a) **46**, K121 (1971)
21.136 J. P. Nougier, M. Rolland, O. Gasquet: Phys. Rev. B **11**, 1497 (1975)
21.137 E. J. Ryder, W. Shockley: Phys. Rev. **81**, 139 (1951)
21.138 W. Shockley: Bell. Syst. Tech. J. **30**, 990 (1951)
21.139 E. J. Ryder: Phys. Rev. **90**, 766 (1953)
21.140 M. Costato, L. Reggiani: Lett. Nuovo Cimento **3**, 728 (1970)
21.141 C. Jacoboni, R. Minder, G. Majni: J. Phys. Chem. Solids **36**, 1129 (1975)
21.142 C. Canali, G. Ottaviani: Phys. Lett. A **32**, 147 (1970)
21.143 J. P. Nougier, M. Rolland: Phys. Rev. B **8**, 5728 (1973)
21.144 G. Persky, D. J. Bartelink: J. Appl. Phys. **42**, 4414 (1971)
21.145 D. L. Scharfetter, H. K. Gummel: IEEE Trans. Electron. Dev. **ED-16**, 64 (1969)
21.146 J. G. Ruch, G. S. Kino: Phys. Rev. **174**, 921 (1968)
21.147 W. Maes, K. de Meyer, R. van Overstraeten: Solid State Electron. **33**, 705 (1990)
21.148 S. S. Li, W. R. Thurber: Solid State Electron. **20**, 609 (1977)
21.149 J. M. Dorkel, P. Leturcq: Solid-State Electron. **24**, 821 (1981)
21.150 M. S. Tyagi, R. van Overstraeten: Solid State Electron. **10**, 1039 (1983)
21.151 H. O. McMachon: J. Opt. Soc. Am. **40**, 376 (1950)
21.152 J. C. Sturm, C. M. Reaves: IEEE Trans. Electron. Dev. **39**, 81 (1992)
21.153 P. J. Timans: J. Appl. Phys. **74**, 6353 (1993)
21.154 P. A. Shumann Jr., W. A. Keenan, A. H. Tong, H. H. Gegenwarth, C. P. Schneider: J. Electrochem. Soc. **118**, 145 (1971)
21.155 R. M. A. Azzam, N. M. Bashara: *Ellipsometry and Polarized Light*, 2nd edn. (Elsevier, Amsterdam 1987)
21.156 D. E. Aspnes, A. A. Studna: Phys. Rev. B **27**, 985 (1983)
21.157 E. Barta, G. Lux: J. Phys. D: Appl. Phys. **16**, 1543 (1983)
21.158 J. Geist: *Handbook of Optical Constants of Solids*, Vol. 3, ed. by D. Palik E. (Academic, New York 1998) p. 519
21.159 D. E. Aspnes, A. A. Studna, E. Kinsbron: Phys. Rev. B **29**, 768 (1984)
21.160 H. R. Philipp, E. A. Taft: Phys. Rev. B **120**, 37 (1960)
21.161 G. E. Jellison Jr.: Opt. Mater. **1**, 41 (1992)
21.162 H. A. Weaklien, D. Redfield: J. Appl. Phys. **50**, 1491 (1979)
21.163 D. F. Edward: *Handbook of Optical Constants of Solids*, Vol. 2 Orlando 1985) p. 547
21.164 R. A. Soref, B. R. Bennett: IEEE J. Quantum Electron. **23**, 123 (1987)

22. Silicon–Germanium: Properties, Growth and Applications

Silicon–germanium is an important material that is used for the fabrication of SiGe heterojunction bipolar transistors and strained Si metal–oxide–semiconductor (MOS) transistors for advanced complementary metal–oxide–semiconductor (CMOS) and BiCMOS (bipolar CMOS) technologies. It also has interesting optical properties that are increasingly being applied in silicon-based photonic devices. The key benefit of silicon–germanium is its use in combination with silicon to produce a heterojunction. Strain is incorporated into the silicon–germanium or the silicon during growth, which also gives improved physical properties such as higher values of mobility. This chapter reviews the properties of silicon–germanium, beginning with the electronic properties and then progressing to the optical properties. The growth of silicon–germanium is considered, with particular emphasis on the chemical vapour deposition technique and selective epitaxy. Finally, the properties of polycrystalline silicon–germanium are discussed in the context of its use as a gate material for MOS transistors.

- 22.1 **Physical Properties of Silicon–Germanium** 482
  - 22.1.1 Critical Thickness 482
  - 22.1.2 Band Structure 483
  - 22.1.3 Dielectric Constant 484
  - 22.1.4 Density of States 484
  - 22.1.5 Majority-Carrier Mobility in Strained $Si_{1-x}Ge_x$ 486
  - 22.1.6 Majority-Carrier Mobility in Tensile-Strained Si on Relaxed $Si_{1-x}Ge_x$ 486
  - 22.1.7 Minority-Carrier Mobility in Strained $Si_{1-x}Ge_x$ 486
  - 22.1.8 Apparent Band-Gap Narrowing in $Si_{1-x}Ge_x$ HBTs........................ 487
- 22.2 **Optical Properties of SiGe** 488
  - 22.2.1 Dielectric Functions and Interband Transitions 488
  - 22.2.2 Photoluminescence 489
  - 22.2.3 SiGe Quantum Wells 490
- 22.3 **Growth of Silicon–Germanium**............... 492
  - 22.3.1 In-Situ Hydrogen Bake 492
  - 22.3.2 Hydrogen Passivation 492
  - 22.3.3 Ultra-Clean Epitaxy Systems 492
  - 22.3.4 $Si_{1-x}Ge_x$ Epitaxy...................... 492
  - 22.3.5 Selective $Si_{1-x}Ge_x$ Epitaxy........... 492
- 22.4 **Polycrystalline Silicon–Germanium**........ 494
  - 22.4.1 Electrical Properties of Polycrystalline $Si_{1-x}Ge_x$ 496
- **References** 497

Silicon–germanium ($Si_{1-x}Ge_x$) alloys have been researched since the late 1950s [22.1], but it is only in the past 15 years or so that these layers have been applied to new types of transistor technology. $Si_{1-x}Ge_x$ was first applied in bipolar technologies [22.2, 3], but more recently has been applied to metal–oxide–semiconductor (MOS) technologies [22.4–7]. This has been made possible by the development of new growth techniques, such as molecular-beam epitaxy (MBE), low-pressure chemical vapour deposition (LPCVD) and ultra-high-vacuum chemical vapour deposition (UHV-CVD). The key feature of these techniques that has led to the development of $Si_{1-x}Ge_x$ transistors is the growth of epitaxial layers at low temperatures (500–700 °C). This allows $Si_{1-x}Ge_x$ layers to be grown without disturbing the doping profiles of structures already present in the silicon wafer. $Si_{1-x}Ge_x$ layers can be successfully grown on silicon substrates even though there is a lattice mismatch between silicon and germanium of 4.2%.

The primary property of $Si_{1-x}Ge_x$ that is of interest for bipolar transistors is the band gap, which is smaller than that of silicon and controllable by varying the germanium content. Band-gap engineering concepts, which were previously only possible in compound semiconductor technologies, have now become viable in silicon technology. These concepts have introduced new degrees of freedom in the design of bipolar transistors that have led to dramatic improvements in transistor performance. In $Si_{1-x}Ge_x$ heterojunction bipolar transistors (HBTs), the $Si_{1-x}Ge_x$ layer is incorporated into the base and the lower band gap of $Si_{1-x}Ge_x$ than Si is used to advantage to dramatically improve the high-frequency performance. $Si_{1-x}Ge_x$ HBTs have been produced with values of cut-off frequency, f_T, approaching 300 GHz [22.8], a value unimaginable in silicon bipolar transistors. Values of gate delay well below 10 ps can be achieved in properly optimised $Si_{1-x}Ge_x$ HBTs [22.9]. In $Si_{1-x}Ge_x$ MOS field-effect transistors (MOSFETs), the $Si_{1-x}Ge_x$ layer is incorporated in the channel and is used to give improved values of mobility.

Initially, strained $Si_{1-x}Ge_x$ layers on silicon substrates were used to give improved hole mobility in p-channel transistors [22.7], but more recently thin, strained silicon layers on relaxed SiGe virtual substrates have been used to give improvements in both electron and hole mobility [22.4–6].

In this chapter, the properties of single-crystal silicon–germanium will first be outlined, followed by a description of the methods used for growing silicon–germanium layers. The properties and applications of polycrystalline silicon–germanium are also discussed later in the article.

22.1 Physical Properties of Silicon–Germanium

Silicon and germanium are completely miscible over the full range of compositions and hence can be combined to form $Si_{1-x}Ge_x$ alloys with the germanium content, x, ranging from 0 to 1 (0–100%). $Si_{1-x}Ge_x$ has a diamond-like lattice structure and the lattice constant is given by Vegard's rule:

$$a_{Si_{1-x}Ge_x} = a_{Si} + x\left(a_{Ge} - a_{Si}\right) , \quad (22.1)$$

where x is the germanium fraction and a is the lattice constant. The lattice constant of silicon, a_{Si}, is 0.543 nm, the lattice constant of germanium, a_{Ge}, is 0.566 nm and the lattice mismatch is 4.2%.

When a $Si_{1-x}Ge_x$ layer is grown on a silicon substrate, the lattice mismatch at the interface between the $Si_{1-x}Ge_x$ and the silicon has to be accommodated. This can either be done by compression of the $Si_{1-x}Ge_x$ layer so that it fits to the silicon lattice or by the creation of misfit dislocations at the interface. These two possibilities are illustrated schematically in Fig. 22.1. In the former case, the $Si_{1-x}Ge_x$ layer adopts the silicon lattice spacing in the plane of the growth and hence the normally cubic $Si_{1-x}Ge_x$ crystal is distorted. When $Si_{1-x}Ge_x$ growth occurs in this way, the $Si_{1-x}Ge_x$ layer is under compressive strain and the layer is described as *pseudomorphic*. In the second case, the $Si_{1-x}Ge_x$ layer is unstrained, or relaxed, and the lattice mismatch at the interface is accommodated by the formation of misfit dislocations. These misfit dislocations generally lie in the plane of the interface, as shown in Fig. 22.1, but dislocations can also thread vertically through the $Si_{1-x}Ge_x$ layer.

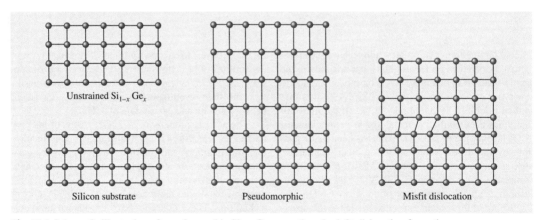

Fig. 22.1 Schematic illustration of pseudomorphic $Si_{1-x}Ge_x$ growth and misfit-dislocation formation

22.1.1 Critical Thickness

There is a maximum thickness of $Si_{1-x}Ge_x$ that can be grown before relaxation of the strain occurs through the formation of misfit dislocations. This is known as the critical thickness of the $Si_{1-x}Ge_x$ layer, and depends strongly on the germanium content, as shown in Fig. 22.2. The original calculations of critical layer thickness were made by *Matthews* and *Blakeslee* [22.11, 12] on the basis of the mechanical equilibrium of an existing threading dislocation. However, measurements of dislocation densities in $Si_{1-x}Ge_x$ showed, in many cases, no evidence of misfit dislocations for $Si_{1-x}Ge_x$ layers considerably thicker than the Matthews–Blakeslee limit. These results were explained by *People* and *Bean* [22.13] who calculated the critical thickness on the assumption that misfit-dislocation generation was determined solely by energy balance. The discrepancy between these two types of calculation can be explained by the observation that strain relaxation in $Si_{1-x}Ge_x$ layers occurs gradually. Layers above the People–Bean curve can be considered to be completely relaxed, whereas layers below the Matthews–Blakeslee curve can be considered to be fully strained. These fully strained layers are termed *stable* and will not relax during any subsequent high temperature processing. Layers lying between the two curves are termed *metastable*; these layers may be free of dislocations after growth, but are susceptible to relaxation during later high-temperature processing.

In practice, a number of additional factors influence the critical thickness of a $Si_{1-x}Ge_x$ layer. Of particular importance to both $Si_{1-x}Ge_x$ HBTs and $Si_{1-x}Ge_x$ MOSFETs, is the effect of a silicon cap layer, which has been shown to increase the critical thickness of the underlying $Si_{1-x}Ge_x$ layer. Figure 22.3 shows a comparison of the calculated critical thickness as a function of germanium percentage for stable $Si_{1-x}Ge_x$ layers with and without a silicon cap. It can be seen that the critical thickness is more than doubled by the presence of the silicon cap.

The presence of misfit dislocations in devices is highly undesirable, since they create generation/recombination centres, which degrade leakage currents when they are present in the depletion regions of devices. Threading dislocations also highly undesirable, as they can lead to the formation of emitter/collector pipes in $Si_{1-x}Ge_x$ HBTs. When designing $Si_{1-x}Ge_x$ devices, it is important that the $Si_{1-x}Ge_x$ thickness is chosen to give a stable layer, so that dislocation formation is avoided. A base thickness below the silicon cap curve in Fig. 22.3 will ensure a stable layer, which will withstand ion implantation and high-temperature annealing without encountering problems of relaxation and misfit-dislocation generation.

Considerable research has been done on the oxidation of $Si_{1-x}Ge_x$ [22.14, 15], and it has been found that the germanium in the $Si_{1-x}Ge_x$ layer does not oxidise, but piles up at the oxide/$Si_{1-x}Ge_x$ interface. This pile-up of germanium makes it difficult to achieve low values of interface state density in oxidised $Si_{1-x}Ge_x$ layers. It is therefore advisable to avoid direct oxidation of $Si_{1-x}Ge_x$ layers, particularly in $Si_{1-x}Ge_x$ MOS technologies. In $Si_{1-x}Ge_x$ MOSFETs, a silicon cap is often

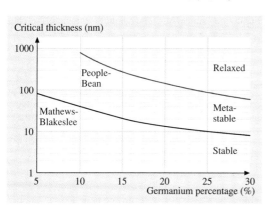

Fig. 22.2 Critical $Si_{1-x}Ge_x$ thickness as a function of germanium percentage (after *Iyer* et al. [22.2], copyright 1989 IEEE)

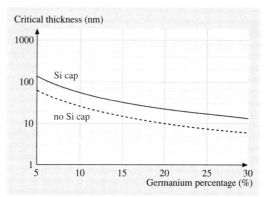

Fig. 22.3 Critical thickness as a function of germanium percentage for stable $Si_{1-x}Ge_x$ layers with and without a silicon cap (after *Jain* et al. [22.10], copyright 1992 Elsevier)

included above the $Si_{1-x}Ge_x$ layer that can be oxidised to create the gate oxide.

22.1.2 Band Structure

$Si_{1-x}Ge_x$ alloys have a smaller band gap than silicon partly because of the larger lattice constant and partly because of the strain. Figure 22.4 shows the variation of band gap with germanium percentage for strained and unstrained $Si_{1-x}Ge_x$. It can be seen that the strain has a dramatic effect on the band gap of $Si_{1-x}Ge_x$. For 10% germanium, the reduction in band gap compared with silicon is 92 meV for strained $Si_{1-x}Ge_x$, compared with 50 meV for unstrained $Si_{1-x}Ge_x$. The variation of band gap with germanium content for strained $Si_{1-x}Ge_x$ can be described by the following empirical equation:

$$E_G(x) = 1.17 + 0.96x - 0.43x^2 + 0.17x^3 \quad (22.2)$$

The band alignment for compressively strained $Si_{1-x}Ge_x$ on unstrained silicon is illustrated schematically in Fig. 22.5. This band alignment is referred to as *type I*, and the majority of the band offset at the hetero-

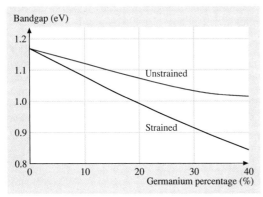

Fig. 22.4 Band gap as a function of germanium percentage for strained [22.16] and unstrained [22.1] $Si_{1-x}Ge_x$ (after *Iyer* et al. [22.2], copyright 1989 IEEE)

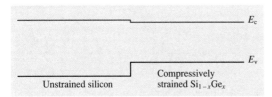

Fig. 22.5 Schematic illustration of the band alignment obtained for a compressively strained $Si_{1-x}Ge_x$ layer grown on an unstrained silicon substrate

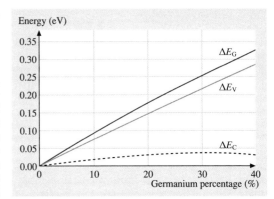

Fig. 22.6 Valence- and conduction-band offsets as a function of germanium percentage for strained $Si_{1-x}Ge_x$ grown on unstrained silicon (after *Poortmans* et al. [22.17], copyright 1993 Elsevier)

junction interface occurs in the valence band, with only a small offset in the conduction band. Different band alignments can be obtained by engineering the strain in the substrate and the grown layer in different ways. For example, *type II* band alignments can be obtained by growing tensile-strained silicon on top of unstrained $Si_{1-x}Ge_x$. This arrangement gives large conduction- and valence-band offsets and is used in strained silicon MOSFETs.

Figure 22.6 shows the variation of valence-band offset, ΔE_V, conduction-band offset, ΔE_C, and band-gap narrowing, ΔE_G, with germanium content. It can be seen that the majority of the band offset occurs in the valence band. For example for 10% germanium, the valence-band offset is 0.073 eV, compared with 0.019 eV for the conduction-band offset. The conduction-band offset can therefore be neglected for most practical purposes.

22.1.3 Dielectric Constant

The dielectric constant of $Si_{1-x}Ge_x$ can be obtained by linear interpolation between the known values for silicon and germanium [22.17] using the following equation:

$$\varepsilon(x) = 11.9(1 + 0.35x) \quad (22.3)$$

22.1.4 Density of States

While, the density of states in the conduction band in $Si_{1-x}Ge_x$ is generally assumed to be the same as that in silicon, there is some evidence in the literature to suggest that the density of states in the valence band

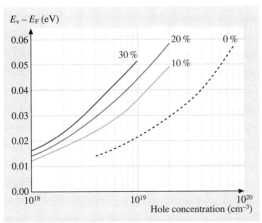

Fig. 22.7 Fermi-level position as a function of hole concentration for $Si_{1-x}Ge_x$ with four different germanium concentrations (after *Iyer* et al. [22.2], copyright 1989 IEEE)

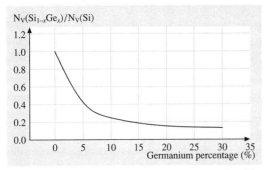

Fig. 22.8 Ratio of density of states in the valence band for $Si_{1-x}Ge_x$ to that for Si as a function of germanium percentage (after *Poortmans* [22.20], copyright 1993 University of Leuven)

is considerably smaller. *Manku* and *Nathan* [22.18, 19] have calculated the E–k diagram for strained $Si_{1-x}Ge_x$ and shown that the density-of-states hole mass is significantly lower, by a factor of approximately three at 30% germanium. There is some experimental evidence to support this calculation. For example, freeze-out of holes in p-type $Si_{1-x}Ge_x$ has been reported to occur at higher temperatures than in p-type silicon [22.20] and enhancements in the majority-carrier, hole mobility have been reported for p-type $Si_{1-x}Ge_x$ [22.21].

Using the calculated values of hole density of states of *Manku* and *Nathan* [22.18, 19], the hole concentration can be calculated as a function of Fermi-level position.

These results are shown in Fig. 22.7 for $Si_{1-x}Ge_x$ with four different germanium contents. It can be seen that the Fermi level moves deeper into the valence band as the germanium concentration increases. Figure 22.8 shows the ratio of the calculated density of states in the valence

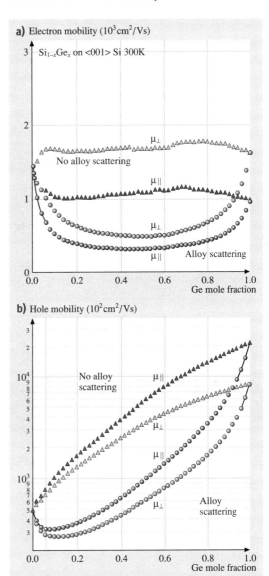

Fig. 22.9a,b Calculated 300-K electron (**a**) and hole (**b**) in-plane and out-of-plane low-field mobilities in strained $Si_{1-x}Ge_x$ grown on (100) Si (after *Fischetti* et al. [22.22], copyright 1996 American Institute of Physics)

band for $Si_{1-x}Ge_x$ to that for silicon as a function of germanium content. It is clear that the density of states in the valence band for $Si_{1-x}Ge_x$ is significantly lower than that for silicon at germanium contents of practical interest.

22.1.5 Majority-Carrier Mobility in Strained $Si_{1-x}Ge_x$

Values of in-plane and out-of-plane low-field mobility in strained $Si_{1-x}Ge_x$ grown on (100) Si have been calculated by *Fischetti* et al. [22.22], and are shown in Fig. 22.9. There is some uncertainty in the chosen values of alloy scattering parameters used in the calculations, but nevertheless the results are representative of current understanding. These results show a large enhancement of low-field hole mobility for $Si_{1-x}Ge_x$ compared with unstrained silicon, but only a modest enhancement of low-field electron mobility. These results indicate that strained $Si_{1-x}Ge_x$ channels can be used to significantly improve the mobility of p-channel MOSFETs, but little benefit is obtained for n-channel MOSFETs. For this reason, industry focus has moved away from channels realised in $Si_{1-x}Ge_x$ to channels realised in tensile-strained silicon, as discussed below.

22.1.6 Majority-Carrier Mobility in Tensile-Strained Si on Relaxed $Si_{1-x}Ge_x$

Tensile-strained silicon can be produced by growing a thin silicon layer on top of a relaxed $Si_{1-x}Ge_x$ virtual substrate. Figure 22.10 shows a typical virtual substrate for a surface-channel MOS transistor. A graded $Si_{1-x}Ge_x$ layer is grown on top of the silicon substrate with the Ge content varying from 0 to 30%. Misfit dislocations will form in this layer, but the majority of dislocations will be in the plane of the $Si_{1-x}Ge_x$ layer and only a small percentage will propagate vertically

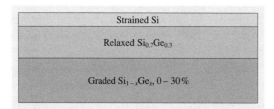

Fig. 22.10 Schematic illustration of a typical tensile-strained Si layer grown on top of a $Si_{1-x}Ge_x$ virtual substrate

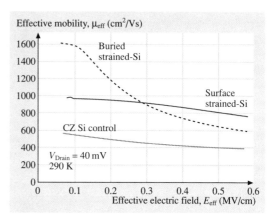

Fig. 22.11 Effective electron mobility as a function of effective electric field for strained Si MOSFETs fabricated on a 30% $Si_{1-x}Ge_x$ virtual substrate (after *Welser* et al. [22.23], copyright 1994 IEEE)

to the surface of the layer. A 30%-relaxed $Si_{1-x}Ge_x$ buffer is then grown followed by a thin tensile-strained $Si_{1-x}Ge_x$ layer in which the channel is fabricated. The key to any virtual substrate growth process is the minimisation of dislocation propagation to the surface of the wafer.

Figure 22.11 shows typical values of effective electron mobility obtained from measurements on n-channel MOS transistors for a $Si_{1-x}Ge_x$ virtual substrate with 30% Ge [22.23]. For the surface-channel strained Si device, the effective mobility is enhanced by 80% compared with the Si control transistor due to the tensile strain in the $Si_{1-x}Ge_x$ layer.

Enhanced hole mobility can also be obtained in tensile-strained Si grown on a $Si_{1-x}Ge_x$ virtual substrate, though higher germanium contents are needed to obtain a significant mobility enhancement. Figure 22.12 shows typical values of effective hole mobility in strained Si for Ge contents between 35 and 50% [22.24]. The effective mobility of the strained Si device is enhanced by 100% compared with the Si control device.

22.1.7 Minority-Carrier Mobility in Strained $Si_{1-x}Ge_x$

$Si_{1-x}Ge_x$ HBTs are minority-carrier devices and hence values of the minority-carrier mobility of more interest than the majority-carrier mobility. Unfortunately very few measurements of minority-carrier mobility have been made in $Si_{1-x}Ge_x$. *Poortmans* [22.20] inferred values of minority-carrier mobility from measure-

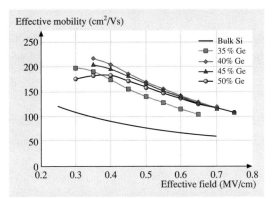

Fig. 22.12 Effective hole mobility as a function of effective electric field for strained Si MOSFETs fabricated on a $Si_{1-x}Ge_x$ virtual substrate with Ge contents in the range 35–50% (after *Leitz* et al. [22.24], copyright 2002 IEEE)

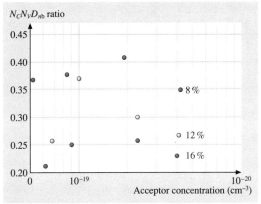

Fig. 22.13 Measured values of the ratio of $N_C N_V D_{nb}$ in $Si_{1-x}Ge_x$ to that in Si as a function of acceptor concentration (after *Poortmans* [22.20], copyright 1993 University of Leuven)

data directly obtained from measurements on $Si_{1-x}Ge_x$ HBTs. The gain enhancement in a $Si_{1-x}Ge_x$ HBT is determined by the ratio of the product $N_C N_V D_{nb}$ in $Si_{1-x}Ge_x$ and Si, together with the band-gap narrowing due to the strained $Si_{1-x}Ge_x$. Figure 22.13 shows a graph of this $N_C N_V D_{nb}$ ratio as a function of acceptor concentration for three values of germanium content. It can be seen that for germanium contents of practical interest, in the range 11–16%, this ratio has a value of around 0.25.

22.1.8 Apparent Band-Gap Narrowing in $Si_{1-x}Ge_x$ HBTs

In $Si_{1-x}Ge_x$ HBTs, the apparent band-gap narrowing is often quoted, which combines the effect of the band-gap reduction and the effect of high doping. *Poortmans* et al. [22.17] have developed a theoretical approach that has been shown to be in reasonable agreement with experiment. Figure 22.14 shows the apparent band-gap narrowing in $Si_{1-x}Ge_x$ as a function of acceptor concentration for three values of germanium content. At low acceptor concentrations, the apparent band-gap narrowing in $Si_{1-x}Ge_x$ is slightly higher than that in silicon, but at acceptor concentrations in the range $1-2 \times 10^{19}$ cm^{-3}, the apparent band-gap narrowing is approximately the same. This latter doping range is the base doping range that is of practical interest for $Si_{1-x}Ge_x$ HBTs.

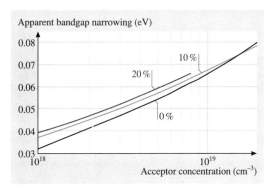

Fig. 22.14 Apparent band-gap narrowing as a function of acceptor concentration for $Si_{1-x}Ge_x$ with three different germanium concentrations (after *Poortmans* et al. [22.17], copyright 1993 Elsevier)

ments on $Si_{1-x}Ge_x$ HBTs and found an enhancement in mobility compared with silicon by a factor of 1.2–1.4 for base doping concentrations in the range $5 \times 10^{18} - 5 \times 10^{19}$ cm^{-3}. Given the scarcity of measured data on minority-carrier mobility and density of states in $Si_{1-x}Ge_x$, the most reliable way of calculating the expected gain improvement in a $Si_{1-x}Ge_x$ HBT is to use

22.2 Optical Properties of SiGe

Interest in the optical properties of SiGe stems from the desire to design silicon-based optoelectronic devices as well as the usefulness of many optical techniques in material analysis. The optical properties of bulk SiGe provides an important starting point for any attempt at in-depth understanding, however, nearly all real applications of SiGe involve the use of thin, strained layers. Beyond this, the formation of SiGe quantum structures within silicon devices has remained one of the most promising methods by which device engineers hope to improve the largely unimpressive optical behaviour of silicon that is brought about by its indirect band gap. The growth of quantum wells, quantum wires and quantum dots in the Si/SiGe system has been extensively explored [22.26–31]. However, there has as yet been little success at using Si/SiGe quantum structures to produce efficient silicon-based light emitters, although there have been impressive attempts [22.31]. Perhaps the most promising devices based on SiGe quantum wells are near-infra-red photodetectors in which the SiGe can be used to enhance sensitivity at the optical-communications wavelengths [22.32–36]. More futuristic applications of SiGe quantum wells include devices based on transitions between the confined energy levels in quantum wells. Devices based on these *inter-subband* transitions include quantum-well infrared photodetectors (QWIPs) [22.37, 38] and quantum cascade lasers [22.39, 40].

22.2.1 Dielectric Functions and Interband Transitions

A range of 1-μm-thick $Si_{1-x}Ge_x$ films grown by MBE on Si(100) substrates have been studied by spectroscopic ellipsometry to yield their complex dielectric functions at room temperature [22.25]. Both the real (ε_1) and imaginary (ε_2) parts of the measured dielectric function for Ge compositions of 0, 0.2, 0.4, 0.6, 0.8 and 1.0 are shown in Fig. 22.15a and b, respectively.

In Fig. 22.15b the absorption structures observed at 3.4 and 4.2 eV in the spectra shown for silicon originate from direct band-to-band transitions at various regions in the Brillouin zone of silicon. The structure seen around 3.4 eV is due to E_0', E_1 and $E_1 + \Delta_1$ interband transition

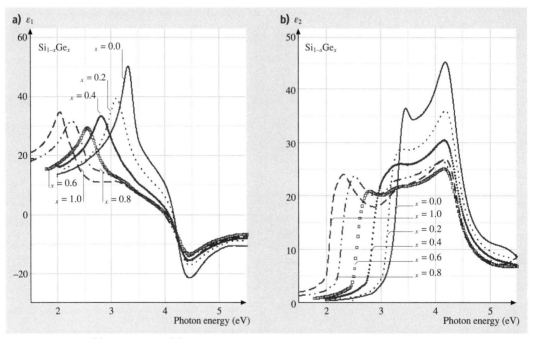

Fig. 22.15a,b Real (**a**) and imaginary (**b**) parts of the dielectric functions of relaxed $Si_{1-x}Ge_x$ alloys with composition x indicated in the legend (after *Bahng* et al. [22.25], copyright 2001, American Physical Society)

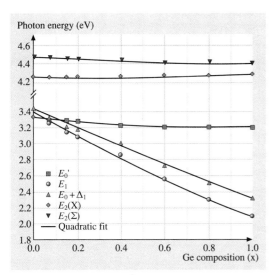

Fig. 22.16 Evolution of E_0', E_1, $E_1 + \Delta_1$, E_2 (X) and $E_2(\Sigma)$ transition energies for relaxed $Si_{1-x}Ge_x$ with composition x (after *Bahng* et al. [22.25], copyright 2001, American Physical Society)

photoluminescence spectra is nearly always sufficient to allow a simple qualitative assessment of material quality; alternatively, detailed analysis can permit a broad range of material or structural parameters to be assessed or determined. No two photoluminescence spectra are the same. In this section and the section that follows on photoluminescence studies of Si/SiSe quantum wells, we will present a range of spectra that represent most of the key features that have been observed.

Weber and *Alonso* [22.41] have provided a very useful study of the near-band-gap photoluminescence of bulk SiGe alloys. In their study bulk SiGe samples are cut from nominally undoped polycrystalline ingots prepared by a zone-levelling technique. Figure 22.17 shows the photoluminescence spectra for a range of compositions. Samples were excited using the 514-nm line of an argon ion laser and the sample temperature was 4.2 K.

edges, whereas the structure at 4.2 eV is due to E_2 (X) and E_2 (Σ) edges [22.25]. The evolution of each of these transition edges for the full range of SiGe compositions is shown in Fig. 22.16.

The quadratic fits shown in Fig. 22.16 are as follows [22.25]:

$$E_0'(x) = 3.337 - 0.348x + 0.222x^2 \quad (22.4)$$
$$E_1(x) = 3.398 - 1.586x + 0.27x^2 \quad (22.5)$$
$$E_1 + \Delta_1(x) = 3.432 - 1.185x + 0.065x^2 \quad (22.6)$$
$$E_2(X)(x) = 4.259 - 0.052x + 0.084x^2 \quad (22.7)$$
$$E_2(\Sigma)(x) = 4.473 - 0.139x + 0.072x^2 \quad (22.8)$$

22.2.2 Photoluminescence

The emission properties of semiconductor structures are of fundamental interest to scientists as well as being an important analytical technique for engineers. In general, the features of low-temperature photoluminescence spectra are very dependent on the specific conditions under which materials are grown and treated. This is because emission energies and emission rates are often sensitive to even small variations of impurity or defect densities, as well as variations in strain or composition. A brief examination of low-temperature

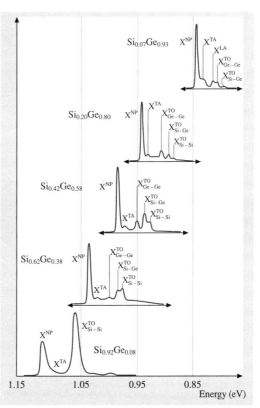

Fig. 22.17 Near-band-gap photoluminescence spectra for several bulk SiGe samples (after *Weber* et al. [22.41], copyright 1989, American Physical Society)

Excitonic emission lines are a strong feature of photoluminescence spectra when the thermal energy of the semiconductor is less than the exciton binding energy. Each spectrum featured in Fig. 22.17, across the full range of SiGe compositions, show similar *excitonic* features.

In most spectra the most pronounced peak is the no-phonon (X^{NP}) line caused by the optical recombination of excitons bound to shallow impurities. In the case of the no-phonon line, momentum is conserved through interaction with the binding impurity. There are many candidate atoms for these shallow impurities and with B, P and As having binding energies of 4.2, 5.0 and 5.6 meV, respectively [22.42]. These bound energy states will tend to dominate luminescence spectra for doped samples and will always tend to be present in nominally undoped samples. The no-phonon line is accompanied by transverse-optical (X^{TO}) or transverse acoustic (X^{TA}) phonon replicas that are created as photon emission is accompanied by the momentum-conserving creation of lattice vibrations in Si–Si, Si–Ge or Ge–Ge bonds.

Figure 22.18 shows how the spectrum from a bulk $Si_{0.915}Ge_{0.085}$ sample develops with increasing temperature [22.41]. With increasing temperature the X^{NP} line thermalises to leave the free-exciton (FE^{NP}) line. Here, emission from nominally free excitons is greatly enhanced for the alloy samples by local fluctuations in composition that provide momentum-conserving scattering centres [22.43].

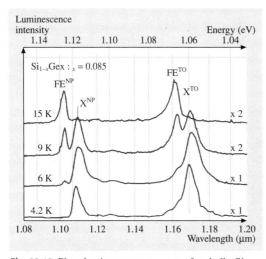

Fig. 22.18 Photoluminescence spectra of a bulk $Si_{0.915}Ge_{0.085}$ sample at different temperatures (after *Weber* et al. [22.41], copyright 1989, American Physical Society)

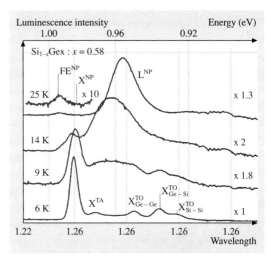

Fig. 22.19 Photoluminescence spectra of a bulk $Si_{0.42}Ge_{0.58}$ sample at different temperatures (after *Weber* et al. [22.41], copyright 1989, American Physical Society)

At higher temperatures the free-exciton line is also thermalised and all fine structure is lost, at temperatures around 25 K broad luminescence bands are commonly observed (Fig. 22.19 [22.41]). These deep luminescence bands are difficult to assign and have been ascribed to impurities, structural defects and, as in the case of the line presented in Fig. 22.19, potential wells formed by alloy fluctuations [22.41].

Weber and *Alonso* [22.41] use their data to provide analytical expressions for both the X^{NP} and L bands for bulk $Si_{1-x}Ge_x$ in the range $0 \leq x \leq 0.85$ as follows:

$$E_{gx}^{(x)}(x) = 1.155 - 0.43x + 0.206x^2 \text{ eV} \tag{22.9}$$

$$E_{gx}^{L}(x) = 2.010 - 1.270x \text{ eV} \tag{22.10}$$

At low temperatures narrow excitonic luminescence features are indicative of defect-free material, and in this way low-temperature photoluminescence becomes a good qualitative tool with which material quality can be assessed.

22.2.3 SiGe Quantum Wells

Figure 22.20 shows the first excitonic luminescence spectra from a Si/SiGe multiple quantum well grown by atmospheric-pressure CVD [22.28]. As we can see, many of the features seen in the photoluminescence spectra of quantum-well samples are similar to

Fig. 22.20 Excitonic photoluminescence spectrum of a SiGe quantum well (after *Grutzmacher* et al. [22.28], copyright 1993, AVS)

those seen from the bulk samples described in the previous section. Again, the most pronounced peak is the no-phonon (NP) line and this is accompanied by phonon replicas, including, impressively, a two-phonon replica of the NP line $(TO + TO^{Si-Si})$. The most significant difference between bulk and quantum well spectra is the energy positions of the excitonic features as these are shifted by quantum confinement effects.

Robbins et al. [22.26], have provided one of the most detailed studies of near-band-gap photoluminescence from pseudomorphic SiGe layers and provide analytical expressions for all factors pertaining to the energy positions of the excitonic energy gap for $Si_{1-x}Ge_x$ quantum wells in the range $0 < x < 0.24$. The effects of alloying, confinement, band offsets, alignment type and exciton binding energy are all taken into account. Samples used in the study were grown by low-pressure CVD at 920 °C; a typical set of photoluminescence spectra are shown in Fig. 22.21.

The exciton band gap at 4.2 K is considered for thick (50-nm) strained layers (E_X^S) where the energies are not affected by quantum shifts, the following expression is derived:

$$E_X^S(x) = 1.155 - 0.874x + 0.376x^2 \text{ eV}, \quad (x < 0.25)$$
(22.11)

Here the presence of strain is responsible for the differences from the expression obtained for bulk samples (22.9). An expression for the exciton binding energy $E_B^C(x)$ is theoretically derived for the cubic alloy and the following quadratic expression is fitted:

$$E_B^C(x) \approx 0.0145 - 0.022x + 0.020x^2 \text{ eV}, \quad (x < 0.25)$$
(22.12)

A strain-corrected expression is also provided but this is found to modify (22.12) only slightly. Thus by adding (22.11) and (22.12) an expression for the band gap can be obtained.

$$E_C - E_V \approx 1.17 - 0.896x + 0.39x^2 \text{ eV}, \quad (x < 0.25)$$
(22.13)

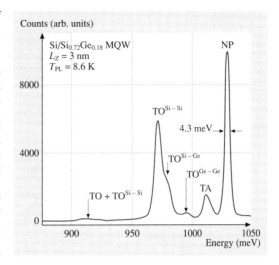

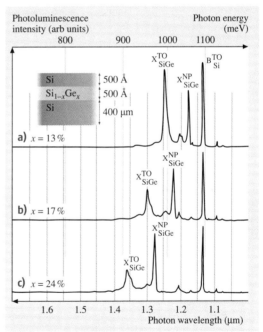

Fig. 22.21 4.2-K photoluminescence spectra from layers with the nominal structure shown in the inset (514-nm Ar-ion laser excitation) (after *Robbins* et al. [22.26], copyright 1992, American Physical Society)

22.3 Growth of Silicon–Germanium

Over the past ten years and more there have been rapid developments in techniques for the growth of Si and $Si_{1-x}Ge_x$ epitaxial layers at low temperatures. This has been made possible by a number of changes in the design of epitaxy equipment and by improvements to growth processes. There are two main prerequisites for the growth of epitaxial layers at low temperature:

- Establishment of a clean surface prior to growth [22.44–47]
- Growth in an ultra-clean environment [22.48–50]

The removal of oxygen and carbon is the main problem in establishing a clean surface prior to growth. A clean silicon surface is highly reactive and oxidises in air even at room temperature. The secret of low-temperature epitaxial growth is therefore the removal of this native oxide layer and the maintenance of a clean surface until epitaxy can begin. Two alternative approaches to pre-epitaxy surface cleaning have been developed, as described below.

22.3.1 In-Situ Hydrogen Bake

The concept that underlies this surface clean is the controlled growth of a thin surface oxide layer, followed by its removal in the epitaxy reactor using a hydrogen bake. The controlled growth of the surface oxide layer is generally achieved using a Radio Corporation of America (RCA) clean [22.44] or a variant [22.45]. The oxide created by the RCA clean is removed in the reactor using an in-situ bake in hydrogen for around 15 min at a temperature in the range 900–950 °C. The temperature required to remove the native oxide depends on the thickness of the oxide, which is determined by the severity of the surface clean.

22.3.2 Hydrogen Passivation

An alternative approach to pre-epitaxy cleaning is to create an oxide-free surface using an ex-situ clean and then move quickly to epitaxial growth before the native oxide can grow. The aim of the ex-situ clean is to produce a surface that is passivated by hydrogen atoms bonded to dangling bonds from silicon atoms on the surface. When the wafers are transferred in the epitaxy reactor, the hydrogen can be released from the surface of the silicon very quickly using a low-temperature bake or even in the early stages of epitaxy without any bake. *Meyerson* [22.46] has reported that hydrogen desorbs at 600 °C at a rate of a few monolayers per second, so the hydrogen passivation approach allows epitaxial layers to be grown at low temperatures without the need for a high-temperature bake. The hydrogen-passivated surface is stable for typically 30 min after completion of the ex-situ cleaning [22.47].

22.3.3 Ultra-Clean Epitaxy Systems

Having produced a clean hydrogen-passivated silicon surface, it is clearly important to maintain the state of this surface in the epitaxy system. This necessitates the use of low-pressure epitaxy systems if epitaxial growth at low temperatures is required. Figure 22.22 summarises the partial pressures of oxygen and water vapour that need to be achieved in an epitaxy system if an oxide-free surface is to be maintained at a given temperature [22.48, 49]. This figure shows that epitaxial growth at low temperature requires low partial pressures of oxygen and water vapour, which of course can be achieved by reducing the pressure in the epitaxy system. Research [22.50] has shown that a pressure below 30 Torr is needed to achieve silicon epitaxial growth below 900 °C.

22.3.4 $Si_{1-x}Ge_x$ Epitaxy

The growth of $Si_{1-x}Ge_x$ epitaxial layers can be achieved over a wide range of temperatures using low-pressure chemical vapour deposition (LPCVD) [22.50] or ultra-high-vacuum chemical vapour deposition (UHV-CVD) [22.51, 52]. The gas used to introduce the germanium into the layers is germane, GeH_4. The influence of germanium on the growth rate is complex, as illustrated in Fig. 22.23. At temperatures in the range 577–650 °C a peak in the growth rate is seen. At low germanium contents, the growth rate increases with germanium content, whereas at high germanium content, the growth rate decreases with germanium content. In the low-temperature regime it has been proposed that hydrogen desorption from the surface is the rate-limiting step. In $Si_{1-x}Ge_x$ this occurs more easily at germanium sites than at silicon sites and hence the growth rate increases with germanium content [22.37]. As the germanium content increases, the surface contains more and more germanium and less and less hydrogen. The rate-limiting step then becomes the adsorption of germane or silane. *Robbins* [22.53] proposed that the sticking coefficient for germane or silane was lower at germanium sites. This would slow the adsorption rate as the ger-

done by adding chlorine or HCl as a separate gas or by using a growth gas that contains chlorine, for example dichlorosilane, SiH_2Cl_2. With chlorine chemistry, selective growth of silicon and $Si_{1-x}Ge_x$ can be achieved to both silicon dioxide and silicon nitride.

Chlorine is reported to have two effects that lead to selective growth. First it increases the surface mobility of silicon and germanium atoms, so that atoms deposited on the oxide or nitride layer are able to diffuse across the surface to the window where the growth is occurring. Second it acts as an etch [22.50] and hence can remove silicon or germanium atoms deposited on the oxide or nitride. The strength of the etching action increases with chlorine content and, if the chlorine content is too high, etching of the substrate will occur instead of epitaxial growth.

A typical growth process for selective silicon epitaxy would use silane and a few percent of HCl [22.50]. The growth rate for this process is shown in Fig. 22.24, and compared with the growth rate for dichlorosilane and silane epitaxy. It can be seen that the activation energy for the silane-plus-HCl process is very similar to that for the dichlorosilane process, indicating that the growth mechanisms are similar. One disadvantage of chlorine-based growth processes over the silane process is a lower growth rate at low temperatures, as can clearly be seen in Fig. 22.24. It is also possible to grow silicon selectively using dichlorosilane and HCl [22.55].

Selective $Si_{1-x}Ge_x$ growth is generally easier to achieve than selective silicon growth, as illustrated in

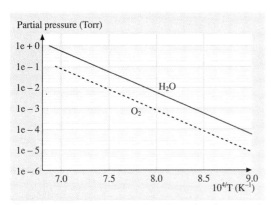

Fig. 22.22 Conditions for oxide formation in an epitaxy system. Note that 1 atm = 1.113 bar = 760 Torr = 1.113×10^5 Pa (after *Smith* and *Ghidini* [22.48, 49], copyright Electrochemical Society 1984)

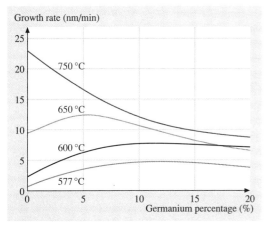

Fig. 22.23 Growth rate of $Si_{1-x}Ge_x$ as a function of germanium percentage for temperatures in the range 577–750 °C (after Racanelli et al. [22.54], copyright 1990, American Institute of Physics)

manium content increased and hence slow the growth rate.

22.3.5 Selective $Si_{1-x}Ge_x$ Epitaxy

Selective epitaxy is the growth of a single-crystal layer in a window, with complete suppression of growth elsewhere, and can be achieved in a number of different ways. The most common method of achieving both selective Si and $Si_{1-x}Ge_x$ epitaxy is by introducing chlorine or HCl into the growth chamber. This can either be

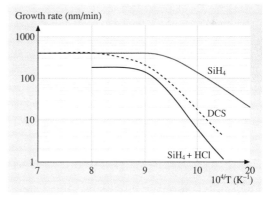

Fig. 22.24 Silicon growth rate as a function of reciprocal temperature for three different growth gases: 40 sccm of SiH_4, 80 sccm of dichlorosilane and 20 sccm of SiH_4 with 2 sccm of HCl. The hydrogen flow was 2 slm (after *Regolini* et al. [22.50], copyright 1989 American Institute of Physics)

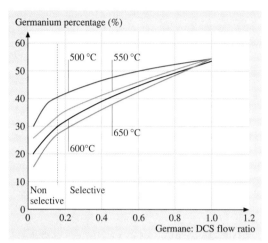

Fig. 22.25 Germanium percentage as a function of germane: dichlorosilane (DCS) flow ratio for temperatures in the range 500–650 °C showing the move from nonselective to selective growth as the proportion of germane in the gas flow increases (after *Zhong* et al. [22.56], copyright 1990, American Institute of Physics)

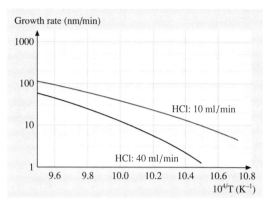

Fig. 22.26 Arrhenius plots for $Si_{1-x}Ge_x$ growth at two different HCl flow rates. The dichlorosilane and germane flow rates were fixed at 100 and 8 ml/min respectively (after *Kiyota* et al. [22.57], copyright 2002, IEEE)

Fig. 22.25 [22.56] for $Si_{1-x}Ge_x$ growth using germane and dichlorosilane. The growth moves from nonselective to selective as the proportion of germane in the gas flow increases.

Arrhenius plots for $Si_{1-x}Ge_x$ growth using germane and dichlorosilane are shown in Fig. 22.26 for $Si_{1-x}Ge_x$ layers grown using germane and dichlorosilane and for two different HCl flows. It can be seen that the growth rate decreases and the activation energy increases with increasing HCl flow. The explanation proposed for this behaviour is that the limiting growth mechanism changes from hydrogen desorption from the growing surface to chlorine or HCl desorption from the surface [22.57]. This decrease in growth rate at high HCl flows is a disadvantage because it leads to increased growth times. High HCl flows can also cause surface roughening when the $Si_{1-x}Ge_x$ layer is heavily boron-doped [22.57]. These considerations demonstrate that the HCl flow should be chosen to be to the smallest value that is consistent with good selective epitaxy.

Silane can be used for selective silicon epitaxy if the growth is performed at a high temperature. This approach relies on the fact that nucleation of growth on oxide is more difficult than that on silicon. This incubation time for growth on an oxide layer is relatively long at high temperatures but much shorter for growth at low temperatures. Selective silicon layers 1 μm thick can be grown using silane at a temperature of 960 °C [22.58], but the achievable layer thickness decreases with decreasing temperature. At 800 °C the maximum selective silicon layer thickness is around 130 nm, at 700 °C it is around 60 nm, and at 620 °C it is around 40 nm. Selective growth to silicon dioxide can be achieved using silane only, but not to silicon nitride.

22.4 Polycrystalline Silicon–Germanium

In the past ten years there has been increasing interest in polycrystalline silicon–germanium for a number of applications that require polycrystalline material deposition at low temperature (around 600 °C). Examples of potential applications are thin film transistors, gates of MOS transistors and polySiGe emitters for SiGe HBTs.

In thin-film transistor technologies [22.59–61], polycrystalline silicon–germanium is compatible with the low-thermal-budget processing that is needed to produce thin-film devices for large-area electronics. The key physical property of polycrystalline silicon–germanium that makes it attractive is its lower melting point than silicon. This means that processes such as deposition,

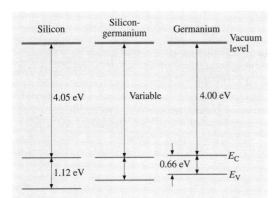

Fig. 22.27 Band-energy levels in silicon, silicon–germanium and germanium

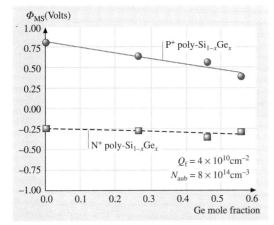

Fig. 22.28 Work-function difference between a polySi$_{1-x}$Ge$_x$ gate and an n-type substrate as a function of germanium content (after *King* et al. [22.60], copyright 1994, IEEE)

electron affinities (4.05 and 4.00 eV respectively), but germanium has a much smaller band gap (0.66 eV compared with 1.12 eV). The energy difference between the valence band and the vacuum level is therefore about 0.5 eV smaller in germanium than in silicon. In silicon–germanium, this energy difference can be varied by varying the germanium content. This allows the threshold voltage of p-channel MOS transistors to be tuned by varying the germanium content in the polySi$_{1-x}$Ge$_x$ gate.

Figure 22.28 shows values of work-function difference between the polySi$_{1-x}$Ge$_x$ gate and the n-type silicon substrate as a function of germanium content in a polySi$_{1-x}$Ge$_x$ gate [22.60]. The work function is defined as the difference in energy between the vacuum level and the Fermi level. In p$^+$ polySi$_{1-x}$Ge$_x$ the Fermi level is near the valence band and hence the work-function difference varies strongly with germanium content. In n$^+$ polySi$_{1-x}$Ge$_x$ the Fermi level is near the conduction band and hence the work-function difference varies little with germanium content.

In Si$_{1-x}$Ge$_x$ HBTs, polySi$_{1-x}$Ge$_x$ has potential as an emitter of a SiGe HBT [22.62]. In bipolar transistors, the breakdown voltage, BV_{CEO}, is inversely proportional to the gain [22.66] and hence transistors with a high gain have lower values of breakdown voltage. Si$_{1-x}$Ge$_x$ HBTs inherently have high values of the gain because the reduced band gap of the Si$_{1-x}$Ge$_x$ base enhances the collector current. The base current is unchanged by the Si$_{1-x}$Ge$_x$ base and hence the gain, which is the

crystallisation, grain growth and dopant activation will occur at a lower temperature than in silicon. Thus lower temperature processes can be used for polySiGe devices and hence it is preferable to polySi in applications with tight thermal-budget requirements.

In MOS transistors, polycrystalline silicon–germanium is attractive as a gate material for future generations of MOS transistor, since the germanium content in the silicon–germanium layer can be varied by 200–300 mV in the direction of a mid-gap gate [22.63–65]. This can be understood from Fig. 22.27, which compares the conduction- and valence-band energy levels in single-crystal silicon, silicon–germanium and germanium. Silicon and germanium have similar

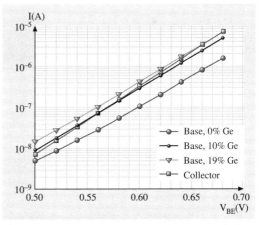

Fig. 22.29 Use of a polySi$_{1-x}$Ge$_x$ emitter to vary the base current of a Si$_{1-x}$Ge$_x$ HBT and hence give the best trade-off between gain and breakdown voltage BV_{CEO}. (after *Kunz* et al. [22.62], copyright 2003, IEEE)

ratio of collector current to base current, is increased. The use of a polySi$_{1-x}$Ge$_x$ emitter instead of a polySi emitter provides a reduced band gap in the emitter, which enhances the base current, and thereby reduces the gain. Typical measured values of base current in a polySi$_{1-x}$Ge$_x$ emitter are shown in Fig. 22.29, where it can be seen that 19% germanium gives a factor of approximately four reduction in gain. A polySi$_{1-x}$Ge$_x$ emitter therefore allows the gain to be tuned to give the best trade-off between gain and breakdown voltage BV_{CEO}.

22.4.1 Electrical Properties of Polycrystalline Si$_{1-x}$Ge$_x$

Figure 22.30 shows the sheet resistance as a function of anneal temperature for boron- and phosphorus-doped Si$_{1-x}$Ge$_x$ for different germanium contents. For boron-doped polySi$_{1-x}$Ge$_x$ the sheet resistance decreases with increasing germanium content, with the decrease being large between 0 and 25% germanium and smaller between 25 and 50% germanium. In contrast, for phosphorus-doped polySi$_{1-x}$Ge$_x$ the sheet resistance

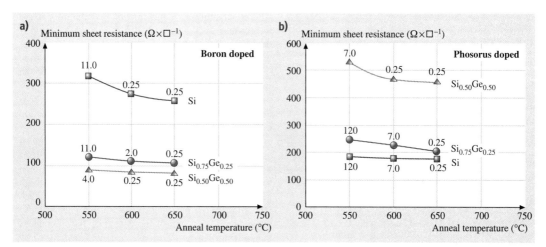

Fig. 22.30 Sheet resistance as a function of anneal temperature for boron- and phosphorus-doped polycrystalline Si$_{1-x}$Ge$_x$ with various germanium contents (after *Bang* et al. [22.67], copyright 1995, American Institute of Physics)

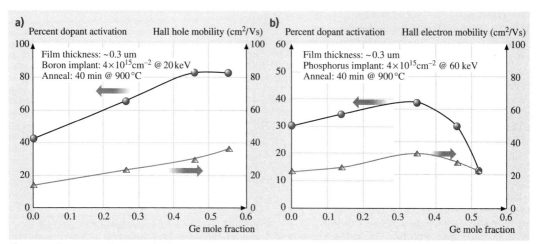

Fig. 22.31 Percentage dopant activation and Hall hole mobility as a function of germanium content for boron- and phosphorus-doped polySi$_{1-x}$Ge$_x$ (after *King* et al. [22.60], copyright 1994, IEEE)

increases with increasing germanium content, with the increase being small between 0 and 25% germanium and large between 25 and 50%. Similar behaviour is seen for arsenic-doped polySi$_{1-x}$Ge$_x$ where higher values of sheet resistance have been reported for polySi$_{1-x}$Ge$_x$ than for polySi [22.64].

The explanation for the sheet-resistance results in Fig. 22.30 can be found in Fig. 22.31, which shows the results of Hall measurements [22.60]. For boron-doped polySi$_{1-x}$Ge$_x$ both the activation and the Hall mobility increase with increasing germanium content, thereby explaining the decrease in sheet resistance with increasing germanium content. For phosphorus-doped polySi$_{1-x}$Ge$_x$ there is little change in activation and electron mobility at low germanium contents, but a sharp decrease in activation at germanium contents above 35%. This explains the sharp increase in sheet resistance seen in Fig. 22.30 for germanium contents between 25 and 50%. The decrease in activation at high germanium contents in the phosphorus-doped polySi$_{1-x}$Ge$_x$ may be due to increased segregation at grain boundaries. Boron does not generally segregate to grain boundaries [22.68], which may explain the different behaviour in boron- and phosphorus-doped polySi$_{1-x}$Ge$_x$.

References

22.1 R. Braunstein, A. R. Moore, F. Herman: Phys. Rev. **109**, 695 (1958)
22.2 S. S. Iyer, G. L. Patton, J. M. C. Stork, B. S. Meyerson, D. L. Harame: IEEE Trans. Electron. Dev. **36**, 2043 (1989)
22.3 C. A. King, J. L. Hoyt, J. F. Gibbons: IEEE Trans. Electron. Dev. **36**, 2093 (1989)
22.4 H. Miyata, T. Yamada, D. K. Ferry: Appl. Phys. Lett. **62**, 2661 (1993)
22.5 T. Vogelsang, K. R. Hofmann: Appl. Phys. Lett. **63**, 186 (1993)
22.6 J. Welser, J. L. Hoyt, J. F. Gibbons: IEEE Electron. Dev. Lett. **15**, 100 (1994)
22.7 A. Sadak, K. Ismile, M. A. Armstrong, D. A. Antoniadis, F. Stern: IEEE Trans. Electron. Dev. **43**, 1224 (1996)
22.8 B. Jagannathan, M. Khater, F. Pagette, J.-S. Rieh, D. Angell, H. Chen, J. Florkey, F. Golan, D. R. Greenberg, R. Groves, S. J. Jeng, J. Johnson, E. Mengistu, K. T. Schonenberger, C. M. Schnabel, P. Smith, A. Stricker, D. Ahlgren, G. Freeman, K. Stein, S. Subbanna: IEEE Electron Dev. Lett. **23**, 258 (2002)
22.9 Z. A. Shafi, P. Ashburn, G. J. Parker: IEEE J. Solid State Circuits **25**, 1268 (1990)
22.10 S. C. Jain, T. J. Gosling, J. R. Willis, R. Bullough, P. Balk: Solid State Electron. **35**, 1073 (1992)
22.11 J. M. Matthews, A. E. Blakeslee: J. Cryst. Growth **27**, 118 (1974)
22.12 J. M. Matthews, A. E. Blakeslee: J. Cryst. Growth **32**, 265 (1975)
22.13 R. People, J. C. Bean: Appl. Phys. Lett. **47**, 322 (1985)
22.14 S. Margalit, A. Bar-lev, A. B. Kuper, H. Aharoni, A. Neugroschel: J. Cryst. Growth **17**, 288 (1972)
22.15 O. W. Holland, C. W. White, D. Fathy: Appl. Phys. Lett. **51**, 520 (1987)
22.16 R. People: Phys. Rev. B **32**, 1405 (1985)
22.17 J. Poortmans, S. C. Jain, D. H. J. Totterdell, M. Caymax, J. F. Nijs, R. P. Mertens, R. Van Overstraeten: Solid State Electron. **36**, 1763 (1993)
22.18 T. Manku, A. Nathan: J. Appl. Phys. **69**, 8414 (1991)
22.19 T. Manku, A. Nathan: Phys. Rev. B **43**, 12634 (1991)
22.20 J. Poortmans: Low temperature epitaxial growth of silicon and strained Si$_{1-x}$Ge$_x$ layers and their application in bipolar transistors; PhD thesis, University of Leuven (1993)
22.21 J. M. McGregor, T. Manku, A. Nathan: *Measured in-plane hole drift mobility and Hall mobility in heavily doped, strained p-type* Si$_{1-x}$Ge$_x$ (Boston 1992) presented at Electronic Materials Conference
22.22 M. V. Fischetti, S. E. Laux: J. Appl. Phys. **80**, 2234 (1996)
22.23 J. Welser, J. L. Hoyt, J. F. Gibbons: IEEE Electron. Dev. Lett. **15**, 100 (1994)
22.24 C. W. Leitz, M. T. Currie, M. L. Lee, Z.-Y. Cheng, D. A. Antoniadis, E. A. Fitzgerald: J. Appl. Phys. **92**, 3745 (2002)
22.25 J. H. Bahng, K. J. Kim, H. Ihm, J. Y. Kim, H. L. Park: J. Phys.: Condens. Matter **13**, 777 (2001)
22.26 D. J. Robbins, L. T. Canham, S. J. Barnett, A. D. Pitt, P. Calcott: J. Appl. Phys. **71**, 1407 (1992)
22.27 N. L. Rowell, J.-P. Noel, D. C. Houghton, A. Wang, D. D. Perovic: J. Vac. Sci. Technol. B **11**, 1101 (1993)
22.28 D. A. Grutzmacher, T. O. Sedgwick, G. A. Northrop, A. Zaslavsky, A. R. Powell, V. P. Kesan: J. Vac. Sci. Technol. B **11**, 1083 (1993)
22.29 J. Brunner, J. Nutzel, M. Gail, U. Menczigar, G. Abstreiter: J. Vac. Sci. Technol. B **11**, 1097 (1993)
22.30 K. Terashima, M. Tajima, T. Tatsumi: J. Vac. Sci. Technol. B **11**, 1089 (1993)
22.31 H. Prestling, T. Zinke, A. Splett, H. Kibbel, M. Jaros: Appl. Phys. Lett. **69**, 2376 (1996)
22.32 L. Masarotto, J. M. Hartmann, G. Bremond, G. Rolland, A. M. Papon, M. N. Semeria: J. Cryst. Growth **255**, 8 (2003)
22.33 J. S. Park, T. L. Lin, E. W. Jones, H. M. Del Castillo, S. D. Gunapall: Appl. Phys. Lett. **64**, 2370 (1994)
22.34 S. S. Murtaza, J. C. Cambell, J. C. Bean, L. J. Peticolas: IEEE Photon. Tech. Lett. **8**, 927 (1996)

22.35 D. J. Robbins, M. B. Stanaway, W. Y. Leong, R. T. Carline, N. T. Gordon: Appl. Phys. Lett. **66**, 1512 (1995)

22.36 A. Chin, T. Y. Chang: Lightwave Technol. **9**, 321 (1991)

22.37 R. People, J. C. Bean, C. G. Bethia, S. K. Sputz, L. J. Peticolas: Appl. Phys. Lett. **61**, 1122 (1992)

22.38 P. Kruck, M. Helm, T. Fromherz, G. Bauer, J. F. Nutzel, G. Abstreiter: Appl. Phys. Lett. **69**, 3372 (1996)

22.39 R. A. Soref, L. Friedman, G. Sun: Superlattices Microstruct. **23**, 427 (1998)

22.40 G. Sun, L. Friedman, R. A. Soref: Superlattices Microstruct. **22**, 3 (1998)

22.41 J. Weber, M. I. Alonso: Phys. Rev. B **40**, 5684 (1989)

22.42 H. Landolt, R. Bornstein: *Numerical data and functional relationships in science and technology*, Vol. III/17a, ed. by O. Madelung (Springer, Berlin Heidelberg New York 1982)

22.43 G. S. Mitchard, T. C. McGill: Phys. Rev. B **25**, 5351 (1982)

22.44 M. Meuris, S. Verhaverbeke, P. W. Mertens, M. M. Heyns, L. Hellemans, Y. Bruynseraede, A. Philipessian: Jpn. J. Appl. Phys. **31**, L1514 (1992)

22.45 A. Ishizaki, Y. Shiraki: J. Electrochem. Soc. **129**, 666 (1986)

22.46 B. S. Meyerson, F. J. Himpsel, K. J. Uram: Appl. Phys. Lett. **57**, 1034 (1990)

22.47 G. S. Higashi, Y. T. Chabal, G. W. Trucks, K. Raghavachari: Appl. Phys. Lett. **56**, 656 (1990)

22.48 F. W. Smith, G. Ghidini: J. Electrochem. Soc. **129**, 1300 (1982)

22.49 G. Ghidini, F. W. Smith: J. Electrochem. Soc. **131**, 2924 (1984)

22.50 J. L. Regolini, D. Bensahel, E. Scheid, J. Mercier: Appl. Phys. Lett. **54**, 658 (1989)

22.51 G. R. Srinivasan, B. S. Meyerson: J. Electrochem. Soc. **134**, 1518 (1987)

22.52 M. Racanelli, D. W. Greve, M. K. Hatalis, L. J. van Yzendoorn: J. Electrochem. Soc. **138**, 3783 (1991)

22.53 D. J. Robbins, J. L. Glasper, A. G. Cullis, W. Y. Leong: J. Appl. Phys. **69**, 3729 (1991)

22.54 M. Racanelli, D. W. Greve: Appl. Phys. Lett. **56**, 2524 (1990)

22.55 A. Ishitani, H. Kitajima, N. Endo, N. Kasai: Jpn. J. Appl. Phys. **28**, 841 (1989)

22.56 Y. Zhong, M. C. Ozturk, D. T. Grider, J. J. Wortman, M. A. Littlejohn: Appl. Phys. Lett. **57**, 2092 (1990)

22.57 Y. Kiyota, T. Udo, T. Hashimoto, A. Kodama, H. Shimamoto, R. Hayami, E. Ohue, K. Washio: IEEE Trans. Electron. Dev. **49**, 739 (2002)

22.58 J. M. Bonar: "Process development and characterisation of silicon and silicon–germanium grown in a novel single-wafer LPCVD system"; *PhD thesis*, University of Southampton (1996)

22.59 T.-J. King, K. C. Saraswat: IEDM Tech. Dig., 567 (1991)

22.60 T.-J. King, K. C. Saraswat: IEEE Trans. Electron. Dev. **41**, 1581 (1994)

22.61 J. A. Tsai, A. J. Tang, T. Noguchi, R. Reif: J. Electrochem. Soc. **142**, 3220 (1995)

22.62 V. D. Kunz, C. H. de Groot, S. Hall, P. Ashburn: IEEE Trans. Electron. Dev. **50**, 1480 (2003)

22.63 T.-J. King, J. R. Pfiester, K. C. Saraswat: IEEE Electron. Dev. Lett. **12**, 533 (1991)

22.64 C. Salm, D. T. van Veen, D. J. Gravesteijn, J. Holleman, P. H. Woerlee: J. Electrochem. Soc. **144**, 3665 (1997)

22.65 Y. V. Ponomarev, P. A. Stolk, C. J. J. Dachs, A. H. Montree: IEEE Trans. Electron. Dev. **47**, 1507 (2000)

22.66 P. Ashburn: *Silicon-germanium heterojunction bipolar transistors* (Wiley, Chichester 2003)

22.67 D. S. Bang, M. Cao, A. Wang, K. C. Saraswat, T.-J. King: Appl. Phys. Lett. **66**, 195 (1995)

22.68 I. R. C. Post, P. Ashburn: IEEE Trans. Electron. Dev. **38**, 2442 (1991)

23. Gallium Arsenide

The history of gallium arsenide is complicated because the technology required to produce GaAs devices has been fraught with problems associated with the material itself and with difficulties in its fabrication. Thus, for many years, GaAs was labelled as "the semiconductor of the future, and it will always be that way." Recently, however, advances in compact-disc (CD) technology, fibre-optic communications and mobile telephony have boosted investment in GaAs research and development. Consequently, there have been advances in materials and fabrication technology and, as a result, GaAs devices now enjoy stable niche markets.

The specialised uses for GaAs in high-frequency and optoelectronic applications result from the physical processes of electron motion that allow high-speed and efficient light emission to take place. In this review, these advanced devices are shown to result from the physical properties of GaAs as a semiconducting material, the controlled growth of GaAs and its alloys and the subsequent fabrication into devices.

Extensive use is made of chapters from "Properties of Gallium Arsenide, 3rd edition" which I edited with the help of *Prof. G. E. Stillman* [23.1]. This book was written to reflect virtually all aspects of GaAs and its devices within a readable text. I believe that we succeeded in that aim and I make no apologies in referring to it. Readers who need specialised data, but not necessarily within an explanatory text, should refer to the Landolt–Börnstein, group III (condensed matter) data collection [23.2, 3]. The sub-volumes A1α (lattice properties) and A2α (impurities and defects) within volume 41 are rich sources of data for all III–V compounds. Although there are no better sources than the original research papers, I have referred to textbooks where possible. This is because the presentation and discussion of scientific data is often clearer than in the original text, and these books are more accessible to students.

Gallium arsenide (GaAs) is one of the most useful of the III–V semiconductors. In this chapter, the properties of GaAs are described and the ways in which these are exploited in devices are explained. The limitations of this material are presented in terms of both its physical and its electronic properties.

23.1 **Bulk Growth of GaAs** 502
    23.1.1 Doping Considerations 502
    23.1.2 Horizontal Bridgman
           and Horizontal Gradient Freeze
           Techniques 503
    23.1.3 Liquid-Encapsulated Czochralski
           (LEC) Technique 504
    23.1.4 Vertical Gradient Freeze (VGF)
           Technique 506

23.2 **Epitaxial Growth of GaAs** 507
    23.2.1 Liquid-Phase Epitaxy (LPE) 507
    23.2.2 Vapour-Phase Epitaxy (VPE)
           Technologies 508
    23.2.3 Molecular-Beam Epitaxy (MBE) ... 509
    23.2.4 Growth of Epitaxial
           and Pseudomorphic Structures ... 511

23.3 **Diffusion in Gallium Arsenide** 511
    23.3.1 Shallow Acceptors 512
    23.3.2 Shallow Donors 513
    23.3.3 Transition Metals 513

23.4 **Ion Implantation into GaAs** 513

23.5 **Crystalline Defects in GaAs** 514
    23.5.1 Defects in Melt-Grown GaAs 514
    23.5.2 Epitaxial GaAs (not Low
           Temperature MBE GaAs) 516
    23.5.3 LTMBE GaAs 517

23.6 **Impurity and Defect Analysis
of GaAs (Chemical)** 517

23.7 **Impurity and Defect Analysis
of GaAs (Electrical)** 518
    23.7.1 Introduction to Electrical Analysis
           of Defects in GaAs 518

23.8 **Impurity and Defect Analysis
of GaAs (Optical)** 521
    23.8.1 Optical Analysis of Defects in GaAs 521

23.9 **Assessment of Complex Heterostructures**................ 522
    23.9.1 Carrier Concentration Measurements in Heterostructures................... 522
    23.9.2 Layer Thickness and Composition Measurements.......................... 522

23.10 **Electrical Contacts to GaAs**.................... 524
    23.10.1 Ohmic Contacts........................ 524
    23.10.2 Schottky Contacts..................... 524

23.11 **Devices Based on GaAs (Microwave)**....... 524
    23.11.1 The Gunn Diode......................... 524
    23.11.2 The Metal−Semiconductor Field-Effect Transistor (MESFET)... 525
    23.11.3 The High-Electron-Mobility Transistor (HEMT) or Modulation Doped FET (MODFET)................................. 526
    23.11.4 The Heterojunction Bipolar Transistor (HBT)....................... 526

23.12 **Devices based on GaAs (Electro-optical)**. 527
    23.12.1 GaAs Emitters 527
    23.12.2 GaAs Modulators 531
    23.12.3 GaAs Photodetectors.................. 531

23.13 **Other Uses for GaAs**............................. 532

23.14 **Conclusions**.. 532

References 533

GaAs is one of the compound semiconductors that occur in the zincblende structure. As can be seen in Fig. 23.1, each Ga atom is tetrahedrally bonded to four As atoms in a structure that is similar to the diamond lattice but with alternating Ga and As atoms.

The presence of two types of atoms introduces a small component of ionic bonding into the structure and increases the bond strength above that of the group IV element (Ge) which Ga and As neighbour. As a result the band-gap energy E_g, is increased to 1.518 eV at $T = 0$ K ($E_g = 1.41$ eV at room temperature).

The lattice constant of GaAs at 300 K is 0.565 36 nm, giving a resultant density of 5.3165 g/cm^3. At atmospheric pressure GaAs melts near 1238 °C but with the loss of As vapour.

Although all ⟨001⟩ atomic planes are equivalent this is not true for ⟨011⟩ and ⟨111⟩ planes. In particular, there are two types of ⟨111⟩ atomic planes; one terminated with Ga atoms only (the ⟨111⟩A planes), the other being terminated with As atoms only (the ⟨111⟩B planes). This makes the ⟨111⟩ directions strongly polar. A further consequence of the reduced symmetry of the lattice compared with the group IV semiconductors is that GaAs exhibits strong piezoelectric effects.

The easily cleaved directions are the ⟨110⟩ and ⟨111⟩ types. The ⟨110⟩ are the more useful as they allow square sections to be cleaved from a ⟨001⟩ section wafer, one of the major reasons for the use of ⟨001⟩ wafers in GaAs device fabrication.

The original interest in GaAs arose from its unusual band structure, compared to that of Si in Fig. 23.2, which demonstrates several attributes that are expected to be interesting for high-speed electronics and optoelectronics.

1. The fundamental band-gap energy of 1.41 eV (300 K) corresponds to a point in the diagram, the Γ point of the Brillouin zone, where both the crystal momentum for holes and electrons is zero. It follows that low-energy electrons and holes can recombine without the moderating influence of phonons that would otherwise be required to conserve momentum. Such direct recombination is expected to result in efficient emission of photons and GaAs would then be the base material for efficient light-emitting diodes (LEDs) and lasers.
2. The effective mass of conduction-band electrons is inversely proportional to the curvature at the bottom of the band and this curvature is considerably greater, and the effective mass is considerably smaller, in GaAs (about $0.063\, m_e$) than in silicon (about $1.1\, m_e$).

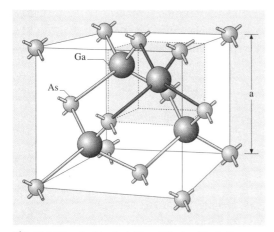

Fig. 23.1 The zincblende lattice of GaAs. The lattice constant a is indicated. (After [23.4])

Table 23.1 Selected important properties of pure GaAs at 300 K

Lattice constant (Å)	5.653
Density (g/cm^3)	5.318
Band-gap energy (eV)	1.424
Band-gap type	Direct
Electron mobility (cm^2/Vs)	8500
Hole mobility (cm^2/Vs)	400
Zero-frequency dielectric constant	13.18
Conduction-band effective density of states (cm^{-3})	4.45×10^{17}
Valence-band effective density of states (cm^{-3})	7.72×10^{18}
Intrinsic carrier concentration (cm^{-3})	1.84×10^6

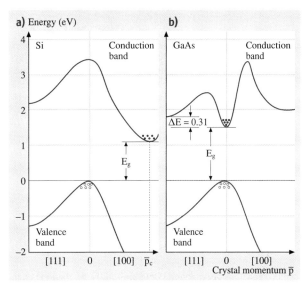

Fig. 23.2a,b The band structures of (a) Si and (b) GaAs. The primary band gap of Si is indirect with an energy E_g, of 1.12 eV. However, that of GaAs is direct (both free electrons and holes of low energy have states that correspond to zero crystal momentum at the Γ point of the Brillouin zone). Secondary conduction-band minima occur at 0.31 eV above the primary minimum near points of maximum crystal momenta along the [111]-directions. These are at the L points. (After [23.4])

Because the mobility of carriers is inversely proportional to their effective mass, it follows that free electrons in GaAs should be very mobile under the influence of an electric field, resulting in fast devices.

3. An additional feature of Fig. 23.2 is the occurrence of subsidiary conduction-band minima near the six points of maximum momentum along the ⟨100⟩ directions. These are at the so-called L points of the Brillouin zone. These minima are degenerate and at an energy of 0.3 eV above the minimum energy of the conduction band at the Γ point. The relatively small curvature of these X minima results in a large effective electron mass of about 1.2 m_e. Electrons at these points have a correspondingly low mobility. It follows that the excitation of low-effective-mass free electrons from the Γ point to the L minima, where they have high effective mass, can result in a *reduction* of electron velocity and hence electron current. This is the *transferred electron effect*. The current/voltage characteristics of such a device exhibit a negative differential resistance region, the Gunn effect, and this can be used to generate microwave radiation.

4. One property of GaAs, not resulting from Fig. 23.2, is that the material can be produced in a high electrical resistivity, semi-insulating (SI) state. This allows devices to be fabricated on a near-insulating substrate. This is advantageous for high-frequency devices as the parasitic capacitances that occur between Si devices and their substrate, reducing their maximum operating frequency, will be absent in GaAs devices fabricated on SI substrates.

5. A further property of great interest is that the lattice constants of GaAs and aluminium arsenide, AlAs, and their alloys, written as $Ga_{1-x}Al_xAs$, where x is the atomic fraction of Al, are very similar. This allows different alloys to be grown consecutively without incurring defects due to lattice mismatch. Such heterostructures are not possible with Si and its known alloys.

The predictions made above will be addressed in later sections but first we consider the growth of GaAs single crystals suitable for device use. Subsequently, we consider the epitaxial growth of thin, GaAs-based layers, which are the basis of modern devices, means of processing and the fabrication and properties of important devices.

23.1 Bulk Growth of GaAs

In the simplest terms, GaAs crystals are synthesised by reacting together high-purity Ga and As. Dopants may be added to produce materials of different conduction type and carrier concentration. However, the growth of compounds from the melt is a thermodynamic process and is characterised by the shape of the phase diagram near the solidus at temperatures just below the melting point of the compound. *Hurle* has published a thorough reassessment of the thermodynamic parameters for GaAs, with and without doping [23.6]. The complete phase diagram for Ga−As, showing the GaAs compound, is shown in Fig. 23.3. However, details near the GaAs solidus are not revealed at this large scale.

The shape of the GaAs solidus is more complicated than indicated in Fig. 23.3. In particular, the congruent point, the maximum melting temperature, corresponds to a GaAs compound containing more As than exists in stoichiometric material (Fig. 23.4).

It follows that growth from a stoichiometric melt results in the growth of a solid phase containing an excess of As of approximately 1×10^{19} cm^{-3}. The solidus is retrograde so that, if thermodynamic equilibrium is maintained, this excess As must be lost from the host as the solid cools. If this As supersaturation cannot be relieved by out-diffusion from the crystal, then it will result in the generation of As-rich second phases (precipitates). Growth from a Ga-rich melt will result in the growth of a stoichiometric crystal. However, the melt will become more Ga-rich as growth proceeds, leading to the growth of solid GaAs that steadily becomes richer in Ga.

Experimental work has shown that the electrical properties of Ga-rich GaAs are poor (undoped Ga-rich GaAs is strongly p-type [23.7]) and this growth regime is rarely used. Although As richness is found in melt-grown GaAs, measured concentrations of excess As do not exceed 10^{17} cm^{-3} [23.8, 9]. It is not known how most of the grown-in excess As atoms are lost.

23.1.1 Doping Considerations

Unlike group IV semiconductors where shallow donors are group V atoms only, donors in III–V compounds can occur on either sublattice, as IV$_{III}$ or VI$_V$, where the

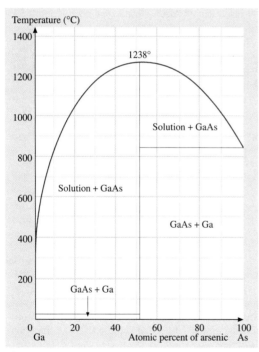

Fig. 23.3 The complete phase diagram of the Ga−As system at an assumed pressure of As vapour of about 1 atm. The *vertical line* near 50% composition is crystalline GaAs. (After [23.5])

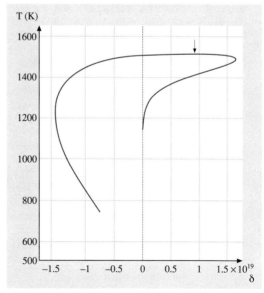

Fig. 23.4 The GaAs solidus. The *arrow* marks the congruent point (After [23.6])

capital letters represent the group in the periodic table from which the donor is selected and the subscript is the sublattice occupied. However, only certain group IV atoms have a suitable solubility on the Ga sublattice for them to be useful and the same situation is true for group VI atoms on As sites.

Similarly, acceptors can be of II_{III} or IV_V types. (In principle, group IV atoms can be both acceptors and donors, IV_{III} and IV_V respectively, a situation known as *amphotericity*. Silicon, and to a similar extent germanium atoms in GaAs, behave like this.) Once again not all of these occur in practice. The choice of dopant depends mainly on its segregation coefficient and solubility (see [23.6] for a discussion of this point). Figure 23.5 shows most of the known shallow donors and acceptors in GaAs with their ionisation energies. Also shown in this figure are two important impurities that generate levels deeper in the band gap. Cu is an unwanted deep acceptor contaminant but Cr has some technological importance and this element is discussed later.

In melt-grown GaAs, donors are usually either Si_{Ga} or Te_{As}, as each has a satisfactory segregation coefficient and solubility, although other group VI species have been used. Donors experience some degree of auto-compensation with the result that, up to a carrier concentration n of about 3×10^{18} cm^{-3}, the carrier concentration is only about 80% of the added dopant concentration. Thermodynamic analysis has demonstrated that for group VI donors this is a result of the formation of (donor–V_{Ga}) pair defects. Auto-compensation for the case of Si doping is a result of the amphotericity of Si with a fraction, about 10%, of the Si atoms taking up As lattice sites and behaving as shallow acceptors. Above $n = 3 \times 10^{18}$ cm^{-3}, the carrier concentration tends to saturate and extra donor atoms are involved in vacancy complexes with a drop in electron mobility, the appearance of dislocation loops and a rapid rise in the lattice constant [23.6].

The p-type doping of melt-grown GaAs appears to be much easier. There is no auto-compensation and the solubility of Zn, the most common acceptor dopant in melt-grown GaAs, is over 10^{20} cm^{-3}. The segregation coefficient of Zn in GaAs is close to unity.

Although the partial pressure of Ga above a Ga–As melt at all reasonable temperatures is extremely small and can be ignored, the partial pressure of As vapour (As_2 and As_4) depends sensitively on the melt composition and this can be used to control the ratio of Ga/As in the melt as the crystal grows. Over a stoichiometric GaAs melt at the melting temperature of 1238 °C this pressure is approximately two atmospheres (1 atm above

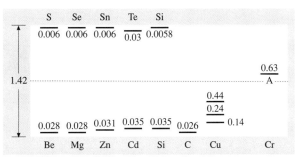

Fig. 23.5 The ionisation energies in eV of the shallow donors and acceptors in GaAs. Also shown are the important deep levels due to Cu and Cr. Cu is an impurity that can be introduced at growth or by processing and causes damaging deleterious effects in devices. Cr was originally used as a deep acceptor to produce semi-insulating GaAs. Its use has fallen since the introduction of LEC and similar growth techniques, see later. (Updated after [23.4])

ambient). It is necessary to control this pressure to retain melt composition and to impede the loss of As from the melt. This has resulted in the development of several growth techniques.

23.1.2 Horizontal Bridgman and Horizontal Gradient Freeze Techniques

The first growth method to be used commercially was the horizontal Bridgman (HB) technique [23.10]. In essence, the growth chamber consists of a horizontal quartz tube, at one end of which is a boat containing the Ga–As melt near the melting temperature (T_m) of 1238 °C while the other end is cooler, near 617 °C, and contains a small quantity of elemental As, Fig. 23.6.

This As produces the overpressure of As_4 that controls the stoichiometry of the Ga–As melt. The Ga–As melt is often formed by reacting metallic Ga in the growth section with As vapour produced from the As reservoir at the other end. In other processes polycrystalline GaAs is used as the source for the molten Ga–As. The growth can proceed in two ways. In the classical HB technique the initial melt exists in a region at a temperature just above T_m but adjacent to a small temperature gradient. A seed crystal, already placed in the crucible with part of it in the melt while the rest is at a temperature below T_m, will initiate growth. The melt is moved mechanically through the temperature gradient so that the entire melt is gradually cooled below T_m. Eventually, all of the GaAs melt solidifies on the seed as a single crystal and with the same orientation as the seed.

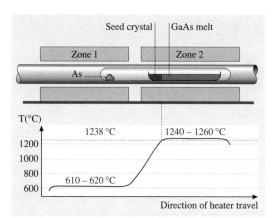

Fig. 23.6 The horizontal Bridgman technique for growing GaAs. The growth takes place in the high-temperature region of the furnace whilst the As vapour pressure is controlled by the presence of elemental As in the low-temperature region. The interface between the melt and the solid is gradually moved until all the GaAs is solidified. (After [23.4])

In a modification to HB, the mechanical withdrawal of the GaAs melt through the gradient is replaced by slowly reducing the temperature of the growth region electronically. This is the horizontal gradient freeze (GF) method [23.11]. It has the advantage of needing less room and is less sensitive to mechanical disturbance.

In both cases, crystals are best grown along ⟨111⟩ to reduce the occurrence of twinning. ⟨001⟩ wafers can be extracted from the ⟨111⟩ boule by accurate sawing. Advantages of the HB and horizontal GF methods include the good visibility of the growth procedure, allowing the operator to make modifications as growth takes place. However, grown crystals have a D-shaped cross section because of the shape of the melt in the boat and subsequent wafering. Considerable loss of material is incurred if the wafers are edge-ground to make them circular.

HB and horizontal GF GaAs are contaminated with silicon atoms from the quartz growth tube and this renders them n-type. This is not a problem if highly doped n-type material is required, as extra Si will be added to the melt anyway. Although there are few uses for p-type bulk GaAs, the over-doping of the melt with an acceptor species (usually Zn) effectively renders the crystal p-type. The reduction in hole mobility by compensation of a minority of the Zn atoms by the Si atoms is of little consequence.

However, the advantages of SI behaviour can only be obtained after the Si donors have been counter-doped by the incorporation of Cr atoms in the melt [23.12, 13]. Chromium atoms act as deep acceptor centers in n-type GaAs and act to pin the Fermi energy just above the mid-gap, see Fig. 23.5. The GaAs so produced exhibits a resistivity that can exceed 10^8 Ωcm. Such Cr-doped GaAs was the mainstay of the high-speed GaAs device industry for over a decade, but growth problems, from the low segregation coefficient and solubility of Cr, resulted in low wafer yields, making the substrates expensive. Device fabrication problems from the rapid out-diffusion of Cr from the substrates into devices, resulting in lower speed, unacceptable device characteristics and device instability, also became evident.

Finally, the combination of poor thermal conductivity and low critically resolved shear stress (CRSS) at elevated temperatures results in inevitable thermal gradients in the cooling crystal producing plastic deformation by the creation of slip dislocations. In HB and horizontal GF material the dislocation densities are around 10^3 cm^{-2} [23.14], although these densities can be reduced by over an order of magnitude due to *impurity hardening* in highly n-type GaAs [23.15]. Dislocations, their generation and their properties will be considered later.

23.1.3 Liquid-Encapsulated Czochralski (LEC) Technique

General Considerations

The Czochralski technique, in which a crystal is pulled from a melt, was originally used for metals and was then modified for the commercial growth of Ge and then Si. The rotation of the seed and crystal results in the immediate advantage of producing boules of circular cross section. Simultaneous rotation of the melt can result in efficient mixing of the host and dopant, a great advantage for doped crystals. In the case of compounds, where the partial pressures of one or both components are large, the melt surface must be protected either by incorporating an independent source of the vapour under dissociation (as in HB or GF growth) or by other means. In LEC growth, the surface is covered by an encapsulant layer of liquefied boric oxide, B_2O_3 [23.16, 17]. The axially rotating crystal is withdrawn through the encapsulant and then cools naturally by heat radiation, and by conduction and convection via the high-pressure ambient gas, see Fig. 23.7.

LEC pullers are high-pressure growth machines (and, therefore, expensive) because the internal pressure can exceed several atmospheres, especially when the Ga and As are compounded. In the growth of a compound

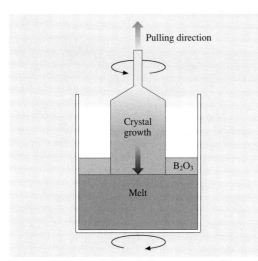

Fig. 23.7 LEC growth. The *dark-grey arrow* indicates the liquid–solid interface where growth takes place. As the growing crystal is pulled through the molten boric oxide it is rotated, resulting in a cylindrical boule. The boat is rotated in the opposite direction to stir the melt efficiently. (Courtesy Wafer Technology plc. UK)

like GaP, the internal pressures can exceed 100 atm. The boric oxide liquefies at temperatures of a few hundred degrees Celsius and floats to the surface of the other materials in the crucible. At all times during the subsequent reaction and growth it protects the surface from As loss. It also acts to purify the melt, probably because of the oxidising behaviour of its water content [23.18]. If the crucible is quartz, the resulting, nominally undoped, crystal is n-type because of Si incorporation. However, if the crucible is made from pyrolytic boron nitride (pBN) the crystal is SI over its entire length even without the introduction of Cr into the melt [23.19] (Sect. 23.1.3).

Nearly all LEC GaAs is contaminated by boron at concentrations up to 10^{18} cm^{-3} [23.20]. In all SI material these B atoms take up Ga lattice sites and are electrically inactive [23.21]. In Si-doped GaAs some of the B atoms appear to be incorporated as an acceptor species [23.22].

The automated growth of near-cylindrical crystals of accurately defined diameter would make LEC the choice of growth if it were not for the high dislocation density found in all but the highest carrier concentration n-type GaAs (this exception being a result of impurity hardening). Uncontrolled cooling of the LEC crystal causes the outside to contract on to the core, creating slip dislocations at densities of 10^4–10^5 cm^{-2} [23.23]. This defect creation occurs at temperatures just below T_m where dislocation motion by slip is easy. In addition, high concentrations of native point defects, which allow dislocation motion by climb, are also present. As a result, the dislocations are able to polygonise into cells, a rearrangement that reduces their strain energies. The final arrangement of a dislocation cell structure is the situation that is usually seen when a wafer is assessed.

As in HB and horizontal GF growth, the crystal can be doped n-type or p-type. Si or Te are normally used for n-type doping while Zn is used for p-type material. As mentioned above, an unfortunate reaction between the melt and the boric oxide has been found to occur with Si doping, leading to a reduction in Si uptake and considerable boron acceptor contamination of the GaAs [23.24]. In many cases, the concentration of B in the final crystal is comparable to the Si concentration. Nevertheless, Si remains one of the preferred donors for LEC GaAs.

LEC growth is controlled by computer and crystals of mass up to 20 kg and diameter up to 200 mm are easily produced. Wafers of 150 mm diameter are routinely supplied to device manufacturers from LEC crystals: at the time of writing this article 200-mm-diameter wafers were being made available to manufacturers for assessment.

Growth of SI LEC GaAs

The growth of nominally undoped GaAs by LEC from a pBN crucible results in SI behaviour. Chemical analysis of this type of GaAs always finds a concentration of carbon that is higher than the total concentrations of all other electrically active impurities [23.21]. The high carbon concentrations are not too surprising because not only is carbon a possible impurity in both Ga and As but there are many components of the LEC puller, namely the heaters and much of the thermal insulation, which are also made of carbon. Much work has shown that the carbon is introduced to the Ga–As melt through the gas phase, probably as carbon monoxide and the control of the partial pressure of this gas can be used to control the uptake of carbon in the crystal [23.25]. Because carbon atoms take up As sites and act as shallow acceptors, the resulting crystal would be expected to be p-type.

Some of the carbon acceptors are compensated by residual concentrations of shallow donors such as silicon and sulphur. The compensation of the rest of the carbon acceptors is performed by a native deep donor species, EL2, which pins the Fermi Energy close to the mid-gap. However, the final resistivity of the GaAs depends mainly on the carbon concentration [23.26]. The atomic identity of EL2 was a hot topic of research for many years and there still remain questions as to its identity [23.27]. However, what is not questioned is that the defect involves the As-antisite defect, As_{Ga}.

After growth, high concentrations of EL2 are found associated with dislocations and this results in nonuniformities in electrical properties [23.28, 29]. Micro-precipitates of hexagonal As are also found in close association with the dislocations [23.30]. Most manufacturers use ingot anneals to render EL2 concentrations more uniform and to improve electrical uniformity [23.31]. Some follow ingot heat treatments with anneals of the individual wafers [23.32]. The schedules of these treatments vary between wafer suppliers but, in general, after these treatments the resistivity is of the order of $10^7\,\Omega\,cm$ with a uniformity of better than $\pm 10\%$.

It must be emphasised that the dislocation density cannot be reduced by heat treatments and the needs of device manufacturers for material of lower dislocation density has led to the development of improved growth techniques, the one now in general use being the vertical gradient freeze (VGF) method.

23.1.4 Vertical Gradient Freeze (VGF) Technique

This is a modification of the horizontal technique where the melt is contained in a vertical crucible above a seed crystal. The crucible, surrounded by the furnace, is pBN if SI GaAs is to be grown; otherwise, it can be quartz. The growth proceeds from the bottom of the melt upwards until the melt is exhausted [23.33]. Because cooling is better controlled, the resulting temperature gradients are much reduced compared to the LEC method, resulting in crystals with lower dislocation densities, typically $10^2 - 10^3\,cm^{-2}$. A schematic of VGF method is shown in Fig. 23.8.

The starting material is polycrystalline GaAs. Often, this can be synthesised from the elements in a LEC puller before being withdrawn rapidly from the crucible. This is a useful way of employing a LEC puller that might otherwise be redundant. The LEC puller also allows the introduction of a controlled amount of carbon, which will be necessary for SI behaviour in the final ingot. The use of polycrystalline GaAs starting material means that the VGF equipment can be easily fabricated in-house, as

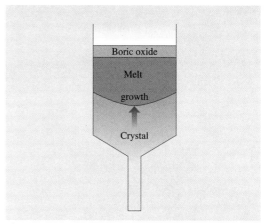

Fig. 23.8 VGF growth. The seed crystal, in the narrow section at the bottom of the crucible, is held below the melt and the growth takes place upwards. The *solid arrow* indicates the solid–liquid interface where growth takes place. The boric oxide surrounds the liquid and solid and helps in the final extraction of the cooled crystal from the crucible. (Courtesy Wafer Technology plc. UK)

high pressures will not be experienced. Boric oxide can be used to encapsulate the melt and to ease crystal removal from the crucible but the sealed system will ensure that the melt remains nearly stoichiometric [23.34].

Initially the major drawback to VGF was the low growth rate coupled with the inability to see the progress of the growth. In other words, if growth was not progressing correctly, this could not be detected until after the entire melt was solidified and this was several days in most circumstances. However, after the method is optimised the yield can approach 100%. The usual dopants can be employed if conducting material is required.

The reduced dislocation density in VGF GaAs has made it the material of choice for most applications and especially in optoelectronics, where dislocations are particularly deleterious to performance as they act as nonradiative recombination paths. At the time of writing, SI wafers of 200 mm diameter have been supplied as test wafers to device manufacturers.

Readers who require a recent and more thorough discussion of the growth of GaAs from the melt are referred to [23.35].

23.2 Epitaxial Growth of GaAs

Very few devices now use melt-grown GaAs in their active parts. Instead, they rely on the substrate to act as support for complex structures that are grown on the surface. These will typically employ ternary or even quaternary alloys. The former are the most important and are exemplified by the $Ga_{1-x}Al_xAs$ alloys, all of which have a very similar lattice constant as GaAs, varying from 5.6533 Å for GaAs to 5.6605 Å for AlAs. Selected electrical and optical properties of these alloys are presented in Fig. 23.9.

These alloys are in the form of thin layers which must have an accurately controlled composition, lattice-match the substrate, have accurate doping and, most of all, a well-defined layer thickness, often of only a few hundred nm. Growth of a new material on to a substrate where the atomic planes accurately line up is *epitaxial*. In this section, we present four approaches to the epitaxial growth of GaAs, the first two being more of historical than practical interest.

23.2.1 Liquid-Phase Epitaxy (LPE)

In LPE, the GaAs layer to be deposited is formed by first dissolving GaAs and the dopant into a liquid, usually Ga, at elevated temperatures, normally around 800 °C [23.37]. The solubility of GaAs in Ga is 3×10^{-3} mass fraction at these temperatures but it falls rapidly if the temperature is reduced. This behaviour can be seen in Fig. 23.3. Thus, if the Ga melt is saturated at a high temperature and is placed over the substrate, GaAs will be deposited as the temperature is reduced. Dopants added to the Ga melt will also be introduced into the growing GaAs layer, allowing junctions to be fabricated. If the melt is replaced by a molten (Ga + Al) alloy, a layer of GaAlAs can be grown, also with doping.

LPE usually uses several pots of molten metal (Ga or Ga + Al, for example) in a high-purity graphite, slider system. These can be moved over the substrate in turn to allow the growth of consecutive layers. A figure of a sliding-boat LPE reactor is shown in Fig. 23.10.

Advantages of LPE include the high crystalline purity of layers especially in terms of low concentrations of native point defects. Disadvantages are associated with the control of growth. It is difficult to produce reliably flat layers as convection currents in the liquid metal tend to produce a rippled surface. The thickness often

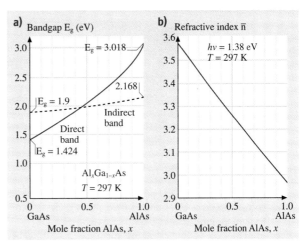

Fig. 23.9a,b Compositional dependence of some important properties of AlGaAs. (**a**) shows the dependence of the band-gap energy on composition. Below 45% Al, the alloy has a direct band-gap energy that increases steadily with Al composition. However, above 45% the band gap becomes indirect and the band-gap energy of pure AlAs is only 2.168 eV. (**b**) The refractive index of the AlGaAs alloys shows no dependence on the band-gap type and decreases steadily from a value of approx. 3.6 (GaAs) to 2.97 (AlAs), both these values being wavelength-dependent. (After [23.4])

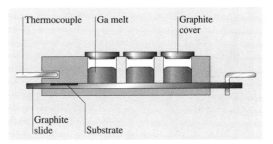

Fig. 23.10 LPE growth using the slider system. The Ga melts saturated with GaAs and dopants are sequentially placed over the GaAs substrate allowing several epitaxial layers to be grown. (After [23.36])

increases towards the periphery of the substrate and it is difficult to produce large areas of constant thickness. LPE was important for the LED and laser industry in early years but it is now rarely used for GaAs.

23.2.2 Vapour-Phase Epitaxy (VPE) Technologies

Chloride and Hydride Growth

In VPE, the components of the layer to be grown are transported to the substrate surface in the form of gases.

In chloride and hydride growth, the As vapours are $AsCl_3$ and AsH_3, respectively, both of which are converted to gaseous As_4 in the growth reactor. The volatile Ga component is GaCl produced by passing HCl gas over a well containing heated Ga metal. This takes place near 850 °C. The volatile Ga and As components are reacted with H_2 over the GaAs substrate, which is held near 750 °C. A two-zone furnace is therefore required. The growth rate can be up to 20 μm per hour. Dopants can be introduced as chlorides. Full details of these growth techniques have been reviewed by *Somogyi* [23.38].

The major commercial use for these growth methods is for the fabrication of $GaAs_{1-x}P_x$ LEDs, see Fig. 23.11.

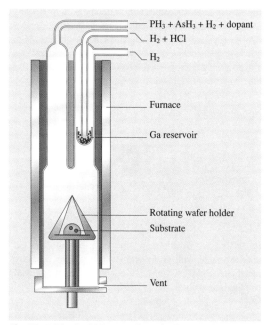

Fig. 23.11 The hydride growth method. A vertical system is shown but horizontal methods are also used. The volatile gallium chloride GaCl is produced near 850 °C, and growth takes place near 740 °C. The rotating sample holder ensures that the growth on several substrates is uniform and reproducible. (After [23.36])

Phosphorus can be introduced as PCl_3 or PH_3 and the growth is sufficiently fast that grading from pure GaAs to the required alloy can be performed in a reasonable time by varying the ratio of $AsCl_3$ to PCl_3 partial pressures. Unfortunately, modern device technologies rely heavily on alloys from the GaAs/AlAs system and growth of AlAs is not possible using these methods because no chloride of Al is sufficiently volatile.

Metalorganic Chemical Vapour Deposition (MOCVD)

In MOCVD, sometimes known as metalorganic vapour-phase epitaxy (MOVPE), the gaseous As precursor is AsH_3 and the volatile metal sources are alkyls (sometimes known as metalorganics) such as trimethyl-gallium (TMG), $Ga(CH_3)_3$, although others have been used. Trimethyl-aluminium (TMA), $Al(CH_3)_3$, is used for the Al component of the alloy and trimethyl-indium (TMI) is used for the In component.

For p-type material the usual acceptor is Zn in the form of diethyl zinc, $Zn(CH_3)_2$. Si is the most commonly used donor in the form of silane, SiH_4.

All the alkyls are volatile liquids at room temperature and are kept refrigerated in order to reduce and control their vapour pressures. AsH_3 and SiH_4 are gases and are supplied diluted in hydrogen.

The alkyls are released from their cylinders by bubbling high-purity hydrogen through them. They are then introduced to all the other constituents with more hydrogen at room temperature. The reaction to produce the layer only occurs when the vapours are passed downstream over the heated substrate. This is held at approximately 750 °C for GaAs growth but is somewhat higher for AlGaAs. Figure 23.12 is a schematic of a MOCVD apparatus.

This technique has been discussed in detail [23.39, 40]. Once the control of the gas flow over the heated substrate has been optimised, layers can be grown with great ease and with great accuracy, the layers being of high crystalline quality, and of good uniformity in both constitution and thickness. The growth of alloys of different composition or material with varying dopant concentration can be simply performed by changing the hydrogen flow rate through the alkyls or dopant sources using mass-flow controllers.

Note that SI epitaxial GaAs cannot be obtained unless Cr is added during growth or lightly p-type GaAs is grown and a post-growth anneal to generate EL2 centres is undertaken. Both methods have been reported, (see [23.41–44] for further details).

Fig. 23.12 MOCVD growth. Separate mass-flow controllers and valves control the hydrogen flow through each of the alkyl liquids, which are held in cooled, sealed flasks. The metalorganic vapours are introduced to the substrate with arsine. The substrate is heated by an induction furnace via a graphite susceptor that supports the substrate as well as coupling to the field of the RF field to produce heating. The gaseous silane is for n-type doping

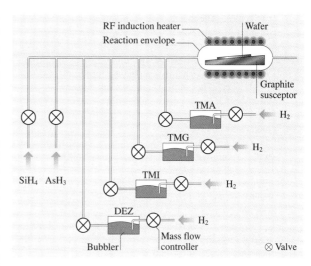

In a research laboratory, the growth equipment can be built in-house and a skilled operator can control the growth of single samples. Commercially the equipment is much more complicated, with several substrates being processed together and the entire growth procedure being computer-controlled. Manufacturers put much effort into ensuring that the gas flow over each substrate is as uniform as possible to ensure reproducible growth. Early problems with difficulties in obtaining alkyls of satisfactory and reproducible quality have been solved.

However, a limitation of MOCVD concerns the rate at which a layer composition can be changed. The abruptness of the interface between layers can be critical in many modern applications and this is often limited in MOCVD by mixing of the gas upstream from the substrate. Thus, even if the mass-flow controllers that control the rate at which hydrogen bubbles through the alkyls change their relative concentrations instantly, there will be some mixing in the upstream gases which will cause this change to be more gradual when the gases flow over the substrate. This causes a slight grading in the grown interface and this can be deleterious to some devices. Nevertheless, most of current GaAs-based devices are produced commercially by MOCVD.

Unfortunately, all the precursors used for MOCVD are poisonous, with AsH_3 being extremely so. Moreover, satisfactory growth requires a surplus of AsH_3 over the alkyls of about ten to one in atomic concentrations to be present. It follows that the exhaust gases

Fig. 23.13 MBE growth takes place in a UHV chamber. Each component of the required epitaxial layer is supplied by an effusion cell (also known as a Knudsen cell) and shutter. The substrate is held on a heated, rotating holder. Cryopanels cooled by liquid nitrogen help to keep the base pressure of the system to about 10^{-11} Torr. The RHEED gun and associated fluorescent screen allow the growth process to be monitored continuously. Substrates are positioned on the sample block via the buffer chamber where they have been previously prepared

are As-rich and very toxic. MOCVD equipment requires considerable exhaust gas cleaning and this makes the technique rather demanding in terms of effluent processing.

23.2.3 Molecular-Beam Epitaxy (MBE)

At its simplest, MBE is a high-purity growth technique where all the components of a required material are evaporated as elemental beams. The term molecular reflects the observation that some elements evaporate as molecules. As an example, under MBE growth conditions arsenic vapour is primarily in the form of As_4

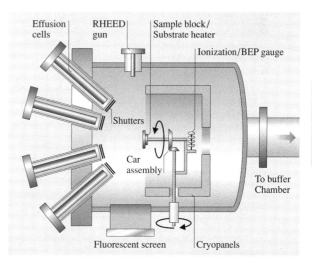

molecules. The beams are incident on a heated, rotating substrate where they react to form a growing layer [23.45].

In practice, the growth takes place in an ultra-high-vacuum (UHV) chamber at a base pressure of about 10^{-11} Torr. The sources are usually high-purity elements held in Knudsen cells (high-vacuum effusion cells). The molecular beams are controlled by simple metal shutters that can be moved rapidly in and out of the beams by external drives. A typical MBE reactor will have up to eight such sources, some for the components of the host semiconductor, Ga, Al, etc. and others for dopants, each held at the relevant temperature to produce a molecular beam of the correct effective pressure, see Fig. 23.13.

The use of UHV technology allows the use of in situ analysis of the growth to be monitored and this has provided much information on details of the growth mechanisms. A favourite tool is reflection high-energy electron diffraction (RHEED) whose use gives important information regarding the quality of the growth with single atomic layer sensitivity.

The advantages of MBE include the use of high-purity elemental sources rather than compounds (whose purity is less controllable). The use of moveable shutters in front of each cell means that each source can be turned on or off in a fraction of a second. As a result, MBE-grown material can be of very high quality with the sharpest interface abruptness. These qualities are essential in several microwave devices and this is one of the major commercial uses for MBE. However, MBE has been a favourite growth technique for semiconductor research laboratories because it is well suited for the small-scale growth of specialised structures.

As an industrial technique, MBE suffers from several problems. The first results from its reliance on UHV technology, meaning that it is expensive to install, requires large quantities of liquid nitrogen to keep cold, considerable power to keep under vacuum, and special clean rooms. It is expensive to operate. Conditions for correct growth include a beam effective pressure ratio of As to group III metals of about six. As a result, the inside of the machine becomes coated with arsenic, making it very unpleasant to clean. Replenishing the cells when they are exhausted can be a slow and difficult process because recovery of the vacuum after the inside of the machine is exposed to atmospheric gases can take several days.

There are three modes for MBE growth: normal, low temperature and gas source.

Normal MBE Growth

In conventional MBE [23.46], the growth normally takes place at around 600 °C with an As_4 flux that is considerable greater than the Ga flux. Often the beam effective pressure (BEP) ratio of As to Ga is around six. The Ga flux controls the growth rate and a Ga flux of 6×10^{14} cm^{-2}s^{-1} produces a growth rate of one monolayer per second.

The GaAs so produced is of excellent structural and electrical quality. It is usually doped n-type with Si but Se and Te are often used. Unlike melt-grown GaAs, there is little auto-compensation and doping levels exceeding 10^{19} cm^{-3} can be obtained; p-type doping is normally achieved using Be, although for some devices carbon is preferable. These acceptors can be incorporated at concentrations exceeding 10^{20} cm^{-3}. AlGaAs alloys are grown by opening the shutter of an Al cell. Growth proceeds at a somewhat higher temperature but with similar ease. Should AlGaAs alloys of different composition be required, it is usual to have two Al cells at different temperatures and these are opened as required. For similar reasons a device structure with different n-type doping regions such as a metal–semiconductor field-effect transistor (MESFET) or high-electron-mobility transistor (HEMT) will require the use of at least two Si cells at different temperatures. These means of changing material composition increase equipment complexity and cost.

Low-Temperature MBE (LTMBE) Growth

Unlike the VPE techniques presented previously, MBE growth can proceed at temperatures as low as 200 °C, in a regime called low-temperature MBE (LTMBE), see [23.47, 48]. One reason for this is that no chemical reactions are needed to release the elemental Ga and As on the GaAs surface from gaseous precursors (although the As_4 must be dissociated) and there is sufficient surface atomic mobility at these low temperatures to ensure uniform, crystalline growth. At these low growth temperatures and under standard As-rich growth conditions with a BEP ratio of greater than 3, undoped GaAs is very As-rich, with up to 1.5% excess As, contains high concentrations of point defects and is high resistivity. This material is heat-treated before use to improve its properties and stabilise it. Growth of LTMBE GaAs under less As-rich conditions, or at higher temperatures, results in nearly stoichiometric material. In all cases, however, SI behaviour is observed.

The real advantage of annealed LTMBE GaAs to device manufacturers is the extremely low minority carrier lifetime of < 0.5 ps. This makes it useful in the fabrication of ultrafast optodetectors [23.49] and as a substrate for high-speed integrated circuits (ICs) [23.50].

Gas-Source and Metalorganic MBE

In modifications known as gas-source MBE (GSMBE) and metalorganic MBE (MOMBE), some sources that are volatile at room temperature are used.

In GSMBE, AsH$_3$ replaces the metallic As source and some dopants are also replaced by volatile sources. One of the advantages of GSMBE is the virtually unlimited source of As (from a high-pressure gas bottle) and the ease of replacing it and similar group V sources. These are exhausted rapidly in normal growth because of the need to use an excess of these elements.

In MOMBE the group III metallic sources are replaced by metalorganics like TMG. By using a combination of triethyl-gallium (TEG) and TMG, MOMBE growth can lead to the controlled incorporation of carbon acceptors at concentrations up to 10^{21} cm^{-3}, necessary for heterojunction bipolar transistors (HBT) fabrication (see later) and for other uses where stable, highly p-type GaAs is required. Similar ease of carbon incorporation in GSMBE uses CCl$_4$ or CBr$_4$ as sources. A thorough review of these techniques has been given by *Abernathy* [23.51].

23.2.4 Growth of Epitaxial and Pseudomorphic Structures

The epitaxial growth of GaAs on GaAs is one of the simplest epitaxial growth processes where, if the growth is performed correctly, analysis of the resultant structure will show no break in the atomic planes at the substrate-layer interface. Because the lattice constants of the ternary GaAlAs alloys match that of GaAs, the growths of these are also simple in principle and are the basis for many devices.

Attempts to grow layers with lattice constants that are dissimilar to the substrate result in the incorporation of misfit dislocations. Although these are initially parallel to the interface, they are deleterious to device performance because their interactions can cause them to orient themselves in the growth direction and to invade the electrically active structure above. They can then act as recombination centres. However, even when a required ternary does not lattice-match the substrate two approaches can be used to give satisfactory device results; grading and pseudomorphic growth.

In grading, the material composition, and hence the lattice constant, is changed gradually from that of the substrate to that of the required alloy. Misfit dislocations are still produced but they are now distributed on many lattice planes and they tend not to interact. This restricts them to the graded region and away from the active regions above. Grading is used for the production of low-cost visible LEDs based on GaAs$_{1-x}$P$_x$. These devices can be formed on either GaP or GaAs substrates according to the required alloy composition (which controls the colour) and growth by the hydride or chloride process allows grading to be performed. Other uses for grading occur in the production of optical confinement layers in certain types of lasers (Sect. 23.13).

Theoretical calculations show that mismatched growth does not result in the immediate creation of misfit dislocations, because there exists a critical thickness below which the strain energy is insufficient to produce them [23.52]. If growth is stopped at this stage, the layer remains dislocation-free. Indeed, if a layer of the substrate compound now covers the layer, the structure is further stabilised. Its enforced lattice match to the substrate strains the mismatched layer; if it would normally be cubic, the epilayer assumes a tetragonal structure, with the c-axis aligned in the growth direction. This distortion produces electrical and optical properties that are no longer isotropic. In other words, properties parallel to the growth interface are different to those perpendicular to it. Often these offer advantages over the cubic material. This type of growth is called pseudomorphic and is very useful in several microwave devices. Pseudomorphic structures can be grown by MOCVD or MBE, although the latter is usually preferred.

23.3 Diffusion in Gallium Arsenide

Diffusion, which refers to the motion of impurity atoms under a concentration gradient, can be a positive or negative process. As an example of the first category, diffusion can be used to introduce impurity atoms into the bulk of a solid. This is the most important process in Si technology, for example. Diffusion is also one of the atomic processes that are thought to be important in making effective Ohmic contacts. However, diffusion is also the mechanism by which large concentration gradients grown in to a structure can relax and cause degradation to the characteristics of the device.

Tuck has written excellent reviews of atomic diffusion in semiconductors [23.53, 54]. For various reasons, the diffusion characteristics of most impurities in GaAs are only poorly understood. In this section, we present a short summary of those diffusion parameters that are well understood or are important in GaAs.

Diffusion is a high-temperature process resulting from the random motion of the impurity atoms within the host matrix. The process is described by the two Fick's laws.

The first law states that the flux of diffusing atoms J in a given direction, say x, is proportional to the concentration gradient along x or $J = -D(\partial C/\partial x)$. The negative sign shows that the net motion of atoms is in the opposite direction to the concentration gradient. This law can easily be proved by atomic models. The constant of proportionality D is the diffusion coefficient and is a strong function of temperature T according to $D = D_0 \cdot \exp(-E/kT)$, where E is an activation energy for the process and D_0 is a pre-exponential function, the diffusion constant.

There are two simple diffusion mechanisms. The first is the *substitutional* mechanism where impurities on a sublattice (Ga or As) move by jumping into adjacent vacant lattice sites. E has two components which add: an energy to move from one site to another, the migration energy, E_m, and a further *energy of formation* to produce the vacancy, E_V. This results in a large value of E and slow diffusion, normally undetectable at diffusion temperatures under a few hundred degrees Celsius below the melting temperature. The second is the *interstitial* mechanism where impurity atoms in interstitial sites move by jumping into adjacent interstices. Here E has a single component, the migration energy, of a value close to E_m. This results in a small value of E and rapid diffusion. Interstitial diffusion can often be observed at low temperatures, in extreme cases at or below room temperature.

The second law represents conservation of mass and states that the rate at which a concentration changes in a given volume, $\partial C/\partial t$, is equal to the difference in input and output fluxes from the volume. Combination of the two laws, for a value of D that does not depend on concentration, results in the diffusion equation, $\partial C/\partial t = D(\partial^2 C/\partial x^2)$. Solutions to this equation have been calculated for many situations including those that are of experimental interest [23.56]. When experimental diffusion profiles do not match theoretical predictions, it must be concluded that D is a function of C.

Experimental investigations usually start with coating the surface of a GaAs sample with the impurity.

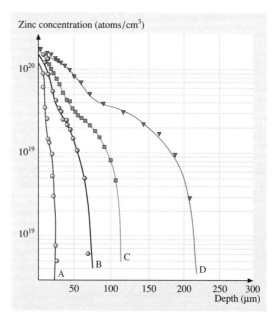

Fig. 23.14 Experimental diffusion profiles for Zn in GaAs at 1000 °C with excess arsenic in the ampoule. Diffusion times: A: 10 min, B: 90 min, C: 3 h, D: 9 h. (After [23.55])

The GaAs is heated rapidly to temperature T and held at T for a known time. The GaAs is then cooled rapidly. Concentrations of the impurity as a function of depth are then determined, either by using radiotracer methods (if the impurity can be tagged with a relatively long half-life isotope), by secondary-ion mass spectrometry (SIMS) if this technique is available, or electrically if the impurity is electrically active at room temperature.

Diffusions at different temperatures can be compared via an Arrhenius plot to derive the activation energy E. Diffusions from a thin layer are expected to result in a smooth, exponential profile. Other diffusion conditions result in profiles of different forms [23.53]. It is necessary to consider the loss of As from the GaAs at elevated temperatures when undertaking this procedure and a little elemental As is often added to the diffusion ampoule to replace loss from the surface. Other complications include possible chemical reactions between the impurity and either Ga or As, or both, which can result in severe changes to the experimental conditions. For these reasons, many diffusion measurements quoted in the literature may be quite inaccurate.

23.3.1 Shallow Acceptors

The first impurity whose diffusion profile in GaAs was carefully analysed was zinc because it was by diffusion of Zn into n-type GaAs to make a p–n junction that early LEDs were fabricated. When Zn is diffused into GaAs (Fig. 23.14) the profile is found to be deeper than expected for a substitutional mechanism. Moreover, it is not exponential but exhibits a concave shape with a very abrupt diffusion front.

Electrical evaluation reveals that, after diffusion, nearly all the Zn atoms act as shallow Zn_{Ga} acceptors. Interpretation suggests two possible *interstitial–substitutional* diffusion mechanisms, both of which involve the diffusion of singly positively charged interstitial Zn atoms, Zn_i^+, which take up Ga lattice positions to become substitutional Zn_{Ga}^- acceptors. Either of these models suggests that D is proportional to $[Zn_{Ga}^-]^2$.

Under certain circumstances, the diffusion of Mn, which is a shallow acceptor, Mn_{Ga}, can be very similar to Zn. However, surface reactions cause experimental problems so that data is not so consistent.

Be_{Ga} is an acceptor often used in MBE growth. Its diffusion behaviour is poorly investigated but it has been concluded that D is concentration-dependent. Rapid ageing characteristics of GaAs/AlGaAs heterojunction bipolar transistors (HBTs) have been attributed to diffusion of Be from the highly p-type base into the emitter and collector regions [23.57].

C_{As} is a shallow acceptor and is important in SI GaAs as well as in MBE growth. It seems to be a very slow diffusing species, probably via purely substitutional diffusion. The diffusion rate depends on whether the GaAs is As- or Ga-rich, being greater in the former case. Arrhenius plots give expressions for D as $0.110 \exp(-3.2\,\text{eV}/kT)$ and $2.8 \times 10^{-4} \exp(-2.7\,\text{eV}/kT)\,\text{cm}^2\text{s}^{-1}$, respectively. Carbon diffusion in GaAs has been reviewed by *Stockman* [23.58].

23.3.2 Shallow Donors

All shallow donors of scientific interest, group IV atoms occupying Ga or group VI atoms occupying As lattice sites, are slow substitutional diffusers whose measurement has posed considerable problems [23.54]. Donor diffusion appears neither to have any direct technical application nor to be involved in any device degradation, although it is expected to be involved with the creation of good Ohmic contacts to n-type GaAs. A recent assessment of Si diffusion in GaAs using SIMS to trace the motion of the Si atoms has been reported in [23.59].

In general, it is observed that atoms occupying the group V sublattice are generally very slow diffusing, possibly a result of the high energy of formation of V_{As}.

23.3.3 Transition Metals

Transition metals normally occupy Ga lattice sites and act as deep acceptors. *Cronin* and *Haisty* [23.12] grew GaAs crystals doped with virtually all the first series of transition metals and were the first to observe their acceptor properties. This investigation showed that Cr doping produced SI material. However, most of these metals are very rapid, interstitial diffusers in GaAs, although their final situation is substitutional, and this has led to their almost total disuse from GaAs technology.

As an example, diffusion of Cr has been investigated by several groups, (unfortunately, with disparate results). Diffusion into GaAs from a surface source (indiffusion) produces diffusion rates that are larger than those from within the host lattice. However, there are clear surface effects that result from a chemical reaction between Cr and GaAs. After diffusion the profile is deep, indicating interstitial diffusion, but nearly all the Cr atoms seem to act as acceptors, i. e. they are substitutional [23.54]. Thus, there are similarities with the earlier case of Zn diffusion. This interstitial–substitutional diffusion is probably common in GaAs, although details are rare. One case where some justification for this mechanism was obtained was the case of Fe diffusion. Here, electron paramagnetic resonance (EPR) studies of the diffused layer showed good agreement between Fe_{Ga}^{3+} and the atomic concentration of Fe determined by radiotracer measurements although the rapid penetration rate strongly indicated interstitial diffusion [23.60].

23.4 Ion Implantation into GaAs

Modification to the electronic properties of the near-surface region of a semiconductor is the basis of nearly all devices. In Si this is routinely performed by diffusion or ion implantation. However, most microwave GaAs devices that require this modification are n-type devices such as metal–semiconductor field-effect transistors (MESFET) and, as mentioned previously, diffusion of shallow donors into GaAs is not technically possible. MESFETs are majority carrier, n-type conduction devices that are fabricated on the surface of SI GaAs. They need abrupt depth profiles of two carrier concentrations (Sect. 23.11.2 for more details). In ion-implanted MESFETs, these regions are produced by direct implantation of donor atoms into the surface. The depth profile of implanted ions can be calculated using computer programs. In general, the greater the energy of the ion, the greater is the mean penetration and the greater the mass of the ion, the lower the penetration.

Unfortunately, damage to the host crystalline structure is associated with the implantation, resulting in zero electrical activity from the implanted ions. This damage must be repaired by thermal annealing. Two types of anneal schedule are used, furnace annealing and rapid thermal annealing (RTA). The first entails heating the GaAs in a conventional furnace at 850–900 °C for at least 15 min. RTA uses a higher temperature of 900–950 °C but for times of only approximately ten seconds. In each case the surface must be protected by an impervious layer to stop the loss of As. This is often silicon nitride (Si_3N_4) but the simple expedient of annealing two wafers face to face will reduce As loss to insignificant amounts. The recovery of electrical activation, being a thermally activated process, has an associated activation energy that is usually less than 1 eV. Because that for diffusion often exceeds 2.5 eV, careful implant annealing does not result in appreciable diffusion. A full list of these parameters is given in the article by *Sealy* [23.61].

In Si, implant anneals remove nearly all this damage; in GaAs only a fraction of the implant damage can be removed, especially after high implant fluences, unless the heat treatments are applied for times and temperatures that are incompatible with device manufacture. Anneals of acceptor-implanted GaAs are more effective with nearly 100% activation of the implant being achievable. The inefficient activation of donor implants in GaAs is not fully understood.

Applications of ion implantation, mostly reserved for MESFETs and monolithic microwave integrated circuits (MMICs) based on MESFETs, are discussed later.

23.5 Crystalline Defects in GaAs

Crystal defects are crucial in determining many electrical and optical properties of GaAs. Most of these defects are incorporated during growth into melt-grown (substrate) material but they can also be introduced in later processing steps.

23.5.1 Defects in Melt-Grown GaAs

Structural Defects

Previously, the problems associated with melt growth were introduced. The growing crystal, cooling from the outside, experiences compressional stresses that cause plastic deformation by the introduction and subsequent motion of dislocations.

In SI GaAs the polygonised dislocation structure is complex. Such structure is best revealed by specialised chemical etching of the surface to reveal dislocation arrangements [23.62, 63] or reflection X-ray topography [23.64]. Each technique reveals classical dislocation cells, *lineage* (the boundary between sections of the crystal that have a small tilt misorientation and caused by dislocation motion) and slip. The lineage occurs preferentially along the ⟨110⟩ diameters. A typical image from a 20-mm² area from a 3-inch-diameter ⟨001⟩ SI LEC GaAs is shown in Fig. 23.15.

The use of selective etching by molten alkalis is commonly used by substrate manufacturers in order

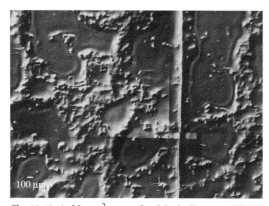

Fig. 23.15 A 20-mm² area of a 3-inch-diameter {001}SI LEC GaAs after etching to reveal dislocations. Many of the dislocations are arranged in a cellular wall formation enclosing regions where the dislocation density is small. The *nearly vertical line*, closely aligned along [110] is lineage where dislocations have interacted to produce a small-angle tilt boundary. (Courtesy of D. J. Stirland)

to reveal dislocations passing through the surface and being revealed as pits. Thus, they obtain an *etch pit density* (EPD), which corresponds to the dislocation density [23.65]. However, transmission electron microscopy (TEM) is a preferable, though much slower, method for investigating dislocations [23.66].

VGF GaAs contains a much lower density of dislocations but even these are found to be in the form of a cell structure [23.68]. n^+–GaAs is much less dislocated and no cell structure is present. It is thought that the slip velocity of dislocations on their glide planes is reduced in highly n-type material, so that n^+-GaAs contains far fewer dislocations than SI or p-type material. However, this cannot be the only reason for this hardening process as the addition of neutral atoms such as In at high concentration can also reduce the dislocation density to near zero [23.69]. The dislocation structures also contain microscopic precipitates, which have been found to be hexagonal, elemental As [23.30, 70]. These are absent in n^+-GaAs.

Point Defects

SI GaAs is a relatively pure material. It contains boron impurities at high concentration but these are neutral (Sect. 23.1.2). Silicon and sulphur, both shallow donors, are found at concentrations of around 10^{15} cm^{-3}. Concentrations of other electrically active impurities are extremely low and can be ignored.

The exception is carbon, a shallow acceptor, which occurs naturally in LEC GaAs, but which is often added intentionally to VGF GaAs, at concentrations of 10^{14}–10^{16} cm^{-3}. For this reason carbon must be treated as a dopant and not an impurity. The accurate measurement of the carbon concentration is challenging. The standard method uses the low-temperature far-infrared absorption due to the localised vibrational modes (LVM) of carbon acceptors, a technique that has a sensitivity of around 10^{14} cm^{-3} [23.71, 72].

However, SI GaAs contains many native defects, and these are listed in Table 23.2.

Most native point defects are deep donors, at concentrations of 10^{13}–10^{14} cm^{-3} with one, EL2, the As antisite defect (As$_{Ga}$), being dominant, existing at a concentration of 1×10^{16}–1.5×10^{16} cm^{-3}. Compensation of the carbon acceptors by EL2 pins the Fermi level near the centre of the band gap, the other point defects being fully ionised by this process. The lack of native acceptors at concentrations above those of donors makes the presence of carbon (or other chemical acceptors) mandatory for SI behaviour.

The EL2 Centre

EL2 is a native deep double donor. Its first ionisation state is at 0.75 eV above the valence band. This is the level that controls the Fermi level in SI GaAs. A second ionisation from + to ++ occurs at an energy at

Table 23.2 Deep electronic levels observed in melt-grown SI GaAs. EL is an electron level and HL is a hole level. In most commercial material only EL2 exists at concentrations exceeding 10^{15} cm^{-3} and is the only deep level assumed to be involved in the compensation mechanism to give SI properties. (From [23.67])

Label	Origin	Concentration (cm^{-3})	Emission energy (eV)	Capture cross section (cm^2)
EL11			$E_c - 0.17$	3×10^{-16}
EL17			$E_c - 0.22$	1.0×10^{-14}
EL14			$E_c - 0.215$	5.2×10^{-16}
EL6	Complex defect	10^{14}–10^{16}	$E_c - 0.35$	1.5×10^{-13}
EL5		10^{14}–10^{16}	$E_c - 0.42$	10^{-13}
EL3		10^{13}–10^{15}	$E_c - 0.575$	1.2×10^{-13}
EL2	Native defect As$_{Ga}$ or [As$_{Ga}$–X]	5×10^{15}–3×10^{16}	$E_c - 0.825$	1.2×10^{-13}
HL10		Below 2×10^{14}	$E_v + 0.83$	1.7×10^{-13}
HL9			$E_v + 0.69$	1.1×10^{-13}
HL7		1.7×10^{15}	$E_v + 0.35$	6.4×10^{-15}
"Ga$_{As}$"	Gallium antisite Ga$_{As}$ or boron antisite B$_{As}$	3×10^{15}–3×10^{16} (dependent on Ga richness of melt)	$E_v + 0.077$ $E_v + 0.203$ (double acceptor)	

0.54 eV above the valence band. Only the singly ionised state is paramagnetic but EPR measurements on this state demonstrated that EL2 contained the As antisite defect [23.73].

EL2 exhibits a broad near-infrared absorption band in SI GaAs, which has been separated into components from neutral and singly ionised EL2 [23.74, 75].

EL2 has come under extreme scrutiny over the past two decades, not only because it results in SI behaviour, but also because it exhibits a remarkable photo-quenching property; at low temperatures EL2 defects can be excited into an inert metastable state by irradiating the GaAs with sub-band-gap light [23.74]. This process is associated with a photoactivated relaxation of As_{Ga} along a $\langle 111 \rangle$ direction into an interstitial site to produce a $[As_i - V_{Ga}]$ complex. Normal deep-donor behaviour is recovered by warming above 140 K [23.76, 77].

The two states of EL2 are shown in Fig. 23.16. Interestingly, a similar deduction about the motion of donors has been made in highly n-type GaAlAs when the Al content exceeds a critical value. The effect, which, results in the loss of normal donor behaviour, is ascribed to the generation of the DX centre. Thus, conduction electrons are lost on irradiating the cooled sample with sub-band-gap light. Like EL2 in GaAs, the recovery from DX-like behaviour is recovered by warming the sample.

High concentrations of EL2 are found in close association with dislocations in as-grown material. Coupled with the presence of As precipitates, this demonstrates that the environments of dislocations are very As-rich.

EL2 concentrations are rendered uniform by ingot anneals which, as expected, also improve the uniformity of electrical properties.

Concentrations of EL2 defects are easily determined by measuring their room-temperature infrared absorption at wavelengths near 1 μm [23.74]. In practice, EL2 concentrations are always $1.0 \times 10^{16} - 1.5 \times 10^{16}$ cm^{-3} in crystals grown from a melt that is slightly As rich. The use of an infrared-sensitive closed-circuit television (CCTV) camera or photographic film can allow mapping of the concentrations to be made [23.78].

Another defect, often referred to as the *reverse contrast* (RC) defect can also be mapped. This must take place at lower sample temperatures but otherwise uses a similar technique [23.79, 80]. RC absorption also reveals dislocations. However, unlike EL2 concentrations, those of RC defects are not homogenised by these heat treatments, indicating that other point defects may also be resistant to standard anneal protocols. RC defects occur at relative concentrations that are the reverse to EL2 defects in unannealed material (they are high where [EL2] is low, and vice versa) and seem to control minority carrier lifetime in SI GaAs and, thus, their low-temperature luminescence. Their identity as being As vacancies was demonstrated by positron-annihilation methods [23.81] and their nonuniform distribution was explained, at least in part, by the influence of dislocations on the As Frenkel reaction on the As sublattice as the crystal cools [23.82]. The identities of other point defects in SI GaAs have not been determined.

23.5.2 Epitaxial GaAs (not Low Temperature MBE GaAs)

Structural Defects

Because they cannot end within the crystal, dislocations that pierce the surface of the substrate must grow into the epilayer and, for devices that rely on minority carriers, this is a problem. The reason for this is that dislocations in GaAs can act as potent minority-carrier recombination centers. This may result from the presence of a high density of electrically active dangling bonds at the dislocation core. However, enhanced recombination from the presence of a relatively high concentration of point defects around the core (Cotterell atmosphere) cannot be totally ruled out. The latter would seem to be less likely in epitaxial GaAs because of the low growth temperatures and the low measured concentrations of these defects. However, the types of atoms that are expected to act as recombination centers are also expected to be rapid diffusers and their concentrations around dislo-

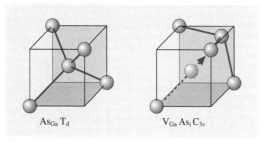

Fig. 23.16 The two states of EL2. The *spheres* represent As atoms. In T_d symmetry the central As atom is tetrahedrally bonded to four nearest neighbour As atoms as As_{Ga}. In the state represented by C_{3v} symmetry this As atom has been excited to move along the $\langle 111 \rangle$ direction where an uncharged metastable site is present. It is now represented by $[V_{Ga} - As_i]$. Heating above 150 K causes the As atom to return to the stable As_{Ga} structure. (After [23.27])

cations could be much greater than is estimated. The deleterious effects of these *threading dislocations* are the prime reason for the development of VGF and similar substrates.

Point Defects

Epitaxial GaAs contains a far lower concentration of intrinsic point defects than melt-grown material because of the low temperatures to which these materials are subjected. EL2 defects are found in epilayers grown by all the VPE techniques at concentrations up to 10^{14} cm^{-3} making it both the most common trap and the one occurring at highest concentration. Higher concentrations may be generated after growth by annealing at temperatures near 900 °C. In this way, lightly p-type VPE GaAs can be rendered SI, see Sect. 23.2.2. A comprehensive list of electron traps has been presented in [23.83] and a similar list of hole traps has been presented in [23.84]. Capture cross-section determinations have been presented in [23.85].

In general, most point defects in epitaxial material are extrinsic and result from impurities present in the precursors or in the growth apparatus. In modern equipment, using purified starting chemicals, residual impurities are typically shallow donors such as Si and S that are difficult to remove. These produce a residual carrier concentration of around 10^{14} cm^{-3} in the best commercial, nominally undoped, epitaxial GaAs. There are notable exceptions to the high purity usually found in MOCVD and MOMBE GaAs when carbon is used as a p-type dopant. Hydrogen can enter the growing layer and passivate the acceptors, see [23.86, 87], a case where intentional doping by one species can result in unintentional incorporation of an impurity, see also the previous case of Si in LEC GaAs. The hydrogen concentration is up to 50% of the carbon concentration in MOMBE GaAs and up to 20% in MOCVD GaAs. Fortunately, it can be removed by a post-growth heat treatment at 600 °C for around 10 min.

23.5.3 LTMBE GaAs

The low growth temperature and an excess of As cause this material to be very As-rich [23.47, 88]. Initially the excess As is in solution and device processing causes it to precipitate. For reasons of reproducibility, an anneal at around 500 °C is applied before processing which stabilises the LTMBE GaAs, which now contains a very high number of As precipitates and increases its resistivity to over 10^8 Ω cm [23.89]. The high resistivity is probably due to high concentrations of point defects, especially EL2, but Schottky barrier effects, resulting from the As/GaAs interfaces, which would result in virtually all the material being depleted, cannot be ruled out [23.90]. The acceptor species, which must be present to ionise the EL2 defects and pin the Fermi level, is widely believed to be the Ga vacancy, V_{Ga} [23.91] and there is some positron-annihilation spectroscopy evidence to support this [23.92].

23.6 Impurity and Defect Analysis of GaAs (Chemical)

Concentrations of impurities in all grades of GaAs, with the exceptions mentioned previously, are extremely low and always less than 1 part per million (ppm) or $\approx 4 \times 10^{16}$ cm^{-3}. Secondary-ion mass spectrometry (SIMS), or similar techniques like glow-discharge mass spectrometry (GDMS), are the chosen method for detecting most impurity elements, see [23.93]. In these methods, the surface is ablated and ionised. The secondary ions from the material under investigation are then passed through a mass spectrometer before detection. Sensitivities of better than 10^{14} cm^{-3} are quoted by most SIMS and GDMS laboratories. Unfortunately, important species like hydrogen, oxygen and carbon are not detected with adequate sensitivity by these techniques unless considerable care is taken to remove these impurities from the vacuum ambient of the mass spectrometer. This normally requires extra pumping often by cryogenic means (cooling a section of the instrument to near liquid-helium temperatures to freeze impurities and remove them from the vacuum), atypical extra components of most spectrometers. Because impurities and their concentrations tend to be specific to manufacturers and GaAs is being improved continuously, an example of a chemical analysis is not given.

At low concentrations, C, H, and O are difficult to measure in epitaxial GaAs because of the low volume of material that is available for analysis. In bulk GaAs, C, and H are best detected using infrared absorption by localised vibrational modes (LVM). This technique exploits the fact that light impurities in a solid have vibrational frequencies that do not couple well to the lattice modes (phonons); they vibrate at frequencies above

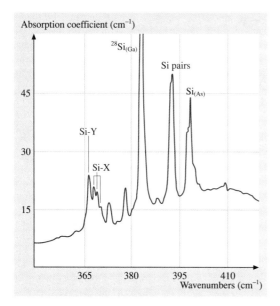

Fig. 23.17 An LVM absorption spectrum of GaAs doped with Si at a concentration of around 5×10^{18} cm^{-3}. Lines due to Si$_{Ga}$ donors are present at 384, 379 and 373 cm^{-1}, corresponding to vibrations of ^{28}Si, ^{29}Si and ^{30}Si on Ga lattice sites, respectively. Lines from Si$_{As}$ shallow acceptors, [^{28}Si$_{Ga}$ − ^{28}Si$_{As}$] nearest-neighbour pairs and complexes labelled Si−X and Si−Y, seen only at high Si concentration, are also seen. (After [23.72])

the maximum optical-phonon frequency (Restrahlen) and lose this energy only slowly to the lattice. This gives them a long lifetime and a narrow bandwidth, resulting in sharp absorption bands which fall in the mid- to far-infrared part of the spectrum at 4−30 μm. Because vibrational frequencies and their absorption strengths depend on the atomic weights of the atoms and their isotopic abundances, respectively, LVM measurements can reveal the atom type unambiguously in many circumstances [23.72].

Measurement of these narrow absorption bands, often employing a Fourier-transform infrared (FTIR) spectrophotometer and low sample temperatures, provides sensitive information not only of concentrations but also of the atomic environments of the impurity atoms.

Unfortunately, absorption from free carriers is very strong in this spectral region and this must be reduced, usually by irradiation of the sample with high-energy electrons. At suitably low fluences, this produces deep levels that compensate the material, apparently without otherwise affecting the impurities. However, this makes the technique difficult for routine measurements. SI GaAs does not need this treatment making it the method of choice for the measurement of carbon concentration (Sect. 23.5).

A LVM absorption spectrum of GaAs containing a high concentration of Si is shown in Fig. 23.17.

The importance of oxygen in commercial GaAs has been questioned for many years. However, the maximum concentrations detected in bulk material are usually below the sensitivity limit for chemical techniques or LVM absorption. The only reports of sizeable concentrations of O in LEC GaAs were based on few, carefully prepared crystals [23.94].

Other methods of chemical analysis are either in disuse or out of the scope of this short section and will not be discussed here.

23.7 Impurity and Defect Analysis of GaAs (Electrical)

23.7.1 Introduction to Electrical Analysis of Defects in GaAs

One of the simplest ways of assessing a sample of GaAs is a measurement of carrier concentration and mobility. This is easily achieved in conducting material by Hall effect analysis. Applications of the Hall effect to GaAs have been discussed in detail by *Look* in [23.95]. This rather simple measurement can give considerable information about concentrations of shallow and deeper electronic levels. However, its sensitivity to the latter is rather limited. Also measurements in SI GaAs are complicated by the difficulty in passing sufficient current through the material to achieve a measurable Hall voltage and the extremely high source impedance of the latter. Fortunately, there exist other highly sensitive electrical assessment techniques that can be used.

However, before these are introduced the importance of mapping needs to be emphasised. It is clearly important to reveal changes of electrical parameters across a wafer and changes that may occur from wafer to wafer. These measurements must be performed rapidly and preferably without contact to the GaAs wafer. Such measurements are often limited to resistivity and assume that this parameter does not vary with depth. For conducting materials, microwave eddy-current loss measurements can be made very rapidly with a resolution of a few mm. These mappers have been available commercially for many years. For SI GaAs the absorption of microwave radiation is very small and another technique, time-domain charge measurement (TDCM), has been developed [23.96]. This method, which measures the dielectric relaxation time in order to give resistivity data, can have a spatial resolution of a few tens of μm [23.97].

Shallow-Level Defects

In p-type GaAs shallow acceptors have an appreciable ionisation energy and in nondegenerate material the resulting free holes can often be frozen out by cooling to temperatures close to 4.2 K. The Hall effect can be very useful here. An Arrhenius plot extracted from carrier concentration measurements taken as a function of temperature can reveal this activation energy and, therefore, the acceptor. The presence of more than one acceptor makes the analysis more difficult; if compensating donors are present they are not revealed directly but mobility data can indicate their concentration.

This method does not work with n-type material because the ionisation energies of shallow donors are very similar and, although total donors concentrations can be established, identification is impossible. Once again, it is assumed that no acceptors act as compensating defects.

However, a more complex technique, photothermal ionisation spectroscopy (PTIS) can identify donors [23.99]. This is based on the fact that the energy, ΔE, to excite the bound s electron to its first excited 2p state depends on the donor type. In PTIS, the sample, with Ohmic contacts, is cooled to temperatures near 10 K. Its conductivity is measured under illumination with light from a monochromator. When the photon energy is equal to ΔE, the 1s electron is excited to the 2p state. It is now ionised very rapidly by thermal excitation to the conduction band where it can be detected as a photocurrent. PTIS is very sensitive (less than 10^{14} cm^{-3} has been cited) but it fails for concentrations much higher than 10^{16} cm^{-3} because overlap of the electron wavefunctions causes broadening of the 1s–2p transitions. Extra information can be obtained by application of a magnetic field, stress or other perturbations to the crystal and these are discussed in [23.100].

Deep-Level Defects

Deep levels are produced by many impurities and intrinsic point defects. In many ways, these are more insidious than shallow levels as their presence is often only detected when fabricated devices fail to perform adequately. Deep levels can result in unexpected noise, parameter drifting, electrical hysteresis, poor light output from optoelectronic devices, etc. These effects arise from carrier trapping and subsequent de-trapping at deep levels, and their role as recombination–generation (RG) centres. Detection and measurement depends on whether the GaAs is SI or conducting.

SI GaAs. Deep levels can trap carriers, produced optically by illumination with above-band-gap light, and then release them thermally. The released carriers, electrons or holes, can produce a current in an external circuit. The temperatures at which different traps release their carriers can be used to determine their ionisation energies. This is the basis of thermally stimulated current (TSC) spectroscopy, a technique that is especially useful for assessing high-resistivity materials such as SI GaAs. See [23.101] for discussions of this and other electrical assessment techniques. Two Ohmic contacts are placed on the sample. After optical filling of the traps at low temperature and at zero applied bias, the temperature is increased at a constant rate with the GaAs under bias and a sensitive current meter in series with the voltage supply. The current in the external circuit shows a series of peaks corresponding to trap emptying, the deeper traps requiring higher temperatures before they can con-

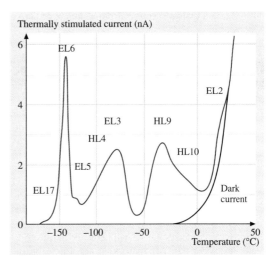

Fig. 23.18 TSC spectra recorded on a 350-μm-thick sample of SI GaAs. The traps have been filled by illumination with white light for 3 min at 80 K. The reverse bias is 25 V. (After [23.98])

tribute. This technique is sensitive (10^{13} cm^{-3}) and can be semiquantitative. It cannot, however, show whether a deep level is an electron or a hole trap, because the emptying of either will produce the same external current. A typical TSC spectrum from a sample of SI GaAs is shown in Fig. 23.18.

Conducting GaAs. Application of TSC to conducting semiconductors with Ohmic contacts is impracticable because of the high temperature-independent background current that would flow. However, the filling and emptying of traps is still a useful approach. If a rectifying junction can be formed, applying a forward bias can fill traps; the subsequent application of a reverse bias produces a depletion layer in which the traps can empty thermally. The emptying of each trap type produces a flow of charge that changes exponentially with time so that each current pulse is a sum of several exponentials with different time constants.

Although the current in an external circuit could be used to perform the subsequent measurement, it is easily shown that more information can be obtained from the change in depletion-layer capacitance because this can reveal whether electrons or holes are being de-trapped. If this capacitance change transient, also a sum of exponentials, is analysed by applying the *rate window* concept [23.103], the data can be presented as a spectrum of capacitance change in that window versus temperature. This is known as deep-level transient spectroscopy (DLTS) and a typical spectrum from a sample of GaAs is shown in Fig. 23.19.

DLTS is quantitative, giving defect concentrations, ionisation energies, capture cross sections and the identification of the type of trap (electron or hole). The labelling of the deep levels in Table 23.2 followed the early application of DLTS to GaAs.

DLTS equipment includes a sensitive and fast capacitance meter, usually furnished by Boonton, Inc., and a control box that applies the forward and reverse bias voltages as well as applying the rate window analyser to the output. The sample must be placed in a variable-temperature cryostat. As a rule, DLTS is controlled by computer. Often the fabrication of a p–n junction is inconvenient and a Schottky contact is used. Because minority carriers cannot be introduced into a semiconductor by this means, DLTS will only be sensitive to majority-carrier traps. If minority traps are to be investigated using the Schottky barrier method, minority

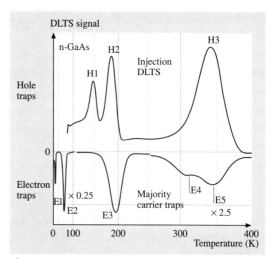

Fig. 23.19 Minority and majority traps introduced by high-energy electron irradiation of a sample of n-type GaAs after the fabrication of a p$^+$–n diode. After [23.102]. For further details [23.101]

carriers must be introduced using a pulse of above-band-gap light instead of forward-bias, optical DLTS (ODLTS), Fig. 23.19. The uses of the many variations of DLTS have been discussed and assessed by several workers [23.91, 95, 101].

The sensitivity of DLTS depends on the carrier concentration, which, for standard operation, must exceed the total deep-level concentration by more than an order of magnitude. Within this restriction, the sensitivity is about two orders of magnitude less than the carrier concentration, i. e. for undoped LPE GaAs with a carrier concentration of 10^{16} cm^{-3}, DLTS will be able to measure defects down to the 10^{14} cm^{-3} range.

Although DLTS is widely used, it has a serious disadvantage: each trap produces a wide peak and several traps of similar emission properties cannot be distinguished. So-called Laplace-transform DLTS has been developed to overcome this [23.104]. A rate window is no longer used but the transient is analysed directly in terms of its exponential components by applying an inverse Laplace transform. However, this is demanding in terms of temperature control and the linearity of capacitance measurement. Older techniques such as photo-capacitance spectroscopy are beyond the scope of this section but are reviewed in the books mentioned above.

23.8 Impurity and Defect Analysis of GaAs (Optical)

23.8.1 Optical Analysis of Defects in GaAs

For obvious reasons, analytical techniques that do not require electrical contacts to be made to the sample, or are otherwise nondestructive, are attractive to device manufacturers. Many of these are optical in nature. The use and limitations of LVM absorption has been addressed in Sect. 23.6 and that for measuring EL2 concentrations was described in Sect. 23.1.3. Other infrared absorption techniques are of limited interest; we concentrate on light emission, or luminescence, methods.

Photoluminescence (PL)

Luminescence in semiconductors results from the radiative recombination of excess carriers. With the exception of direct recombination via free excitons, luminescence takes place via the mediation of defects [23.105, 106]. The energies of luminescence bands give information of the defects, the intensities are related (albeit in a rather indirect way) to concentrations and the total luminescence intensity is related to concentrations of parallel, nonradiative recombination paths. If the excess carriers are produced optically, often by illumination of the sample with a laser beam, the technique is photoluminescence. Excitation using a beam of high-energy electrons in a scanning electron microscope (SEM) results in cathodoluminescence (CL). Although the scanning ability of the SEM allows CL to map the luminescence of the material, CL is rarely applied to GaAs.

In either method, excess carriers are produced within the top few μm of the surface. However, this is not the only volume that is probed by the technique, as these carriers will diffuse about one diffusion length, L_D, before they recombine. L_D will be greater in material where the minority carrier lifetime is greater; in general, this will be associated with better, more luminescent material. It follows that there are restrictions on the spatial information that can be obtained, better resolution being obtained with material of lower L_D. Moreover, excess carriers that diffuse to the surface are lost to the measurement as they recombine there without emission of light. This process is quantified by a *surface recombination velocity*, which increases strongly as the temperature is raised. PL and CL are usually performed at low sample temperature because this reduces thermal broadening of the emission lines and reduces the diffusion of excess carriers to the sample surface. The emitted light is passed through a high-resolution spectrometer before detection by either a suitable photomultiplier tube or semiconductor (usually Si or Ge) detector. A Ge detector must be cooled, although cooling of all detectors generally results in less noise.

Low-Temperature Luminescence from Point Defects

Low-temperature luminescence from shallow acceptors often dominates the spectra of SI GaAs, the energy of the lines being approximated by $E_G - E_A$, where E_A is the energy of the acceptor relative to the top of the valence band. Because E_A depends on the acceptor species, the presence of a particular acceptor can be determined. Similar luminescence from donors is less easy to differentiate as they have similar values of E_D (where E_D is the energy level relative to the bottom of the conduction band). Luminescence from intrinsic deep levels is weak because recombination is mostly nonradiative. However, that from EL2 centres can be resolved both at low and at room temperature [23.98, 107]. The situation from impurity levels, especially transition metals, usually resulting from contamination is different with strong luminescence resulting from many. Of interest is the line due to Cu at 1.05 eV, which is particularly strong and well resolved and allows this common contaminant to be detected. The PL spectra from GaAs depend very sensitively on the method of growth, subsequent treatments and the presence of impurities. Readers interested in this technique are directed towards the relevant chapters of [23.1].

Room-Temperature Luminescence Measurements

The broadening effects of temperature and the increased influence of the surface greatly reduces the use of room-temperature luminescence as an analytical tool. However, mapping at room temperature is often attractive because of its relative ease compared to low-temperature studies. It is then a truly nondestructive method giving spatial information of surface properties (that control the surface recombination velocity) and it is often used for assessing substrates and the uniformity and quality of quantum wells (QW). A review of these techniques has been given by [23.108].

Mapping of Surface Properties. The surface quality of GaAs substrates is critical to successful epi-growth. Manufacturers perform special cleaning and oxidation procedures to ensure reproducible surface properties, those being suitable for immediate epi-growth being supplied as *epitaxial-ready*. A valuable check on these properties is scanning PL as changes to the surface result in changes to the surface recombination velocity. Of course, this assumes that the material is well annealed and that residual nonuniformities in point-defect concentrations do not otherwise affect the recombination kinetics, a situation that is never fully achieved. There is no need for a spectrometer in this measurement as the total PL efficiency is altered by the surface and, in any case, PL line widths are thermally broadened. Collection of all the luminescence means that a relatively high signal is detected, with a concomitant increase in the measurement rate. Small-scale structures are often associated with residual volume nonuniformities (see, for example, [23.109]) whilst large-scale effects are most often due to polishing and subsequent contamination of the surface [23.110].

Mapping of Quantum-Well Properties. Quantum wells act as very efficient collectors of excess carriers and are, in general, highly efficient radiative centers. This is the reason for their use in lasers, LEDs, etc. The reproducibility of QW characteristics, especially in terms of QW widths and the quality of the barriers, across large substrates and from wafer to wafer is a key to successful device manufacture and a challenge to all epitaxial growth techniques. Two-dimensional mapping of room-temperature PL using a spectrometer to resolve the spectrum is used to assess the uniformity of these properties. Although the time to map a wafer depends on the spatial and spectral resolution that is required, the efficient radiative recombination at QWs and the bright luminescence makes this a relatively rapid process [23.111].

23.9 Assessment of Complex Heterostructures

Most electrically active GaAs, as opposed to substrate material, is used in complex device structures consisting of several layers. In many optical and microwave devices these layers will be of GaAs and $Ga_{1-x}Al_xAs$, where x may take several values and the carrier concentration and thickness of each layer is critical.

23.9.1 Carrier Concentration Measurements in Heterostructures

Although Hall-effect measurements accompanied by a process where the surface is gradually removed by chemical dissolution can be used to assess material of more than a single layer, more complex structures normally use capacitance–voltage (CV) profiling. This method is sensitive and has good thickness resolution. However, it does not give carrier mobility information. Some point-defect information can be derived from the frequency dependence of the data, although expert analysis is required. CV measurements are normally a part of DLTS instrumentation and suitable software is then available to analyse the data see [23.93, 101]. In some situations, where individual layer thicknesses exceed a micron or so, the wafer can be cleaved (normally to reveal a ⟨110⟩ surface) and etched in a solution that stains the layers according to doping type and material. Such a cleaved and stained surface can be viewed in an optical microscope and probed electrically for rapid assessment [23.93].

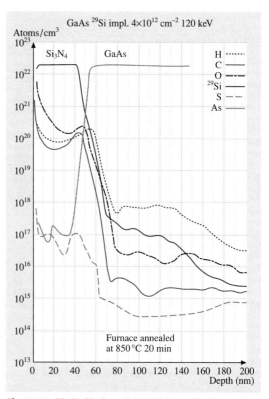

Fig. 23.20 H, O, Si, S, and As depth profiles in reactive-ion-beam sputtered Si_3N_4 film on GaAs as determined by SIMS. (After [23.112])

Fig. 23.21a,b Experimental and theoretical rocking curves from a 165-Å-thick layer of $In_{0.18}Ga_{0.82}As$ on GaAs, **(a)** at the wafer centre and **(b)** near the wafer periphery. The *vertical scales* are logarithmic. Note the complexity of these curves for this relatively simple system

23.9.2 Layer Thickness and Composition Measurements

Layer thickness and composition are normally measured together. As an example, dynamic SIMS, where the atomic composition and layer thickness is measured as the ablation of the surface takes place, gives this information effectively, at least for the first few microns of depth. At greater depths, the flatness of the floor of the ablated material degrades. Except that SIMS is destructive, it is often the method of choice [23.93]. Spatial mapping is possible in principle but the increase in machine time makes this prohibitively expensive as a routine procedure. A SIMS profile of a Si_3N_4 film on GaAs, demonstrating the depth resolution and large sensitivity range, is shown in Fig. 23.20.

A nondestructive approach is that of X-ray diffraction, normally double-crystal X-ray diffraction (DXD). X-rays are emitted from a tube and then *conditioned* by diffraction from a first crystal of high-structural-quality GaAs. The radiation is then diffracted by the sample, which can be tilted to move it in and out of the diffraction condition. A scintillation counter or semiconductor detector detects the diffracted X-rays.

Only the symmetrically diffracted beam is normally used for routine assessment, although other diffraction spots can be more sensitive to the surface structure as the penetration of the beam can be less in these cases. The diffracted energy as a function of tilt is presented as a *rocking curve*, which reveals the small dispersion of the total diffracted energy resulting from lattice strain and other effects. Penetration of the X-rays into the top few microns of a complex structure results in a more complicated rocking curve. However, in virtually all cases, most of this diffracted energy comes from the substrate and diffraction from the epilayers results in small additions to the wings of the rocking curve. It follows that the curve has a large dynamic range of several orders of magnitude and that it can take many hours to obtain the entire curve. Analysis of the rocking curve requires

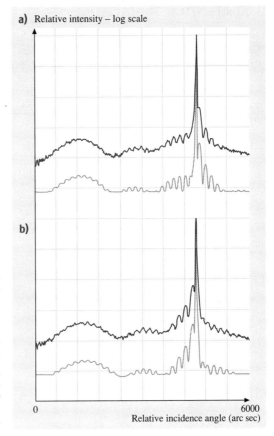

considerable computation so that the entire process is time-consuming. Often the rocking curve is compared to a computer simulation so that the calculation has to be performed only once. For reviews of these techniques, see [23.113, 114].

Spatial mapping of the rocking curves from substrates to map changes in lattice constant and residual strain is possible, as the rocking curves are simple. However, mapping from complex structures is extremely time-consuming. Figure 23.21 shows rocking curves from a 165-Å-thick layer of $In_{0.18}Ga_{0.82}As$ on GaAs, (a) at the wafer centre and (b) near the wafer periphery. The lower curve of each set of data is a theoretical fit to the experimental curve above.

23.10 Electrical Contacts to GaAs

It is clearly necessary to make electrical contacts to a GaAs device. These are of two types, those with a small internal barrier voltage and which do not affect the passage of a current, *Ohmic* contacts, and those whose internal electrical fields result in a sizeable potential difference and which modify the current flow, non-Ohmic or Schottky contacts.

23.10.1 Ohmic Contacts

A true Ohmic contact would exploit the constancy of the Fermi energies in the semiconductor and the metal. Such a situation rarely occurs and the true nature of practical Ohmic contacts often remains unclear. As an example, the most commonly used Ohmic contact to n-type GaAs is the eutectic alloy of Au–12% Ge. This is evaporated onto the surface and then heated to about 400 °C to produce the required electrical properties. It is believed that the heat treatment results in the in-diffusion of Ge to produce an extremely thin n^+ region, although this may not be the only mechanism; electron microscopy shows that complex intermetallic reactions between the Au–As and Ge take place. If Ni is added to the Au–Ge alloy to produce a ternary, a greatly improved contact results, certainly as a result of the Ni reacting with the GaAs surface to produce intermetallic phases with As and possibly by removing the oxide layer. Other contact materials are sometimes used and many of these have been reviewed in [23.95].

Ohmic contacts to p-type GaAs often use evaporated Au–Zn alloys followed by a heat treatment similar to that used for n-type contacts. The in-diffusion of Zn atoms clearly takes place but the removal of oxides on the GaAs surface by the Zn is also important.

23.10.2 Schottky Contacts

The barrier height between most metals and n-type GaAs is found to be nearly constant at 0.6–0.7 V. This high and constant value is a result of the Fermi energy at the surface being controlled by surface electronic states rather than the properties of the metal. Titanium is the most important Schottky contact. It is usually introduced via a TiPtAu alloy. This exploits the high sticking coefficient of Ti to GaAs whilst the Pt acts as a diffusion barrier to the low-electrical-resistance Au metallisation that would otherwise enter the GaAs. TiAl is sometimes a preferred alternative.

Interestingly, Al also produces a Schottky barrier on p-type GaAs with a barrier height of about 0.6 V. Gold was used as a Schottky contact on n-type GaAs with a barrier height of about 0.9 V but, as discussed above, this is degraded by in-diffusion at high temperatures. Contacts used specifically for high-temperature use include some silicides, with WSi_x being commonly used in some microwave devices. Metals for Schottky contacts have been reviewed in [23.115].

23.11 Devices Based on GaAs (Microwave)

23.11.1 The Gunn Diode

The occurrence of secondary conduction-band minima in the E–k diagram of GaAs leads to the possibility of high-energy electrons being excited from the Γ point where they have low effective mass to the X points where their mass is much greater. This is the transferred-electron effect. Although on applying a greater electric field their kinetic energy is increased, their increased mass leads to a fall in velocity. As a result, the current falls. This appears as a negative differential resistance (NDR) region in the current–voltage characteristics [23.116–118]. If an applied bias keeps the current flow in the NDR region, it can be shown that the electron flow will break up into pulses and the current in the external circuit will also be pulsed. Under certain circumstances, this results in the production of power at a particular microwave frequency determined mainly by the thickness of the GaAs [23.119]. The device optimised for microwave power production is called the *Gunn diode*, after J. B. Gunn who first observed this effect experimentally. This device is shown in Fig. 23.22.

The device is very simple, consisting solely of a few microns thickness of high-purity n-type ($\approx 10^{15}$ cm^{-3}) GaAs with more highly doped contacting regions to improve connection to the outside world [23.120]. The Gunn diode can produce microwave emission at frequencies up to 10 GHz and with powers of a few Watts. These devices have uses as portable microwave sources in radar speed detectors, for example.

23.11.2 The Metal–Semiconductor Field-Effect Transistor (MESFET)

As discussed in the introduction, the increased mobility of electrons in GaAs over those in Si seem to offer the prospect of faster devices, especially improved microwave transistors. This turns out to be based on an over simplistic interpretation as to how these devices really work. However, consider the first, three-terminal GaAs microwave device, the metal–semiconductor field-effect transistor (MESFET) whose structure is presented in Fig. 23.23.

The MESFET is similar to the Junction Field Effect (JFET) in Si technology with the p–n junction gate being replaced by a simple Schottky gate. Ti-based alloys are normally used for the Schottky contact as these adhere well to the surface. Application of a voltage to the gate modulates the width of the conducting channel and hence its conductivity. The channel is connected to two highly doped contact regions where ohmic contacts are made. These are the source and drain.

Initially all MESFETs were made by donor ion implantation into SI GaAs substrates using the methods described in Sect. 23.4. However, more recent methods including epitaxial growth by MOCVD and MBE result in higher effective doping levels, more abrupt interfaces, especially at the rear of the channel where a rapid change in carrier concentration is critical and the possibility of depositing alloys such as GaAlAs. Ion implantation is still used throughout the industry, however, especially for digital circuits.

In operation, the gate bias causes the channel to be pinched off. The drain–source voltage produces an extremely high field near the pinched-off channel and the channel current is saturated. Figure 23.24 shows that under these circumstances the electron velocity is expected to be saturated at $\approx 5 \times 10^6$ cm/s, a result of velocity saturation and the transferred-electron effect, the latter producing the reduction in velocity at higher fields.

This value is actually *less* than the saturated electron velocity in Si at similar electric fields and it is clear that the advantages of low electron effective mass at low fields in GaAs are not important. However, microwave measurements show an unexpected extension of high-frequency amplification especially with MESFETs with a small gate length. This is caused by the extremely short

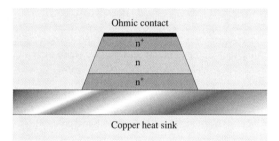

Fig. 23.22 The Gunn diode. The active region is the low-doped n-type GaAs in the centre. (After [23.120])

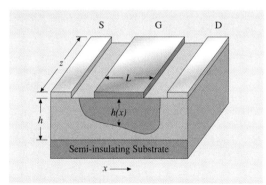

Fig. 23.23 The MESFET. Current flows only in the undepleted region. L is the gate length and determines the speed of the device; z is the gate width and controls the maximum current capabilities of the device; h is the channel thickness. (After [23.121])

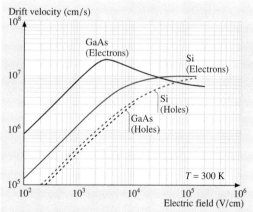

Fig. 23.24 Drift velocity of electrons and holes in Si and GaAs. At low electric fields the carrier velocities increase linearly with electric field (Ohm's law). However, at high fields saturation takes place. The reduction in velocity of electrons in n-type GaAs is a result of the transferred-electron effect. (After [23.4])

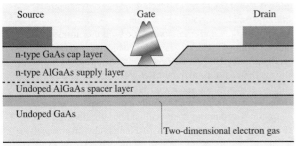

Fig. 23.25 The GaAs/AlGaAs HEMT. (After [23.125])

time that electrons are in the channel, which does not allow them to lose their kinetic energy to the lattice. This effect, known as *velocity overshoot*, allows electrons to reach a velocity near 2.5×10^7 cm/s in short-channel devices and is the first true reason for the high-frequency capability of the GaAs MESFET; the other is the use of a SI substrate. The latter reduces the parasitic capacitances that exist between the device and its contact pads and ground. The cut-off frequency of GaAs MESFETs is inversely proportional to the gate length and is 110 GHz for a gate length of 0.15 μm. The MESFET is discussed in detail in [23.122–124].

23.11.3 The High-Electron-Mobility Transistor (HEMT) or Modulation Doped FET (MODFET)

The channel of a GaAs MESFET must be doped with donor atoms to produce the necessary free carriers. Scattering by the positively charged donor ions results in a reduction in electron mobility and increased noise. In the HEMT the free carriers in the channel are produced by diffusion of electrons from an n-type layer of greater band-gap energy such as GaAlAs.

GaAs cap	40 nm	4×10^{18} cm^{-3}
Al$_{0.23}$Ga$_{0.77}$As	25 nm	2×10^{18} cm^{-3}
Al$_{0.23}$Ga$_{0.77}$As	4 nm	Undoped
In$_{0.2}$Ga$_{0.8}$As channel	12 nm	Undoped
GaAs buffer layer	0.5 μm	Undoped
Semi-insulating GaAs substrate		

Fig. 23.26 The layer structure of the AlGaAs/InGaAs pHEMT. (After [23.125])

A diagram of a GaAs/GaAlAs HEMT is shown in Fig. 23.25.

The GaAlAs supply layer of 50 nm thickness, doped n-type to a concentration of about 2×10^{18} cm^{-3}, supplies the free electrons. Alternatively, this layer can be delta-doped with a suitable donor. The undoped GaAlAs layer of about 2 nm thickness reduces Coulombic effects between the donors in the GaAlAs and the electrons in the channel which would otherwise reduce their mobility. The thickness of the undoped GaAs channel layer is not critical and is typically less than 1 μm. Advantages of this HEMT structure are best revealed at low temperature where effects due to phonon scattering are also minimised. However, even at 300 K this device is faster and less noisy than a comparable MESFET. This is because the real improvement in performance comes from the heterojunctions which act to constrain the carriers close to the gate with a resultant increase in high-frequency performance.

One disadvantage of the simple HEMT is the poor transfer of electrons from the GaAlAs to the GaAs channel resulting from the rather small conduction-band discontinuity at the interface. This results in a relatively low electron concentration in the channel and a large residual electron concentration in the supply layer, which acts as a slow GaAlAs MESFET in parallel with the HEMT. For this and other reasons, the simple GaAs/AlGaAs HEMT has been replaced by the pseudomorphic HEMT (pHEMT).

The pHEMT exploits the greater conduction-band discontinuity that exists between GaAlAs and InGaAs. Although there is a large lattice mismatch between GaAlAs and InGaAs, the latter can be grown as a thin, (12-nm) pseudomorphic layer on GaAs. The resulting structure is shown in Fig. 23.26.

Cut-off frequencies up to 300 GHz have been reported for these devices, which can also be designed for use as power devices. There is a strong commercial market for MMICs employing pHEMTs in many microwave applications. These include mobile phones, domestic satellite TV antennas and radar and space use. Again, [23.122–124] offer an excellent description of these device types.

23.11.4 The Heterojunction Bipolar Transistor (HBT)

The current gain of a bipolar homojunction transistor is optimised by having an emitter doping that is much higher than the base doping and a base width

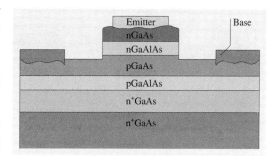

Fig. 23.27 Schematic diagram of a heterojunction GaAs/AlGaAs bipolar transistor. The collector contact is made to the rear of the device. (After [23.121])

that is as small as possible [23.121]. Unfortunately, high-frequency operation requires a high doping level in the base to reduce the base resistance. This must be accompanied, therefore, by very heavy emitter doping. However, such heavy doping leads to band-gap shrinkage, resulting in high base injection current and a reduction of the current gain. This argument is true both for npn and pnp homojunction bipolar transistors.

An elegant solution to this problem uses different materials in the device; for an npn device, a heavily doped, n-type, wide-band-gap emitter, a heavily doped p-type base and a lightly doped, n-type wide-band-gap collector. This is the HBT.

The base–emitter heterojunction offers a different barrier to hole injection from the base and electron injection from the emitter, resulting in efficient minority-carrier injection into the base [23.126, 127]. The base–collector junction need not be of different materials but a heterojunction may increase the breakdown voltage at the collector. The GaAs/GaAlAs system can be used for these devices, as shown in Fig. 23.27, although HBTs based on the alternative GaAs/GaInP system are superior and have a large market in applications like mobile phones.

Typically, HBTs have f_t values of around 40 GHz, although devices with $f_t > 160$ GHz have been reported. They can produce more power per unit area of device than a MESFET or MODFET. However, catastrophic breakdown can occur at the collector–base junction. High-temperature operation can result in rapid ageing effects, resulting from diffusion of acceptors from the base into the acceptor region. Because of its low diffusion coefficient the use of carbon as the acceptor in the base has reduced these problems. Readers are referred to [23.122, 124] for further reading. HBTs find extensive use in mobile telephony, where they are efficient microwave power sources.

23.12 Devices based on GaAs (Electro-optical)

The range of GaAs based electro-optical devices can be separated into emitters, modulators and detectors, the former being of much greater commercial interest than the others.

23.12.1 GaAs Emitters

Light-Emitting Diodes

The direct band gap of GaAs has made it popular for the fabrication of LEDs for several decades. The simplest LED is a p–n junction in GaAs, which in the early days, was fabricated by diffusing Zn into n-type material. The internal efficiency of this early device was poor at only a few percent but improvements in both material and processing showed that it could be surprisingly high, near 100%, even in such a simple structure. However, the external efficiency is always poor, with most of the emitted radiation being reflected back into the GaAs where eventually it is reabsorbed. This is because of the high refractive index of GaAs ($n \approx 3.5$), which, for a planar p–n junction, gives a maximum external efficiency from a single face of 2%. There are several approaches to extracting more light from the LED and these have been discussed by *Schubert* [23.128]. The use of lenses, antireflection coatings and epoxy domes are only three of the many commercial practices that improve the external efficiency of a simple GaAs p–n junction.

GaAs can be used as a basis for LEDs that cover the wavelength range from green (560 nm) through to IR wavelengths of over 1000 nm. Table 23.3 is a list of commercial LEDs based on GaAs.

The first entry [$(Al_xGa_{1-x})_{0.5}In_{0.5}P$] is a series of quaternary alloys, all of which have direct band gaps, and lattice-match to GaAs for any value of x. This gives a wide range of colours in the visible part of the optical spectrum. These alloys are grown by MOCVD and, although much more expensive than LEDs based on simpler materials, have the advantage of being very

Table 23.3 Light-emitting-diode materials and their properties. (After [23.125])

Active region	Gap	Substrate	Lattice match	Wavelength (nm)	Color
$(Al_xGa_{1-x})_{0.5}In_{0.5}P$	D	GaAs	Yes	560–640	Green, yellow, orange, red
$[Al_xGa_{1-x}As$–$GaAs]$ MQW ($x < 0.45$)	D	GaAs	Yes	630–870	Red to IR
$Al_xGa_{1-x}As$ ($x < 0.45$)	D	GaAs	Yes	630–870	Red to IR
$GaAs_{0.6}P_{0.4}$	D	GaAs	No	650	Red
GaAs	D	GaAs	Yes	870	IR
GaAs: Zn	D	GaAs	Yes	870–940	IR
GaAs doping SL	I-RS	GaAs	Yes	870–1000	IR
$Ga_{1-x}In_xAs$	D	GaAs	Yes	> 870	IR

MQW–Multiple quantum well; SL–superlattice; D–direct gap in k space; I-RS–indirect in real space; IR–infrared

bright and efficient. Light emission from the rear of the device is not possible because the GaAs substrate absorbs it. If a thick layer of GaP is grown as the top layer, a variation of this device can be fabricated where the entire GaAs substrate is removed by etching and the GaP becomes a new substrate. Because the

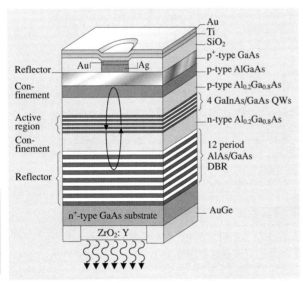

Fig. 23.28 The structure of a GaAs-based RCLED. The active region is a multiple-quantum-well structure. The lower reflector is a distributed Bragg reflector formed by growing a series of AlAs and GaAs layers so that each has an optical thickness of $\lambda/4$ giving near 100% reflectivity at the chosen wavelength. The *upper* reflector is the simple SiO_2 interface. (After [23.129])

$(Al_xGa_{1-x})_{0.5}In_{0.5}P$ is now the material with the lowest band-gap energy, all the light is available for emission from the device and the external efficiency is markedly improved.

The second entry represents multiple quantum wells (MQW) based on GaAs wells with GaAlAs walls and whose confined electron states depend on the well widths. The recombination of excess carriers in the wells efficiently produces light emission, the wavelength of which is engineered by modifying the well widths. These LEDs can emit red light efficiently.

The third entry is for bulk $Al_xGa_{1-x}As$ where the emitted light wavelength depends on the value of x. In both this and the previous case, x must be less than 45% because at higher fractions of Al the band gap becomes indirect and the efficiency for optical emission falls rapidly. Both these structures allow emission from 630–870 nm to be achieved.

The fourth entry has been mentioned in relation to grading in epitaxial growth. $GaAs_{1-x}P_x$ alloys have been the basis for low-cost visible LEDs for many years, those suitable for growing on n-type GaAs, resulting in red emission. Green- and yellow-emitting LEDs are based on GaP substrates. Grading the alloy composition from n-type GaAs until the required n-type $GaAs_{1-x}P_x$ alloy is grown produces the correct composition for emission. The next growth is of p-type material of the same composition to produce the p–n junction. The p-type contact is made to the top of this structure to complete the LED.

The next entry relates to the simple GaAs LED while the next demonstrates the use of heavy Zn doping which leads to band-gap narrowing and the emission of light of wavelengths longer than 870 nm. A similar effect has been found by using doping superlattices.

The final entry is an attempt to lower the bandgap energy of GaAs by the addition of In to make $Ga_{1-x}In_xAs$ alloys. These alloys are not lattice-matched to GaAs and the introduction of misfit dislocations at the growth interface results in nonradiative processes and a reduction in the efficiency.

Simple IR GaAs LEDs are mass-produced for use in infrared remote controls, optoisolators, emitters for low-cost fibre-optic networks, etc. High-efficiency visible LEDs have a very large market in portable and permanent displays. The former uses include mobile telephones and laptop computers. The second includes advertising and other public displays.

Highly efficient LED structures include the resonant-cavity LED (RCLED) where the emission takes place in a cavity tuned to the emission wavelength. The spontaneous emission along the axis of the resonant cavity is strongly enhanced and the increase in light output in this direction can be more than 10 times that of a more conventional LED. The structure of a RCLED is shown in Fig. 23.28. It should be compared to that of a vertical-cavity surface-emitting laser in the next section.

The recent work dedicated to light-emitting diodes by *Schubert* is recommended for further reading [23.128].

Lasers

The major difference between an LED and a laser is the existence of a resonant cavity and the confinement of the light in the cavity, which encourages stimulated rather than spontaneous emission in the laser. The use of GaAlAs is very important for the optical confinement as its lower refractive index produces total internal reflection along the axis of the laser. The optical cavity is often produced by simply cleaving parallel ⟨110⟩ facets on the ends of the cavity, the laser being fabricated on a ⟨001⟩ substrate. The high current density that is required to ensure population inversion in the laser is accomplished by careful engineering of the contact metallisation.

There are several types of GaAs laser structure but these can be separated into those where the lasing takes place in bulk material and those where this region is modified by using quantum wells.

Laser-diode structures have been reviewed in [23.130–132] and in many other general textbooks on semiconductors.

The GaAs/AlGaAs Heterojunction Laser. This type of laser is typical of those used in compact disc players, etc. The structure and operating conditions are shown in Fig. 23.29a and b.

Stimulated emission takes place in the GaAs active region, where excess carriers are injected from the surrounding p- and n-type regions. Different lasers have either GaAs or GaAlAs as these adjacent layers. We dis-

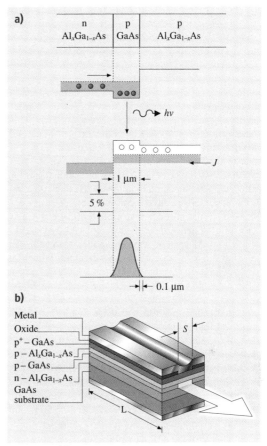

Fig. 23.29 (a) Band-gap diagram of a simple double heterojunction, GaAs/AlGaAs laser. The injected carriers are effectively trapped in the GaAs active region where they recombine to produce light. The increase of refractive index in the active region also constrains the light by total internal reflection. Thus both electrical and optical confinement is achieved. (b) Final structure of a GaAs/ AlGaAs *ridge* laser. The n^+-GaAs substrate and upper p^+-GaAs layers are used for the contacts as metallisation to AlGaAs is less satisfactory. The SiO$_2$ oxide layer provides further electrical confinement. The optical cavity is defined by the {110} end facets. (After [23.4])

cuss only those lasers with GaAlAs on either side of the active layer, the double heterojunction (DH) laser, as these are most commonly used.

Electrons and holes introduced from the adjacent GaAlAs layers are effectively confined to the GaAs active layer because of the conduction- and valence-band offsets that exist at the heterojunctions. However, this is not optimal for optical confinement and further cladding by $Ga_{0.4}Al_{0.6}As$ is often necessary. This modification results in the device known as the separate confinement heterojunction (SCH) laser. However it is achieved, GaAlAs cladding layers must always surround the active layer to produce optical confinement.

In some designs, the Al content of the GaAlAs cladding layers is graded with the Al content increasing linearly with the distance from the active layer. This graded refractive index (GRIN) structure has optical and electrical advantages and obviates the use of two separate confinement layers. The final structure is often called a GRINSCH laser. A layer of insulating SiO_2 with a thin axial opening to allow electrical connection to the upper p-type layer achieves the electrical confinement. Emission occurs at around 800 nm if GaAs is used in the active layer. If it is replaced by GaAlAs with a low Al content, the emission wavelength can be reduced. Laser structures were originally grown by LPE but large-scale production uses MOCVD or even MBE, although GRINSCH lasers are usually grown by MOCVD. There is a rather high dependence of emission wavelength on temperature with these simple lasers. The use of distributed optical feedback stabilises the output. Though outside the confines of this review, details of these lasers can be found in [23.134].

Although many millions of small GaAs/GaAlAs lasers (less than 20 mW output) are sold annually for domestic applications such as CD players, these lasers can be designed to produce powers exceeding 50 W and are used for pumping Er-doped quartz amplifier fibres for communications use. High-power lasers are not as rugged as smaller units and can suffer from catastrophic failure at the reflecting facets. Special treatments have to be used to prevent these failures.

The Red-Emitting, GaAs/GaAlAs Multi-Quantum-Well (MQW) Laser. The efficient radiative recombination that can occur in quantum wells has been exploited in the design of red lasers based on the GaAs/GaAlAs system. Attempts to produce a red-emitting laser using a conventional active layer fail because GaAlAs with the necessary Al content has an indirect band gap and, therefore, low emission efficiency. However, replacing

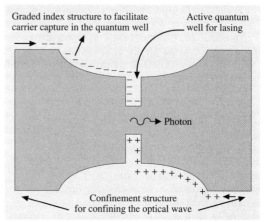

Fig. 23.30 The band structure of a GRINSCH single-QW laser. The large-band-gap material is $Al_{0.6}Ga_{0.4}As$. This is graded to $Al_{0.3}Ga_{0.7}As$ towards the centre of the device, providing electrical and optical confinement. The active region is the $Al_{0.3}Ga_{0.7}As/GaAs/Al_{0.3}Ga_{0.7}As$ quantum well at the centre. The emission wavelength depends on the width of this well, which is typically 1–4 nm. In a GRINSCH MQW laser there is a series of several identical quantum wells. Barriers of typically 5 nm thickness separate them. (After [23.133])

this layer with a series of quantum wells increases the energies of the trapped electrons and holes with respect

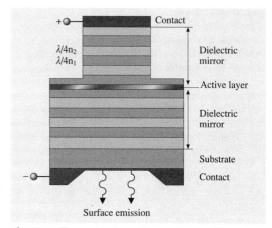

Fig. 23.31 The vertical cavity surface emitting laser (VCSEL). The optical cavity is vertical being defined by the two dielectric mirrors (Distributed Bragg Reflectors). The active region is a multi-quantum well structure at the centre of the device

to the conduction band and the valence band, respectively [23.133]. This effect is the same as used in the MQW LED. The structure of a red-emitting GRINSCH QW laser is shown in Fig. 23.30.

These lasers are used in digital versatile disc (DVD) players, visible pointing devices, light pens, and several other devices where a visible narrow light beam is important, and where a larger and more expensive He/Ne would have been used previously. The mass production of this complex structure has resulted in the price of these lasers being reduced to the single-dollar range. The MQWs in these lasers are very narrow, involving layers that are only a few lattice spacings wide. This is the regime where pseudomorphic growth is possible. Thus, the use of GaAs/GaInAs MQWs within GaAlAs cladding layers has produced lasers emitting in the range $0.85-1.0\,\mu$m.

The Vertical-Cavity Surface-Emitting Laser (VCSEL). The standard heterojunction laser suffers from several disadvantages from a fibre-optic systems point of view. It is difficult to coat the exit facets with antireflection coatings, it is very difficult to integrate this type of device with microwave amplifiers such as MESFETs and the output beam has a cylindrical cross section that is not optimised for injection into a fibre. The VCSEL attempts to overcome these problems (Fig. 23.31).

The light is amplified by a QW active region as only the QW has sufficient gain over such a thin layer to allow lasing. Even then, the use of simple cleaved mirrors would reduce the overall gain below that necessary for laser operation. Highly reflective multilayer dielectric reflectors (distributed Bragg reflectors often made from a series of GaAs/AlGaAs layers below and above the active layer) are necessary. However, the final device is truly planar, providing a beam of circular cross section. Output powers of several mW have been obtained [23.136] together with integration within a MESFET integrated circuit [23.137]. Advanced semiconductor diode lasers and their incorporation within integrated circuits are thoroughly described in [23.137, 138].

23.12.2 GaAs Modulators

It is sometimes necessary to modulate the intensity of a light signal electrically after it has been produced. The near-band-edge absorption of quantum wells can be moved to longer wavelengths by the application of a reverse bias. This is the quantum-confined Stark effect [23.139] and this is the basis of the MQW modulator. The modulator structure and the change of near-band-edge absorption with applied voltage are shown in Fig. 23.32.

This device can operate at frequencies up to 40 GHz and has spawned a series of more-advanced devices that are discussed in [23.135].

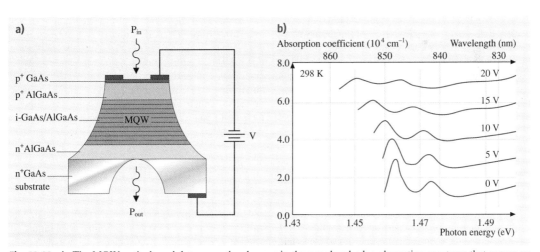

Fig. 23.32a,b The MQW optical modulator uses the changes in the near-band-edge absorption spectrum that occurs on the application of an electric field. The device is shown in (**a**) where it can be noted that the layer structure is very similar to a MQW LED. It is, however, operated under reverse bias. (**b**) Changes in absorption that occur in a MQW when a bias is applied. (After [23.135])

23.12.3 GaAs Photodetectors

Solar Cells

As a bulk material GaAs has found great use in solar cells. There are many reasons for this.

1. The band-gap energy is well matched to the solar spectrum and the high optical absorption above the band edge means that most light is absorbed in a thin layer.
2. In extraterrestrial applications, where large numbers of high-energy atomic particles exist, GaAs solar panels damage less easily than Si panels and therefore retain efficiency for longer.
3. In terrestrial use the higher band-gap energy allows GaAs solar cells to operate in concentrator systems (at the focus of a large Fresnel lens which is mechanically aimed at the Sun) where high temperatures are inevitable and Si devices would lose efficiency.

The simplest structure of this device is a simple p–n junction of high-quality GaAs. It is normally covered with a GaAlAs top layer to passivate the surface and to reduce reflection losses. These devices can have an efficiency of over 20% [23.140] Structures that are more complex involve up to three solar cells grown in series. In a typical two-cell structure, the light sequentially passes through a GaAs solar cell and then a similar GaSb device. Much of the sub-band-gap light that is transmitted by the GaAs is absorbed in the GaSb solar cell ($E_G = 0.72$ eV). Because these solar cells are effectively in series with reducing band-gap energies, more of the sunlight is absorbed than if a single cell were used. The efficiency of these devices can reach 30%. Research and development of multi-junction structures, often using Ge as a substrate and with different layers including GaAs, is very active. Because of their size, especially for space application, solar cells represent a sizeable part of the overall market for GaAs.

Other Photodetectors

The large band-gap energy makes GaAs unsuitable for most optical detector applications; the light generated by GaAs emitters being detected by Si or low-band-gap ternary detectors. However, the ionisation of electrons from bound QW states into continuum states can result in tunable mid- and far-IR detectors. The GaAs/GaAlAs system has been used in this way to make imaging arrays for wavelengths up to 20 µm, see for example [23.141].

23.13 Other Uses for GaAs

There have been attempts to exploit GaAs in other fields. High-energy particle detectors, including those for imaging X-rays, measure the ionisation produced as the particle traverses the detector material. The high stopping power, the large band-gap energy, which reduces leakage current, and the increased radiation hardness of GaAs in comparison with Si would appear to make it suitable for detectors for both scientific and medical use. The ready availability of SI GaAs wafers has encouraged this work. Unfortunately, these detectors suffer from effects resulting from the presence of deep levels, especially EL2 [23.142]. One application that has shown promise is in high-temperature sensors, amplifiers and processors in automotive, aeronautical and other uses. This exploits the high band-gap energy of GaAs, which allows devices to work at temperatures up to 300 °C. An excellent review has been given in [23.143].

23.14 Conclusions

GaAs technology is now mature with devices occupying important niches in the semiconductor marketplace. I have presented the fundamental properties of GaAs and have demonstrated how its unique set of characteristics has allowed it to be exploited in several microwave and optoelectronic applications. Future developments will probably concentrate on further integration of known devices and improvements to speed, power and reliability rather than the development of novel ones.

References

23.1 M. R. Brozel, G. E. Stillman: *Properties of Gallium Arsenide*, 3rd edn. (INSPEC, London 1996)
23.2 W. Martienssen: *Landolt-Börnstein*, New Series Group III/ 41A1α, 41Aα (Springer, Berlin Heidelberg New York 2001, 2002)
23.3 W. Martienssen, H. Warlimont: *Springer Handbook of Condensed Matter and Materials Data* (Springer, Berlin Heidelberg New York 2005) Chap. 4.1, p. 621
23.4 S. M. Sze: *Semiconductor Devices, Physics and Technology* (Wiley, Chichester 1985)
23.5 W. Koster, B. Thoma: Z. Met. Kd. **46**, 291 (1995)
23.6 D. T. J. Hurle: A comprehensive thermodynamic analysis of native point defect and dopant solubilities in gallium arsenide, J. Appl. Phys. **85**(10), 6957–7022 (1999)
23.7 K. R. Elliott: Appl. Phys. Lett. **42**(3), 274–276 (1983)
23.8 K. Kurusu, Y. Suzuki, H. Takami: J. Electrochem. Soc. **136**, 1450–1452 (1989)
23.9 K. Terashima et al.: Jpn. J. Appl. Phys. **79**, 463–468 (1984)
23.10 L. R. Weisberg, F. D. Rosi, P. G. Herkart: *Properties of Elemental and Compound Semiconductors*, Vol. 5, ed. by H. C. Gatos (Interscience, New York 1960) pp. 25–67
23.11 J. M. Woodall: Electrochem. Technol. **2**, 167–169 (1964)
23.12 G. R. Cronin, R. W. Haisty: J. Electrochem. Soc. **111**, 874–877 (1964)
23.13 G. Martinez, A. M. Hennel, W. Szuszkiewicz, M. Balkanski, B. Clerjaud: Phys. Rev. B **23**, 3920 (1981)
23.14 T. P. Chen, T. S. Huang, L. J. Chen, Y. D. Gou: J. Cryst. Growth **106**, 367 (1990)
23.15 F. Moravec, B. Stepanek, P. Doubrava: Cryst. Res. Technol. **26**, 579–585 (1991)
23.16 E. P. A. Metz, R. C. Miller, R. Mazelsky: J. Appl. Phys. **33**, 2016 (1962)
23.17 J. B. Mullin, R. J. Heritage, C. H. Holliday, B. W. Straughan: J. Cryst. Growth **3-4**, 281–285 (1968)
23.18 J. R. Oliver, R. D. Fairman, R. T. Chen: Electron. Lett. **17**, 839–841 (1981)
23.19 H. M. Hobgood, L. B. Ta, A. Rohatgi, G. W. Eldridge, R. N. Thomas: Residual impurities and defect levels in semi-insulating GaAs grown by liquid encapsulated Czochralski, Proc. Conf. Semi-Insulating III–V Materials, Evian, France 1982, ed. by S. Makram-Ebeid, B. Tuck (Shiva, Nantwich 1982) 30
23.20 D. E. Holmes, R. T. Chen, K. R. Elliott, C. G. Kirkpatrick: IEEE Trans. Electron. Dev. **29**, 1045 (1982)
23.21 D. E. Holmes, R. T. Chen, K. R. Elliott, C. G. Kirkpatrick: Appl. Phys. Lett **40**, 46–48 (1982)
23.22 K. Laithwaite, R. C. Newman, J. F. Angress, G. A. Gledhill: Inst. Phys. Conf. Ser. **33**, 133 (1977)
23.23 A. S. Jordan, R. Caruso, A. R. von Neida: Bell Syst. Tech. J. **59**, 593 (1980)
23.24 M. R. Brozel, J. B. Clegg, R. C. Newman: J. Phys. D **11**, 1331 (1978)
23.25 P. J. Doering, B. Friedenreich, R. J. Tobin, P. J. Pearah, J. P. Tower, R. M. Ware: Proc. 6th Conf. Semi-Insulating III–V Materials, Toronto, Canada 1990, ed. by A. G. Milnes, C. I. Miner (IOP, London 1990) 173–181
23.26 U. Lambert, G. Nagel, H. Rufer, E. Tomzig: Proc. 6th Conf. Semi-Insulating III–V Materials, Toronto, Canada 1990, ed. by A. G. Milnes, C. I. Miner (IOP, London 1990) 183–188
23.27 J. M. Baranowski, P. Trautman: *Properties of Gallium Arsenide*, 3 edn., ed. by M. R. Brozel, G. E. Stillman (INSPEC, London 1996) pp. 341–357
23.28 I. Grant, D. Rumsby, R. M. Ware, M. R. Brozel, B. Tuck: Proc. Conf. on Semi-Insulating III–V Materials, Evian, France 1982, ed. by S. Makram-Ebeid, B. Tuck (Shiva, Nantwich 1982) 98–106
23.29 M. S. Skolnick, M. R. Brozel, L. J. Reed, I. Grant, D. J. Stirland, R. M. Ware: J. Electron. Mater. **13**, 107–125 (1984)
23.30 A. G. Cullis, P. G. Augustus, D. J. Stirland: J. Appl. Phys. **51**, 2256 (1980)
23.31 D. Rumsby, I. Grant, M. R. Brozel, E. J. Foulkes, R. M. Ware: Electrical behaviour of annealed LEC GaAs, Proc. Conf. on Semi-Insulating III–V Materials, Kah-Nee-Ta, ed. by D. C. Look, J. S. Blakemore (Shiva, Nantwich 1984) 165–170
23.32 O. Oda: *Properties of Gallium Arsenide*, 3rd edn., ed. by M. R. Brozel, G. E. Stillman (INSPEC, London 1996) pp. 591–595
23.33 E. M. Monberg, H. Brown, C. E. Bormer: J Cryst. Growth **94**, 643–650 (1989)
23.34 E. D. Bourret, E. C. Merk: J. Cryst. Growth **110**, 395–404 (1991)
23.35 M. R. Brozel, I. R. Grant: Growth of gallium arsenide. In: *Bulk Crystal Growth*, ed. by P. Capper (Wiley, Chichester 2005) pp. 43–71
23.36 M. S. Tyagi: *Introduction to Semiconductor Materials and Devices* (Wiley, New York 1991)
23.37 M. G. Astles: *Liquid Phase Epitaxial growth of III–V Semiconductor Materials and their Device Applications* (IOP, Bristol 1990)
23.38 K. Somogyi: *Properties of Gallium Arsenide*, 3rd edn., ed. by M. R. Brozel, G. E. Stillman (INSPEC, London 1996) pp. 625–638
23.39 M. Razeghi: *The MOCVD Challenge* (Hilger, Bristol 1989)
23.40 G. B. Stringfellow: *Organometallic Vapor Phase Epitaxy: Theory and Practice* (Academic, Boston 1989)
23.41 K. F. Jensen: *Handbook of Crystal Growth*, Vol. 3b, ed. by D. T. J. Hurle (Elsevier Science, Amsterdam 1994) p. 3541

23.42 D. W. Kisker: *Handbook of Crystal Growth*, Vol. 3b, ed. by D. T. J. Hurle (Elsevier Science, Amsterdam 1994) p. 393
23.43 G. B. Stringfellow: *Handbook of Crystal Growth*, Vol. 3b, ed. by D. T. J. Hurle (Elsevier Science, Amsterdam 1994) p. 349
23.44 L. Samuelson, W. Seifert: *Handbook of Crystal Growth*, Vol. 3b, ed. by D. T. J. Hurle (Elsevier Science, Amsterdam 1994) p. 745
23.45 M. A. Herman, H. Sitter: *Molecular Beam Epitaxy: Fundamentals and Current Status* (Springer, Berlin Heidelberg New York 1989)
23.46 K. R. Evans: *Properties of Gallium Arsenide*, 3rd edn., ed. by M. R. Brozel, G. E. Stillman (INSPEC, London 1996) pp. 655–662
23.47 F. W. Smith, A. R. Calawa, C. L. Chen, M. J. Manfra, L. J. Mahoney: IEEE Electron. Dev. Lett. **9**, 77 (1988)
23.48 M. Missous: *Properties of Gallium Arsenide*, 3rd edn., ed. by M. R. Brozel, G. E. Stillman (INSPEC, London 1996) pp. 679–683
23.49 J. F. Whitaker: *Properties of Gallium Arsenide*, 3rd edn., ed. by M. R. Brozel, G. E. Stillman (INSPEC, London 1996) pp. 693–701
23.50 N. X. Nguyen, U. K. Mishra: *Properties of Gallium Arsenide*, 3rd edn., ed. by M. R. Brozel, G. E. Stillman (INSPEC, London 1996) pp. 689–692
23.51 C. R. Abernathy: *Properties of Gallium Arsenide*, 3rd edn., ed. by M. R. Brozel, G. E. Stillman (INSPEC, London 1996) pp. 663–671
23.52 J. W. Matthews, A. E. Blakeslee: J. Cryst. Growth **27**, 118 (1974)
23.53 B. Tuck: *Introduction to Diffusion in Semiconductors* (Peregrinus, Stevenage 1974)
23.54 B. Tuck: *Atomic Diffusion in III–V Semiconductors* (A. Hilger, Bristol 1988)
23.55 K. A. Khadim, B. Tuck: J. Mater. Sci. **7**, 68–74 (1972)
23.56 J. Crank: *The Mathematics of Diffusion* (Clarendon, Oxford 1975)
23.57 K. P. Roenker: Microelectron. Reliab. **35**, 713 (1995)
23.58 S. A. Stockman: *Properties of Gallium Arsenide*, 3rd edn., ed. by M. R. Brozel, G. E. Stillman (INSPEC, London 1996) pp. 101–116
23.59 T. Ahlgren, J. Likonen, J. Slotte, J. Raisanen, M. Rajatora: Phys. Rev. B **56**, 4597–4603 (1997)
23.60 M. R. Brozel, E. J. Foulkes, B. Tuck, N. K. Goswami, J. E. Whitehouse: J. Phys. D **16**, 1085–1092 (1983)
23.61 B. J. Sealy: *Properties of Gallium Arsenide*, 3rd edn., ed. by M. R. Brozel, G. E. Stillman (INSPEC, London 1996) pp. 765–782
23.62 M. S. Abrahams, C. J. Buiocchi: J. Appl. Phys. **36**, 2855 (1965)
23.63 R. T. Blunt, S. Clarke, D. J. Stirland: IEEE Trans. Electron. Dev. **29**, 1039 (1982)
23.64 A. R. Lang: Recent Applications of X-Ray Topography. In: *Modern Diffraction and Imaging Technique in Materials Science*, ed. by S. Amelinckx, G. Gevers, J. Van Landuyt (North Holland, Amsterdam 1978) pp. 407–479
23.65 G. T. Brown, C. A. Warwick: J. Electrochem. Soc. **133**, 2576 (1986)
23.66 P. Hirsch, A. Howie, R. B. Nicholson, D. W. Pashley, M. J. Whelan: *Electron Microscopy of Thin Crystals* (Krieger, Malabar 1977)
23.67 M. R. Brozel: *Properties of Gallium Arsenide*, 3rd edn., ed. by M. R. Brozel, G. E. Stillman (INSPEC, London 1996) p. 377
23.68 L. Breivik, M. R. Brozel, D. J. Stirland, S. Tuzemen: Semicond. Sci. Technol. **7**, A269–A274 (1992)
23.69 G. Jacob: Proc. Conf. on Semi-Insulating III–V Materials, Evian, France 1982, ed. by S. Makram-Ebeid, B. Tuck (Shiva, Nantwich 1982) 2
23.70 T. Ogawa, T. Kojima: Mater. Sci. Monogr. **44**, 207–214 (1987)
23.71 R. C. Newman: *Infra-Red Studies of Crystal Defects* (Taylor Francis, London 1973)
23.72 R. Murray: *Properties of Gallium Arsenide*, 3rd edn., ed. by M. R. Brozel, G. E. Stillman (INSPEC, London 1996) pp. 227–234
23.73 R. J. Wagner, J. J. Krebs, G. H. Strauss, A. M. White: Soild State Commun. **36**, 15 (1980)
23.74 G. M. Martin: Appl. Phys. **39**, 747 (1981)
23.75 M. Skowronski, J. Lagowski, H. C. Gatos: J. Appl. Phys. **59**, 2451 (1986)
23.76 J. Dabrowski, M. Scheffler: Phys. Rev. Lett. **60**, 2183 (1988)
23.77 T. J. Chadi, K. J. Chang: Phys. Rev. Lett. **60**, 2187 (1988)
23.78 M. R. Brozel, I. Grant, R. M. Ware, D. J. Stirland: Appl. Phys. Lett. **42**, 610–12 (1983)
23.79 M. S. Skolnick, L. J. Reed, A. D. Pitt: Appl. Phys. Lett. **44**, 447–449 (1984)
23.80 M. R. Brozel, M. S. Skolnick: Near band edge "Reverse Contrast" images in GaAs, Proc. Conf. Semi-Insulating III–V Materials, Hakone 1986, ed. by H. Kukimoto, S. Miyazawa (Shiva, Nantwich 1986) 109
23.81 C. Le Berre et al.: Appl. Phys. Lett. **66**, 2534 (1995)
23.82 M. R. Brozel, S. Tuzemen: Mater. Sci. Eng. (B) **28**, 130–133 (1994)
23.83 G. M. Martin, A. Mitonneau, A. Mircea: Electron. Lett. **13**, 191–193 (1977)
23.84 A. Mitonneau, G. M. Martin, A. Mircea: Electron. Lett. **13**, 666–668 (1977)
23.85 A. Mitonneau, A. Mircea, G. M. Martin, D. Pons: Revue de Phys. Appl. **14**, 853–861 (1979)
23.86 S. A. Stockman, A. W. Hanson, S. L. Jackson, J. E. Baker, G. E. Stillman: Appl. Phys. Letts. **62**, 1248 (1992)
23.87 N. Watanabe, T. Nittono, H. Ito: J. Cryst. Growth **145**, 929 (1994)
23.88 M. R. Melloch, N. Otsuka, J. M. Woodall, A. C. Warren, J. L. Freeouf: Appl. Phys. Lett. **57**, 1631 (1990)
23.89 Z. Liliental-Weber, A. Claverie, J. Washburn, F. Smith, A. R. Calawa: Appl. Phys. A **53**, 141 (1991)

23.90 A. C. Warren, J. M. Woodall, J. L. Freeouf, D. Grischkowsky, D. T. Melloch: Appl. Phys. Lett. **57**, 1331–1333 (1990)

23.91 D. E. Bliss, W. Walukiewicz, J. W. AgerIII, E. E. Haller, K. T. Chan, S. J. Tamigawa: Appl. Phys. **71**, 1699 (1992)

23.92 D. J. Keeble, M. T. Umlor, P. Asoka-Kumar, K. G. Lynn, P. W. Cooke: Appl. Phys. Lett. **63**, 87 (1993)

23.93 D. K. Schroder: *Semiconductor Material and Device Characterization*, 2nd edn. (Wiley Interscience, New York 1998)

23.94 Ch. H. Alt: Appl. Phys. Lett. **54**, 1445 (1989)

23.95 D. C. Look: *Electrical Characterization of GaAs materials and Devices*, Design Meas. Electron. Eng. Ser. 1989 (Wiley, Chichester 1989)

23.96 R. Stibal, J. Windscheif, W. Jantz: Semicond. Sci. Technol. **6**, 995–1001 (1991)

23.97 M. Wickert, R. Stibal, P. Hiesinger, W. Jantz, J. Wagner: High resolution EL2 and resistivity topography of SI GaAs wafers, Proc. SIMC-X, Berkeley, CA 1998, ed. by Z. Liliental-Weber, C. Miner (IEEE, 1999) 21–24

23.98 G. M. Martin, J. P. Farges, G. Jacob, J. P. Hallais, G. Poiblaud: J. Appl. Phys. **51**, 2840–2852 (1980)

23.99 G. E. Stillman, C. M. Wolfe, J. O. Dimmock: *Semicond. Semimet.*, Vol. 21 (Academic, New York 1977) p. 169

23.100 M. N. Afsar, K. J. Button, G. L. McCoy: Inst. Phys. Conf. Ser. **56**, 547–555 (1980)

23.101 P. Blood, J. W. Orton: *The Electrical Characterization of Semiconductors: Majority Carriers and Electron States* (Academic, London 1992)

23.102 D. V. Lang, L. C. Kimmerling: IOP Conf. Ser. **23**, 581 (1975)

23.103 D. V. Lang: J. Appl. Phys. **45**, 3023 (1974)

23.104 L. Dobaczewski, P. Kaczor, I. D. Hawkins, A. R. Peaker: J. Appl. Phys. **76**, 194 (1994)

23.105 H. B. Bebb, E. W. Williams: Photoluminescence I: Theory. In: *Semicond. Semimet.*, Vol. 8, ed. by R. K. Willardson, A. C. Beer (Academic, New York 1972) pp. 181–320

23.106 P. J. Dean: Prog. Cryst. Growth Charact. **5**, 89–174 (1982)

23.107 M. Tajima, T. Iino: Jpn. J. Appl. Phys. **28**, L841–844 (1989)

23.108 C. J. Miner, C. J. L. Moore: *Properties of Gallium Arsenide*, 3rd edn., ed. by M. R. Brozel, G. E. Stillman (INSPEC, London 1996) pp. 320–332

23.109 O. Oda, H. Yamamoto, M. Seiwa, G. Kano, T. Inoue, M. Mori, H. Shimakura, M. Oyake: Semicond. Sci. Technol. **7**, A215 (1992)

23.110 B. J. Skromme, C. J. Sandroff, E. Yablonovitch, T. Gmitter: Appl. Phys. Lett. **51**, 24 (1987)

23.111 C. J. Miner: Semicond. Sci. Technol. **7**, A10 (1992)

23.112 A. M. Huber, C. Grattepain: SIMS Analysis of III–V compound microelectronic materials. In: *Analysis of Microelectronic Materials and Devices*, ed. by M. Grasserbauer, H. W. Werner (Wiley, New York 1991) p. 305

23.113 B. K. Tanner, D. K. Bowen: *Characterization of Crystal Growth Defects by X-Ray Methods* (Plenum, New York 1980)

23.114 B. K. Tanner: *X-Ray Topography and Precision Diffractometry of Semiconductor Materials*, ed. by T. J. Shaffner, D. K. Schroder (Electrochem. Soc., Pennington 1988) pp. 133–149

23.115 S. P. Kwok: J. Vac. Sci. Tech. B **4**, 6 (1986)

23.116 B. K. Ridley, T. B. Watkins: Proc. Phys. Soc. **78**, 293 (1961)

23.117 C. Hilsum: Proc. IRE **50**, 185 (1962)

23.118 J. B. Gunn: Solid State Commun. **1**, 88 (1963)

23.119 M. P. Shaw: *The Physics and Instabilities of Solid State Electron Devices* (Kluwer Academic/Plenum, Dordrecht 1992) pp. 830–835

23.120 C. G. Discus et al.: *Properties of Gallium Arsenide*, 3rd edn., ed. by M. R. Brozel, G. E. Stillman (INSPEC, London 1996)

23.121 J. Singh: *Physics of Semiconductors and Their Heterostructures* (McGraw-Hill, New York 1994)

23.122 Y. Chang, F. Kai: *GaAs High-Speed Devices* (Wiley, New York 1994)

23.123 J. M. Golio: *Microwave Metal Semiconductor Field Effect Transistors and High Electron Mobility Transistors* (Artech House, London 1991)

23.124 W. Liu: *Fundamentals of III–V Devices: HBTs, MESFETs and HFETs/HEMTs* (Wiley, New York 1999)

23.125 R. H. Wallis: *Properties of Gallium Arsenide*, 3rd edn., ed. by M. R. Brozel, G. E. Stillman (INSPEC, London 1996) pp. 811–819

23.126 W. Shockley: US Patent 2 569 347 (1951)

23.127 H. Kroemer: Proc. IRE **45**, 1535 (1957)

23.128 E. F. Schubert: *Light Emitting Diodes* (Cambridge Univ. Press, Cambridge 2003)

23.129 E. F. Schubert: *Properties of Gallium Arsenide*, 3rd edn., ed. by M. R. Brozel, G. E. Stillman (INSPEC, London 1996) pp. 874–886

23.130 E. Kapon (ed): *Semiconductor Lasers I. Fundamentals* (Academic, New York 1998)

23.131 T. Numai: *Fundamentals of Semiconductor Lasers* (Springer, Berlin Heidelberg New York 2004)

23.132 L. A. Coldren, S. W. Corzine: *Diode Lasers and Photonic Integrated Circuits* (Wiley Interscience, New York 1995)

23.133 R. M. Kolbas: *Properties of Gallium Arsenide*, 3rd edn., ed. by M. R. Brozel, G. E. Stillman (INSPEC, London 1996) pp. 887–905

23.134 G. Morthier, P. Vankwikelberge: *Handbook of Distributed Feedback Laser Diodes* (Artech House, London 1997)

23.135 P. K. Bhattacharya: *Properties of Gallium Arsenide*, 3rd edn., ed. by M. R. Brozel, G. E. Stillman (INSPEC, London 1996) pp. 861–873
23.136 T. E. Sale: *Vertical Cavity Surface Emitting Lasers*, Electron. El. Res. Stud. Optoelectronics S (Research Studies, 2003)
23.137 Y. J. Yang: Appl. Phys. Lett. **62**, 600–602 (1993)
23.138 L. A. Coldren, S. W. Corzine: *Diode Lasers and Photonic Integrated Circuits* (Wiley, New York 1995)
23.139 H. Yamamoto, M. Asada, Y. Suematsu: Electron. Lett. **21**, 579 (1985)
23.140 J. M. Woodall, H. J. Hovel: Appl. Phys. Lett. **30**, 492 (1977)
23.141 L. J. Kozlowski et al.: IEEE Trans. Electron. Dev. **38**, 1124 (1991)
23.142 C. M. Buttar: GaAs detectors and related compounds, Nucl. Inst. Phys. Res. A **395**, 1–8 (1997)
23.143 L. P. Sadwick, R. J. Hwu: *Properties of Gallium Arsenide*, 3 edn., ed. by M. R. Brozel, G. E. Stillman (INSPEC, London 1996) pp. 948–962

24. High-Temperature Electronic Materials: Silicon Carbide and Diamond

The physical and chemical properties of wide-band-gap semiconductors make these materials an ideal choice for device fabrication for applications in many different areas, e.g. light emitters, high-temperature and high-power electronics, high-power microwave devices, micro-electromechanical system (MEM) technology, and substrates for semiconductor preparation. These semiconductors have been recognized for several decades as being suitable for these applications, but until recently the low material quality has not allowed the fabrication of high-quality devices. In this chapter, we review the wide-band-gap semiconductors, silicon carbide and diamond.

Silicon carbide electronics is advancing from the research stage to commercial production. The commercial availability of single-crystal SiC substrates during the early 1990s gave rise to intense activity in the development of silicon carbide devices. The commercialization started with the release of blue light-emitting diode (LED). The recent release of high-power Schottky diodes

24.1	**Material Properties and Preparation**	540
	24.1.1 Silicon Carbide	540
	24.1.2 Diamond	544
24.2	**Electronic Devices**	547
	24.2.1 Silicon Carbide	547
	24.2.2 Diamond	551
24.3	**Summary**	557
	References	558

was a further demonstration of the progress made towards defect-free SiC substrates.

Diamond has superior physical and chemical properties. Silicon-carbide- and diamond-based electronics are at different stages of development. The preparation of high-quality single-crystal substrates of wafer size has allowed recent significant progress in the fabrication of several types of devices, and the development has reached many important milestones. However, high-temperature studies are still scarce, and diamond-based electronics is still in its infancy.

The electronic revolution of the 20th century is mainly based on silicon, which can be regarded as the first-generation semiconductor. Around the turn of the 21st century gallium arsenide and indium phosphide have evolved as second-generation semiconductors, constituting the base for the wireless and information revolution. Now at the start of the 21st century, the wide-band-gap semiconductors silicon carbide and gallium nitride are on the rise and may be regarded as third-generation semiconductors used in the electronic and optoelectronic industries. Moreover given diamond's superior properties and the recent surge of research on diamond preparation and fabrication of diamond-based electronic devices, one might speculate that diamond may be a future-generation semiconductor.

The effects of temperature on materials and devices have been of great interest throughout the history of semiconductor research. The aim has been to investigate the high-temperature limits of materials and to enhance high-temperature semiconductor device performance. The development of semiconductor devices for reliable operation for an extended period at high temperatures is a complex process in which a number of physical effects connected with increasing temperature [24.1, 2] have to be considered. The term high temperature is not defined in a unique way in the literature and has a different meaning depending on the semiconductor under consideration and the area of application of semiconductor devices. The definition of high temperature often cited in the literature is temperatures above 125 °C [24.2, 3], since 125 °C is frequently specified as the upper limit at which standard commercial silicon devices function properly, although tests on standard commercial components indicate that even 150 °C may be applicable for selected silicon components [24.3].

Silicon is still the dominant semiconductor and silicon devices are still being developed. The most common and cost-effective integrated circuit technology is now silicon complementary metal–oxide–semiconductor (CMOS), which is able to operate up to 200 °C. The silicon-on-insulator (SOI) technology extended the operational temperature of CMOS circuits to 300 °C [24.4–8]. In addition, devices based on gallium arsenide and related alloys, which are commercialized to a lesser degree than silicon, are also candidates for high-temperature operation beyond 300 °C. The short-term operation of GaAs devices at temperatures as high as 500 °C has been reported [24.9, 10].

A survey of the literature indicates that 300 °C can be regarded as a dividing point from several standpoints, e.g. packaging, wiring, connecting, etc. [24.1, 2, 4]. This temperature is approximately the maximum temperature at which low-power silicon or conventional gallium arsenide devices can function reliably. The intrinsic carrier concentration for several semiconductors as a function of temperature is shown in Fig. 24.1. The control of the free-carrier concentration is vital for the performance of all semiconductor devices. The intrinsic carrier concentration (n_i) is exponentially dependent on the temperature:

$$n_i = \sqrt{N_C N_V}\, e^{-E_g/2k_B T}, \quad (24.1)$$

where E_g is the band gap, k_B is Boltzmann's constant and T is the temperature in Kelvin. Evidently, at temperatures above 300 °C, SiC, GaN and diamond and AlN have much lower intrinsic carrier concentrations than Si and GaAs. This implies that devices designed for higher temperatures should be fabricated from wide-band-gap semiconductors to avoid the deteriorating effects of thermally generated carriers.

The wide-band-gap third-generation semiconductors, SiC and GaN (including the III–nitride systems e.g. AlGaN), have been recognized for over three decades as materials which are well suited for high-temperature electronics and for light emitters, but until recently, the low material quality has not allowed the production of high-quality devices. The availability of single-crystal SiC wafers at the start of the 1990s initiated a great deal of activity towards the development of SiC-based devices, and their commercialization started with the release of blue light-emitting diode (LED). The availability of commercial high-quality substrates meant that more research has been carried out on SiC than on the GaN and III–nitride systems. The SiC devices have the advantages of a more mature semiconductor material growth and device fabrication technology. Furthermore, GaN and III–nitride crystals have mostly been grown by heteroepitaxy on e.g. sapphire and SiC, since a viable GaN substrate technology does not exist. Unfortunately, GaN crystals always contain more defects than SiC and the current aim is to reduce the surface defect densities in GaN from current densities of the order 10^8 cm^{-2} to 10^5 cm^{-2}. The unavailability of low-defect-density substrates and defect-free material limits the ability to fabricate high-quality GaN devices. The discussion of wide-band-gap semiconductors must mention AlN since it has one of the largest band gaps (wurtzite: 6.23 eV, and zincblende: 6.0 eV [24.11]). The growth of defect-free AlH crystals (as with the GaN) is an outstanding issue.

The reduction of the defect density and the effects of specific defects of third-generation semiconductors are the most urgent current problems that must be solved. Diamond is a future-generation semiconductor which is at a different stage of research than the third-generation semiconductors, particularly SiC, which is far more developed then diamond.

The research in wide-band-gap semiconductors has been driven by the need for light emitters, high-temperature and high-power industrial applications, and microwave power applications. A variety of applications e.g. in aircraft and space systems, automotive electronics, deep-well drilling, energy production centers etc., would benefit from power devices that function at high temperatures [24.12–16]. When the ambient temperature is too high, the performance-enhancing electronics presently used beneficially to monitor and control crucial hot sections must reside in cooler areas, this is achieved

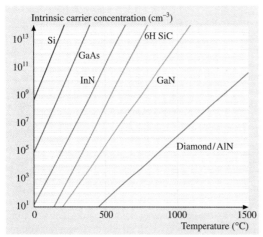

Fig. 24.1 Intrinsic carrier concentration as a function of temperature of several semiconductors. (After [24.9])

by their remote location or actively by cooling with air or liquids. These thermal management approaches introduce additional overheads that can have a negative impact relative to the desired benefits when considering the overall system performance. The additional overhead, in the form of longer wires, more connectors and plumbing for the cooling system, can add undesired size and weight to the system, and increased complexity that corresponds to an increased potential for failure. The economic benefits of high-temperature electronics for various systems are likely to be orders of magnitude greater than the total market for actual high-temperature electronics. The world market for high-temperature electronics between 2003–2008 is predicted to increase from 400 to 900 million US dollars, which is substantially lower than the world's total semiconductor electronic market [24.1]. The situation can be dramatically described as follows, a mere handful of high-temperature electronic chips that may cost a few hundred dollars, can optimize the performance of a very large number of systems, thus saving many millions of dollars, e.g. deep-well drilling [24.5].

A survey of the potential industrial users of high-temperature electronics revealed that the majority of applications for high-temperature electronics operate in the range 150–300 °C [24.1, 2, 4]. The recent development of silicon and gallium arsenide electronics and their cost (silicon technology is much cheaper than SiC), indicates that wide-band-gap semiconductor devices are unlikely in the near future to be used in low-power electronics applications for temperatures up to 300 °C. These devices maybe used for application which cannot be satisfied by available technologies such as SOI, and for temperatures above 300 °C. However, in order to realize viable low-power SiC devices for the temperature range 300–600 °C, long-term reliability of electronic circuits must be achieved [24.1].

The performances of silicon power devices have almost reached their theoretical limits [24.17]. The practical operation of Si power devices at ambient temperatures higher than 200 °C appears problematic, as self-heating due to current flow at higher power levels results in high internal junction temperatures and leakage. The overall goal for high-temperature power-electronic circuits is to reduce power losses, volume, weight, and at least the costs of the system. The continuous progress in high-temperature electronics creates a demand for unique material properties, novel processing technologies and electronic devices. The physical and chemical properties required for meeting the demands of the high-temperature and high-power applications can only be found in wide-band-gap semiconductors, which offer a number of advantages over corresponding devices fabricated from silicon. These include higher temperature stability, higher chemical stability, higher thermal conductivity, and higher breakdown field. Various device implementations not only use these standard semiconductor parameters, but also the special peculiarities these materials exhibit, e.g. aluminium nitride and gallium nitride, unlike diamond and silicon carbide, have a direct band gap and have complete miscibility with each other and with indium nitride. This is important for the implementation of optoelectronic device since it allows the band gap to be controlled, and thus the wavelength of the spectral characteristic maximum [24.4]. The wide-bad-gap silicon carbide and diamond are next discussed in this review.

The properties of silicon carbides make it an excellent material for high-power devices operating at temperatures up to 600 °C and above, and at frequencies around 20 GHz. Within power electronics, SiC has the potential to replace Si-based diodes and insulated gate bipolar transistors (IGBTs), and Si gate turn-off (GTO) thyristors, which are part of the mass market of discrete power devices in general and in converter systems in particular. The power losses in SiC switches are two orders of magnitude lower compared with Si devices, thus SiC devices have a large potential for applications in e.g. uninterrupted power systems (UPS), motor controls, etc. The maximum operating temperature of a Schottky diode in SiC may be limited by an increasing leakage currents, but active power devices for operation at high temperature has been presented. U-shaped-trench metal–oxide–semiconductor field-effect transistors (UMOSFET) made from SiC that operate up to 450 °C and thyristors (6 A, 700 V) that operate at 350 °C have been presented. Furthermore, SiC MOSFETs have been reported to operated even at 650 °C, and devices based on n-type-channel metal–oxide–semiconductor (NMOS) technology, which is an integrated operational amplifier, have been reported to work at 300 °C [24.18–22]. The properties and preparation of SiC are elucidated in the next section.

Among the wide-band-gap semiconductors, diamond has the best physical, chemical and electrical properties [24.23], unmatched by any other material. The properties of interest relevant to high-temperature high-frequency power electronics are the large band-gap energy (5.5 eV), the breakdown electric field (10 MV/cm), the carrier mobilities (≈ 2200 and $\approx 1600\,\text{cm}^2/\text{V s}$ for electrons and holes, respectively), the thermal conductivity (10–20 W/cmK), the low dielectric constant (5.5), and the excellent resistance to radiation. Diamond can be found naturally or must be synthesized. In nature diamond occurs as single crystals only, whereas synthetic diamond can be prepared as single crystals, or as a polycrystalline or as a nanocrystalline material.

The discovery that diamond can be grown by the chemical vapor deposition (CVD) technique has opened up some of the expected applications of diamond. However, the utilization of diamond's many unique properties in electronics has so far been limited among others by the unavailability of large-area high-quality diamond and that only p-type (acceptor-type impurity) diamond with high hole densities are available today. The n-type (donor-type impurity) diamond with high electron densities would find many applications, apart from the fundamental interest to realize pn-junctions and other electronic devices in diamond. The n-type diamond is expected to be a better electron emitter for field emission, photo emission, and ion or electron impact-induced emission, and may also serve as a better inert electrode for electrochemical applications.

Nevertheless, many studies have been reported with natural, high-pressure high-temperature (HPHT) synthesized and polycrystalline CVD diamonds [24.24, 25]. The pn-junctions were formed from boron- and phosphorus-doped diamond films, and from boron- and nitrogen-doped diamond films, respectively. The diamond films with high crystalline perfection were grown epitaxially on diamond single crystals. The current–voltage (I–V) characteristic of the boron/nitrogen pn-junction diode was studied up to 400 °C. The combination of two boron/nitrogen pn-junctions, a bipolar junction transistor (BJT) which can operate in direct-current (DC) mode up to 200 °C was fabricated. The fabrication of many types of field-effect transistors (FETs) for both DC and radio-frequency (RF) modes has crossed many important milestones. The cutoff frequency of 1.7 GHz and a maximum drain current of 360 mA/mm were measured for a metal–semiconductor field-effect transistor (MESFET) with a gate length of 0.2 μm. Recently, a FET functioning up to 81 GHz was reported by a collaboration between Nippon Telegraph and Telephone Corp. and the University of Ulm in Germany. The research groups fabricated T-shaped gates on a diamond layer with a carrier mobility of 130 cm^2/V s [24.26]. In addition, Schottky diodes that function up to 1000 °C were fabricated from either single-crystal or polycrystalline diamond. Low-resistance thermostable resistors deposited on ceramic substrates have been investigated for temperatures up to 800 °C. The temperature dependence of the field emission of nitrogen-doped diamond films has been investigated at temperatures up to 950 °C.

There has been much progress in the fabrication of diamond-based electronic devices and several types of devices have reached an important stage in their development. However, despite these developments, diamond-based electronics is still in its infancy.

24.1 Material Properties and Preparation

24.1.1 Silicon Carbide

The properties of silicon carbide makes it an excellent material for devices operating at high temperatures (600 °C and higher), high power (4H-SiC transistor: presently RF output power on the order of 5 W/mm), and high frequency [RF through X band (5.2–10.9 GHz) potentially to K band (20–40 GHz)]. The large band gap of silicon carbides (2.2, 3.26 and 3.0 eV for 3C-SiC, 4H-SiC and 6H-SiC, respectively) compared to the band gap of silicon (1.1 eV) enables devices to function at temperatures beyond 600 °C. The very high breakdown electric field of these materials ($\approx$ 1.8, 3.5 and 3.8 MV/cm for 3C-SiC, 4H-SiC and 6H-SiC, respectively) which are approximately 10 times higher than that of Si (0.3 MV/cm), allows a reduction of the thickness of the conduction regions (for constant doping), which results in very low specific conduction resistance. The 4H-SiC junctions exhibit a negative temperature coefficient, with a breakdown voltage that decreases by about 8% within the temperature range from room temperature to 623 °C [24.27]. The high thermal conductivity ($\approx$ 4–4.5 W/cmK) permits a power density increase which facilitates a more compact or much higher power per area. The high saturation velocity of all three types of silicon carbide is high ($\approx 2 \times 10^7$ cm/s) compared to the value for silicon (1×10^7 cm/s). The low carrier mobilities of silicon carbide is a disadvantage which limits RF performance at frequencies above the X band. The electron mobilities are of the order 900, 500 and 200 cm^2/V s for 3C-SiC, 4H-SiC and 6H-SiC, respectively. The hole mobilities are of the order of 50 cm^2/V s for all three types of SiC (for Si: $\approx$ 1350 and $\approx$ 500 cm^2/V s for electrons and holes, respectively). The carrier mobilities

Table 24.1 Approximate values of physical properties for some semiconductors

Name	Bandgap E_g(eV)	Maximum electric field (V/cm)	Dielectric constant ε_s	Thermal conductivity (W/cmK)	Carrier mobility (cm^2/Vs)
Si	1.1	3×10^5	11.8	1.5	1350
					480
GaAs	1.4	3.5×10^5	10.9	0.8	8600
					250
SiC	3.3	2.5×10^6	9.8	4.9	980
					200
GaN	3.4	2×10^6	7.8	1.4	2000
Diamond	5.5	1×10^7	5.5	10–20	1800
					1600

of SiC are adequate however, for high-power devices in the X band. The properties of silicon carbide and diamond relevant for electronics are given in Table 24.1.

Noteworthy, is that silicon carbide has a close lattice match with III–nitrides, which makes it a preferred substrate material for nitride-based electronic and optoelectronic devices. The commercial production of large substrates which have improved electronic and optoelectronic properties constitutes a milestone in their application. These materials have been used (among others) for large-scale production of green, blue and ultraviolet light-emitting diodes. Unfortunately, unavailability of high-quality defect-free SiC substrates is slowing down the pace of transition from research and development (R&D) to production of SiC devices, which may include high-power solid-state switches or diodes for electrical power control, and high-power-density microwave transistors.

Silicon carbide occurs in a large number of polytype structures. The number of polytypes in the literature

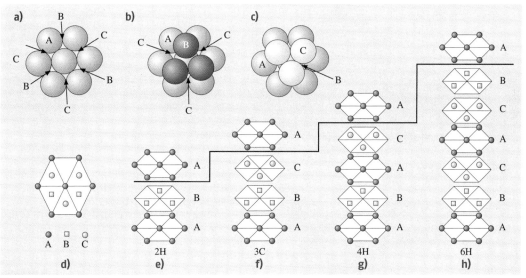

Fig. 24.2a–d The stacking sequence of double layers of the four most common SiC polytypes, (a) locations of C atoms, labeled A, in the first biatomic layer in the {0001} plane; (b), (c) optional positions of C atoms, labeled B and C, respectively, in the next biatomic layer above the first layer; (d) first biatomic layer with six C atoms and the optional positions of the three C atoms in the next biatomic layer, (e)–(h) stacking sequence of the most common SiC polytypes. The *solid line* indicates the completion of the unit cell in the [0001] direction (After [24.28])

varies between 150 and 250. These polytypes are differentiated by the stacking sequence of the biatomic close-packed layers. A detailed study of silicon carbide's polytypism was done in [24.29]. The most famous polytypes are the hexagonal 4H and 6H, cubic 3C, and rhombohedral 15R structures. Not all types are easy to grow; only 4H and 6H polytypes are available as substrate materials. In a single bilayer of SiC each C atom is tetrahedrally bonded to four Si atoms – to three ones within the layer and to one in the next layer. Looking at a bilayer from the top in the direction of the c-axis [0001], the C atoms form a hexagonal structure, as shown in Fig. 24.2a and Fig. 24.2d. These are labeled 'A'. The C atoms of the next biatomic layer have the option to be positioned at the lattice sites 'B' or 'C', as shown in Fig. 24.2b and Fig. 24.2c, respectively. This is the stacking sequence defining a polytype. Figure 24.2d shows schematically the first bilayer, with six C atoms forming the hexagonal structure and the optional positions for the three C atoms in the next layer beyond the first layer. Figures 24.2e–h shows the stacking sequence for the most common SiC polytypes. The change in sequence has an impact on the properties of the material, for example the band gap changes from 3.4 eV for 2H polytype to 2.4 eV for 3C polytype. Figures 24.3a–e shows the structure of the most common SiC polytypes viewed in the $\{11\bar{2}0\}$ plane, i. e. in the [0001] direction.

The leading manufacturer of substrates is Cree Inc., though new manufacturers have recently appeared. The crystalline quality in terms of a low defect density, and a specially low micropipe density the substrates of Cree Inc. are still ahead. The preparation of silicon carbide is complicated by the fact that it does not melt, it sublimates at temperature above 2000 °C, thus standard growth techniques, e.g. the Czochralski process by which large single-crystal ingots are produced by pulling a seed crystal from the melt, cannot be used. Silicon carbide crystals are grown by a sublimation method first developed by *Lely* in 1955 [24.30], and later extended to a seed sublimation technique by *Tairov* and *Tsvetkov* in 1978 [24.31]. This method is also termed physical vapor transport (PVD) growth. The crystals are grown by SiC deposition derived from Si and C molecular species provided by a subliming source of SiC placed in close proximity to the seed wafer.

A new high-temperature CVD (HTCVD) technique was developed in 1999 [24.28], where the growth rate can be tuned in such a way that a high-quality thick epitaxial layer with precisely controlled doping levels can be grown in a few hours, which is fast compared to conventional CVD, which takes days to grow a similar structure. As an alternative to CVD, sublimation epitaxy has also been demonstrated for the growth of thick epitaxial layers. The growth rate in sublimation epitaxy is also very high, of the order of a few hundred micrometers an hour. In devices where the control of the doping level is an important issue, sublimation epitaxy has been shown to work successfully.

The commercial availability of SiC substrates with increasing diameter and quality has been a prerequisite for advances in SiC device technology. The SiC substrates are available in two different polytypes, namely 6H- and 4H-SiC. The latter is relevant for electronic application due to its higher carrier mobility and wider band gap than 6H-SiC. SiC substrates have been in the market for over a decade, but the absence of defect-free growth is slowing the pace of transition from research and development to the production of power devices such as high-power solid-state switches or diodes for electrical power control and high-power-density microwave transistors. Three-inch 4H-SiC substrates have been commercially available since 2001, but it was only recently that their defect concentration have been reduced to levels that allows for the fabrication of com-

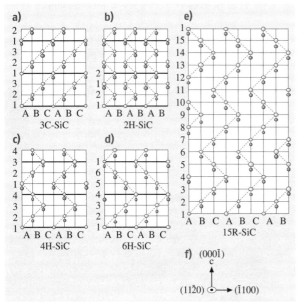

Fig. 24.3a–f Schematic representation of the structure of the most common SiC polytypes viewed in the $\{11\bar{2}0\}$ plane. The *black dots* represent C atoms and the *open circles* represent Si atoms. The *bold line* indicates the completion of the unit cell in the [0001] direction (After [24.28])

mercially viable high power switches. During 2003 Infineon and Cree Inc. released 10 Amps devices, which is clear evidence that substrates have reached an acceptable level of quality.

The important defects of SiC are different types of dislocations. The open core screw dislocations called micropipes are of particular concern for SiC due to their detrimental effects on power devices. Micropipes cause diodes to fail for voltages that are much smaller than the voltage at which avalanche breakdown occurs [24.33]. Progress in the development of the physical vapor deposition (PVD) technique during the last four years, has resulted in a significant reduction of the micropipe density in 3-inch 4H-SiC wafers, from a previously typical value above $100\,\text{cm}^{-2}$ to a value as low as $0.22\,\text{cm}^{-2}$ in R&D samples. The micropipe densities in commercially available substrates are $< 30\,\text{cm}^{-2}$ and $< 80\,\text{cm}^{-2}$ for n-type and semi-insulating materials, respectively. 100-mm 4H-SiC wafers are now under development. The micropipe densities for such wafers are $\approx 22\,\text{cm}^{-2}$ and $\approx 55\,\text{cm}^{-2}$ for n-type and semi-insulating crystals, respectively [24.34, 35].

Low-angle grain boundaries, also known as domain walls, are an other class of defect that has to be reduced, since they are associated with leakage currents and failure in devices, and may cause wafers to crack during epitaxial processing. This type of defect can be observed through whole-wafer X-ray topography, though it is difficult to obtain quantifiable numbers on wafer quality by such measurements. This defect seems to be intimately related to the growth method used [24.36]. A class of defects known as threading screw dislocations are suggested to have an impact on the leakage behavior of Schottky diodes. The evidence suggests that these defects are introduced at the seed/growth interface by seed subsurface damage. The application of seed treatment reduces the density of dislocations in a 3-inch 4H-SiC wafer from a value of the order $3 \times 10^4\,\text{cm}^{-2}$ to $3 \times 10^3\,\text{cm}^{-2}$ [24.35]. The so-called basal plane dislocation is another important and significant defect. It has been shown that PiN device structures are susceptible to severe degradation of the forward voltage characteristics due to the presence of these defects in the active layer of the device [24.37]. The presence of these dislocations increases the resistance of the active layer of the device. It is critical to reduce the density of these dislocations in epitaxial layers for stable device production. A level of basal plane dislocations in the substrate which may be acceptable in order to allow reasonable yields of PiN diodes, is on the order of $100\,\text{cm}^{-2}$. The average density of the basal plane dislocations in 3-inch 4H-SiC wafers is $1.5 \times 10^3\,\text{cm}^{-2}$ [24.35]. Recently it was reported that this type and even other types of dislocations and defects can be reduced by growing the material along the a-face direction [24.32]. Single crystals of SiC are usually grown by the method termed c-face growth, where crystals are grown along the $\langle 0001 \rangle$ c-axis direction using a seed of $\{0001\}$ substrate. In the new method, known as a repeated a-face growth process, single crystals are grown along the a-axis [$11\bar{2}0$] or [$1\bar{1}00$] (both axis are called the a-axis in the report) direction in several steps. The a-face growth process is shown in Fig. 24.4.

The electrical properties of SiC substrates are related to the purity of the as-grown crystals. The substrate purity is dominated by the presence of residual nitrogen and boron impurities. The level of these impurities is critical for the production of undoped high-purity semi-insulating substrates with uniform and stable semi-insulating properties. High-purity 3-inch and 100-mm

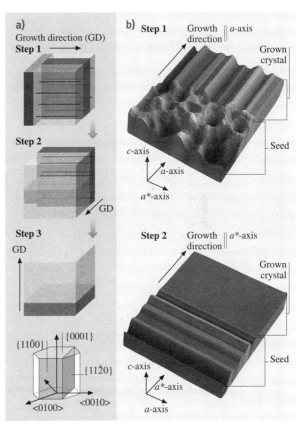

Fig. 24.4 Schematic picture of the a-face growth of SiC. (After [24.32])

4H-SiC substrates with low micropipe densities and uniform semi-insulating properties ($> 10^9\,\Omega\,\mathrm{cm}$) over the full wafer diameter have been produced. These wafers had typical residual contamination densities $5\times 10^{15}\,\mathrm{cm}^{-3}$ and $3\times 10^{15}\,\mathrm{cm}^{-3}$ for nitrogen and boron, respectively [24.38].

Although most doping of SiC is obtained by an in situ method during epitaxial growth, additional selected-area doping is often required during fabrication of devices such as MOSFETs and lateral bipolar transistor. Due to the extremely low diffusion coefficient of dopant atoms in SiC even at very high temperatures ($\approx 2000\,^\circ\mathrm{C}$), ion implantation is the only viable doping technique during device fabrication. The critical parameters of ion implantation of dopants in SiC are the temperature of SiC during implantation (from room temperature up to $900\,^\circ\mathrm{C}$), as well as the subsequent annealing required to activate the dopant, performed at around $1700\,^\circ\mathrm{C}$. Nitrogen is typically used as the n-type dopant, while Al is often the p-type dopant widely used during epitaxy. The element boron is a lighter element than Al and subsequently causes less lattice damage during implantation, and may eventually replace Al. However, a small amount of B also diffuses into the lightly doped drift-layer side during the annealing process, thereby degrading the junction.

The high bond strength of SiC means that room-temperature wet etches for this material do not exist, and so reactive-ion etching (RIE) is the standard method used. Frequently fluorine-based chemistries are used in which the silicon forms a volatile SiF_4 molecule and C is removed either as CO_2 or CF_4. However, RIE is not regarded as a limitation since, as feature sizes decrease, dry etching processes are actually preferred to wet etching.

A unique advantage of SiC compared to other wide-band-gap semiconductors, is its ability to oxidize and form SiO_2 exactly as in Si technology. The oxidation rates are much lower for SiC than for Si, and are very dependent on if a silicon- or carbon-terminated face is exposed to the growing SiO_2. The fabrication of high-quality thermal oxides with low interface state and oxide-trap densities has proven to be a great challenge. Finally, the reliability of oxides is a major issue for SiC devices since, at high electric fields and high temperatures, oxides have poor longevity. This issue needs further research to reduce the leakage current in the devices that operate at elevated temperatures.

An important issue in high-temperature electronics is the type of metallization used, where examples include ohmic, Schottky, heat-sinking and capping. It is necessary to have reasonable thermal expansion matching and good adhesion between the metal and SiC. The wide band gap of silicon carbide makes it difficult to control the electrical properties at the metal–semiconductor interface of devices. In addition, stable noncorrosive contacts are also key issues in high-temperature electronics. The main parameter of concern for SiC high-frequency devices is a stable Schottky barrier for good rectification and a low reverse leakage current while operating at elevated temperatures. Several groups have tried different combinations of transition metals that form good Schottky contact on n- and p-type SiC with barrier heights in the range $0.9–1.7\,\mathrm{eV}$ [24.20, 39–41]. The rectifying properties either change to ohmic or degrade severely while operating at temperatures above $600\,^\circ\mathrm{C}$. Among the ohmic contact the most widely used material for n-type is Ni_2Si which is generally formed by deposition of Ni film and silicidation is obtained by annealing at above $900\,^\circ\mathrm{C}$. The Ni_2Si ohmic contact has been shown to be stable at very high temperatures [24.39, 42]. The formation of low-resistance ohmic contacts to p-type SiC is still difficult since metals with sufficiently large work functions are not available to offset the wide band gap and electron affinity of SiC. Aluminum is typically used to form p-type ohmic contacts. A major drawback of Al however is its relatively low melting point, which prohibits its use for high-temperature applications. Several other combinations of different metals have also been reported in the literature these have poor contact resistivities compared to Al [24.43, 44]. A special effort is required to develop stable contacts for SiC devices operating at higher temperatures, and metals with a high melting temperature and their silicides and carbides should be studied in the future towards this goal.

The packaging of SiC devices for high-power and high-frequency applications and operation at elevated temperature is an issue that has been neglected compared to material growth and device processing technology. It is highly desirable to find suitable packaging for high-temperature electronics which can endure high thermal stress and high power without the extra effort of cooling.

24.1.2 Diamond

Among the wide-band-gap semiconductors, diamond has the best properties, unmatched by any other material [24.45–49]. Most electrical, thermal and optical properties of diamond are extrinsic, i.e. strongly dependent on the impurity content [24.23, 46], the most common impurity being nitrogen. Diamond has a large band gap (5.5 eV), high breakdown electric field (10 MV/cm), low dielectric constant (5.66–5.70), high carrier mobilities (≈ 1800 and $\approx 1600 \text{ cm}^2/\text{Vs}$ for electrons and holes, respectively [24.45]), high saturated carrier velocity (2.7×10^7 cm/s and 1×10^7 cm/s for electrons and holes, respectively), high thermal conductivity (10–20 W/cmK), high resistivity (10^{13}–$10^{16} \, \Omega \, \text{cm}$), low thermal expansion coefficient (1.1 ppm/K at room temperature), the highest sound velocity (1.833×10^6 cm/s), exceptional hardness (10 000 kg/mm^2) and wear resistance, low friction coefficient (0.05, dry), broad optical transparency [from 225 nm to far infrared (IR)], excellent resistance to radiation, chemical and thermal stability. A unique feature of diamond is that some of its surfaces can exhibit a very low or negative electron affinity. Obviously diamond is the material of choice for many applications, including electronics. The properties of diamonds make it the most suitable semiconductor for power electronics at high (RF) frequencies and high temperatures [24.50–53]. Since diamond-like silicon is a single-element semiconductor it is less susceptible to have the high density of structural defects that are usually present in compound semiconductors. However to date, diamond is regarded as one of the most difficult semiconductors to synthesize for the fabrication of electronic devices.

Diamond is a cubic semiconductor with lattice constant $a = 3.566$ Å. The covalent bonding of the carbon atoms (sp^3 bonds) is extremely strong and short, which gives diamond its unique physical, chemical and mechanical properties [24.46–49]. Diamond is available naturally and can also be synthesized. The natural form of diamond occurs as single crystals, whereas synthetic diamond can be prepared as single crystals, or as polycrystalline or nanocrystalline material. Usually natural diamond single crystals have a high nitrogen content and cannot be used for the fabrication of electronic components. The natural form of diamond has been classified according to several criteria, a detailed description of the classification of diamonds has been given by *Walker* [24.46] and *Zajtsev* [24.54].

Diamond melts at approximately 3827 °C [24.45]. It is stable at elevated temperatures, but the stability depends on the ambient. In hydrogen ambient diamond is stable up to 2200 °C [24.55], but it is graphitized in vacuum [24.56, 57] or in an inert gas [24.56]. Diamond does not have a native oxide, but it oxides in air at elevated temperatures. This is a critical point for the application of diamond for high-temperature devices. The oxidation of natural and synthetic diamond has been studied since the beginning of the 1960s; despite this more research is needed for a complete understanding of the oxidation process. The activation energy for the oxidation of CVD-grown films in air was 213 kJ/mol for temperatures of 600–750 °C and the oxidation proceeded by etching pits into the CVD film, thus creating a highly porous structure [24.58]. The results of several studies indicated that diamond oxidized preferentially. The oxidization of natural diamond and CVD-grown diamond films in oxygen has been observed to be dependent on the crystallographic orientation, here the (111) plane oxidized more easily than the (100) and (220) planes, and also the CVD films were less resistant to oxidation than natural diamond [24.56, 59]. *Sun* et al. [24.56] observed that the oxidation of synthetic diamond started in air at 477 °C when oxygen is able to impinge into the densely packed (111) planes and they suggested that the oxidation of diamond occurs by the same mechanism as the corrosion of metals, whereby oxygen penetrates into the bulk by bonding and rebonding, leaving behind weakly interacting dipoles which are eroded away during processing. *Lu* et al. [24.60] reported that the oxidation in air of diamond films prepared by DC arc plasma jet started at 650 °C which was about 100 °C lower than the temperature of oxidation of natural diamond. Furthermore, it was reported that the oxidation rate of CVD diamond depended on the diamond's growth condition [24.61].

There are several etchants for diamond; the most commonly used method is oxidative etching. The effects of dry oxygen and a mixture of oxygen and water in the temperature range 700–900 °C has been studied and compared with the effect of molten potassium nitrate [24.62, 63].

Diamond-based electronic devices have now been fabricated from natural and synthesized single crystals, high-purity single-crystal films (homoepitaxial diamond), and from polycrystalline films (heteroepitaxial diamond). Single crystals can be synthesized artificially by the high-pressure high-temperature method (HPHT), which mimics the process used by nature. The drawback of this method is that it produces single crystals limited in size. The largest crystals prepared by this method have a dimension of the order of millimeters, and the processing time to produce such crystal is very long [24.64]. These crystals have been used for the fabrication of discrete electronic devices and as substrates for homoepitaxial growth of diamond films by CVD technique. Diamond films have been epitaxially deposited on diamond single-crystal substrates, this demonstrates that single-crystal diamond deposition is possible by low-pressure processing [24.65, 66].

The discovery that diamond can be grown homoepitaxially and heteroepitaxially by the chemical vapor deposition (CVD) technique has opened up some of the expected applications of diamond. The history of this technique goes back to the late 1960s. During the 1980s, researchers [24.67, 68] made a series of discoveries which enabled them to grow, at significant growth rates, diamond films of high quality on non-diamond substrates by using hot-filament CVD and subsequently by microwave plasma chemical vapor deposition (MPCVD). This started a worldwide interest in diamond CVD for both research and technology. Since then a number of low-pressure CVD techniques have been developed [24.69, 70] and the volume of research on the preparation of large-area diamond films has been immense.

It has been shown recently [24.71, 72] that homoepitaxial growth of diamond films on high-quality HPHT diamonds can produce single-crystal films of a purity that exceeds the purity of the purest diamonds found in nature. The measurements of the carrier mobilities in these films have revealed interesting results, such as mobilities of 4500 and 3800 cm^2/Vs for the electrons and holes, respectively. These values are the highest ever reported for diamond and are approximately twice as high as those found in pure natural diamond. The carrier mobility measurements were performed on homoepitaxial diamond deposited by a microwave plasma-assisted CVD technique and a HPHT diamond single crystal of dimensions $4 \times 4 \times 0.5\, mm$ was used as the substrate. The homoepitaxial diamond film was found to be of exceptional purity and was found to contain a low concentration of intrinsic and extrinsic defects. The total measured nitrogen concentration was less than $1 \times 10^{15}\, cm^{-3}$ and the dislocation density was less than $1 \times 10^6\, cm^{-2}$. The exceptionally high values of the carrier mobilities were attributed to the low defect and dislocation densities.

The disadvantage of diamond homoepitaxy is that only small-area single crystals can be fabricated, and substrates typically have a size of the order of millimeters. In order to exploit diamond's superior properties for the fabrication of electronic devices, thin diamond films are required, i.e. a method for the production of large-area, inexpensive single-crystal films with a low defect density.

Despite the progress made, the available diamond homoepitaxy methods cannot solve the technological problem of producing large-area diamond wafers for the fabrication of electronic devices. During the last 10 years film preparation has focused on diamond heteroepitaxy. The aim has been to produce films of homoepitaxial diamond's quality by avoiding the formation of grain boundaries and other defects. The research has focused on finding suitable substrates, conditions for achieving high diamond nucleation densities on various substrates, and the optimization of textured growth procedures. To date many substrates have been investigated, e.g. Ni, Co, Pt, Si, BeO, SiO_2, cubic BN, β-SiC, GaN, etc. [24.73–77]. Although some of these materials are suitable substrates, e.g. BN [24.77], all attempts to grow large diamond single-crystal films have hitherto failed. Most of the CVD diamond films reported to date have been grown on Si, mainly due to the availability of large-area single-crystal wafers and the low cost of Si as well as the favorable properties of Si [24.78–84]. These films are still polycrystalline but highly oriented (HOD) with respect to the substrate and have found application in many fields, e.g. electrochemical electrodes, field-emitter arrays, radiation detectors, micro-electromechanical systems (MEM), etc. In the field of MEMS a large number of devices for various applications have already been built, demonstrating thus the excellent properties of these films [24.85–87]. Despite their high quality, these films are not suitable for the fabrication of electronic devices since their attractive properties are deteriorated by structural imperfections, particularly by grain boundaries. The performance of electronic devices fabricated using such low-quality films is significantly reduced. Furthermore, due to their mosaic spread, these films cannot be used as substrates for the homoepitaxial growth of diamond.

A significant advance in diamond heteroepitaxy was made by the application of substrates with a multilayer structure. It was discovered that iridium single-crystal films grown as a buffer layer on MgO could serve as a substrate for the nucleation and growth of low-pressure microwave plasma-enhanced CVD diamond [24.88, 89]. The substrate MgO was later replaced by $SrTiO_3$ [24.90, 91], which decreased the mosaic spread of the epitaxial iridium and of the resulting heteroepitaxial diamond. The diamond layer had single-crystal quality and was used for the fabrication of field-effect transistors [24.92]. A further advance in the large-scale heteroepitaxial growth of diamond has been made recently when $SrTiO_3$ was replaced successfully by sapphire [24.93, 94]. Diamond produced in this way had the same high quality as that prepared on $SrTiO_3$. However, since sapphire is a relatively inexpensive large-area substrate, this development is a further step towards the wafer-scale production of heteroepitaxial diamond.

When cooling from the growth temperature the diamond film experiences significant compressive stresses due to the difference in the thermal expansion coefficient of the materials present. These stresses can cause delamination, which is a serious obstacle for the development of diamond wafers. Moreover, calculations have shown [24.95] that films grown on MgO, SrTiO₃ and α-Al₂O₃ substrates are exposed to a significant amount of stress, −8.30, −6.44 and −4.05 GPa, respectively. These large stresses make the treatment of thick films difficult. From the thermal stress point of view, Si with a stress of −0.68 GPa is the best substrate so far.

Further step towards the fabrication of large-area diamond single-crystal films for electronic applications has been made recently [24.95] by introducing a new concept for the substrate multilayer. The multilayer was prepared in two steps, where yttrium-stabilized zirconia (YSZ) was first deposited on Si, then an iridium thin film was deposited on the YSZ. This process decreases the lattice misfit between consecutive layers and such a substrate multilayer is shown in Fig. 24.5. The diamond was then grown on the iridium film as before. The quality of the diamond was the same as of that grown on SrTiO₃. The advantage of this concept is in the type and the combination of the substrate materials, which minimizes thermal stress thereby avoiding delamination.

In order to exploit diamond's superior properties for the fabrication of electronic devices, thin diamond films are required, i.e. a method for the production of large-area inexpensive single-crystal films with low

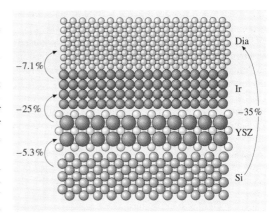

Fig. 24.5 Schematic representation of the layer system diamond/Ir/YSZ/Si(001). In the YSZ crystal the *large spheres* correspond to the oxygen ions. The *numbers* indicate the lattice mismatch between consecutive layers. (After [24.95]) (Figure provided by Mathias Schreck, University of Augsburg)

defect and dislocation densities is needed. Diamond device technology has similar problems to third-generation semiconductor technology, namely, the availability of inexpensive large-area diamond crystal with a low defect concentration; this is the prerequisite for a successful application of diamond for the fabrication of electronic devices.

24.2 Electronic Devices

24.2.1 Silicon Carbide

The ability of silicon carbide to operate at high temperature, high power, and high frequencies enables considerable enhancement of the performance of devices used within a wide variety of applications. In particular, SiC power devices can outperform equivalent Si devices, but this would require a mature and viable SiC semiconductor technology. The fundamental physical limitations of Si operation are the strongest motivation for switching to a wide-band-gap semiconductor such as SiC for high-temperature applications. The replacement of Si by SiC for power switches [24.20] is extremely advantageous, since the avalanche breakdown voltage for SiC is about ten times higher than that of Si (see Fig. 24.6 and Fig. 24.7). Moreover, silicon-carbide-based power switches also have a faster response with a lower parasitic resistance, so that the physical size of SiC devices

will be much smaller than equivalent silicon devices. In addition, the faster switching speed enhances the efficiency of power system conversion, and allows the use of smaller transformers and capacitors, which reduce significantly the overall size and weight of the system. The cooling requirements in power electronics, which are a considerable portion of the total size and cost of power conversion and distribution systems, can be significantly reduced by the high-temperature capabilities of SiC.

Power switching devices based on SiC such as Schottky barrier diodes, PiN junction diodes, metal–oxide–semiconductor field-effect transistors (MOSFETs), junction field-effect transistors (JFETs), bipolar junction transistors (BJTs) have already been demonstrated in the research stage. To date, SiC Schottky barrier diodes used for high-voltage applications have evolved from the research stage to limited commercial production [24.96]. Although Schottky barrier diodes offer rectification with lower switching losses compared to PiN diodes (when switching from the conducting state to the blocking state) Si-based Schottky barrier diodes are still not used for high-voltage applications. The reason for this is the comparatively lower barrier height between ordinary metals and Si (typically less than 0.5 eV), which is further reduced in reverse-biased mode. The electron injection current from metal to semiconductor increases exponentially as the barrier height reduces. As a result very large reverse currents are observed at relatively low voltages, in the case of Si. The high Schottky barrier height of about 1.5 eV for SiC reduces the leakage current such that SiC Schottky diodes can operate at high voltages, and a blocking voltage of up to 5 kV has been demonstrated [24.39].

Diodes, Schottky and PiN Diodes

In contrast to Si diodes, SiC rectifiers exhibit ultimate low switching losses, no reverse current peak, and therefore an extreme soft recovery behavior. In Schottky diodes, a reduced forward-voltage drop and hence a reduced power loss would be desirable, and this can be achieved by lowering the Schottky barrier height. On the other hand, even at high temperatures, the barrier must be high enough to ensure a certain blocking voltage with a reasonably low reverse-leakage current. There is a strong dependency of the measured reverse current on the electric field for a Ti/SiC Schottky diode, with the temperature as a parameter [24.98].

Schottky barrier diodes with Ta and TaC on p- and n-type SiC have been investigated by a group at Carnegie Mellon University [24.99]. The rectifying behavior of both Ta and TaC was observed on p-type 6H-SiC to 250 °C, while on n-type, TaC showed ohmic behavior above 200 °C [24.99]. The maximum operating temperature of SiC Schottky diodes is restricted because of the increasing leakage current and therefore, junction devices can approach and perform better at higher temperature.

The pn-junction diode characteristics at temperatures up to 400 °C have been reported both for epitaxially grown and ion-implanted SiC. The characteristics of the 6H-SiC pn-diode were obtained where nitrogen was implanted into p-type substrate. The rectifying ratio was measured to be 10^9 at room temperature and 10^5 at 400 °C. PiN diodes of 4H-SiC have been successfully fabricated and this device has a blocking voltage of up to 19 kV [24.100]. The high-temperature operation of a 8-kV diode indicates that the reverse leakage current density at 5 kV increases by only an order of magnitude between room temperature and 300 °C [24.101]. The reverse recovery of these diodes showed an extremely fast switching, which only increased by a factor of two between room temperature and 275 °C. This is small compared to Si diodes, which have an increase of four times between room temperature and 120 °C [24.101]. In the 5.5-kV blocking range of a PiN diode, a forward voltage drop of less than 5 V was observed at 500 A cm^{-2} between room temperature and 225 °C, in addition to a 50% increase in the peak reverse recovery current [24.102]. The reverse leakage current in a large-area (6×6 mm^2) diode which blocked 7.4 kV showed a small increase in the current up to 200 °C after packaging when measured up to 4.5 kV [24.103].

Another diode, the so-called junction barrier Schottky (JBS), is also considered an attractive device for power switching applications. The JBS device was first demonstrated in Si [24.104, 105]. The device structure is a combination of a Schottky barrier and a pn-junction, which allows a reduction in the power loss of the pn-junction under forward conduction and the utilization of the Schottky barrier. In the reverse direction the Schottky region is pinched off by the pn-junction, thus exhibiting a smaller leakage current. The spacing between the p^+ regions should be designed so that pinch-off is reached before the electric field at the Schottky contact increases to the point where excessive leakage currents occur due to tunneling, this complicates the device design and several attempts have been reported [24.106–108]. The major problem concerning SiC may be the poor ohmic contact with p-doped samples, which requires very-high-temperature

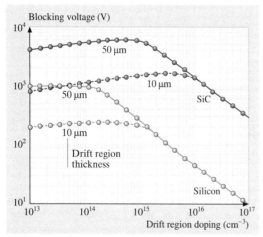

Fig. 24.6 Comparison of silicon carbide and silicon dielectric strength. (After [24.97])

annealing (above 850 °C) that causes severe damage to the Schottky contact; this then increases the excessive leakage current compared to an undamaged Schottky contact.

Though all these results on PiN diodes are very encouraging, there still remain problems in their practical operation. The most important perhaps, is the degradation of the current with time when operating under forward bias. One of the most attractive applications of SiC PiN diodes is in high-voltage DC (HVDC) electrical transmission systems. The current degradation is presumed to be related to the generation and extension of stacking fault defects in the basal plane. These defects lie in the crystalline plane perpendicular to the current flow direction. Recently it has been claimed [24.109] that this problem may be solved by growing the material along another direction. In order to get a defect-free crystalline structure, the material was grown on several a-face surfaces in at least 2–3 steps. The PiN diodes fabricated from this material were stressed and measured for 4 h at constant voltage. Growth using a few steps might be not feasible and more research on the operation of these PiN diodes is needed.

High-frequency MESFETs, SITs and BJTs

Concerning high-frequency devices, several groups from the USA, Japan and Europe have demonstrated two types of SiC transistors, metal–semiconductor field-effect transistor (MESFETs) and static induction transistors (SITs). These transistors suffer from large gate leakages, so that the possible application may be restricted to less than 350 °C. Static induction transistors (SITs) shows the best performance for high-peak-power applications up to 4 GHz. Under class C operation, the transistor produces over 350 W output power with 50% efficiency and a gain greater than 10 dB [24.110]. At ultra-high frequency (UHF) 900 W, pulsed power was obtained from a single chip packaged with a 51-cm gate periphery at L-band, and the same power was obtained for a 54-cm periphery package [24.110]. No investigations of the thermal limit of the high-temperature operation have been reported for such high-power SITs. For MESFETs, power densities as high as 5.2 W/mm² with 63% power added efficiency (PAE) have been demonstrated at 3.5 GHz while single device with 48 mm of gate periphery yielded 80 W continuous wave (CW) at 3.1 GHz with 38% PAE [24.111, 112]. Though these results are impressive, the transistor performance suffers severely from self-heating due to current flow in the device. Thus the majority of published articles are related to the small-gate periphery since thermal prob-

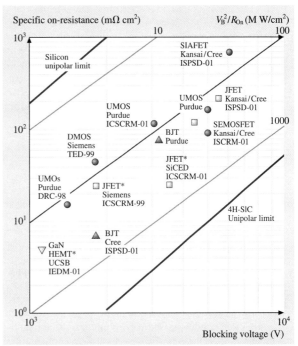

Fig. 24.7 Comparison of silicon and silicon carbide operating voltage and conduction resistance. (After [24.20])

lems are limited and the power densities are recorded to be very high. When the gate periphery increases, the power densities decrease very quickly, this is related to the self-heating effect. For a wide transistor, the temperature becomes very high and, therefore, the electron mobility decreases together with the drain current, and finally the RF power decreases. In a recent investigation, a channel temperature as high as 340 °C was observed for a large 31.5-mm transistor when only DC biases without any RF excitation was applied; the estimated dissipated power was 58.6 W in class AB [24.113]. In another investigation, the saturated DC drain current was reduced from 80 mA to 50 mA when the temperature was increased to 250 °C for a 6H-SiC MESFET [24.114].

Very little work has been published so far on bipolar transistors and specially within microwave applications. Due to the wide band gap, junctions of the transistors can control the reverse and forward current, and hence can be very attractive for high-ambient-temperature operations. However, the transistor suffers from some severe materials properties such as low minority-carrier lifetime, a very small hole mobility that restricts the cutoff (f_T), and a maximum frequency of oscillation (f_{max}).

The transistors have been fabricated and exhibited typical emitter breakdown at 500 V while the f_T was only 1.5 GHz. A transistor with an emitter width of 2.5 μm and an emitter periphery of 2.62 cm has demonstrated an output power of 50 W using a 80 V power supply in common-emitter class AB mode. The pulse width was 100 μs and the duty cycle was only 10%. The collector efficiency was 51% with a power gain of 9.3 dB [24.115]. Since SiC junctions can withstand high junction temperatures, the transistor can function efficiently without any external cooling, resulting in significant system advantages.

JFETs

SiC-based JFETs are very attractive for high-temperature electronics for the reasons described for BJTs. JFETs are more attractive as they are unipolar devices and thus do not suffer from a low value of the hole mobility. However, JFETs are normally depletion-mode (normally on) devices, and the gate must be kept at a negative voltage to keep the transistor off. Most power control systems require enhancement-mode (normally off) transistors so that the system can be switched off in a safe condition. One way to circumvent this problem is to connect a JFET in a cascode configuration with an enhancement-mode control device such as a Si or SiC MOSFET [24.116]. High-temperature operation of SiC-based JFETs has been reported in the literature since the early 1990s along with the evolution of SiC as a semiconductor for microelectronics.

Transistors with high blocking voltage up to 5.5 kV have been successfully fabricated. This transistor showed a specific on-resistance of 69 mΩ cm^2 and the turn off time was 47 ns [24.117]. The material 4H-SiC is more attractive due to high carrier mobilities which results in some favorable properties. Specifically, in a vertical JFET with the drain on the back side of the wafer, the forward current density reached 249 A/cm^2 at a drain voltage of only 1.2 V at room temperature, while at 600 °C the current dropped to 61 A/cm^2 with an increase of the specific on-resistance from 4.8 to 19.6 mΩ cm^2. In addition, the breakdown voltage of the transistor was 1644 V at room temperature, while it increased to 1928 V at 600 K [24.118]. In a thermal stress study of a 6H-SiC JFET, a decrease of about 40% in the drain current was observed at 300 °C [24.119].

High performance of SiC JFET has been reported by Siemens. The buried gate is permanently connected to the source, and the device blocked 1800 V at a specific on-resistance of 24 mΩ cm^2. Recently, Kansai Electric and Cree Inc. reported the first enhancement-mode SiC JFET [24.120]. The device structure consisted of two gates; the buried gate was connected to the top gate, providing a gating effect from both sides. The transistor with a 50-μm-thick drift layer blocked 4.4 kV and the specific on-resistance was 121 mΩ cm^2.

MOSFETs

In contrast to Si MOSFETs, the 6H-SiC MOSFETs transconductance and channel mobility increases with rising temperature up to 225 °C. The high interface state density [24.121] may be the reason for these trends. MOSFETs fabricated from 6H-SiC have operated at temperatures around 400 °C with a very low drain leakage current in air [24.122]. In a recent report, a large-area device (3.3 mm × 3.3 mm) blocked 1.6 kV with an on-resistance of 27 mΩ cm^2. This device exhibited a peak channel mobility of 22 cm^2/V s and a threshold voltage of 8.8 V, which decreased to 5.5 V at around 200 °C [24.123].

The transistors were fabricated both in the lateral double-diffused MOSFET (DMOSFET) and the vertical U-shaped-trench MOSFET (UMOSFET) directions, these blocked several kilovolts. The Northrop Grumman group determined that both transistor types have similar channel mobilities and the same temperature effect due to the interface traps. In addition, they operated at up to 300 °C where the current and transconductance increased with rising temperature [24.124]. A group at Purdue University demonstrated a novel DMOSFET which could block 2.6 kV, where the drain current at 155 °C was four times the value at room temperature [24.125]. Recently, Purdue and Auburn University fabricated a UMOSFET that could block more than 5 kV, with a specific on-resistance of only 105 mΩ cm^2, for a 100-μm-thick low-doped drift layer (see e.g. Fig. 24.7). This device has not been characterized at elevated temperatures [24.126]. The transistor maintained a low on-resistance to current densities above 100 A/cm^2, while the maximum reported current density was 40 A/cm^2. Cree Research has already reported a record blocking of more than 6 kV.

Due to the problems caused by a high interface-state density, which results in a very low channel mobility, a new type of transistor, the accumulation-mode MOSFET (ACCUFET), with output characteristics that are similar to MOSFETs has been introduced and fabricated in both 6H- and 4H-SiC [24.127]. The 4H-SiC transistor exhibited a high specific on-resistance, 3.2 Ω cm^2 at a gate bias of 5 V, which reduced to 128 mΩ cm^2 at 450 K. The transistors were not designed for high blocking voltages thus the unterminated breakdown voltage was only 450 V.

Recently, encouraging results from the continuing research into 4H-SiC MOSFET has been reported [24.128, 129]. Growing the gate oxide in N_2O ambient results in a significant enhancement of the inversion channel mobility in lateral n-channel Si face 4H-SiC MOSFETs. A mobility of $150\,cm^2/V\,s$ was obtained, whereas a value below $10\,cm^2/V\,s$ was obtained when growing the gate by the conventional process: in a wet or dry oxygen ambient.

Thyristors, GTOs and IGBTs

One of the most significant developments in device technology during the last 15–20 years is the insulated gate bipolar transistor (IGBT), which blocks high voltages but at the same time has a high conduction current. The basic advantage of the IGBTs is conductivity modulation due to carrier injection and a MOS-driven gate. The device operates between 600 and 6.5 kV where the current varies over 1–3500 A. The device has a big potential for further development in the areas of higher currents and voltages, and for higher frequencies and lower dissipated power.

A large amount of effort is being made on the development of SiC thyristors and IGBTs. The main advantage is the high electric breakdown field strength, which leads to very thin drift layers, and consequently much faster switching behavior.

The thyristor is the most popular controllable device used in high-power systems with controllable turn-on, and large-area Si thyristors are produced today as a single device on a wafer with a 4-inch diameter. The current capability is more than 1000 A with a blocking voltage approaching 10 000 V. These devices are used in high-voltage DC (HVDC) transmission systems, and so far no other device can match its performance. A 6H-SiC polytype thyristor was first demonstrated in early 1993. Due to the high resistivity of the p-type substrate, a high specific on-resistance of $128\,m\Omega\,cm^2$ has been obtained for a 100 forward blocking voltage device [24.130]. However, at a higher temperature of 633 K, the thyristor performance improved with a specific on-resistance reduced to only $11\,m\Omega\,cm^2$. The opposite polarity thyristor has been realized on the same structure with an improvement in all the characteristic parameters, and a very low specific on-resistance of about $3.6\,m\Omega\,cm^2$ that increased to $10\,m\Omega\,cm^2$ at 623 K temperature. Furthermore, early 4H devices exhibited a blocking voltage of −375 V. The low on-resistance of the 4H devices resulted in a lower voltage drop. Subsequently, the Northrop Gramman group has demonstrated a gate turn-off (GTO) with 1-kV blocking, but the device was turned-off by an external MOSFET. The device has been operated successfully at 390 °C [24.131]. Recently remarkable effort has been put into this area and 4H-SiC-based GTOs have been demonstrated with a 3.1-kV forward blocking capability with a 12-A conduction current [24.132].

24.2.2 Diamond

The many exceptional properties of diamond make it a very attractive material in several fields of electronics, optoelectronics and micro-electromechanical devices. Since an exhaustive review on the application of diamond in electronics and optoelectronics would be beyond the scope of this relatively short review, here the survey is limited to the application of diamond for pn-junctions, BJTs, FETs, passive components, heat-spreading elements, field emission, and resistors. For the application of diamond in other areas, e.g. sensors and detectors [24.133–139], microwave filters [24.140], acoustic wave filters [24.69,141] and electro-mechanical microdevices [24.86, 142], the interested reader is directed to the above references.

pn-Junction

In order to exploit the superior properties of diamond for high-power/high-temperature electronic devices, high-quality inexpensive diamond must be available. The preparation of heteroepitaxial diamond is viable by bias-enhanced nucleation of iridium single-crystal film on sapphire or $SrTiO_3$ followed by low-pressure plasma-enhanced chemical vapor deposition. The crucial factor in the fabrication of electronic devices is the ability to prepare pn-junction, i.e. the ability to prepare in a reliable way p- and n-type materials with high carrier densities.

The p-type diamond has up to now been created by doping with boron. The boron is introduced from a gas or a solid [24.143–147] during diamond growth. The activation energy of electrical resistivity decreases with increasing boron concentration, being in the range 0–0.43 eV [24.143]. The energy is 0.35 eV for a boron concentration of $1\times 10^{18}\,cm^{-3}$ and it becomes zero for boron concentrations greater than $1.7\times 10^{20}\,cm^{-3}$ [24.143]. The p-type diamond can also be created by exposing the material to hydrogen plasma. The hydrogenation creates a layer with p-type conductivity with a fairly low resistivity $\approx 10^6\,\Omega\,cm$ [24.148, 149]. The discovery that both single-crystal and polycrystalline diamond can be treated by the plasma has led to successful use of these materials for the fabrication of electronic devices, e.g. the fabrication of FETs.

The growth of n-type diamond is one of the most challenging issues in the diamond field. Many theoretical predictions as well as experimental attempts to obtain n-type diamond have recently been published. Unfortunately, all experimental attempts have up to now delivered n-type diamond with low electron densities; this restricts the use of n-type diamond for the fabrication of transistor. Nitrogen is often used to create n-type diamond, but nitrogen-doped diamond is an electrical insulator at room temperature due to the deep level of 1.7 eV of the nitrogen impurity. Nitrogen is thus not a practical dopant atom to use. Phosphorus would be an obvious candidate but it has an impurity level of ≈ 0.59 eV, which is still rather high. Several reports on phosphorus doping have been published reporting activation energies in the range 0.32–0.59 eV [24.143, 150–152]. Reproducible results on phosphorus doping have been reported recently [24.150, 151]. At room temperature a reasonably high electron mobility of 250 cm^2/V was measured in the p-doped films [24.150]. Attempts to use other elements for n-type doping have also been made, e.g. sulphur (0.32 eV [24.143, 152, 153]), and lithium (0.16 eV [24.143, 154]). A survey of p-type doping of diamond has been given by *Kalish* [24.155].

Despite the experimental difficulties in n-type doping several research groups have succeeded to create a well function pn-junction. *Koizumi* et al. [24.150] created a pn-junction by epitaxial growth using a MPCVD technique to fabricate a phosphorus doped n-type diamond film on a boron doped p-type diamond film, on the [24.112] oriented surface of a diamond single crystal. The substrate contained over 100 ppm boron and exhibited high electrical conductivity. The electrical and optical properties of the junction have been investigated. The pn-junction exhibited clear diode characteristics, where the rectification ratio was over five orders in magnitude and the turn-on voltage was 4–5 V. The temperature dependence of the hole and electron concentrations in doped diamond films has been studied up to 1000 K. At room temperature the carrier mobilities of the boron and the phosphorus doped film were 300 and 60 cm^2/V s, respectively, and ultraviolet (UV) light emission has been observed at 235 nm.

A pn-junction has also been formed by using boron-doped p-type and nitrogen-doped n-type diamond films [24.156]. The active diamond films were grown by an MPCVD technique on heavily boron-doped HTHP diamond single crystals. The forward and reverse I–V characteristics of the junction were studied as a function of temperature up to 400 °C. At room temperature the resistivity of the n-layer is extremely high, ≈ 10 GΩ cm. The activation energy of the pn-junction diode saturation current obtained from these measurements was 3.8 eV. This value is in good agreement with theoretical predictions that assumes that the energy of the nitrogen level 1.7 eV. A pn-diode was also formed from a thick lithium-doped layer grown on a highly boron-doped substrate [24.143]. At room temperature the rectification ratio was $\approx 10^{10}$ for ± 10 V, but the series resistance was very high ≈ 200 kΩ, whereas at higher temperatures the series resistance decreased to 30 kΩ but the rectification ration decreased to $\approx 10^6$.

A single-crystal diamond pnp-type BJT consisting of boron/nitrogen pn-junctions has been fabricated [24.147, 156]. The transistor characteristics for common-base and common-emitter configurations have been measured. The temperature-activated leakage current limited the operation of the transistor up to 200 °C. The study showed that it was possible to fabricate a pnp-type BJT, but the high resistance of the base ≈ 10 GΩ m at 20 °C limited the operation of the BJTs to the DC mode, small currents (nA), and moderate temperatures.

Diodes as electronic devices must have a high reverse voltage and a low resistance in the forward direction. It has been shown that lightly boron-doped (10^{17} cm^{-3}) Schottky diodes on oxygen-terminated surfaces exhibit good performance at high temperatures. Stable contacts at high temperatures have been developed to create diodes that may operate up to 1000 °C [24.157–159]. The diodes showed breakdown behavior at reverse bias, but the reverse current was higher than theoretical values. The breakdown electric field for CVD diamond has been estimated to be 2.106 V/cm [24.143], a value well below that of natural diamond.

FETs

While the development of the diamond bipolar junction transistor is hindered by the lack of n-type diamond with high electron densities, there has been some promising results towards the development of diamond field-effect transistors, although studies at high temperatures are absent. Nevertheless, a short review of the present situation regarding diamond FETs is given here. There are presently two concepts for obtaining high-performance devices: boron δ-doped p-channel FETs and hydrogen-induced p-type surface-channel FET. Since a shallow diamond dopant is absent, δ-doping and two-dimensional (2D) conduction are essential parts of the diamond device concept.

Hydrogenated surfaces of natural diamond and films grown by plasma-assisted chemical vapor deposition technique exhibit substantial conductivity. Hydrogenation is carried out by treating the diamond surface with hydrogen plasma and cooling to room temperature in hydrogen ambient. The hydrogen-terminated surfaces exhibits the following, a p-type conductivity (2D conduction) of the order of 10^{-4}–$10^{-5}\,\Omega^{-1}$, a shallow acceptor level which is less than 50 meV, a high carrier concentration $10^{13}\,\text{cm}^{-2}$, and a hole mobility in the range 100–150 cm^2/Vs [24.148, 149, 161–165]. The thickness of the conductive layer is estimated to be less than 10 nm. It has been suggested that the surface conductivity depends on the details of the hydrogen termination together with the coverage of physisorbed adsorbates [24.166–168]. The high conductivity of the layer can be destroyed by dehydrogenation of the surface, by exposing the surface to oxygen or by heating in air at a temperature above 200 °C [24.148, 169]. The latter consideration may complicate the fabrication of devices. In order to exploit the conducting layer for the fabrication of high-temperature diamond devices, a way of protecting the layer has to be developed [24.169]. The opposite of hydrogenation is obtained by treating the diamond surface with oxygen plasma. The oxygen-terminated surface becomes highly insulating due to the surface potential pinning at ≈ 1.7 eV above the valence band [24.170]. This effect is exploited to isolate parts of a device when fabricating devices using the planar process. The hydrogen-terminated areas are transformed into oxygen-terminated ones. The hydrogen-terminated surface has a low surface-state density thus the barrier height between a metal and the surface is almost determined by the work function of the metal only. Usually Al and Au are used as ohmic contacts. Furthermore, when hydrogenation is used to create surface p-type conductivity, alternative cheaper types of diamond can be used [24.171].

The high hole conductivity allows for the fabrication of high-performance FETs, e.g. [24.160, 162, 172–175]. Two types of FET have been investigated, the metal–insulator–semiconductor field-effect transistor (MISFET) and the metal–semiconductor field-effect transistor (MESFET). The high-frequency performance and DC output characteristics of a MISFET with a 0.7-μm gate length has been investigated at room temperature [24.175]. The measurements show that the cutoff frequency f_T and the maximum frequency (MAG – maximum available gain) f_{max} are 11 and 18 GHz, respectively. In addition, a transconductance, g_m, of 100 mS/mm, has been obtained from DC measurements. The DC output characteristics of a circular FET with a gate length of 3 μm has been investigated at room temperature [24.176]. The maximum current was 90 mA/mm at a gate bias -6 V, and the maximum transconductance was 25 mS/mm. Breakdown voltage at pinch-off was -200 V. The high-frequency performance of MESFETs has also been also investigated at room temperature [24.177]. The cutoff frequency and the maximum frequency for a transistor with a 2-μm gate length were 2.2 and 7.0 GHz, respectively. The highest frequencies at room temperature were obtained for a MESFET with a gate length of 0.2 μm [24.174, 178, 179], and the following values were reported: maximum drain current, I_{Dmax}, 360 mA/mm, the transconductance, g_m, 150 mS/mm, the maximum drain voltage, V_{DSmax}, 68 V, and the maximum estimated power capability, P_{Rfmax}, 3.0 W/mm for class A.

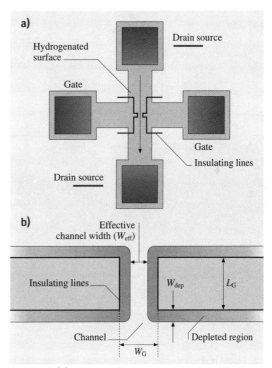

Fig. 24.8 (a) Schema of an in-plane transistor fabricated on a H-terminated diamond surface, using oxidized lines to define channel and gate, (b) channel region with oxidized lines and depletion regions along the *lines*. L_G and W_G are the gate length and gate width, respectively. (After [24.160])

The cutoff and maximum frequencies were measured to be 11.7 GHz and 31.7 GHz, respectively [24.178]. In a later investigation even higher frequencies were obtained, 21 GHz and 63 GHz, respectively [24.179]. These results indicate that a high gain can be obtained and that this transistor can operate at RF frequencies. Unfortunately, the instabilities of the hydrogen-terminated surface prevented large signal measurements on the device. The structure of a FET fabricated using the hydrogen layer is shown in Fig. 24.8.

The hole concentration of homoepitaxially prepared diamond films was measured for temperatures up to 500 °C. By altering the manner in which the temperature is increased and decreased, it has been shown that adsorbates from the environment can change the hole concentration in the conducting layer. Furthermore, the influence of the crystalline and surface properties of diamond homoepitaxial layers on device properties of hydrogen-terminated surface-channel FETs has been investigated [24.180].

The δ-channel FET approach is based on boron δ-doping, where the doping profile with peak concentration of 10^{20} cm^{-3} must be confined to within a few nanometers [24.181,182]. The DC output characteristics of δ-doped MESFETs were investigated at room temperature and at 350 °C [24.183]. The maximum drain current I_{Dmax} at room temperature and 350 °C were 35 μA/mm and 5 mA/mm, respectively. The maximum usable drain–source voltage V_{DSmax} at room temperature and 350 °C were 100 V and 70 V, respectively. A δ-channel JFET, with a δ-doping profile width of 1 nm, has been fabricated and the DC output performance has been investigated at a temperature of 250 °C and −125 °C [24.174]. The mobility of such δ-doped channels would be $\approx$ 350 cm^2/V s. At temperature 250 °C all the carriers are activated and a maximum drain current I_{Dmax} of 120 mA/mm has been measured for a transistor with a gate length of 2 μm and a gate width of 100 μm. Figure 24.9 shows δ-doping profile and fabricated recessed gate δ-channel JFET.

All FETs were fabricated from homoepitaxial films deposited on HTHP single-crystal diamond substrates, since the use of polycrystalline diamond films significantly reduced the FETs' performance. In addition, encouraging results which can be regarded as an important breakthrough in the development of high-power diamond electronics on the wafer scale, have been reported recently. The fabrication of high-performance p-type surface-channel MESFETs with sub-micron gate length from a heteroepitaxially grown diamond film. The diamond film was grown on a SrTiO$_3$ ceramic substrate with an intermediate iridium buffer layer [24.92]. The components exhibited similar performance as structures fabricated from homoepitaxial diamond films. The devices have been analyzed at room temperature under DC, small signal and large-signal RF power conditions. For devices with the smallest gate length of 0.24 μm the following values were obtained, $I_{Dmax} \approx$ 250 mA/mm, transconductance $\approx$ 97 mS/mm and V_{Dmax} − 90 V. The cutoff f_T and maximum frequencies $f_{max(MAG)}$ were measured to be 9.6 and 16.3 GHz, respectively.

In connection with the solid-state diamond FETs, the diamond vacuum field-effect transistor (VFET) ought to be mentioned. This component is part of vacuum electronics using diamond electrodes [24.184]. The diamond

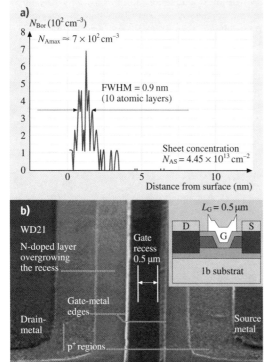

Fig. 24.9 (a) Elastic recoil detection (ERD) analysis of a δ-doped layer showing a full-width at half-maximum (FWHM) of 0.9 nm (10 atomic layers); (b) fabricated recessed gate δ-channel JFET with roughness visible at the bottom of the epitaxially overgrown recess trench. The gate length of 0.5 μm is given by the width of the gate recess. (After [24.174])

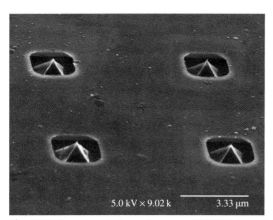

Fig. 24.10 Scanning electron microscopy (SEM) picture of a 2×2 array of gated diamond emitters. (After [24.185])

up to 500 °C. The capacitor functioned up to 450 °C with low loss and constant capacitance, whereas for higher temperatures the capacitance was temperature-dependent. The micro-switch was fabricated using the same type of diamond. The properties of diamond make it an ideal material for switch fabrication. Switches often operate in a vacuum at temperatures around 650 °C in a kHz frequency range. Moreover, no change in the switching threshold voltage indicates that there has been no change in the device's mechanical properties.

field emitter has been integrated with silicon-based micro-electromechanical system processing technology in order to obtain a monolithic diamond field-emitter triode on a Si wafer [24.185] (see Fig. 24.10). The monolithic VFET with a self-aligned n^{++} gate has been operated in a vacuum in the triode configuration, with an external anode placed 50 μm above the gate. The low turn-on gate voltage of 10 V, the high anode emission current of 4 μA at an applied gate voltage of 20 V, and the high gain factor of 250, indicate that high-speed diamond VFETs for low and high power can be realized.

Passive Components

Highly oriented heteroepitaxial diamond films grown on wafer-size Si, HOD films, are polycrystalline and cannot therefore be used for the fabrication of transistors or as substrates for the homoepitaxial growth of diamond. Nevertheless, these films possess properties that approach the properties of ideal diamond, and may be the most suitable material for MEMS technologies, heavy-duty, high-frequency and high-temperature electronic components for integrated high-power microwave circuits, and passive components like e.g. capacitors, switches, and planar waveguides. This should also enable the integration with Si- and 3C-SiC electronics.

In connection with high-temperature devices two passive components can be mentioned here, capacitors [24.187] and switches [24.188]. The capacitor has been realized using a diamond membrane as the dielectric and Au as the contacts. The diamond films were grown on large-area Si substrate by the MPCVD technique. The dielectric loss of the capacitor has been measured up to 600 °C and the C–V characteristics for

Heat-Spreading Elements

The properties of diamond suggest that its main application will be in power electronics at high frequencies and high temperature. Natural diamond has the highest known thermal conductivity, 2200 W/mK at room temperature. The highest-quality synthetic diamond grown by CVD technique has an identical thermal conductivity. This implies that power devices can be built on an integrated heat spreader. The high thermal conductivity and thermal stability of diamond will allow ideal heat sinks to be combined with high-power microelectronic and optoelectronic devices operating at high temperatures. Although the thermal conductivity of CVD diamond reduces with decreasing layer thickness, it still exceeds the conductivity of gold, even when the thickness is as low as 2 μm [24.189, 190].

GaN-based FETs have, owing to their excellent characteristics, great potential for high-frequency power

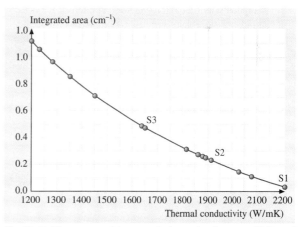

Fig. 24.11 The measured integrated absorption between 2760 and 3030 cm^{-1} plotted against the thermal conductivity determined using the laser flash technique. Samples S1–S3 are marked, the other data points are measured from a range of material with as-grown thickness of 100–800 μm. (After [24.186])

amplifiers [24.21]. A limitation of the performance of GaN FETs is the rise in operating temperature caused by heat dissipation, which leads to large leakage currents and reduced channel mobility [24.191, 192]. It has been demonstrated that a heat-spreading diamond film can be deposited on GaN FETs, without any degradation of the transistors characteristics [24.190]. The films of thickness 0.7 and 2 μm were deposited by the CVD technique using a new seeding process at low temperatures (less than 500 °C). The GaN FETs were fabricated by a standard process for GaN transistors.

A correlation between the integrated absorption in the IR range of $2700-3030\,\text{cm}^{-1}$, associated with stretch modes of CH_4, and the measured thermal conductivity at 300 K were reported for CVD diamond samples of different optical quality [24.186]. This provides a quick and reliable method of ascertaining a film's thermal quality. The measured integrated absorption as a function of the thermal conductivity is shown in Fig. 24.11.

Field Emission

The emission of electrons from an electrically active surface by means of tunneling into vacuum is commonly referred to as field emission. The emission properties of diamond, diamond-like carbon (DLC) and carbon-based materials are very robust [24.184]. Thin diamond film deposited by the CVD technique is an excellent cold electron emitter. The electron density yield is around $1\,\text{mA/cm}^2$ for an applied field of $10-100\,\text{V/μm}$. There is much scientific and technological interest in electron field emission from carbon-based materials for applications in many fields, e.g. vacuum field-effect transistors, diodes and triodes, ion sources, electron guns, flat-panel displays, scanning microscopes, energy conversion, and many others. A very large number of reports concerning the field emission from carbon-based materials has been published since the first reported electron emission from diamond in 1991 [24.193]. The physical reason behind the outstanding emission properties of these materials however is still under discussion [24.194].

Depending on the surface termination, diamond can exhibit a low or even negative electron affinity (NEA) [24.165]. The surfaces of diamond and DLC prepared by the CVD technique are naturally terminated by physisorbed hydrogen, making diamond the only semiconductor for which a true negative electron affinity can be obtained. This means that the vacuum level can be below the conduction-band minimum at the surface barrier [24.165]. The hydrogen layer lowers the electron affinity and reduces the turn-on voltage required to achieve field emission. Many reports treating different aspects of the field-emission process has been published to date, e.g. [24.195–197]. However, there are several problems connected with a direct application of diamond, DLC or other forms of carbon emitters, for instance nonuniform emitter microstructures that result in inconsistent emission and poor long-term stability. In diamond there are no free electrons to be easily emitted. A three-step emission model was proposed by *Cutler* et al. [24.198]. According to this model the electrons are injected through the back contact, transported through the diamond, and emitted at the diamond–vacuum surface. This model was extended by a proposition [24.199] that: (i) electrons are transferred by the applied electric field from the substrate to the nitrogenated DLC conduction band through a band-to-band tunneling process; (ii) electrons are transported across the nitrogenated DLC film through its conduction band, and (iii) electrons are emitted into the vacuum at the surface of nitrogenated DLC film by tunneling through the barrier. The barrier was crated by the strong upward band-bending at the surface.

Hydrogen has been the most commonly used element for inducing NEA on diamond surfaces. It has been reported that other materials e.g. a Ti layer, can also induce NEA on diamond surfaces [24.200]. High-temperature thermionic electron emission has been studied for several decades [24.201]. Fowler–Nordheim theory, the theoretical foundation of field emission, neglects the effects of temperature on emission characteristics. The temperature dependence of field emission from diamond film surfaces has been reported recently. The electron emission properties of nitrogen-doped diamond films were studied as a function of temperature for up to 950 °C [24.202]. The films were prepared by a microwave-assisted CVD technique for application as a low-temperature thermionic field-emission cathode. The film surfaces were terminated with hydrogen or titanium. Measurements at elevated temperatures showed the importance of a stable surface passivation. Hydrogen-passivated films showed enhanced electron emission, but measurements at elevated temperatures, e.g. 725 °C, showed that the hydrogen layer had degraded rapidly. The titanium-terminated films showed a similar enhanced emission as the hydrogen-terminated ones but the titanium-passivated surfaces were stable up to 950 °C. The electron emission increased with increasing temperature, implying that the field emission is strongly temperature-dependent. At temperatures below 500 °C no emission was detected, while increasing the temperature above 700 °C a strong contribution to the electron emission was detected. At constant temperature the emission current was nearly constant at low anode voltages, whereas it increased exponentially for anode voltages above 15 kV. At low anode voltages, below 10 kV, the activation energy was approximately 1 eV. This energy value indicates contributions from nitrogen donor levels and defect states to the emission current. It was suggested, according to the emission characteristics, that the emission at low anode voltages could be attributed to thermionic emission of electrons in the diamond's conduction band, whereas the exponential increase at higher voltages indicated a tunneling through or thermionic emission over the potential barrier.

The temperature dependence of electron field-emission characteristics of nitrogen-doped polycrystalline diamond films has also been investigated by [24.203]. The films were grown by the PECVD technique for high-temperature applications. The measurements show that maximum current density increased and the turn-on voltage decreased for temperatures between 300 and 500 °C. The experimental results were interpreted in terms of the Fowler–Nordheim theory and by means of an alternative model that included size and temperature effects explicitly. Furthermore, analysis of the temperature dependence of emission was carried through by parameter estimation of the effective emitting area, field-enhancement factor, and work function. All of the estimates indicated that the emission characteristics exhibit a strong dependence on temperature. From these results it was suggested that the thermally excited electrons are responsible for improved emission at high temperature.

Resistors

An attractive feature of diamond films is their ability to be an excellent dielectric when undoped, or as an interesting resistors when doped. Even polycrystalline or nanocrystalline diamond films, depending on the process deposition conditions, exhibit breakdown strength (as a dielectric) and power-density capability (as a resistor) with values particularly interesting for high-temperature and high-power applications. Diamond resistors were fabricated on a ceramic substrate of aluminum nitride [24.204], by etching in oxygen plasma from the CVD-deposited layer of boron-doped diamond films. The resistors were intended for investigation of high-power-density characteristics of boron-doped diamond. The resistor exhibits ohmic behavior at low to medium current levels, e.g. up to ≈ 3 mA. At higher currents, thermal excitations are important, and the carrier-density enhancement and conductivity increases. The component enters a thermal runaway situation, but unlike conventional devices, the resistor by virtue of diamonds tolerance for high power, continues to operate at very-high-power and temperature conditions. Operating the resistor under load at high voltages is not unlike operating certain devices in the reverse breakdown mode, except for the much higher power-density levels. For example, the resistance at low power is typically 315 kΩ. The power density just before entering the thermal runaway region is 480 kW/cm^2. At a maximum current of 14.2 mA and voltage of 413.2 V the resistance fell to 29 kΩ. At the end of the test the power was 5.87 W and the current density was 4730 A/cm^2 [24.204].

Technological applications of micro-electromechanical systems (MEMs) and optoelectronic system can be broadened by the existence of thermally stable thin-film resistive microheaters. Thermally stable resistors working in a wide temperature range were presented recently [24.205]. Metal containing amorphous carbon like diamond (DLC) films, CrSiDLC and MoSiDLC, were prepared by the CVD technique and doping with Mo and Cr. The resistors, in the form of strips, were deposited on glass ceramic substrates. The films were studied at $700–800$ °C. The average resistivity was 2×10^{-3} Ω. The thermal coefficient of resistance was 7×10^{-4} Ω/K and the activation energy was 30 meV.

24.3 Summary

This chapter reviewed the advances in research and technology of the wide-band-gap semiconducting material SiC and diamond for high-temperature electronics. The hexagonal 4H-SiC structure is the most appealing for high-temperature power electronics. Basic SiC devices for use in power electronics that can operate at elevated temperature were presented. Four major device concepts: Schottky and pn-diode, high-frequency devices, power switching devices and high-current devices were discussed. PiN high-blocking-voltage power diodes can operate above 300 °C with an order of magnitude increase in the reverse leakage-current density. JFET is supposed to be the most promising power device. Transistors with blocking voltage up to 5 kV were characterized up to 300 °C. MOSFETs are still under development, and their operation has been tested up to 400 °C. The operation of high-frequency transistors are limited by to Schottky barrier height and possible applications may be limited to below 350 °C. The most probable near-future devices are thyristors and GTOs since their junctions can withstand higher temperatures, and exhibit large blocking voltages and high conduction currents.

Concerning high-temperature/high-power electronics, diamond has the best properties and is at different stage of research than SiC, since the technology of diamond electronic devices is in its infancy. The foremost progress in diamond research has been in the preparation of high-quality heteroepitaxial diamond, which is the prerequisite for wafer-scale fabrication.

The state of the art concerning high-temperature devices are pn-diodes and Schottky diodes that operate up to 400 °C and 1000 °C, respectively, and passive component capacitors and switches that operate up to 450 and 650 °C, respectively. Moreover, the development of FETs have reached important milestones for both DC and RF modes. The performance of these components, however, are still below standard, and components that can operate at high temperatures and high powers are still far away. The development of the BJT is hindered by the unavailability of n-type diamond with high electron concentrations. The application of diamond for field emission has made a lot of progress, since vacuum FETs have been fabricated and field emission from diamond has been studied at high temperatures. Diamond as a heat sink for high-power electronic devices can soon find wide applications due to the recent progress in the preparation of large-area high-quality diamond films. Carbon-based resistors exhibit excellent thermal stability, and with the continuing progress in the preparation of carbon-based materials, will find increasing application in the fabrication of high-temperature devices.

References

24.1 P. G. Nuedeck, R. S. Okojie, L-Y. Chien: Proc. IEEE **90**, 1065–1076 (2002)
24.2 P. L. Dreike, D. M. Fleetwood, D. B. King, D. C. Sprauer, T. E. Zipperian: IEEE Trans. A **17**, 594–609 (1994)
24.3 C. S. White, R. M. Nelms, R. W. Johnson, R. R. Grzybowski: High temperature electronic systems using silicon semiconductors, Industry Applications Conference, 1998. Thirty-Third IAS Annual Meeting, Proc. IEEE **2**, 967–976 (1998)
24.4 M. Willander, H. L. Hartnagel (Eds.): *High Temperature Electronics* (Chapman Hall, London 1997)
24.5 S. Cristoloveanu, G. Reichert: High-Temp. Electron. Mater. Dev. Sensors Conf., 86–93 (1998)
24.6 B. Gentinne, J. P. Eggermont, D. Flandre, J. P. Colinge: Mater. Sci. Eng. B **46**, 1–7 (1997)
24.7 L. Demeus, P. Delatte, V. Dessard, S. Adriaensen, A. Viviani, C. Renaux, D. Flandre: The art of high temperature FD-SOI CMOS, IEEE Cat. No. 99EX372, IEEE Conference Proceedings (1999), High Temperature Electronics, 1999. HITEN 99. The Third European Conference on 4-7 July 1999, pp. 97–99
24.8 S. Cristoloveanu: Circ. Dev. Mag. **15**, 26–32 (1999)
24.9 J. C. Zolner: Solid-State Electron. **42**, 2153–2156 (1998)
24.10 P. Schmid, K. L. Lipka, J. Ibbetson, N. Nguyen, U. Mishra, L. Pond, C. Weitzel, E. Kohn: IEEE Electron. Dev. Lett. **19**, 225–227 (1998)
24.11 I. Vurgaftman, J. R. Meyer, L. R. Ram-Mohan: J. Appl. Phys. **89**, 5815–5875 (2001)
24.12 C. M. Carlin, J. K. Ray: The requirements for high-temperature electronics in a future high speed civil transport, Trans. 2nd Int. High-Temperature Electronics Conf., Charlotte, NC June 1994, ed. by D. B. King, F. V. Thome (Phillips Laboratory, Sandia National Laboratories, Wright Laboratory, 1994) I.19–I.26
24.13 Z. D. Gastineau: High-temperature smart actuator development for aircraft turbine engines, Proc. 5th Int. High-Temperature electronics Conf., Albuquerque, NM, June 2000, pp. X.1.1–X1.5
24.14 S. P. Rountree, S. Berjaoui, A. Tamporello, B. Vincent, T. Wiley: High-temperature measure while drilling: Systems and applications, proc. 5th Int. High-Temperature Electronics Conf., Albuquerque, NM, June 2000, pp. X.2.1–I.2.7
24.15 M. Gerber, J. A. Ferreira, I. W. Hofsajer, N. Seliger: High temperature, high power density packaging for automotive applications, Power Electronics Specialist Conference, 15–19 June, 2003. PESC '03. 2003 IEEE 34th Annual IEEE Vol **1**, 425–430 (2003)
24.16 S. Lande: Supply and demand for high temperature electronics, High Temperature Electronics, 1999. HITEN 99. The Third European Conference on 4-7 July, 1999, pp. 133–135
24.17 J. B. Baliga: IEEE Electron. Dev. Lett. **10**, 455–457 (1989)
24.18 A. R. Powell, L. B. Rowland: Proc. IEEE **90**, 942–955 (2002)
24.19 A. Elasser, T. P. Chow: Proc. IEEE **90**, 969–986 (2002)
24.20 J. A. Cooper, A. Agarwal: Proc. IEEE **90**, 956–968 (2002)
24.21 R. J. Trew: Proc. IEEE **90**, 1032–1047 (2002)

24.22 J. A. Cooper: Mater. Sci. Forum **389-393**, 15–20 (2002)

24.23 J. F. H. Custers: Physika **18**, 489–496 (1952)

24.24 L. S. Pan, D. R. Kania (Eds.): *Diamond: Electronic Properties and Applications* (Kluwer Academic, Boston 1995)

24.25 J. Walker: Rep. Prog. Phys. **42**, 1605–1660 (1979)

24.26 http://www.eetimes.com/story/0EG20030822S0005

24.27 J. Palmour, R. Singh, R. C. Glass, O. Kordina, C. H. Carter: (1997), Silicon carbide for power devices, Power Semiconductor Devices and ICs 1997, ISPSD '97, 1997 IEEE International Symposium on, 26-29 May 1997, pp. 25–32

24.28 A. Ellison: PhD Thesis; Disertation No. 599, Linköping Studies in Science and Technologies, Linköping University (1999)

24.29 G. R. Fisher, P. Barnes: Philos. Mag. **B61**, 217–236 (1990)

24.30 J. A. Lely: Ber. Dt. Keram. Ges. **32**, 299 (1955)

24.31 Y. M. Tariov, V. F. Tsvetkov: J. Cryst. Growth **43**, 209–212 (1978)

24.32 D. Nakamura, I. Gunjishima, S. Yamagushi, T. Ito, A. Okamoto, H. Kondo, S. Onda, K. Takatori: Nature **430**, 1009–1012 (2004)

24.33 P. G. Nuedeck, J. A. Powell: IEEE Electron. Dev. Lett. **15**, 63–65 (1994)

24.34 H. McD. Hobgood, M. F. Brady, M. R. Calus, J. R. Jenny, R. T. Leonard, D. P. Malta, S. G. Müller, A. R. Powell, V. F. Tsvetkov, R. C. Glass, C. H. Carter: Mater. Sci. Forum **457-460**, 3–8 (2004)

24.35 A. R. Powell, R. T. Leonard, M. F. Brady, S. G. Müller, V. F. Tsvetkov, R. Trussel, J. J. Sumakeris, H. McD. Hobgood, A. A. Burk, R. C. Glass, C. H. Carter: Mater. Sci. Forum **457-460**, 41–46 (2004)

24.36 R. C. Glass, L. O. Kjellberg, V. F. Tsvetkov, J. E. Sundgren, E. Janzén: J. Cryst. Growth **132**, 504–512 (1993)

24.37 J. P. Bergman, H. Lendermann, P. A. Nilsson, U. Lindefelt, P. Skytt: Mater. Sci. Forum **353-356**, 299–302 (2001)

24.38 J. R. Jenny, D. P. Malta, M. R. Calus, S. G. Müller, A. R. Powell, V. F. Tsvetkov, H. McD. Hobgood, R. C. Glass, C. H. Carter: Mater. Sci. Forum **457-460**, 35–40 (2004)

24.39 H. M. McGlothin, D. T. Morisette, J. A. Cooper, M. R. Melloch: 4 kV silicon carbide Schottky diodes for high-frequency switching applications, Dev. Res. Conf. Dig. 1999 57th Annual, 28-30 June 1999, Santa Barbara, California, USA, pp. 42–43

24.40 S. Liu, J. Scofield: Thermally stable ohmic contacts to 6H- and 4H- p-type SiC, High Temperature Electronics Conference, HITEC 1998, 4th International, 14-18 June 1998, pp. 88–92

24.41 Q. Wahab, A. Ellison, J. Zhang, U. Forsberg, E. A. Duranova, L. D. Madsen, E. Janzén: Mat. Sci. Forum **338-342**, 1171–1174 (1999)

24.42 R. Raghunathan, B. J. Baliga: IEEE Electron. Dev. Lett. **19**, 71–73 (1998)

24.43 L. Zheng, R. P. Joshi: J. Appl. Phys. **85**, 3701–3707 (1999)

24.44 K. J. Schoen, J. P. Henning, M. Woodall, J. A. Cooper, M. R. Melloch: IEEE Electron. Dev. Lett. **19**, 97–99 (1998)

24.45 Landolt-Börnstein: *Condensed Matter Group III/41A2, Impurities and Defects in Group IV Elements, IV-IV and III-V Compounds. Part α: Group IV Elements*, ed. by W. Martiensen (Springer, Berlin, Heidelberg 2001)

24.46 J. Walker: Rep. Prog. Phys. **42**, 1605–1654 (1979)

24.47 A. T. Collins, B. C. Lightowlers: *The Properties of Diamond* (Academic, London 1979) p. 87

24.48 J. E. Field (Ed.): *The Properties of Natural and Synthetic Diamond* (Academic, New York 1992)

24.49 M. H. Nazaré, A. J. Neves (Eds.): *Properties, Growth and Applications of Diamond* (INSPEC, London 2001)

24.50 R. J. Trew, J. B. Yan, P. M. Mock: Proc. IEEE **79**, 598–620 (1991)

24.51 M. W. Geis, N. N. Efremow, D. D. Rathman: J. Vac. Sci. Technol. **A6**, 1953–1954 (1988)

24.52 K. Shenai, R. S. Scott, B. J. Baliga: IEEE Trans. Electron. Dev. **36**, 1811–1823 (1989)

24.53 B. J. Baliga: J. Appl. Phys. **53**, 1759 (1982)

24.54 A. M. Zaitsev: *Handbook of Diamond Technology* (Trans Tech, Zurich 2000) p. 198

24.55 S. Kumar, P. Ravindranathan, H. S. Dewan, R. Roy: Diamond Rel. Mater. **5**, 1246–1248 (1996)

24.56 Ch. Q. Sun, H. Xie, W. Zang, H. Ye, P. Hing: J. Phys. D: Appl. Phys. **33**, 2196–2199 (2000)

24.57 J. Chen, S. Z. Deng, J. Chen, Z. X. Yu, N. S. Su: Appl. Phys. Lett. **74**, 3651 (1999)

24.58 C. E. Jonhson, M. A. S. Hasting, W. A. Weimar: J. Mater. Res. **5**, 2320–2325 (1990)

24.59 Ch. Q. Sun, M. Alam: J. Electrochem. Soc. **139**, 933–936 (1992)

24.60 F. X. Lu, J. M. Liu, G. C. Chen, W. Z. Tang, C. M. Li, J. H. Song, Y. M. Tong: Diamond Rel. Mater. **13**, 533–538 (2004)

24.61 R. R. Nimmagadda, A. Joshi, W. L. Hsu: J. Mater. Res. **5**, 2445–2450 (1990)

24.62 F. K. de Theije, O. Roy, N. J. van der Laag, W. J. P. van Enckevort: Diamond Rel. Mater. **9**, 929–934 (2000)

24.63 F. K. de Theije, E. van Veenendaal, W. J. P. van Enckevort, E. Vlieg: Surface Sci. **492**, 91–105 (2001)

24.64 R. C. Burns, G. J. Davies: *The Properties of Natural and Synthetic Diamond* (Academic, London 1992)

24.65 B. V. Spitsyn, L. L. Bouilov, B. V. Derjaguin: J. Cryst. Growth **52**, 219–226 (1981)

24.66 H. Okushi: Diamond Rel. Mater. **10**, 281–288 (2001)

24.67 M. Kamo, Y. Sato, S. Matsumoto, N. J. Setaka: J. Cryst. Growth **62**, 642 (1983)

24.68 Y. Saito, S. Matsuda, S. Nogita: J. Mater. Sci. Lett. **5**, 565–568 (1986)

24.69 B. Dischler, C. Wild: *Low-Pressure Synthetic Diamond* (Springer, Berlin Heidelberg New York 1998)

24.70 P.W. May: Phil. Trans. R. Soc. Lond. **A 358**, 473 (2000)
24.71 J. Isberg, J. Hammarberg, E. Johansson, T. Wikströ, D.J. Twitchen, A.J. Whitehead, S.E. Coe, G.A. Scarsbrook: Science **297**, 1670–1672 (2002)
24.72 J. Isberg, J. Hammarberg, D.J. Twitchen, A.J. Whitehead: Diamond Rel. Mater. **13**, 320–324 (2004)
24.73 B.R. Stoner, J.T. Glass: Appl. Phys. Lett. **60**, 698–700 (1992)
24.74 T. Tachibana, Y. Yokota, K. Nishimura, K. Miyata, K. Kobashi, Y. Shintani: Diamond Rel. Mater. **5**, 197–199 (1996)
24.75 H. Karawada, C. Wild, N. Herres, R. Locher, P. Koidl, H. Nagasawa: J. Appl. Phys. **81**, 3490–3493 (1997)
24.76 M. Oba, T. Sugino: Jpn. J. Appl. Phys. **39**, L1213–L1215 (2000)
24.77 S. Koizumi, T. Murakami, T. Inuzuka, K. Suzuki: Appl. Phys. Lett. **57**, 563–565 (1990)
24.78 S.D. Wolter, B.R. Stoner, J.T. Glass, P.J. Ellis, D.S. Buhaenko, C.E. Jenkins, P. Southworth: Appl. Phys. Lett. **62**, 1215–1217 (1993)
24.79 X. Jiang, C.P. Klages, R. Zachai, M. Hartweg, H.J. Füsser: Appl. Phys. Lett. **62**, 3438–3440 (1993)
24.80 C. Wild, R. Kohl, N. Herres, W. Müller-Sebert, P. Koidl: Diamond Rel. Mater. **3**, 373–381 (1994)
24.81 X. Jiang, C.P. Klages: *New Diamond and Diamond-Like Films* (Techna, Srl. 1995) pp. 23–30
24.82 S.D. Wolter, T.H. Borst, A. Vescan, E. Kohn: Appl. Phys. Lett. **68**, 3558 (1996)
24.83 A. Flöter, H. Güttler, G. Schulz, D. Steinbach, C. Lutz-Elsner, R. Zachai, A. Bergmaier, G. Dollinger: Diamond Rel. Mater. **7**, 283–288 (1998)
24.84 X. Jiang, K. Schiffmann, C.P. Klages, D. Wittorf, C.L. Jia, K. Urban, W. Jäger: J. Appl. Phys. **83**, 2511–2518 (1998)
24.85 P. Gluche, M. Adamschik, A. Vescan, W. Ebert, F. Szücs, H.J. Fecht, A. Flöter, R. Zachai, E. Kohn: Diamond Rel. Mater. **7**, 779–782 (1998)
24.86 P. Gluche, M. Adamschik: Diamond Rel. Mater. **8**, 934–940 (1999)
24.87 E. Kohn, A. Aleksov, A. Denisenko, P. Schmid, M. Adamschick, J. Kusterer, S. Ertl, K. Janischowsky, A. Flöter, W. Ebert: Diamond and Other Carbon Materials, Diamond in electronic applications, CIMTEC 2002 – 3rd Forum on New Materials 3rd International Conference, Florence 2002, ed. by P. Vincenzini, P. Ascarelli (Techna, 2003) 205–216
24.88 T. Tsubota, M. Ohta, K. Kusakabe, S. Morooka, M. Watanabe, H. Maeda: Diamond Rel. Mater. **9**, 1380–1387 (2000)
24.89 K. Ohtsuka, K. Suzuki, A. Sawabe, T. Inuzuka: Jpn. J. Appl. Phys. **35**, 1072 (1996)
24.90 M. Schreck, H. Roll, B. Stritzker: Appl. Phys. Lett. **74**, 650–652 (1999)
24.91 M. Schreck, A. Schury, F. Hörmann, H. Roll, B. Stritzger: J. Appl. Phys. **91**, 676–685 (2002)
24.92 M. Kubovic, A. Aleksov, M. Schreck, T. Bauer, B. Strizker, E. Kohn: Diamond Rel. Mater. **12**, 403–407 (2003)
24.93 C. Bednarski, Z. Dai, A.P. Li, B. Golding: Diamond Rel. Mater. **12**, 241–245 (2003)
24.94 Z. Dai, C. Bednarski-Meinke, B. Golding: Diamond Rel. Mater. **13**, 552–556 (2004)
24.95 S. Gsell, T. Bauer, J. Goldfuss, M. Schreck, B. Stritzker: Appl. Phys. Lett. **84**, 4541–4543 (2004)
24.96 Infineon Technologies (2001), SDP04S60, SPD04S60, SDP06S60, SDB06S60, SDP10S30, SDB10S30, and SDB20S30 preliminary data sheets, homepage http://www.infineon.com/cgi/ecrm.dll/ecrm/scripts/search/advanced_search_result.jsp?queryString=SiC&x=22&y=7 (Jan. 2001)
24.97 M. Mazzola: *SiC high-temperatuer wideband gap mateials*, Combat hybrid power system component technologies (National Academic, Washington D.C. 2001) pp. 31–40
24.98 W. Wondrak, E. Niemann: Proc. IEEE International Symposium (ISIE '98), 1998, p. 153
24.99 T. Jang, L.M. Porter: Electrical characteristics of tantalum and tantalum carbide Schottky diodes on n- and p-type silicon carbide as a function of temperature, Proc. 4th International High Temperature Electronics Conference (HITEC 98) **4**, 280–286 (1998) Cat. No. 98EX145
24.100 Y. Sugawara, D. Takayama, K. Asano, R. Singh, J. Palmour, T. Hayashi: Int. Symp. on Power Semiconductor Devices & ICs, Osaka, Japan **13**, 27–30 (2001)
24.101 R. Singh, J.A.Jr. Cooper, R. Melloch, T.P. Chow, J. Palmour: IEEE Trans. Electron. Dev. **49**, 665 (2002)
24.102 R. Singh, K.G. Irvine, D. Cappel, T. James, A. Hefner, J.W. Palmour: IEEE Trans. Electron. Dev. **49**, 2308 (2002)
24.103 R. Singh, D.C. Capell, K.G. Irvine, J.T. Richmond, J.W. Palmour: Electron. Lett. **38**, 1738 (2002)
24.104 B.M. Wilamowski: Solid State Electron. **26**, 491 (1983)
24.105 B.J. Baliga: IEEE Elect. Dev. Lett. **5**, 194 (1984)
24.106 C.M. Zetterling, F. Dahlquist, N. Lundberg, M. Ostling: Solid State Electron. **42**, 1757 (1998)
24.107 F. Dahlquist, C.M. Zetterling, M. Ostling, K. Rutner: Mater. Sci. Forum **264-268**, 1061 (1998)
24.108 F. Dahlquist, H. Lendenmann, M. Ostling: Mater. Sci. Forum **389-393**, 1129 (2002)
24.109 D. Nakamura, I. Gunjishima, S. Yamagushi, T. Ito, A. Okamoto, H. Kondo, S. Onda, K. Takatori: Nature **430**, 1009–1012 (2004)
24.110 R.C. Clarke, A.W. Morse, P. Esker, W.R. Curtice: Proc. IEEE High Performance Devices, 2000 IEEE/Cornell Conference, 141–143 (2000)
24.111 S.T. Sheppard, R.P. Smith, W.L. Pribble, Z. Ring, T. Smith, S.T. Allen, J. Milligan, J.W. Palmour: Proc. Device Research Conference, 60th DRC. Conference Digest, 175–178 (2002)

24.112 S.T. Allen, W.L. Pribble, R.A. Sadler, T.S. Alcorn, Z. Ring, J.W. Palmour: IEEE MTT-S Int. **1**, 321–324 (1999)

24.113 F. Villard, J.P. Prigent, E. Morvan, C. Brylinski, F. Temcamani, P. Pouvil: IEEE Trans. Microw. Theory Technol. **51**, 1129 (2003)

24.114 J.B. Casady, E.D. Luckowski, R.W. Johnsson, J. Crofton, J.R. Williams: IEEE Electronic Components and Technology Conference **45**, 261–265 (1995)

24.115 A. Agarwal, C. Capell, B. Phan, J. Miligan, J.W. Palmour, J. Stambaugh, H. Bartlow, K. Brewer: IEEE High Performance Devices, 2002, Proceedings, IEEE Lester Eastman Conference, p. 41

24.116 B.J. Baliga: Proc. 6th Int. Conf. on SiC & Related Mat. Ser. 142, 1–6 (1996)

24.117 K. Asano, Y. Sugawara, T. Hayashi, S. Ryu, R. Singh, J.W. Palmour, D. Takayama: IEEE Power Semiconductor Devices and ICs, 2002. Proceedings of the 14th International Symposium, p. 61

24.118 J.H. Zhao, X. Li, K. Tone, P. Alexandrov, M. Pan, M. Weiner: IEEE Semiconductor Device Research Symposium, 2001 International, p. 564

24.119 C.J. Scozzie, C. Wesley, J.M. McGarrity, F.B. Mclean: Reliability Physics Symposium, 1994. 32nd Annual Proceedings., IEEE International, p. 351

24.120 K. Asano, Y. Sugawara, S. Ryu, J.W. Palmour, T. Hayashi, D. Takayama: Proc. 13th Int. Symp. Power Semiconductor Devices and ICs, Osaka, Japan, 2001

24.121 N.S. Rebello, F.S. Shoucair, J.W. Palmour: IEEE Proc. Circuits Dev. Syst. **143**, 115 (1996)

24.122 T. Billon, T. Ouisse, P. Lassagne, C. Jassaud, J.L. Ponthenier, L. Baud, N. Becourt, P. Morfouli: Electron. Lett. **30**, 170 (1994)

24.123 S.H. Ryu, A. Agarwal, J. Richmond, J. Palmour, N. Saks, J. Williams: IEEE-Power Semiconductor Devices and ICs, 2002. Proceedings of the 14th International Symposium, 65–68 (2002)

24.124 J.B. Casady, A. Agarwal, L.B. Rowland, S. Shshadri, R.R. Siergiej, D.C. Sheridan, S. Mani, P.A. Sanger, C.D. Brandt: IEEE-Compound Semiconductors, International Symposium, 1998, p. 359

24.125 J. Spitz, M.R. Melloch, J.A. Jr. Cooper, M.A. Capano: IEEE Electron. Dev. Lett. **19**, 100 (1998)

24.126 I.A. Khan, J.A. Jr. Cooper, M. Capano, T. Isaacs-Smith, J.R. Williams: IEEE – Power Semiconductor Devices and ICs, [2002] Proceedings of the 14th International Symposium, 157–160 (2002)

24.127 R.K. Chilukuri, M. Praveen, B.J. Baliga: IEEE Trans. Ind. Appl. **35**, 1458 (1999)

24.128 G.I. Gudjonsson, H.O. Olafsson, E.O. Sveinbjornsson: Mater. Sci. Forum **457-460**, 1425–1428 (2004)

24.129 H.O. Olafsson, G.I. Gudjonsson, P.O. Nillson, E.O. Sveinbjornsson, H. Zirath, R. Rodle, R. Jos: Electron. Lett. **40**, 508–509 (2004)

24.130 L.A. Lipkin, J.W. Palmour: Mater. Sci. Forum **338**, 1093 (2000)

24.131 R.C. Clarke, C.D. Brandt, S. Sriram, R.R. Siergiej, A.W. Morse, A. Agarwal, L.S. Chen, V. Balakrishna, A.A. Burk: Proc. IEEE High Temperature Electronic Materials, Devices and Sensors Conference, 1998, p. 18

24.132 S.H. Ryu, A. Agarwal, R. Singh, J.W. Palmour: IEEE Trans. Electron. Dev. Lett. **22**, 124 (2001)

24.133 S.F. Kozlov: IEEE Trans. Nucl. Sci **222**, 160 (1975)

24.134 S.F. Kozlov, E.A. Komorova, Y.A. Kuznetsov, Y.A. Salikov, V.I. Redko, V.R. Grinberg, M.L. Meilman: IEEE Trans. Nucl. Sci **24**, 235–237 (1977)

24.135 M. Krammer, W. Adam, E. Berdermann, P. Bergonzo, G. Bertuccio, F. Bogani, E. Borchi, A. Brambilla, M. Bruzzi, C. Colledani: Diamond Rel. Mater. **10**, 1778–1782 (2001)

24.136 A. Mainwood: Semicond. Sci. Technol. **15**, 55 (2000)

24.137 M. Adamschik, M. Müller, P. Gluche, A. Flöter, W. Limmer, R. Sauer, E. Kohn: Diamond Rel. Mater. **10**, 1670–1675 (2001)

24.138 A. Denisenko, A. Aleksov, E. Kohn: Diamond Rel. Mater. **10**, 667–672 (2001)

24.139 M.D. Whitfield, S.P. Lansley, O. Gaudin, R.D. McKeag, N. Rizvi, R.B. Jackman: Diamond Rel. Mater. **10**, 715–721 (2001)

24.140 J. Gondolek, J. Kocol: Diamond Rel. Mater. **10**, 1511–1514 (2001)

24.141 P.R. Chalker, T.B. Joyce, C. Johnston: Diamond Rel. Mater. **8**, 309–313 (1999)

24.142 E. Kohn, M. Adamschik, P. Schmid, S. Ertl, A. Flöter: Diamond Rel. Mater. **10**, 1684–1691 (2001)

24.143 T.H. Borst, O. Weis: Diamond Rel. Mater. **4**, 948–953 (1995)

24.144 B.A. Fox, M.L. Hartsell, D.M. Malta, H.A. Wynands, G.J. Tessmer, D.L. Dreifus: Mater. Res. Soc. Symp. Proc. **416**, 319 (1996)

24.145 S. Yamanaka, H. Watanabe, S. Masai, D. Takeuchi, H. Okushi, K. Kajimura: Jpn. J. Appl. Phys. **37**, 1129 (1998)

24.146 D. Saito, E. Tsutsumi, N. Ishigaki, T. Tashiro, T. Kimura, S. Yugo: Diamond Relat. Mater. **11**, 1804–1807 (2002)

24.147 A. Aleksov, A. Denisenko, M. Kunze, A. Vescan, A. Bergmeir, G. Dollinger, W. Ebert, E. Kohn: Semicond. Sci. Technol. **18**, S59–S66 (2003)

24.148 M.I. Landstrass, K.V. Ravi: Appl. Phys. Lett. **55**, 975–977 (1989)

24.149 M.I. Landstrass, K.V. Ravi: Appl. Phys. Lett. **55**, 1391–1393 (1989)

24.150 S. Koizumi, K. Watanabe, M. Hasegawa, H. Kanda: Diamond Rel. Mater. **11**, 307–311 (2002)

24.151 M. Katagiri, J. Isoya, S. Koizumi, H. Kanda: Appl. Phys. Lett. **85**, 6365–6367 (2004)

24.152 E. Gheeraerf, A. Casanova, A. Tajani, A. Deneuville, E. Bustarret, J.A. Garrido, C.E. Nebel, M. Stutzmann: Diamond Rel. Mater. **11**, 289–295 (2000)

24.153 J. F. Prins: Diamond Rel. Mater. **10**, 1756–1764 (2001)

24.154 H. Sternschulte, M. Schreck, B. Strizker, A. Bergmaier, G. Dollinger: Diamond Rel. Mater. **9**, 1046–1050 (2000)

24.155 R. Kalish: Diamond Rel. Mater. **10**, 1749–1755 (2001)

24.156 A. Aleksov, A. Denisenko, E. Kohn: Solid-State Electron. **44**, 369–375 (2000)

24.157 P. Gluche, A. Vescan, W. Ebert, M. Pitter, E. Kohn: Processing of High Temperature Stable Contacts to Single Crystal Diamond, *Transient Thermal Processing Techniques in Electronic Materials*, Proc. 12th TMS Annual Meeting, Anaheim 1996, ed. by N. M. Ravindra, R. K. Singh (Warrendale, PA 1996) 107–110

24.158 A. Vescan, I. Daumiller, P. Gluche, W. Ebert, E. Kohn: IEEE Electron. Dev. Lett. **18**, 556–558 (1997)

24.159 A. Vescan, I. Daumiller, P. Gluche, W. Ebert, E. Kohn: Diamond Relat. Mater. **7**, 581–584 (1998)

24.160 J. A. Garrido, C. E. Nebel, R. Todt, G. Rösel, M.-C. Amann, M. Stutzmann, E. Snidero, P. Bergonzo: Appl. Phys. Lett. **82**, 988–990 (2003)

24.161 T. Maki, S. Shikama, M. Komori, Y. Sakagushi, K. Sakura, T. Kobayashi: Jpn. J. Appl. Phys. **31**, 1446–1449 (1992)

24.162 H. Karawada: Surface Sci. Rep. **26**, 205–259 (1996)

24.163 K. Hayashi, S. Yamanaka, H. Okushi, K. Kajimura: Appl. Phys. Lett. **68**, 376–378 (1996)

24.164 K. Hayashi, S. Yamanaka, H. Watanabe, T. Sekiguchi, H. Okushi, K. Kajimura: J. Appl. Phys. **81**, 744–753 (1997)

24.165 J. Ristein, F. Maier, M. Riedel, J. B. Cui, L. Ley: Phys. Status Solidi **A 181**, 65 (2000)

24.166 F. Maier, M. Riedel, B. Mantel, J. Ristein, L. Ley: Phys. Rev. Lett. **85**, 3472–3475 (2000)

24.167 B. F. Mantel, M. Stammer, J. Ristein, L. Ley: Diamond Rel. Mater. **10**, 429–433 (2001)

24.168 J. Ristein, F. Maier, M. Riedel, M. Stammer, L. Ley: Diamond Rel. Mater. **10**, 416–422 (2001)

24.169 T. Yamada, A. Kojima, A. Sawaba, K. Suzuki: Passivation of hydrogen terminated diamond surface conductive layer by hydrogenated amorphous carbon, Diamond Rel. Mater. **13**, 776–779 (2004)

24.170 Y. Otsuka, S. Suzuki, S. Shikama, T. Maki, T. Kobayashi: Jpn. J. Appl. Phys. **34**, 551 (1995)

24.171 O. A. Williams, R. B. Jackman, C. Nebel, J. S. Foord: Diamond Rel. Mater. **11**, 396–399 (2002)

24.172 B. A. Fox, M. L. Hartsell, D. M. Malta, H. A. Wynands, C.-T. Kao, L. S. Plano, G. J. Tessmer, R. B. Henard, J. S. Holmes, A. J. Tessmer, D. L. Dreifus: Diamond Rel. Mater. **4**, 622–627 (1995)

24.173 K. Tsugawa, H. Kitatani, A. Noda, A. Hokazono, K. Hirose, M. Tajima: Diamond Rel. Mater. **8**, 927–933 (1999)

24.174 A. Aleksov, M. Kubovic, N. Kaeb, U. Spitzberg, A. Bergmaier, G. Dollinger, Th. Bauer, M. Schreck, B. Stritzker, E. Kohn: Diamond Rel. Mat. **12**, 391–398 (2003)

24.175 H. Ishizaka, H. Umezawa, H. Taniuchi, T. Arima, N. Fujihara, M. Tachiki, H. Kawarada: Diamond Rel. Mat. **11**, 378–381 (2002)

24.176 P. Gluche, A. Aleksov, V. Vescan, W. Ebert, E. Kohn: IEEE Electron. Dev. Lett. **18**, 547–549 (1997)

24.177 H. Taniuchi, H. Umezawa, T. Arima, M. Tachiki, H. Kawarada: IEEE Electron. Dev. Lett. **22**, 390–392 (2001)

24.178 A. Aleksov, A. Denisenko, U. Spitzberg, T. Jenkins, W. Ebert, E. Kohn: Diamond Rel. Mat. **11**, 382–386 (2002)

24.179 A. Aleksov, M. Kubovic, M. Kasu, P. Schmid, D. Grobe, S. Ertl, M. Schreck, B. Stritzker, E. Kohn: Diamond Rel. Mater. **13**, 233–240 (2004)

24.180 M. Kasu, M. Kubovic, A. Aleksov, N. Teofilov, Y. Taniyasu, R. Sauer, E. Kohn, T. Makimoto, N. Kobayashi: Diamond Rel. Mater. **13**, 226–232 (2004)

24.181 M. Kunze, A. Vescan, G. Dollinger, A. Bergmaier, E. Kohn: Carbon **37**, 787–791 (1999)

24.182 A. Aleksov, A. Vescan, M. Kunze, P. Gluche, W. Ebert, E. Kohn, A. Bergmeier, G. Dollinger: Diamond Rel. Mater. **8**, 941–945 (1999)

24.183 A. Vescan, P. Gluche, W. Ebert, E. Kohn: IEEE Electron Dev. Lett. **18**, 222–224 (1997)

24.184 W. P. Kang, T. S. Fischer, J. L. Davidson: New Diamond Frontier Carbon Technol. **11**, 129 (2001)

24.185 J. L. Davidson, W. P. Kang, A. Wisitsora-At: Diamond Rel. Mater. **12**, 429–433 (2003)

24.186 D. J. Twitchen, C. S. J. Pickles, S. E. Coe, R. S. Sussmann, C. E. Hall: Diamond Rel. Mater. **10**, 731–735 (2001)

24.187 W. Ebert, M. Adamschik, P. Gluche, A. Flöter, E. Kohn: Diamond Rel. Mater. **8**, 1875–1877 (1999)

24.188 E. Kohn, M. Adamschik, P. Schmid, S. Ertl, A. Flöter: Diamond Rel. Mater. **10**, 1684–1691 (2001)

24.189 E. Worner: *Low Pressure Synthetic Diamond* (Springer, Berlin Heidelberg New York 1998) p. 137

24.190 M. Seelmann-Eggebert, P. Meisen, F. Schaudel, P. Koidl, A. Vescan, H. Leier: Diamond Rel. Mater. **10**, 744–749 (2001)

24.191 Y. F. Wu, B. P. Keller, S. Keller: IEICE Trans. Electron. **E82**(11), 1895 (1999)

24.192 C. E. Weitzel: IOP Conf. Series **142**, 765 (1996)

24.193 B. C. Djubua, N. N. Chubun: IEEE Trans. Electron. Dev. **38**, 2314 (1991)

24.194 J. Robertson: J. Vac. Sci. Technol. **B 17**, 659–665 (1999)

24.195 J. B. Cui, M. Stammler, J. Ristein, L. Ley: J. Appl. Phys. **88**, 3667–3673 (2000)

24.196 W. Choi, Y. D. Kim, Y. Iseri, N. Nomura, H. Tomokage: Diamond Rel. Mater. **10**, 863–867 (2001)

24.197 K. M. Song, J. Y. Shim, H. K. Baik: Diamond Rel. Mat. **11**, 185–190 (2002)

24.198 P. H. Cutler, N. M. Miskovsky, P. B. Lerner, M. S. Ching: Appl. Surf. Sci. **146**, 126–133 (1999)

24.199 X.-Z. Ding, B. K. Tay, S. P. Lau, J. R. Shi, Y. J. Li, Z. Sun, X. Shi, H. S. Tan: J. Appl. Phys. **88**, 5087–5092 (2000)
24.200 J. van der Weide, R. J. Nemanich: J. Vac. Sci. Technol. **B 10**, 1940–1943 (1992)
24.201 T. S. Fischer, D. G. Walker: J. Heat Trans. **124**, 954–962 (2002)
24.202 F. A. M. Köck, J. M. Garguilo, B. Brown, R. J. Nemanich: Diamond Rel. Mater. **11**, 774 (2002)
24.203 S. H. Shin, T. S. Fisher, D. G. Walker, A. M. Strauss, W. P. Kang, J. L. Davidson: J. Vac. Sci. Technol. **B 21**, 587–592 (2003)
24.204 J. L. Davidson, W. Kang, K. Holmes, A. Wisitsora-At, P. Taylor, V. Pulugurta, R. Venkatasubramanian, F. Wells: Diamond Rel. Mater. **10**, 1736–1742 (2001)
24.205 V. K. Dimitrev, V. N. Inkin, G. G. Kirpilenko, B. G. Potapov, E. A. Ylyichev, E. Y. Shelukhin: Diamond Rel. Mater. **10**, 1007–1010 (2001)

25. Amorphous Semiconductors: Structure, Optical, and Electrical Properties

This chapter is devoted to a survey of the structural, optical and electrical properties of amorphous semiconductors on the basis of their fundamental understanding. These properties are important for various types of applications using amorphous semiconductors.

First, we review general aspects of the electronic states and defects in amorphous semiconductors, i.e., a-Si:H and related materials, and chalcogenide glasses, and their structural, optical and electrical properties.

Further, we survey the two types of phenomena associated with amorphous structure, i.e., light-induced phenomena, and quantum phenomena associated with nanosized amorphous structure. The former are important from the viewpoint of amorphous-silicon solar cells. The latter phenomena promise novel applications of amorphous semiconductors from the viewpoint of nanotechnology.

25.1	Electronic States	565
25.2	Structural Properties	568
	25.2.1 General Aspects	568
	25.2.2 a-Si:H and Related Materials	568
	25.2.3 Chalcogenide Glasses	569
25.3	Optical Properties	570
	25.3.1 General Aspects	570
	25.3.2 a-Si:H and Related Materials	571
	25.3.3 Chalcogenide Glasses	572
25.4	Electrical Properties	573
	25.4.1 General Aspects	573
	25.4.2 a-Si:H and Related Materials	574
	25.4.3 Chalcogenide Glasses	575
25.5	Light-Induced Phenomena	575
25.6	Nanosized Amorphous Structure	577
References		578

Amorphous semiconductors are promising electronic materials for a wide range of applications such as solar cells, thin-film transistors, light sensors, optical memory devices, vidicons, electrophotographic applications, X-ray image sensors, europium-doped optical-fibre amplifications etc, particularly, hydrogenated amorphous silicon (a-Si:H) for solar cells, thin-film transistors, X-ray image sensors, and chalcogenide glasses for optical memory devices including digital video/versatile disk (DVD). In this chapter, we emphasize the basic concepts and general aspects of the electronic properties of amorphous semiconductors such as their electrical and optical properties as well as their structural properties [25.1–3]. Furthermore, some basic and important results of these properties are described to understand these applications and to consider their further development. Light-induced phenomena in amorphous semiconductors, which have been considered to be associated with amorphous structure, are also described.

Nanosized amorphous structures exhibit quantum effects associated with two-dimensional (quantum well), one-dimensional (quantum wire) and zero-dimensional (quantum dot) structures, so they have received significant attention on both the fundamental and application sides. These topics are briefly described.

25.1 Electronic States

Long-range disorder in amorphous network breaks down the periodic arrangement of constituent atoms, as shown in Fig. 25.1. In the figure, the structures of amorphous and crystalline silicon are shown. The periodic arrangement of atoms makes it easy to treat the electronic states mathematically, i.e., the so-called Bloch the-

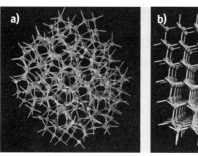
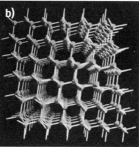

Fig. 25.1a,b Structural models of (**a**) amorphous silicon and (**b**) crystalline silicon

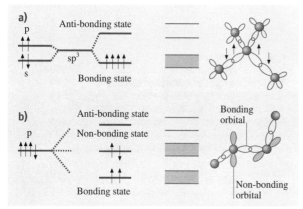

Fig. 25.2a,b Schematic diagram of the energy levels of atomic orbitals, hybridized orbitals and bands for (**a**) tetrahedrally bonded semiconductors, e.g., a-Si and for (**b**) selenium

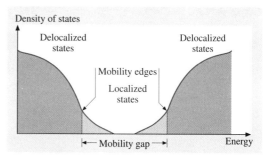

Fig. 25.3 Schematic diagram of the density of states of the conduction band (delocalized states) and the valence band (delocalized states) shown by the *brown-colored*. The mobility edge and mobility gap are shown

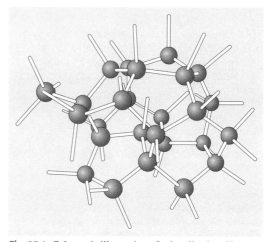

Fig. 25.4 Schematic illustration of a dangling bond in tetrahedrally bonded amorphous semiconductors

ory can be applied to crystalline solids. On the other hand, it becomes difficult to treat the electronic states in amorphous solids mathematically. However, electronic states in amorphous semiconductors are simply described in terms of tight-binding approximations and Hartree–Fock calculations. Using these approaches, spatial fluctuations of bond length, bond angle and dihedral angle lead to broadening of the edges of the conduction and valence bands, constructing the band tail. This is due to spatial fluctuations of bond energy between constituent atoms in tetrahedrally coordinated semiconductors such as Si and Ge, in which the bonding state and the antibonding state constitute the valence band and the conduction band, respectively, as shown in Fig. 25.2a. In chalcogenide glasses, the conduction band arises from the antibonding state, while the valence band arises from the nonbonding state, as shown in Fig. 25.2b. The broadening of those bonds also occurs as a result of potential fluctuations associated with the amorphous network of constituent atoms. The static charge fluctuation associated with bond-length and bond-angle variations in the amorphous network has been discussed theoretically [25.4] and experimentally [25.5, 6] in amorphous silicon.

In the band tails, the electronic states have a localized character and their nature changes from localized to delocalized at a critical boundary called the mobility edge, as shown in Fig. 25.3. The energy separation between the two mobility edges of the conduction and valence bands is called the mobility gap. The nature of the conduction and valence bands has been elucidated by means of photoemission spectroscopy [25.7]. The structure of actual samples of amorphous semiconduc-

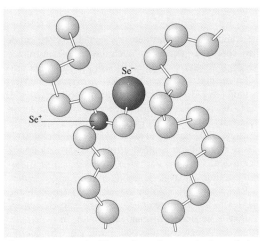

Fig. 25.5 Schematic illustration of two selenium chains with positively charged threefold-coordinated selenium and negatively charged onefold-coordinated selenium

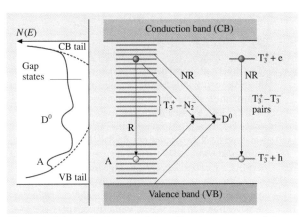

Fig. 25.6 Schematic diagram of the tail and gap states and the recombination processes in a-Si:H. R and NR designate radiative and nonradiative recombination, respectively. A diagram of the density of states spectrum is illustrated on the *left-hand side*. For the definition of other symbols, see the text. (After [25.9])

tors deviates from the ideal random network, namely, coordination of constituent atoms deviates from the normal coordination following the $8 - N$ rule [25.1], where N designates the number of valence electrons. For instance, the normal coordinations of Si and Se are four and two, respectively. However, in actual samples Si atoms with a threefold coordination are present in a-Si (Fig. 25.4), and Se atoms with onefold and threefold coordinations are present in a-Se (Fig. 25.5). These atoms are generally called structural defects (gap states) whose electronic energy levels are located within the band-gap region. In the case of chalcogenide glasses, onefold- and threefold-coordinated Se atoms are called valence-alternation pairs (VAP) [25.8]. The detailed properties of structural defects have been elucidated by means of optical spectroscopies such as photoinduced absorption (PA), photothermal deflection spectroscopy (PDS) and the constant-photocurrent method (CPM), capacitance measurements such as deep-level transient spectroscopy (DLTS) and isothermal capacitance transient spectroscopy (ICTS), and resonance methods such as electron spin resonance (ESR) and electron–nuclear double resonance (ENDOR) [25.2].

In the following, we take a-Si:H as an example and describe the nature of its electronic states. Figure 25.6 shows a schematic diagram of the electronic states involved in the band-gap region of a-Si:H proposed by the authors' group [25.9]. The tails of the conduction and valence bands are expressed by an exponential function with widths of 25 meV [25.10] and 48–51 meV [25.11–13], respectively.

In this figure, the processes of radiative recombination and nonradiative recombination are shown, along with the energy levels of the gap states involved in the recombination processes. The radiative recombination occurs between tail electrons and tail holes at higher temperatures and between tail electrons and self-trapped holes (A centres) at lower temperatures, particularly below 60 K. Such recombinations contribute to the principal luminescence band peaked at 1.3–1.4 eV. These have been elucidated by optically detected magnetic resonance (ODMR) measurements [25.14–16], and are briefly mentioned in Sect. 25.3. The radiative recombination center participating in the low-energy luminescence (defect luminescence) may be attributed to distant $T_3^+ - N_2^-$ pair defects [25.17, 18], in which T_3^+ and N_2^- are the positively charged threefold-coordinated silicon center and the negatively charged twofold-coordinated nitrogen center, respectively. The N_2^- centers are created from contaminating nitrogen atoms introduced during sample preparation. On the other hand, close $T_3^+ - N_2^-$ pair defects act as nonradiative recombination centers. A typical nonradiative recombination center in a-Si:H is the neutral silicon dangling bond, i.e., T_3^0. The details of the recombination processes in a-Si:H are described in Sect. 25.3.

25.2 Structural Properties

25.2.1 General Aspects

The structural properties of amorphous semiconductors have been investigated by means of X-ray, electron and neutron diffraction, transmission electron microscope (TEM) and scanning electron microscope (SEM), extended X-ray absorption fine structure (EXAFS), small-angle X-ray scattering (SAXS), Raman scattering, infrared absorption (IR) and nuclear magnetic resonance (NMR). The absence of long-range order in amorphous semiconductors is manifested in diffraction techniques, e.g., electron diffraction for a-Si results in a halo pattern as shown in Fig. 25.7a, while Laue spots are seen for c-Si, as shown in Fig. 25.7b. The medium-range order of 0.5–50 nm has been discussed based on SAXS measurements.

25.2.2 a-Si:H and Related Materials

The radial distribution function (RDF) in a-Si has been obtained from an analysis of curves of scattered electron intensity versus scattering angle from electron diffraction measurements, as shown in Fig. 25.8 [25.19]. From a comparison of RDF curves of a-Si and c-Si, it was concluded that the first peak of a-Si coincides with that of c-Si, the second peak of a-Si is broadened compared to that of c-Si, the third peak of c-Si almost disappears and a small peak appears, shifted from the third peak of c-Si, as shown in Fig. 25.8. These results indicate that the bond length of a-Si is elongated by 1% over that of c-Si (= 2.35 Å), and that the bond angle is tetrahedral (= 109.47°) with a variation of 10% from that of c-Si. Using the neutron diffraction technique, the RDF was also deduced from the measured structure factor, $S(Q)$, as a function of Q (the momentum transfer) in a-Si [25.20]. The RDF curve of a-Si exhibiting these features was well simulated by the continuous random network model [25.21]. Since then, computer-generated models have been constructed (see, e.g., *Kugler* et al. [25.20]), using the Monte Carlo (MC) method [25.22], the molecular dynamics (MD) method [25.23] and the reverse MC method [25.24, 25]. Here, the reverse MC method, i.e., a MC simulation repeatedly carried out to reach consistency with the measured RDF curve and curve of structure factor $S(Q)$ versus Q from the neutron diffraction data, is briefly mentioned. First we start with an initial set of Cartesian coordinates (particle configuration) and calculate its RDF and $S(Q)$. Comparing the calculated RDF and $S(Q)$ with the measured ones, the new particle configuration is generated by random motion of a particle, being consistent with the constraint, e.g., the coordination number. From the reverse MC simulation, the bond-angle distribution is derived, i.e., it shows a peak at the tetrahedral angle, 109.5° and small peaks at $\approx 60°$ and $\approx 90°$ [25.26].

In a-Si:H, knowledge of the hydrogen configuration is important to understand its electronic properties. This is obtained by NMR [25.27, 28] and IR measurements [25.29]. The bonding modes of Si and H atoms are $Si-H$, $Si-H_2$, $(Si-H_2)_n$ and $Si-H_3$ bonds. The concentrations of these hydrogen configurations depend on the preparation conditions, particularly on the substrate temperature, i.e., the deposition temperature. a-Si:H films prepared at 250 °C mostly contain $Si-H$ bonds in incorporated hydrogen (hydrogen content of ≈ 10 at. %), while those prepared at lower temperatures such as room temperature contain all these bonds in incorporated hydrogen (hydrogen content of

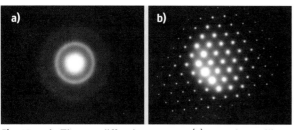

Fig. 25.7a,b Electron diffraction patterns. (a) amorphous silicon, (b) crystalline silicon

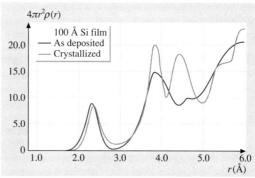

Fig. 25.8 Radial distribution function of amorphous (evaporated) and crystalline silicon obtained from electron diffraction patterns. (After [25.19])

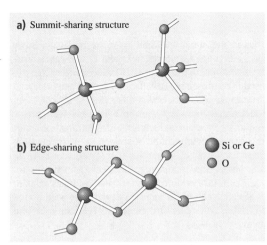

Fig. 25.9a,b Schematic illustration of (a) summit-sharing structure and (b) edge-sharing structure. (After [25.32])

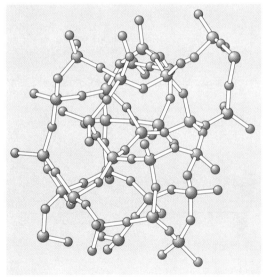

Fig. 25.10 Structural model of germanium disulfide. (After [25.32])

≈ 30 at. %). The hydrogen configuration has also been investigated by neutron scattering measurements using the isotope-substitution method [25.30, 31]. Partial pair-correlation functions of Si−Si, Si−H and H−H are

obtained from those measurements. The Si−H bond length is 1.48 Å.

Raman scattering measurements provide useful information about microcrystallinity, i.e., the volume fraction of the amorphous phase in microcrystalline silicon [25.33]. The Raman shift associated with transverse optical (TO) phonons is observed at $520\,\text{cm}^{-1}$ for c-Si, while it is observed at $480\,\text{cm}^{-1}$ for a-Si.

25.2.3 Chalcogenide Glasses

The chalcogenide glasses are composed of chalcogenide atoms and other constituent atoms such as As, Ge etc. Their crystalline counterpart has a structure consisting of threefold-coordinated As atoms and twofold-coordinated chalcogens, i.e., c-Se and c-As$_2$S$_3$ exhibit a chain structure and a layered structure, respectively. The structure of the chalcogenide glasses is fundamentally built from these constituent atoms, keeping their coordination in the crystalline counterpart. However, actual glasses contain wrong bonds, i.e., homopolar bonds such as Se−Se, As−As, Ge−Ge etc. In the following, we take two examples of chalcogenide glasses, As$_2$X$_3$ (X = S, Se) and GeX$_2$ (X = S, Se).

1. As$_2$X$_3$ (X = S, Se)
   This type of glass, e.g., As$_2$S$_3$, is mainly composed of AsS$_3$ pyramid units, i.e., a random network of these units sharing the twofold-coordinated S site [25.34]. There is no correlation between As atoms, but interlayer correlation seems to hold even for the amorphous structure.

2. GeX$_2$ (X = S, Se)
   Silica glass (SiO$_2$) is a well-known material, which is mainly composed of SiO$_4$ tetrahedron units sharing the twofold-coordinated O site (the summit-sharing structure. See Fig. 25.9a). On the other hand, SiS$_2$ and GeSe$_2$ glasses have a different structure from SiO$_2$ glass: the tetrahedron units are connected through two X sites (the edge-sharing structure) along with one X site (the summit-sharing structure. See Fig. 25.9b). For example, in GeSe$_2$, the edge-sharing structure occurs at a level of more than 30% and the summit-sharing structure with less than 70%. Actually, wrong bonds (homopolar bonds) Ge−Ge, Se−Se exist in GeSe$_2$ glass. A structural model of GeS$_2$ [25.32] is shown in Fig. 25.10.

25.3 Optical Properties

25.3.1 General Aspects

Optical absorption and luminescence occur by transition of electrons and holes between electronic states such as conduction and valence bands, tail states, and gap states. In some cases, electron–phonon coupling is strong and, as a result, self-trapping occurs. Exciton formation has also been suggested in some amorphous semiconductors.

The absorption of photons due to interband transition, which occurs in crystalline semiconductors, is also observed in amorphous semiconductors. However, the absorption edge is not clear since interband absorption near the band gap is difficult to distinguish from tail absorption in the absorption spectra. Figure 25.11 schematically illustrates typical absorption spectra of amorphous semiconductors. The absorption coefficient, α, due to interband transition near the band gap is known to be well described by the following equation [25.35]

$$\alpha \hbar \omega = B(\hbar \omega - E_g)^2 , \quad (25.1)$$

where $\hbar\omega$ and E_g denote the photon energy and optical gap, respectively. In most amorphous semiconductors, the optical gap E_g is determined by a plot of $(\alpha\hbar\omega)^{1/2}$ versus $\hbar\omega$, which is known as Tauc's plot. The photon energy at which the absorption coefficient is 10^4 cm^{-1}, E_{04}, is also used for the band gap in a-Si:H. The absorption coefficient at the photon energy just below the optical gap (tail absorption) depends exponentially on the photon energy, $E = \hbar\omega$, as expressed by

$$\alpha(E) \propto \exp\left(\frac{E}{E_U}\right) , \quad (25.2)$$

where E_U is called the Urbach energy and also the Urbach tail width. In addition, optical absorption by defects appears at energies lower than the optical gap.

Photoluminescence (PL) occurs as a result of the transition of electrons and holes from excited states to the ground state. After interband excitation, electrons relax to the bottom of the conduction band by emitting phonons much more quickly than the radiative transition. Similarly, the holes also relax to the top of the valence band. In the case of crystalline semiconductors without defects or impurities, there is no localized state in the band gap and PL occurs by transition between the bottom of the conduction band and the top of the valence band. In this case the k-selection rule, $k_{photon} = k_i - k_f$, must be satisfied, where k_{photon}, k_i and k_f denote the wavenumbers of photons, and electrons in the initial and final states, respectively. Since k_{photon} is much smaller than k_i and k_f, we can rewrite the selection rule as $k_i = k_f$. Semiconductors satisfying this condition are called direct-gap semiconductors. Crystalline silicon is one semiconductor in which the direct transition is not allowed by the k-selection rule (indirect-gap semiconductors), but the transition is allowed by either absorption of phonons or their emission. On the other hand, strong PL is observed in a-Si:H. In amorphous semiconductors, the k-selection rule is relaxed and, furthermore, the electrons and holes relax to localized states in the gap before radiative recombination. Thus, PL occurs by transitions between localized states.

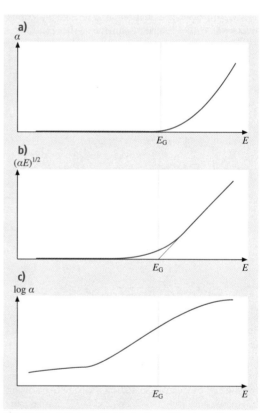

Fig. 25.11a–c Schematic illustration of the absorption spectra of amorphous semiconductors. (**a**) The absorption coefficient α plotted as a function of photon energy E. (**b**) $(\alpha E)^{1/2}$ versus E (Tauc's plot). (**c**) $\log \alpha$ versus E. E_G is the optical gap determined from (**b**)

The lifetime measurement is important for the identification of the origin of the PL since the PL spectra in most amorphous semiconductors are featureless and do not provide much information. The probability of radiative transition between localized electrons and holes depends exponentially on their separation, R. The lifetime is estimated by $\tau = \tau_0 \exp(2R/R_0)$ where R_0 denotes the radius of the most extended wavefunction [25.36]. When the Coulomb interaction is strong enough, excitons are formed. Exciton formation is also suggested in some amorphous semiconductors. The total spin of excitons, S, is either 0 or 1. Singlet excitons of $S = 0$ have a lifetime in the nanosecond region, while triplet excitons with $S = 1$ have a lifetime much longer than that of singlet excitons, e.g. $\tau = 1$ ms, since the transition is forbidden by the spin selection rule.

PL measurements combined with magnetic resonance are a powerful means to study recombination processes [25.14, 15]. This is called optically detected magnetic resonance (ODMR) measurements. PL from triplet excitons has been suggested from ODMR measurements, as described in Sects. 25.3.2 and 25.3.3.

The photoconductivity measurements provide us with useful information about the processes of carrier transport and recombination. Their results on a-Si:H and chalcogenide glasses are briefly reviewed in Sects. 25.3.2 and 25.3.3, respectively.

25.3.2 a-Si:H and Related Materials

The band gap of a-Si:H determined from Tauc's plot is 1.7–1.9 eV, depending on the preparation condition, particularly the deposition temperature, T_s, i. e. hydrogen content, [H], [25.37–39, 48] as shown in Table 25.1. The absorption coefficient of visible light for a-Si:H is of the order 10^5–10^6 cm^{-1}, which is large compared to that of crystalline silicon. A thickness of 1 μm is enough to absorb visible light. Thus a-Si:H is suitable for application to solar cells especially when a thin film is desired.

The optical gap, E_g, and the Urbach tail width, E_U, of a-Si:H, a-Ge:H and Si-based alloys are shown in Table 25.2. Some a-Si:H-based alloys such as a-Si$_{1-x}$N$_x$:H and a-Si$_{1-x}$C$_x$:H have a band gap wider than that of a-Si:H depending on the composition, x. In the preparation of these alloys, x can be varied continuously over a certain range. Thus, the band gap can be varied arbitrarily. It is possible to prepare multilayer films consisting of layers with different band gaps, such as a-Si:H/a-Si$_{1-x}$N$_x$:H. Quantum size effects in multilayer films will be described in Sect. 25.6.

It is difficult to obtain the absorption spectra below the band gap from transmittance measurements in a-Si:H, which is normally prepared as a thin film. Such low absorption in a-Si:H has been measured by PDS and CPM.

PL from a-Si:H was observed by Engemann and Fischer [25.49] for the first time. The quantum efficiency has been found to be of the order of unity [25.50], although it decreases with increasing density of dangling bonds [25.51]. PL spectra from a-Si:H films generally consist of two components. The first is observed as a peak at 1.3–1.4 eV with a FWHM of 0.3 eV. The other component, which is called low-energy PL (defect PL), is observed at 0.8–0.9 eV.

The origin of the main peak has not been fully understood. Electrons and holes in the tail states may give

Table 25.1 Optical gap of a-Si:H prepared at various deposition temperatures

T_s (°C)	[H] (at.%)	E_g (eV)	References
300	7	1.70	[25.37, 38]
200	18	1.75	[25.37, 38]
120	28	1.9	[25.37, 38]
75	33	2.0	[25.38, 39]

T_s: Deposition temperature, [H]: Hydrogen content

Table 25.2 Band gap energies, Urbach tail widths and dark conductivities at room temperature for a-Si:H, a-Ge:H and related materials

Material	E_g (eV)	E_U (meV)	$\sigma(300\,\text{K})$ ($\Omega^{-1}\text{cm}^{-1}$)	References
a-Si:H	1.75	48	10^{-11}	[25.40, 41]
a-Ge:H	1.05	50	10^{-4}	[25.40, 41]
a-Si$_{0.7}$C$_{0.3}$:H	2.28	183.4		[25.42]
a-Si$_{0.8}$C$_{0.2}$:H	2.2		10^{-15}	[25.43]
a-Si$_{0.4}$N$_{0.6}$:H	3.0	≈ 200	≈ 10^{-8}	[25.44–46]
a-Si$_{0.74}$N$_{0.26}$:H	≈ 2.0	≈ 100	1.4×10^{-8}	[25.44, 45, 47]

rise to this PL. Dunstan and Boulitrop [25.52] have shown that the PL spectra of a-Si:H can be understood without considering electron–phonon coupling. However, the electron–phonon coupling has still been discussed. Morigaki proposed a model on the basis of ODMR measurements [25.14] in which self-trapped holes and tail electrons are the origin of the PL at low temperatures [25.53], while tail holes and tail electrons participate in the PL at high temperatures.

In the case of radiative recombination of localized electron–hole pairs the lifetime depends exponentially on the separation of the electron and the hole, as described in Sect. 25.3.1. When the spatial distribution of electrons and holes is random, a broad lifetime distribution is predicted. This model describes some properties well, e.g., the generation-rate dependence of the lifetime. However, it has been pointed out that we have to consider the PL from specific electronic states such as excitonic PL to understand the lifetime distribution at low generation rate [25.54–56]. PL due to triplet excitons in a-Si:H has been suggested by optically detected magnetic resonance measurements [25.57, 58]. The PL from singlet excitons, which is expected to have lifetime of about 10 ns, has also been reported [25.59–61].

The low-energy PL is emission from deep gap states created by defects. However the origin of low-energy PL is not neutral dangling bonds (T_3^0) of Si since they act as nonradiative centers, as has been suggested from optically detected magnetic resonance [25.62]. Yamaguchi et al. [25.17, 18] proposed a model in which the origin of the low-energy PL is $T_3^+ - N_2^-$ pairs.

The drift mobility of carriers, i.e., electrons and holes, in a-Si:H has been measured by the method of time of flight (TOF), using blocking electrodes [25.63]. In the measurement, carriers are created by an optical pulse near one of the electrodes and run as a sheet-like shape against another electrode. The transient photocurrent i_p associated with a pulsed optical excitation exhibits a dispersive behavior, as shown in Fig. 25.12 [25.63, 64], which is given by

$$i_p \propto t^{-(1-\alpha)} \quad \text{for} \quad t < t_r \tag{25.3}$$

and

$$i_p \propto t^{-(1+\alpha)} \quad \text{for} \quad t > t_r, \tag{25.4}$$

where t_r designates the transit time, as shown in Fig. 25.12, given by

$$t_r = \frac{L}{\mu_d F}, \tag{25.5}$$

in which L, μ_d and F are the separation of the two electrodes, the drift mobility and the magnitude of the electric field, respectively. In dispersive conduction, the drift mobility depends on the thickness of the samples. This is due to dispersive carrier processes, which are caused by dispersion of the flight time. Band conduction with multiple trapping of carriers also exhibits dispersive conduction, as was observed in a-Si:H and chalcogenide glasses. For undoped a-Si:H, the drift mobility was on the order of $1 \text{ cm}^2/\text{V}$ at room temperature. $\alpha = 0.51$ was obtained as 160 K [25.11]. This transient behavior depends on temperature, i.e., it becomes more dispersive at lower temperatures. α is related to the exponential tail width (normally equal to E_U) $E_c \equiv k_B T_c$ as follows:

$$\alpha = \frac{T}{T_c}. \tag{25.6}$$

From the value of α, the tail width of the conduction band has been obtained to be 25 meV. For hole transport, dispersive behavior has also been obtained and a drift mobility of the order of $10^{-2} \text{ cm}^2/\text{V}$ has been obtained at room temperature.

For the steady-state photoconductivity of undoped a-Si:H, it has been generally accepted that photoconduction occurs through electrons in the conduction band at temperatures above ≈ 60 K and through hopping of tail electrons at temperatures below ≈ 60 K. The detailed processes of trapping and recombination of carriers involved in photoconduction have been discussed in many literatures (see, e.g., *Morigaki* [25.2]; *Singh* and *Shimakawa* [25.3]). In relation to this issue, spin-dependent photoconductivity measurements are very useful, and have therefore been extensively performed on a-Si:H [25.65–67].

25.3.3 Chalcogenide Glasses

As mentioned in Sect. 25.3.1, the absorption coefficient for the interband transition in many chalcogenide

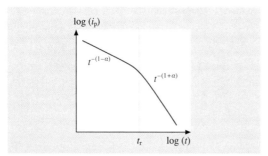

Fig. 25.12 Schematic illustration of the transient photocurrent curve for the dispersive transport

Table 25.3 Band gap energies, drift mobilities, and dark conductivities at room temperature for chalcogenide glasses [25.2]

Material	E_g [25.68] (eV)	μ_d (cm^2/Vs) [25.1,69] n-type	p-type	σ (300 K) [25.68] (Ω^{-1}cm^{-1})
Se	2.05	$3-6 \times 10^{-3}$	$1-2 \times 10^{-1}$	10^{-16}
As$_2$S$_3$	2.32			10^{-17}
As$_2$Se$_3$	1.76		10^{-3}	10^{-12}
As$_2$Te$_3$	0.83		10^{-2}	10^{-4}
GeS$_2$	3.07			10^{-14}
GeSe$_2$	2.18	1.4×10^{-1}	4×10^{-2}	10^{-11}
Sb$_2$Se$_3$	0.70			
Cd–In–S [25.70]	2.2	30 *)		9×10^{-1}

*) Hall mobility

glasses is well described by Tauc's relationship (25.1). The optical gap E_g is shown in Table 25.3 for various chalcogenide glasses. The effect of light on optical absorption in chalcogenide glasses is known as photo-darkening. This effect will be briefly reviewed in Sect. 25.5.

PL has been observed in a-Se, arsenic chalcogenide glasses and germanium chalcogenide glasses. PL in chalcogenide glasses is known to have a large Stokes shift. The peak energy in the PL spectra is approximately half of the band gap. This stokes shift has been attributed to strong electron–phonon coupling [25.71]. The electronic states responsible for the PL have not been well understood. There has been some experimental evidence for PL from self-trapped excitons. The observation of triplet excitons has been reported in ODMR measurements for chalcogenide glasses such as As$_2$S$_3$ and As$_2$Se$_3$ [25.72–74].

Dispersive transport was observed in the transient photocurrent for the first time for chalcogenide glasses such as a-As$_2$Se$_3$, in which the hole transport dominates over the electron transport [25.63]. The drift mobility of holes ranges between $10^{-3}-10^{-5}$ cm^2/V, depending on the temperature and electric field. The zero-field hole mobility at room temperature is about 5×10^{-5} cm^2/V. The steady-state photoconduction in a-As$_2$Se$_3$ has been considered to be governed by charged structural defects such as VAPs.

25.4 Electrical Properties

25.4.1 General Aspects

Electrical Conductivity

Electrical conduction in amorphous semiconductors consists of band conduction and hopping conduction. Band conduction in undoped amorphous semiconductors is characterized by

$$\sigma = \sigma_0 \exp\left(-\frac{E_a}{k_B T}\right), \quad (25.7)$$

where σ and σ_0 are the electrical conductivity and a prefactor, respectively, and E_a, k_B and T are the activation energy, the Boltzmann constant and the temperature, respectively. E_a is given by either $E_c - E_F$ or $E_F - E_V$, depending on whether electrons or holes are considered, where E_c, E_V and E_F are the mobility edges of the conduction band and the valence band, and the Fermi energy, respectively. Hopping conduction in amorphous semiconductors consists of nearest-neighbor hopping and variable-range hopping [25.1]. Nearest-neighbor hopping is well known in crystalline semiconductors, in which electrons (holes) hop to nearest-neighbor sites by emitting or absorbing phonons. Variable-range hopping is particularly associated with tail states, in which electrons (holes) in tail states hop to the most probable sites. This type of hopping conductivity σ_p is characterized by the following temperature variation:

$$\sigma_p = \sigma_{p0} \exp\left(-\frac{B}{T^{1/4}}\right). \quad (25.8)$$

The Hall Effect

The Hall effect is used for the determination of carrier density and its sign in crystalline semiconductors. For amorphous semiconductors, however, the sign of the Hall coefficient does not always coincide with that of the carriers [25.76]. Such a sign anomaly has been observed in a-Si:H, as will be shown in Sect. 25.4.2.

Thermoelectric Power

The thermoelectric power S associated with band conduction of electrons is given by

$$S = -\frac{k_B}{e}\left(\frac{E_c - E_F}{k_B T} + A\right), \quad (25.9)$$

where A is a quantity depending on the energy dependence of the relaxation times associated with electrical conduction. For hole conduction, S is given by (25.9) except that $E_c - E_F$ is replaced by $E_F - E_V$ and that the sign of S is positive. The sign of S coincides with that of the carriers. The thermoelectric power S (25.9) is related to the electrical conductivity σ (25.7) as follows [25.77]:

$$\ln\sigma + \left|\frac{e}{k_B}S\right| = \ln\sigma_0 + A \equiv Q. \quad (25.10)$$

The temperature-dependent quantity Q is defined as above. The activation energies of σ and S, E_σ and E_S, are equal to each other. Then, Q is defined by Q_0 as follows:

$$Q = \ln\sigma_0 + A \equiv Q_0. \quad (25.11)$$

However, if E_σ is not equal to E_S, Q is generally expressed by

$$Q = Q_0 - \frac{E_Q}{k_B T} \quad (25.12)$$

$$E_Q = E_\sigma - E_S. \quad (25.13)$$

It has been generally observed that E_σ is greater than E_S, i.e., $E_\sigma > E_S$. This has been accounted for in terms of long-range fluctuation of the band edge. The electrical conduction is due to those carriers that are thermally excited into the valley of the band-edge fluctuation and cross over its barrier, while the thermoelectric power, i.e., transport of phonon energy by carriers, is governed by carriers in the valleys. Thus, a relationship of $E_\sigma > E_S$ can be accounted for in terms of long-range fluctuation of the band edge [25.77].

Drift Mobility

The drift mobility of carriers is measured by the TOF method, as mentioned in Sect. 25.3.2. The drift mobility μ_d is estimated from (25.5).

In the following, the electrical properties of a-Si:H and related materials and chalcogenide glasses are described.

25.4.2 a-Si:H and Related Materials

A typical example of band conduction and variable-range hopping conduction is shown in Fig. 25.13, which is the temperature dependence of the electrical conductivity for a-Si_xAu_{1-x} films prepared by vapor evaporation [25.75]. Band conduction occurs at high temperatures and variable-range hopping conduction at low temperatures. The $T^{-1/4}$ law of temperature variation of electrical conductivity is clearly seen in Fig. 25.14. The values of the dark conductivities of a-Si:H, a-Ge:H and related materials at room temperature are shown in Table 25.2. Figure 25.15 shows the temperature dependencies of S for a-Ge:H, in which two slopes of the curve of S versus T^{-1} are seen with 0.43 eV in the high-temperature range and 0.17 eV in the low-temperature range [25.78]. This crossover of two slopes is accounted for in terms of two different mechanisms of electrical conduction, i.e., band conduction at high temperatures and hopping conduction in the band-tail states at low temperatures [25.79]. In amorphous semiconductors, doping control is generally difficult, because the constituent

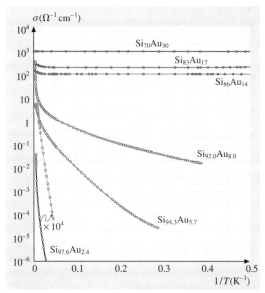

Fig. 25.13 Temperature dependence of the electrical conductivity for an a-Si_xAu_{1-x} film. (After [25.75])

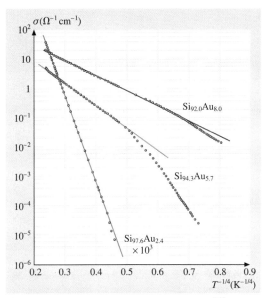

Fig. 25.14 Electrical conductivity versus $T^{-1/4}$ for a-Si_xAu_{1-x} films. (After [25.75])

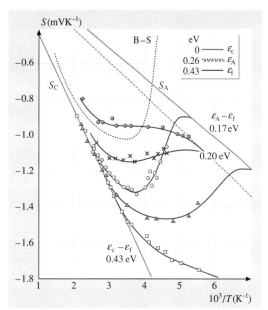

Fig. 25.15 Thermoelectric power versus T^{-1} for various a-Ge:H films. The *solid lines* are the calculated results. The *insert* shows the model on which the calculations are based. Carriers in states just above E_A move in extended states, those at E_A move by hopping. The *dotted curve* marked B–S was obtained by *Beyer* and *Stuke* [25.81] for a slowly evaporated sample annealed at 310 °C [25.78]

atoms obey the so called $8-N$ rule to satisfy the coordination of covalent bonds with their neighboring atoms. In a-Si:H, however, n-type doping and p-type doping were performed by using phosphorus (group V element) and boron (group III element), respectively [25.80].

The Hall coefficient of a-Si:H has been measured [25.76]. In crystalline semiconductors, when the carriers are electrons, its sign is negative, and when the carriers are holes, its sign is positive. The sign of the Hall coefficient is generally negative irrespective of that of the carriers in amorphous semiconductors. This has been accounted for in terms of the random phase model [25.82]. However, for the Hall coefficient of a-Si:H, this is not the case, i.e., when the carriers are electrons, the sign is positive, and when the carriers are holes, the sign is negative. Such an anomalous Hall coefficient in a-Si:H is called double reversal of the Hall coefficient and has been accounted for, taking into account antibonding orbitals for electrons and bonding orbitals for holes [25.83].

25.4.3 Chalcogenide Glasses

The values of the dark conductivities of chalcogenide glasses at room temperature are shown in Table 25.3 along with those of μ_d at room temperature. In chalcogenide glasses, the Fermi level is pinned near the midgap and this has been considered to be due to VAP defects with negative correlation energy. Thus, doping seems difficult, except for a few cases, e.g., the incorporation of Bi into a-Ge:S and a-Ge:Se increases the direct current (DC) conductivity by 6–7 orders of magnitude [25.84, 85]. This doping effect is due to chemical modification instead of conventional doping. Furthermore, Cd–In–S chalcogenide glasses exhibit $\sigma \approx 1 \times 10^{-2}\,\Omega^{-1}\mathrm{cm}^{-1}$ at room temperature [25.70].

25.5 Light-Induced Phenomena

Light-induced phenomena in amorphous semiconductors were first observed for chalcogenide glasses as photo-darkening, PL fatigue and photostructural changes associated with illumination whose energy

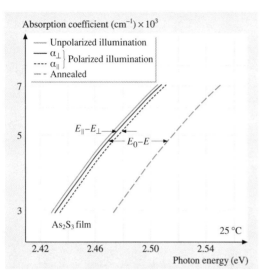

Fig. 25.16 Optical absorption edge of As_2S_3 films annealed or illuminated with unpolarized or linearly polarized light ($\alpha_\perp$: *solid line* and $\alpha_\parallel$: *dashed line*). $\alpha_\parallel$ and $\alpha_\perp$ refer to polarization along the plane of illumination and perpendicular to it, respectively. (After [25.87])

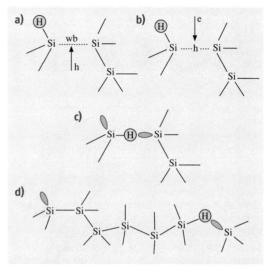

Fig. 25.17 (**a**) Self-trapping of a hole by a weak Si–Si bond adjacent to a Si–H bond. (**b**) Nonradiative recombination of a self-trapped hole with an electron. (**c**) A Si–H bond is switched toward the weak Si–Si bond and a dangling bond is left behind. (**d**) Formation of two separate dangling bonds after hydrogen movements and repeating of processes shown in (**a**)–(**c**)

equals to or exceeds the band-gap energy (see, e.g., Shimakawa et al. [25.86]; Singh and Shimakawa, [25.3]). Photo-darkening is an effect in which the absorption edge shifts towards lower photon energy under illumination, as shown in Fig. 25.16 [25.87]. PL fatique is the decrease of the PL intensity associated with illumination. Photostructural changes are observed as changes of the volume (generally expansion) of the sample with illumination. PL fatique may be due to light-induced creation of charged structural defects acting as nonradiative recombination centers. For the photo-darkening and the photostructural change, it has been discussed whether they are independent of each other or not. Thus, the origins of these phenomena are still unclear.

Light-induced phenomena in a-Si:H were observed first in the dark conductivity and photoconductivity, i.e., drops in their values after prolonged illumination of band-gap light (the so-called Staebler–Wronski effect [25.88]). Subsequently, such phenomena have been observed in PL, optical absorption and ODMR etc. [25.2]. This effect also gives rise to degradation of the performance of amorphous-silicon solar cells, so that this has received great attention from the viewpoint of application. After prolonged illumination, dangling bonds were found to be created [25.89, 90], so that light-induced creation of dangling bonds has been considered as the origin of these phenomena. Several models have been proposed for mechanisms for light-induced creation of dangling bonds [25.53, 91–96], but this issue is still controversial (see, e.g., *Morigaki* [25.2]; *Singh and Shimakawa* [25.3]). Very recently, the creation of a number of dangling bonds such as 1×10^{19} cm^{-3} after pulsed optical excitation has been observed in high-quality a-Si:H films [25.97, 98]. This result has been accounted for in terms of the authors' model [25.99], based on a combination of the following processes occurring during illumination. Self-trapping of holes in weak Si–Si bonds adjacent to Si–H bonds triggers these weak bonds to break using the phonon energy associated with nonradiative recombination of electrons with those self-trapped holes (Fig. 25.17a and b). After Si–H bond switching and hydrogen movement, two types of dangling bonds are created, i.e., a normal dangling bond and a dangling bond with hydrogen at a nearby site, i.e. so-called hydrogen-related dangling bonds (Fig. 25.17c and d). In the latter, hydrogen is dissociated from the Si–H bond as a result of nonradiative recombination at this dangling bond site. Dissociated hydrogen can terminate two types of dangling bonds (Fig. 25.18c and d) or can be in-

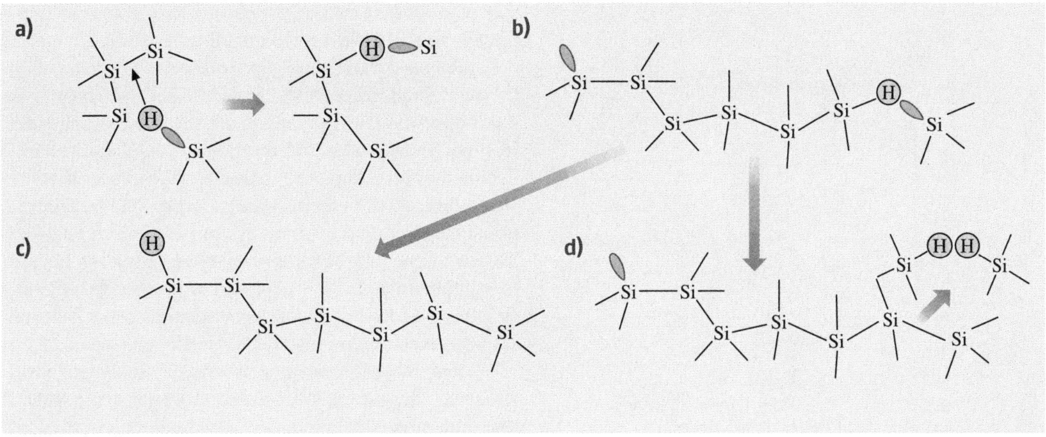

Fig. 25.18 (**a**) Dissociation of hydrogen from a hydrogen-related dangling bond and insertion of hydrogen into a nearby weak Si–Si bond (formation of a new hydrogen-related dangling bond). (**b**) Two separate dangling bonds, i.e. a normal dangling bond and a hydrogen-related dangling bond. (**c**) Dissociation of hydrogen from a hydrogen-related dangling bond and termination of a normal dangling bond by hydrogen. (**d**) Termination of a hydrogen-related dangling bond by hydrogen

serted into a nearby weak Si–Si bond (Fig. 25.18a). The dependencies of the density of light-induced dangling bonds on illumination time and generation rate have been calculated using the rate equations governing these processes. The results have been compared with experimental results obtained under continuous illumination and pulsed illumination with good agreement. Light-induced structural changes, i.e., volume change [25.100] and local structural changes around a Si–H bond [25.101] have also been observed in a-Si:H and the origins for these changes have been discussed [25.102–104].

25.6 Nanosized Amorphous Structure

Recent technologies have enabled us to control the properties of semiconductors by introducing artificial structures of nanometer size such as quantum wells, quantum wires and quantum dots. The quantum well is formed by preparing multilayers consisting of two semiconductors. Figure 25.19a illustrates the conduction and valence band edges in such multilayers consisting of two semiconducting materials with different band gaps, plotted as functions of the distance from the substrate, z. The curves of the conduction and valence bands in Fig. 25.19a are considered to be the potentials for the electrons and holes, respectively. The motion of the electrons and holes is assumed to be described by effective mass equations which are similar to the Schrödinger equation. When the barriers of the potentials are high enough to prevent the carriers from moving to the adjacent well, the quantum levels (for motion along the z-axis) are as illustrated in Fig. 25.19a. The band gap of the multilayer, E_g, is equal to the separation of the quantum levels of the lowest energy in the conduction and valence bands, as shown in Fig. 25.19a. E_g increases with decreasing thickness of the well layer, L_w; this is well known as the quantum-size effect. When the height of the barrier is infinite, $\Delta E = E_g - E_{gw}$ is proportional to L_w^{-2}, where E_{gw} denotes the band gap of the material of the well layer.

Multilayers of various amorphous semiconductors have been prepared. Observation of the quantum size effect has been reported in some amorphous semiconducting multilayers. For example, the shift of the optical gap of an a-Si:H/a-Si$_{1-x}$N$_x$:H multilayer from that of a-Si:H, ΔE, is proportional to L_w^{-2}, where L_w is the thickness of the a-Si:H layers (e.g. Morigaki [25.2]; see Yamaguchi and Morigaki [25.105]; and references therein). The result, $\Delta E \propto L_w^{-2}$, is consistent with quantum size effect in a square-well potential.

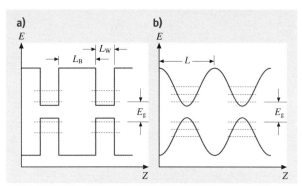

Fig. 25.19a,b The edges of the conduction and valence bands in (**a**) multilayer films and (**b**) band-edge modulated films, as functions of z, where z denotes distance from the substrates. The *dashed lines* indicate quantum levels. The band gap E_g, which increases due to the quantum size effect, is also shown

In the case of amorphous semiconductors, it is easy to introduce a potential well of arbitrary shape, because the optical gap of the amorphous semiconducting alloy, which depends on the composition, can be varied continuously within a certain range. The authors have prepared a new type of a-Si:H-based film called band-edge modulated (BM) a-Si$_{1-x}$N$_x$:H, in which the potentials for electrons and holes are sinusoidal functions of z [25.106, 107]. The band-edges for the BM films are illustrated in Fig. 25.19b. In this case, the quantum size effect is also expected, similarly to the case of multilayer films. However, the quantum levels and the size dependencies of the band gap are different from those in the case of multilayer films. In the case of BM, the potential is approximately parabolic at the bottom of the well. In this case the quantum levels have the same separation as that illustrated in Fig. 25.19b. We expect that the shift of the optical gap in BM films, ΔE, will be proportional to L^{-1}, where L denotes the modulation period. The band gap of BM films observed in experiments are in agreement with that expected from the above consideration [25.106]. Parabolic potentials in crystalline semiconductors have also been introduced (see Gossard et al. [25.108]). However it is difficult to obtain such structures that are small enough to observe significant quantum effects in crystalline semiconductors.

Recently, preparation of amorphous silicon quantum dots in silicon nitride has been reported [25.109]. Theoretical calculations for silicon nanostructures [25.110] have predicted that the oscillator strength in low symmetry is larger than that in high symmetry. A large quantum efficiency for PL is expected in the case of amorphous quantum dots. Thus, amorphous quantum wires and dots have potential applications in light-emitting devices with high efficiency.

Chalcogenide glasses multilayers have been prepared [25.111–113], whose optical properties such as optical absorption and PL, and electrical properties have been measured.

References

25.1 N. F. Mott, E. A. Davis: *Electronic Processes in Non-crystalline Materials*, 2nd edn. (Clarendon, Oxford 1979)
25.2 K. Morigaki: *Physics of Amorphous Semiconductors* (World Scientific, Singapore; Imperial College Press, London 1999)
25.3 J. Singh, K. Shimakawa: *Advances in Amorphous Semiconductors* (Taylor Francis, London 2003)
25.4 S. Kugler, P. R. Surján, G. Náray-Szabó: Phys. Rev. B **37**, 9069 (1988)
25.5 L. Ley, J. Reichardt, R. L. Johnson: Phys. Rev. Lett. **49**, 1664 (1982)
25.6 L. Brey, C. Tejedor, J. A. Verges: Phys. Rev. Lett. **52**, 1840 (1984)
25.7 L. Ley, S. Kawalczyk, R. Pollak, D. A. Shirley: Phys. Rev. Lett. **29**, 1088 (1972)
25.8 M. Kastner, D. Adler, H. Fritzsche: Phys. Rev. Lett. **37**, 1504 (1976)
25.9 K. Morigaki, M. Yamaguchi, I. Hirabayashi, R. Hayashi: *Disordered Semiconductors*, ed. by M. A. Kastner, G. A. Thomas, S. R. Ovshinsky (Plenum, New York 1987) p. 415
25.10 G. D. Cody: In: *Semiconductors and Semimetals*, Vol. 21, Part B ed. by J. I. Pankove (Academic, Orlando 1984) p. 1
25.11 T. Tiedje: In: *Semiconductors and Semimetals*, Vol. 21, Part C ed. by J. I. Pankove (Academic, Orlando 1984) p. 207
25.12 K. Winer, I. Hirabayashi, L. Ley: Phys. Rev. B **38**, 7680 (1988)
25.13 K. Winer, I. Hirabayashi, L. Ley: Phys. Rev. Lett. **60**, 2697 (1988)
25.14 K. Morigaki: In: *Semiconductors and Semimetals*, Vol. 21, Part C ed. by J. I. Pankove (Academic, Orlando 1984) p. 155
25.15 K. Morigaki, M. Kondo: Solid State Phenomena **44–46**, 731 (1995)
25.16 K. Morigaki, H. Hikita, M. Kondo: J. Non-Cryst. Solids **190**, 38 (1995)

25.17 M. Yamaguchi, K. Morigaki, S. Nitta: J. Phys. Soc. Jpn. **58**, 3828 (1989)
25.18 M. Yamaguchi, K. Morigaki, S. Nitta: J. Phys. Soc. Jpn. **60**, 1769 (1991)
25.19 S. C. Moss, J. F. Graczyk: *Proc. 10th Int. Conf. on Physics of Semiconductors*, ed. by J. C. Hensel, F. Stern (US AEC Div. Tech. Inform., Springfield 1970) p. 658
25.20 S. Kugler, G. Molnár, G. Petö, E. Zsoldos, L. Rosta, A. Menelle, R. Bellissent: Phys. Rev. B **40**, 8030 (1989)
25.21 D. E. Polk: J. Non-Cryst. Solids **5**, 365 (1971)
25.22 F. Wooten, K. Winer, D. Weaire: Phys. Rev. Lett. **54**, 1392 (1985)
25.23 R. Car, M. Parrinello: Phys. Rev. Lett. **60**, 204 (1988)
25.24 S. Kugler, L. Pusztai, L. Rosta, P. Chieux, R. Bellissent: Phys. Rev. B **48**, 7685 (1993)
25.25 L. Pusztai: J. Non-Cryst. Solids **227-230**, 88 (1998)
25.26 S. Kugler, K. Kohary, K. Kádas, L. Pusztai: Solid State Commun. **127**, 305 (2003)
25.27 J. A. Reimer: J. Phys. (Paris) **42**, C4-715 (1981)
25.28 K. K. Gleason, M. A. Petrich, J. A. Reimer: Phys. Rev. B **36**, 3259 (1987)
25.29 G. Lucovsky, R. J. Nemanich, J. C. Knights: Phys. Rev. B **19**, 2064 (1979)
25.30 R. Bellissent, A. Menelle, W. S. Howells, A. C. Wright, T. M. Brunier, R. N. Sinclair, F. Jansen: Physica B **156, 157**, 217 (1989)
25.31 A. Menelle: Thése Doctorat (Université Piere et Marie Curie, Paris 1987)
25.32 T. Uchino: Kotai Butsuri (Solid State Physics) **37**, 965 (2002)
25.33 J. H. Zhou, K. Ikuta, T. Yasuda, T. Umeda, S. Yamasaki, K. Tanaka: J. Non-Cryst. Solids **227-230**, 857 (1998)
25.34 G. Lucovsky, F. L. Galeener, R. H. Geils, R. C. Keezer: In: *Proc. Int. Conf. on Amorphous and Liquid Semiconductors*, ed. by W. E. Spear (University of Edinburgh, Edinburgh 1977) p. 127
25.35 J. Tauc: In: *Amorphous and Liquid Semiconductors*, ed. by J. Tauc (Plenum, New York 1974) p. 159
25.36 C. Tsang, R. A. Street: Phys. Rev. B **19**, 3027 (1979)
25.37 K. Morigaki, Y. Sano, I. Hirabayashi: J. Phys. Soc. Jpn. **51**, 147 (1982)
25.38 K. Morigaki, Y. Sano, I. Hirabayashi: *Amorphous Semiconductor Technologies and Devices—1983*, ed. by Y. Hamakawa (Ohomsha, Tokio 1983) Chap. 3.2
25.39 I. Hirabayashi, K. Morigaki, M. Yoshida: Sol. Ener. Mat. **8**, 153 (1982)
25.40 G. H. Bauer: Solid State Phenomena **44-46**, 365 (1995)
25.41 W. Paul: In: *Amorphous Silicon and Related Materials*, ed. by H. Fritzsche (World Scientific, Singapore 1989) p. 63
25.42 Y. Tawada: In: *Amorphous Semiconductors—Technologies and Devices*, ed. by Y. Hamakawa (Ohmusha and North Holland, Tokyo and Amsterdam 1983) Chap. 4.2

25.43 F. Demichelis, C. F. Pirri: Solid State Phenomena **44-46**, 385 (1995)
25.44 K. Maeda, I. Umezu: J. Appl. Phys. **70**, 2745 (1991)
25.45 M. Yamaguchi, K. Morigaki unpublished
25.46 B. Dunnett, D. I. Jones, A. D. Stewart: Philos. Mag. B **53**, 159 (1986)
25.47 M. Hirose: Jpn. J. Appl. Phys. **21**(suppl. 21-1), 297 (1981)
25.48 M. Yamaguchi, K. Morigaki: Philos. Mag. B **79**, 387 (1999)
25.49 D. Engemann, R. Fischer: In: *Amorphous and Liquid Semiconductors*, ed. by J. Stuke, W. Brenig (Taylor & Francis, London 1974) p. 947
25.50 D. Engemann, R. Fischer: In: *Proceedings of the 12th International Conference on the Physics of Semiconductors*, ed. by M. H. Pilkuhn (B. G. Teubner, Stuttgart 1974) p. 1042
25.51 R. A. Street, J. C. Knights, D. K. Biegelsen: Phys. Rev. B **18**, 1880 (1978)
25.52 D. J. Dunstan, F. Boulitrop: Phys. Rev. B **30**, 5945 (1984)
25.53 K. Morigaki: J. Non-Cryst. Solids **141**, 166 (1992)
25.54 R. Stachowitz, M. Schubert, W. Fuhs: J. Non-Cryst. Solids **227-230**, 190 (1998)
25.55 C. Ogihara: J. Non-Cryst. Solids **227-230**, 517 (1998)
25.56 T. Aoki, T. Shimizu, S. Komedoori, S. Kobayashi, K. Shimakawa: J. Non-Cryst. Solids **338-340**, 456 (2004)
25.57 M. Yoshida, M. Yamaguchi, K. Morigaki: J. Non-Cryst. Solids **114**, 319 (1989)
25.58 M. Yoshida, K. Morigaki: J. Phys. Soc. Jpn. **58**, 3371 (1989)
25.59 B. A. Wilson, P. Hu, T. M. Jedju, J. P. Harbison: Phys. Rev. B **28**, 5901 (1983)
25.60 C. Ogihara, H. Takemura, H. Yoshida, K. Morigaki: J. Non-Cryst. Solids **266-269**, 574 (2000)
25.61 H. Takemura, C. Ogihara, K. Morigaki: J. Phys. Soc. Jpn. **71**, 625 (2002)
25.62 M. Yoshida, K. Morigaki: J. Non-Cryst. Solids **59 & 60**, 357 (1983)
25.63 G. Pfister, H. Scher: Adv. Phys. **27**, 747 (1978)
25.64 H. Scher, E. W. Montroll: Phys. Rev. B **12**, 2455 (1975)
25.65 I. Solomon: In: *Amorphous Semiconductors*, ed. by M. H. Brodsky (Springer, Berlin Heidelberg New York 1979) p. 189
25.66 K. Lips, C. Lerner, W. Fuhs: J. Non-Cryst. Solids **198-200**, 267 (1996)
25.67 M. Stutzmann, M. S. Brandt, M. W. Bayerl: J. Non-Cryst. Solids **266-269**, 1 (2000)
25.68 S. R. Elliott: In: *Material Science and Technology*, Vol. 9, ed. by R. W. Cahn et al. (VCH, Weinheim 1991) p. 376
25.69 A. Feltz: *Amorphous Inorganic Materials and Glasses* (VCH, Weinheim 1993)
25.70 H. Hosono, H. Maeda, Y. Kameshima, H. Kawazoe: J. Non-Cryst. Solids **227-230**, 804 (1998)
25.71 R. A. Street: Adv. Phys. **25**, 397 (1976)

25.72 B. C. Cavenett: J. Non-Cryst. Solids **59 & 60**, 125 (1983)
25.73 J. Ristein, P. C. Taylor, W. D. Ohlsen, G. Weiser: Phys. Rev. B **42**, 11845 (1990)
25.74 D. Mao, W. D. Ohlsen, P. C. Taylor: Phys. Rev. B **48**, 4428 (1993)
25.75 N. Kishimoto, K. Morigaki: J. Phys. Soc. Jpn. **46**, 846 (1979)
25.76 P. G. LeComber, D. I. Jones, W. E. Spear: Philos. Mag. **35**, 1173 (1977)
25.77 H. Overhof, W. Beyer: Philos. Mag. B **44**, 317 (1983)
25.78 D. I. Jones, W. E. Spear, P. G. LeComber: J. Non-Cryst. Solids **20**, 259 (1976)
25.79 N. F. Mott: J. Phys. C **13**, 5433 (1980)
25.80 W. E. Spear, P. G. LeComber: Philos. Mag. **33**, 935 (1976)
25.81 W. Beyer, J. Stuke: In: *Proc. Int. Conf. on Amorphous and Liquid Semiconductors, 1973*, ed. by J. Stuke (Taylor & Francis, London 1974) p. 251
25.82 L. Friedman: J. Non-Cryst. Solids **6**, 329 (1971)
25.83 D. Emin: Philos. Mag. **35**, 1189 (1977)
25.84 N. Tohge, T. Minami, Y. Yamamoto, M. Tanaka: J. Appl. Phys. **51**, 1048 (1980)
25.85 L. Tichy, H. Ticha, A. Triska, P. Nagels: Solid State Commun. **53**, 399 (1985)
25.86 K. Shimakawa, A. Kolobov, S. R. Elliott: Adv. Phys. **44**, 475 (1995)
25.87 K. Kimura, K. Murayama, T. Ninomiya: J. Non-Cryst. Solids **77, 78**, 1203 (1985)
25.88 D. L. Staebler, C. R. Wronski: Appl. Phys. Lett. **31**, 292 (1977)
25.89 I. Hirabayashi, K. Morigaki, S. Nitta: Jpn. J. Appl. Phys. **19**, L357 (1980)
25.90 H. Dersch, J. Stuke, J. Beichler: Appl. Phys. Lett. **38**, 456 (1981)
25.91 M. Stutzmann, W. B. Jackson, C. C. Tsai: Phys. Rev. B **32**, 23 (1985)
25.92 C. Godet, P. Roca i Cabarrocas: J. Appl. Phys. **80**, 97 (1996)
25.93 H. M. Branz: Phys. Rev. B **59**, 5498 (1999)
25.94 K. Morigaki, H. Hikita: Solid State Commun. **114**, 69 (2000)
25.95 K. Morigaki, H. Hikita: J. Non-Cryst. Solids **266-269**, 410 (2000)
25.96 K. Morigaki, H. Hikita: *Proc. Int. Conf. on Physics of Semiconductors*, ed. by T. Ando N. Miura (Springer, Berlin Heidelberg New York 2000) p. 1485
25.97 C. Ogihara, H. Takemura, T. Yoshimura, K. Morigaki: J. Non-Cryst. Solids **299-302**, 637 (2002)
25.98 K. Morigaki, H. Hikita, H. Takemura, T. Yoshimura, C. Ogihara: Philos. Mag. Lett. **83**, 341 (2003)
25.99 K. Morigaki, H. Hikita: J. Non-Cryst. Solids **299-302**, 455 (2002)
25.100 T. Gotoh, S. Nonomura, M. Nishio, S. Nitta, M. Kondo, A. Matsuda: Appl. Phys. Lett. **72**, 2978 (1998)
25.101 Y. Zhao, D. Zhang, G. Kong, G. Pan, X. Liao: Phys. Rev. Lett. **74**, 558 (1995)
25.102 H. Fritzsche: Solid State Commun. **94**, 953 (1995)
25.103 R. Biswas, Y. P. Li: Phys. Rev. Lett. **82**, 2512 (1999)
25.104 K. Morigaki: Res. Bull. Hiroshima Inst. Tech. **35**, 47 (2001)
25.105 M. Yamaguchi, K. Morigaki: Phys. Rev. B **55**, 2368 (1997)
25.106 C. Ogihara, H. Ohta, M. Yamaguchi, K. Morigaki: Philos. Mag. B **62**, 261 (1990)
25.107 M. Yamaguchi, C. Ogihara, K. Morigaki: Mat. Sci. Eng. B **97**, 135 (2003)
25.108 A. C. Gossard, M. Sundaram, P. F. Hopkins: In: *Semiconductors and Semimetals*, Vol. 40, ed. by A. C. Gossard (Academic, Boston; Tokio 1994) Chap. 2
25.109 N.-M. Park, C.-J. Choi, T.-Y. Seong, S.-J. Park: Phys. Rev. Lett. **86**, 1355 (2001)
25.110 J. Koga, K. Nishio, T. Yamaguchi, F. Yonezawa: J. Phys. Soc. Jpn. **70**, 3143 (2001)
25.111 R. Ionov, D. Nesheva, D. Arsova: J. Non-Cryst. Solids **137&138**, 1151 (1991)
25.112 R. Ionov: Ph. D. Thesis. Ph.D. Thesis (Technical Univ., Sofia 1993)
25.113 H. Hamanaka, S. Konagai, K. Murayama, M. Yamaguchi, K. Morigaki: J. Non-Cryst. Solids **198-200**, 808 (1996)

26. Amorphous and Microcrystalline Silicon

Processes used to grow hydrogenated amorphous silicon (a-Si:H) and microcrystalline silicon (μc-Si:H) from SiH$_4$ and H$_2$/SiH$_4$ glow discharge plasmas are reviewed. Differences and similarities between growth reactions of a-Si:H and μc-Si:H in a plasma and on a film-growing surface are discussed, and the process of nucleus formation followed by epitaxial-like crystal growth is explained as being unique to μc-Si:H. The application of a reaction used to determine the dangling-bond defect density in the resulting a-Si:H and μc-Si:H films is emphasized, since it can provide clues about how to improve the optoelectronic properties of those materials for device applications, especially thin-film silicon-based solar cells. Material issues related to the realization of low-cost and high-efficiency solar cells are described, and finally recent progress in this area is reviewed.

26.1 Reactions in SiH$_4$ and SiH$_4$/H$_2$ Plasmas .. 581
26.2 Film Growth on a Surface 583
    26.2.1 Growth of a-Si:H 583
    26.2.2 Growth of μc-Si:H 584
26.3 Defect Density Determination
    for a-Si:H and μc-Si:H 589
    26.3.1 Dangling Bond Defects 589
    26.3.2 Dangling Bond Defect Density
          in μc-Si:H 590
26.4 Device Applications 590
26.5 Recent Progress in Material Issues
    Related to Thin-Film Silicon Solar Cells .. 591
    26.5.1 Controlling Photoinduced
          Degradation in a-Si:H 591
    26.5.2 High Growth Rates
          of Device-Grade μc-Si:H 592
26.6 Summary 594
References 594

Hydrogenated amorphous silicon (a-Si:H) and microcrystalline silicon (μc-Si:H) are recognized as being useful materials for constructing devices related to optoelectronics, such as solar cells, thin-film transistors, etc. [26.1, 2]. Several methods have been proposed for the preparation of device-grade a-Si:H and μc-Si:H. These include: reactive sputtering of a crystalline silicon target with Ar+H$_2$ plasma [26.3]; mercury-sensitized photochemical vapor deposition (CVD) utilizing a decomposition reaction of silane (SiH$_4$) molecules with photoexcited Hg (Hg*) [26.4]; a direct photo CVD method where high-energy photons from a Xe-resonance lamp or a low-pressure Hg lamp are used for the direct excitation of SiH$_4$ molecules to excited electronic states [26.5, 6]; a hot-wire CVD method for decomposing SiH$_4$ by means of catalytic reactions on a heated metal surface [26.7]; and a plasma-enhanced CVD method (PECVD). The PECVD method is the most widely used of these due to its ability to consistently prepare uniform, high-quality materials on a large-area substrate.

In this chapter, the PECVD method is highlighted, details regarding the processes used to grow a-Si:H and μc-Si:H from reactive plasmas are explained, and the determination reaction (which is used to obtain the dangling-bond defect density in the films: one of the most important structural properties that influences device performance) is interpreted in order to obtain clues about how to control the optoelectronic properties of those materials for device applications.

26.1 Reactions in SiH$_4$ and SiH$_4$/H$_2$ Plasmas

The initial event required for the growth of a-Si:H and μc-Si:H is the decomposition of the source gas material in SiH$_4$ or SiH$_4$/H$_2$ glow discharge plasma. Figure 26.1 shows a schematic of the dissociation pathway of SiH$_4$ and H$_2$, during which the molecules are excited to higher electronic states due to inelastic collisions with high-energy electrons in the plasma [26.8]. As the electrons in the plasma usually have a wide variety of energies,

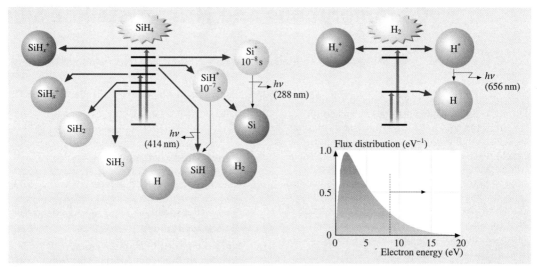

Fig. 26.1 Schematic showing the dissociation of SiH_4 and H_2 molecules to a variety of chemical species in the plasma via excited electronic states. The electron energy distribution function in the plasma is also shown

from zero to several tens of electron volts (eV), ground-state electrons of source gas molecules are excited into their electronic excited states almost simultaneously due to inelastic collisions with energetic electrons. Excited electronic states of complicated molecules like SiH_4 are usually dissociating states, from which dissociation occurs spontaneously to SiH_3, SiH_2, SiH, Si, H_2 and H, as shown in Fig. 26.1, depending on the stereochemical structure of the excited state. Hydrogen molecules are also decomposed to atomic hydrogen. Excitation of ground-state electron to vacuum-state gives rise to ionization events, generating new electrons and ions, which maintains the plasma.

Reactive species produced in the plasma also experience secondary reactions, mostly with parent SiH_4 and H_2 molecules, as shown in Fig. 26.2, resulting in a steady state. Reaction rate constants for each reaction are summarized in the literature [26.9]. Steady-state densities of reactive species are basically determined by the balance between their rate of generation and their annihilation rate. Therefore, highly reactive species such as SiH_2, SiH and Si (short-lifetime species) have much smaller densities than SiH_3 in the steady-state plasma, although the generation rates of those species are not very different from that of SiH_3, which shows low reactivity along with SiH_4 and H_2 (long-lifetime species).

Steady-state densities of reactive species have been measured using various gas-phase diagnostic techniques [26.10–16], such as optical emission

Ion exchanging	$SiH_x^+ + SiH_4$	$\rightarrow SiH_x + SiH_4^+$
Ion–Molecule	$SiH_x + SiH_4$	$\rightarrow SiH_3 + SiH_3$
Neutral–Molecule	$SiH + SiH_4$	$\rightarrow Si_2H_5$
Disproportionation	$Si + SiH_4$	$\rightarrow SiH_3 + SiH$
Insertion	$SiH_2 + SiH_4$	$\rightarrow Si_2H_6$
Recombination	$SiH_2 + H_2$	$\rightarrow SiH_4$
Abstraction	$SiH_3 + SiH_4$	$\rightarrow SiH_4 + SiH_3$
	$H + SiH_4$	$\rightarrow H_2 + SiH_3$
Ion–Radical Radical–Radical	less probable	

Fig. 26.2 Representative secondary reactions of the chemical species produced in the plasma with SiH_4 and H_2 molecules. Their reaction rate constants are available in the literature

spectroscopy (OES) [26.10], laser-induced fluorescence (LIF) [26.12], infrared laser absorption spectroscopy(IRLAS) [26.13], and ultraviolet light absorption spectroscopy (UVLAS) [26.16]. Figure 26.3 shows the steady state number densities of the chemical species, including both emissive and ionic species, in the SiH_4 and SiH_4H_2 plasmas used to prepare device-grade a-Si:H and μc-Si:H. It is clear from Fig. 26.3 that the SiH_3 radical is the dominant chemical species in the growth of both a-Si:H and μc-Si:H, although the density ratio of short-lifetime species to SiH_3 changes depending on the conditions used to generate the plasma. For instance, when high electric power is supplied to the plasma (high electron density) under low SiH_4 flow rate conditions, SiH_4 is rapidly dissociated and then depleted, giving rise to reduced probabilities of gas-phase reactions of short lifetime species with SiH_4 molecules. This leads to an increased contribution from short-lifetime species to the film growth, which causes a deterioration of the structural properties of the resulting films.

The steady-state density of atomic hydrogen (H) varies widely in the plasma, as shown in Fig. 26.3. This is mainly due to the change in the hydrogen dilution ratio R (H_2/SiH_4 in the source gas material, i. e., the density of atomic hydrogen increases with increasing R. Noting the fact that more μc-Si:H is formed with increasing R at constant electron density in the plasma and constant substrate (surface) temperature, it is clear

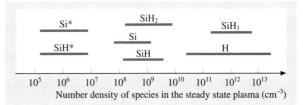

Fig. 26.3 Number densities of chemical species in realistic steady-state plasmas measured or predicted by various diagnostic techniques

that atomic hydrogen plays an important role in μc-Si:H growth [26.17] although SiH_3 is the dominant film precursor for both a-Si:H and μc-Si:H growth [26.18].

26.2 Film Growth on a Surface

26.2.1 Growth of a-Si:H

Upon reaching a film-growing surface, SiH_3 radicals begin to diffuse across it. During this diffusion, the SiH_3 abstracts bonded hydrogen from the surface, forming SiH_4 and leaving dangling bonds on the surface (this known as growth site formation). Other SiH_3 molecules then diffuse across the surface to find the site containing the dangling bonds, whereupon Si–Si bond formation (film growth) occurs, as shown schematically in Fig. 26.4. This surface reaction scheme for film growth has been proposed on the basis of two experimental results [26.18].

Figure 26.5 shows the general concept of the surface reaction process. Some the flux of SiH_3 is reflected off the surface (the proportion of molecules reflected is given by the reflection probability). The remaining SiH_3 is adsorbed onto the surface and it changes its form as follows:

1. SiH_3 abstracts bonded H from the surface, forming SiH_4, or two of the SiH_3 radicals interact on the surface, forming Si_2H_6 (with a recombination probability γ);
2. Surface-diffusing SiH_3 sticks to the site containing the dangling bond, forming Si–Si bond (with a sticking probability s).

The total loss probability (β) is given by the sum of recombination probability and the sticking probability ($\gamma + s$), and the reflection probability is therefore $1 - \beta$. Among the reaction probabilities mentioned above, the

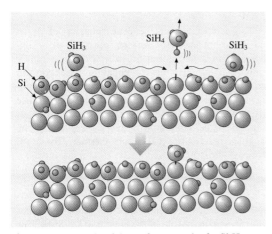

Fig. 26.4 Schematic of the surface growth of a-Si:H

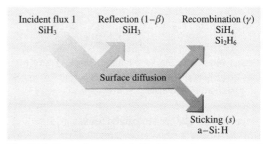

Fig. 26.5 General concepts behind the surface reactions of incoming SiH_3 radicals

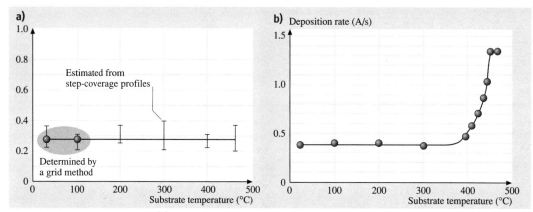

Fig. 26.6a,b Loss probability β (**a**) of SiH$_3$ radicals reaching the film-growing surface, as measured using the grid method and the step-coverage method, and the deposition rate (corresponding to the sticking probability s for SiH$_3$) (**b**) as a function of substrate temperature during the formation of a-Si:H

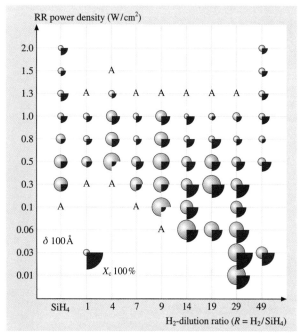

Fig. 26.7 Crystal size (*light*: δ) and volume % (*dark*: X_c) of microcrystallites in the resulting films, mapped out on the RF power density/hydrogen dilution ratio plane

loss probability has been measured using a grid method (GM) as well as a step-coverage method (SCM). Figures 26.6a and 26.6b show the loss probabilities obtained by GM and SCM, as well as the sticking probability predicted from the film deposition rate as a function of the substrate temperature, respectively.

The loss probability β, related to the reflectivity $(1-\beta)$, is simply dependent on the nature of the site where SiH$_3$ from the plasma lands. Therefore, the temperature-independent $1-\beta$ seen in Fig. 26.6 suggests that almost all of the surface sites are covered with bonded H over the whole temperature range used in the experiment, although a few dangling bonds are created thermally above 350 °C, as shown by in situ infrared reflection absorption spectroscopy (IR-RAS) [26.19]. On the other hand, SiH$_3$ diffusing across the surface is easily captured by even just a few dangling bonds, because a mobile species on the surface can find a specific site at a distant location from the landing site.

The temperature-independent behavior of $1-\beta$ and the temperature-dependent behavior of s observed in Figs. 26.6a and 26.6b have been perfectly reproduced by theoretical simulations based on the surface-reaction scheme shown in Fig. 26.4, assuming reasonable activation energies for the surface diffusion of SiH$_3$, the abstraction reaction of H with SiH$_3$, the saturation of the site containing the dangling bond with SiH$_3$, and thermal H-removal processes [26.20].

26.2.2 Growth of μc-Si:H

Atomic hydrogen reaching the film-growing surface plays an important role in the growth of μc-Si:H. This has been confirmed in the μc-Si:H-formation map draw-

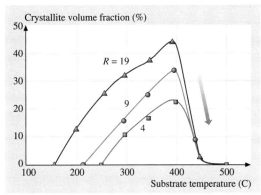

Fig. 26.8 Volume % (X_c) of microcrystallites in the resulting films plotted against the substrate temperature during film growth

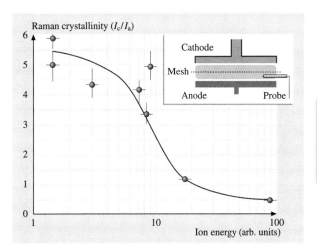

Fig. 26.9 Raman crystallinity in the resulting μc-Si:H as a function of the energy of the ions impinging on the film-growing surface (which is controlled by the application of the bias voltage to the mesh electrode during film growth)

ing on the RF power density/hydrogen diffusion ratio plane as shown in Fig. 26.7. As seen in the figure, large-crystallite μc-Si:H is prepared under high hydrogen dilution conditions, indicating the importance of atomic H in the growth of μc-Si:H. Figure 26.8 shows the crystalline volume fraction (determined by the X-ray diffraction peak area) in the resulting μc-Si:H as a function of the substrate temperature during film growth for three different hydrogen dilution ratios [26.17]. The crystalline volume fraction increases with increasing substrate temperature, reaching a maximum at around 350 °C, and then suddenly drops to zero above 500 °C. The lack of μc-Si:H formation above 500 °C suggests that surface hydrogen coverage is a requirement for crystallite formation in the resulting film [26.17].

Figure 26.9 shows the negative effect of ionic species impinging on the film-growing surface on the formation of μc-Si:H, studied using a triode reactor, as shown in the inset of the figure [26.21]. The Raman crystallinity I_c/I_a, defined as the ratio of the peak intensity from the crystalline phase at around $520\,\text{cm}^{-1}$ to that from the amorphous phase at $480\,\text{cm}^{-1}$ in the Raman scattering spectrum, deteriorates as the energy of the ionic species impinging on the film-growing surface (controlled by the bias voltage applied to the mesh electrode) is increased.

The surface reaction behavior of SiH_3 reaching the film-growing surface has also been investigated for μc-Si:H growth using GM and SCM, in contrast to a-Si:H growth [26.18]. Figures 26.10a and 26.10b show the loss probability (β) and the deposition rate (corresponding to the sticking probability s) of SiH_3 reaching the

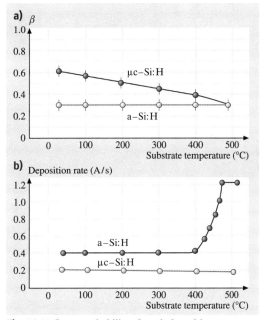

Fig. 26.10 Loss probability β and deposition rate as a function of substrate temperature for μc-Si:H growth in comparison to those for a-Si:H growth

film-growing surface as a function of the substrate temperature for the growth of both μc-Si:H and a-Si:H.

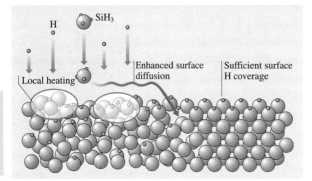

Fig. 26.11 Surface diffusion model for μc-Si:H formation. The *large spheres* and *small spheres* represent Si and H, respectively

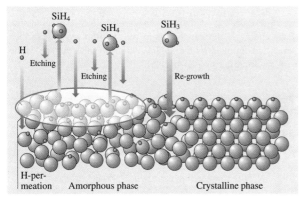

Fig. 26.12 Etching model for μc-Si:H formation

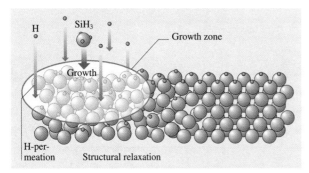

Fig. 26.13 Chemical annealing model for μc-Si:H formation

Unlike in the case of a-Si:H growth, the loss probability β shows a significant dependence on substrate temperature and the sticking probability s shows no temperature dependence in the case of μc-Si:H growth.

Based on these experimental results, the following properties of the formation process of μc-Si:H can be specified:

1. The film precursor is SiH_3, the same as in the case of a-Si:H growth;
2. Atomic hydrogen reaching the film-growing surface plays an important role in the formation of μc-Si:H;
3. The film becomes amorphous when the substrate temperature is higher than 500 °C;
4. High-energy ions impinging on the surface result in crystallinity deterioration;
5. The surface loss probability shows a temperature dependence whereas the sticking probability does not show a temperature dependence in the case of μc-Si:H growth.

Growth Models for μc–Si:H

In an attempt to explain the specific phenomena observed during the formation of μc-Si:H, three models have been proposed:

1. The surface diffusion model [26.17];
2. The etching model [26.22];
3. The chemical annealing model [26.23].

The surface diffusion model is depicted schematically in Fig. 26.11. Here, a high atomic H flux from the plasma results in full bonded hydrogen surface coverage and also local heating through hydrogen exchange reactions on the film-growing surface. These two actions enhance the surface diffusion of film precursors (SiH_3). As a consequence, the SiH_3 adsorbed on the surface can find energetically favorable sites, leading to the formation of an atomically ordered structure (nucleus formation). After the formation of the nucleus, epitaxial-like crystal growth takes place with enhanced surface diffusion of SiH_3 [26.17, 24].

The etching model has been proposed due to the observation that the rate of film growth decreases with increasing hydrogen dilution ratio R. The concept behind the etching model is shown schematically in Fig. 26.12. Atomic H reaching the film-growing surface breaks Si—Si bonds, preferentially weak bonds involved in the amorphous network structure, leading to the removal of Si atoms weakly bonded to other Si atoms. This site is replaced with a new film precursor SiH_3, creating a rigid, strong Si—Si bond, which gives rise to an ordered structure [26.22, 24].

The chemical annealing model has been proposed in order to explain the observation that crystal formation is observed during hydrogen plasma treatment;

growth occurs layer-by-layer via an alternating sequence of thin amorphous film growth and hydrogen plasma treatment. Several monolayers of amorphous silicon are deposited, and these layers are exposed to hydrogen atoms produced in the hydrogen plasma. This procedure is repeated several tens of times in order to fabricate the proper thickness to be able to evaluate the film structure. The absence of any significant reduction in film thickness during the hydrogen plasma treatment is difficult to explain using the etching model, and so the chemical annealing model was proposed, as schematically shown in Fig. 26.13. During the hydrogen plasma treatment, many atomic hydrogens permeate through the subsurface (the growth zone), giving rise to the crystallization of an amorphous network through the formation of a flexible network without any significant removal of Si atoms [26.23, 24].

These three models have been carefully examined, and the merits and drawbacks of each model have been discussed [26.24, 25].

More microscopic observations have recently been reported, based on the use of in situ diagnostic techniques, and a detailed mechanism for the formation process of μc-Si:H has been proposed.

Formation of the Nucleus

Figure 26.14 shows the evolution in surface roughness during film growth, as obtained by spectroscopic ellipsometry (SE), for three hydrogen dilution ratios R of 0, 10 and 20 [26.26]. As is seen in the figure, after the formation of an island, the enforced coalescence of islands takes place, which results in a smooth surface under

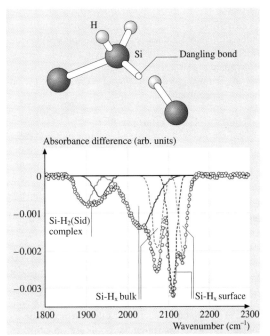

Fig. 26.15 Surface infrared absorption spectrum from the film just before nucleus formation, showing the appearance of the Si−H$_2$−d complex whose structure is also shown

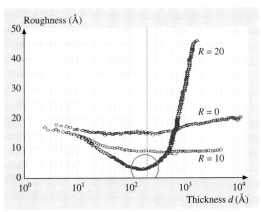

Fig. 26.14 Evolution in surface roughness, as measured using spectroscopic ellipsometry, during film growth for three different R_s

μc-Si:H-growth conditions ($R = 20$). After the smooth surface is obtained (formation of the nucleus is confirmed at this point in time), surface roughness is then enhanced due to an orientation-dependent crystal growth rate [26.27]. As soon as the smooth surface appears, a particular surface absorption band is observed in the infrared absorption spectrum, as measured using the in situ attenuated total reflection technique (ATR) during film growth [26.27, 28]. Figure 26.15 shows the surface infrared absorption spectrum, showing the presence of specific bands at 1897 cm^{-1} and 1937 cm^{-1} together with the usual Si−H$_x$ surface and bulk absorption bands (which occur between 2000 cm^{-1} and 2150 cm^{-1}). This new absorption band is assigned to the SiH$_2$(Sid) complex, which is also sketched in Fig. 26.15. Note that the number density (absorption intensity) of the SiH$_2$(Sid) complex is found to be proportional to the magnitude of the internal stress in the film just before these complexes appear. A nucleation model has been proposed based on the experimental facts mentioned above. The enforced island coalescence due to the enhanced surface diffusion of SiH$_3$ gives rise to an internal stress involving many strained Si−Si bonds in the amorphous incubation layer [26.27]. Atomic hydrogen attacks the strained Si−Si bonds, forming $SiH_2(Sid)$ complexes on the film-growing surface. These complexes provide structural flexibility, which enables structural order to be obtained via successive Si−SiH_3 bond formation at this site; in other words it acts as a prenucleation site on the film-growing surface.

Epitaxial−Like Crystal Growth

Figure 26.16 shows a cross-sectional transmission electron microscope (TEM) image of typical μc-Si:H films deposited on a glass substrate [26.29]. In the figure, epitaxial-like crystal growth is clearly observed to occur from the nucleus. It is a well-known fact that epitaxial crystal growth occurs only when the surface diffusion length of the film precursor is sufficiently long. In order to investigate the origin of the enhanced surface diffusion of SiH_3 during the formation of μc-Si:H, an isotope labeling experiment has been carried out [26.30]. D_2 was used as the source gas material (D_2SiH_4) instead of H_2SiH_4 during film growth under constant substrate temperature conditions, and the number densities of both D and H incorporated into the resulting film are measured by infrared absorption spectroscopy in order to estimate the degree of H-to-D exchange reactions, which mostly occur on the film-growing surface.

Figure 26.17 shows the Raman crystallinity I_c/I_a plotted against the number density ratio of the D/H incorporated in the resulting films. As is clearly from the figure, μc-Si:H is obtained only when D/H exceeds a critical value, indicating that the D/H exchange reaction is required to some extent for the formation of μc-Si:H. The D/H exchange reaction involves two steps:

1. Atomic D reaching the film-growing surface abstracts bonded H on the surface (the Eley-Rideal reaction), forming HD and providing a Si dangling bond site;
2. A recombination reaction occurs between the dangling bond and another atomic D adsorbed onto the surface, forming a Si−D bond.

These abstraction and recombination reactions are known to be strong exothermal reactions, with 1.4 eV and 3.1 eV of energy released, respectively. Considering the highly exothermal nature of the D/H exchange reaction, and the existence of a threshold value for the degree of D/H exchange reactions required for the formation

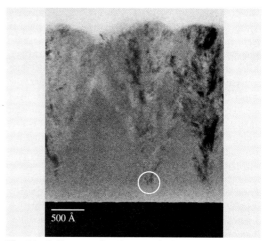

Fig. 26.16 Cross-sectional transmission electron microscope image of typical μc-Si:H deposited on a glass substrate

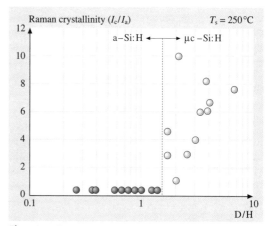

Fig. 26.17 Raman crystallinity plotted against the number density ratio of D/H incorporated into the resulting films prepared from D_2SiH_4 plasmas

of μc-Si:H, as shown in Fig. 26.17, "local heating" on the film-growing surface is believed to play a major role in the enhanced surface diffusion of the film precursor (SiH_3) during both nucleus formation (strained bond formation and $SiH_2(Sid)$-complex formation) and the epitaxial-like crystal growth associated with μc-Si:H film growth [26.17, 24, 28, 31].

26.3 Defect Density Determination for a-Si:H and μc-Si:H

One of the most important structural properties of a-Si:H and μc-Si:H for device applications is their dangling bond defect densities, because each dangling bond creates a localized deep state in the band gap of the material, which acts as a recombination center for photoexcited electrons and holes, although the free carrier mobility is another important property in semiconductors.

26.3.1 Dangling Bond Defects

Figure 26.18 shows the dangling bond defect density in the resulting a-Si:H as a function of substrate temperature. This dependency of the dangling bond density on the substrate temperature has been explained by taking into account the steady-state dangling-bond density on the film-growing surface. The structural properties of the resulting film are generally largely determined by the steady-state surface properties during the thin film growth process, because the surface formed at any given instant is incorporated into the bulk in the next instant due to the successive layering nature of the film growth [26.32, 33]. The steady state number density of surface dangling bonds is determined by the balance between the rate of generation of dangling bonds and their annihilation rate.

At low substrate temperatures, surface dangling bonds are produced by abstraction reactions of surface H atoms with SiH_3; the reaction rate for this is almost independent of substrate temperature, since the reaction rate is affected by both the residence time of SiH_3 at the H-covered site (longer residence times occur at lower temperatures) and by the abstraction reaction rate (slower rates occur at lower temperatures). The surface dangling bond is saturated by SiH_3 diffusing across the surface (this is dangling bond annihilation), which exhibits a slower rate at low temperatures due to the slower surface diffusion of SiH_3, which results in high dangling bond density on the steady-state film-growing surface. As a consequence, the dangling-bond density in a-Si:H grown at low substrate temperatures can be as high as $10^{19}\,cm^{-3}$. When the substrate temperature is increased during film growth, the steady state number density of surface dangling bonds is reduced drastically due to the combination of the temperature-independent H-abstraction reaction and the thermally enhanced surface diffusion of SiH_3, which gives rise to a minimum defect density of $10^{15}\,cm^{-3}$ in the resulting a-Si:H at a substrate temperature of $\approx 250\,°C$. The increased dangling bond density of the a-Si:H prepared at substrate temperatures higher than $350\,°C$ is explained by the increased rate of generation of steady-state dangling bonds due to the addition of a new term in the generation rate of these bonds associated with the thermal removal of surface H, although its annihilation rate strictly increases with increasing substrate temperature [26.33, 34].

We note here that a remarkable increase in the contribution from short-lifetime species such as SiH_2, SiH and Si, which show high reactivity and no diffusion on the film-growing surface, is observed upon the depletion of SiH_4-parent molecules, which causes an increase in the

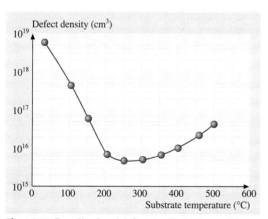

Fig. 26.18 Dangling bond defect density in the resulting a-Si:H films as a function of the substrate temperature during film growth

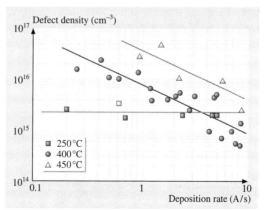

Fig. 26.19 Dangling bond defect density in a-Si:H films plotted against their deposition rates for three different substrate temperatures

number density of dangling bond defects in the resulting films through an enhancement of the dangling bond generation rate and a reduction of the dangling bond annihilation rate on the film-growing surface [26.20, 32].

Based on our understanding of the defect density determination reaction during film growth, several trials have attempted to control the defect density in a-Si:H. The steady state defect density on the film-growing surface could be reduced when the growth rate was much faster than the thermal H removal rate in the substrate temperature range above 350 °C, where the steady state defect density is mainly determined by the thermal H removal process. Figure 26.19 shows the number density of dangling bond defects in a-Si:H films plotted against their growth rate. As expected, the defect density in a-Si:H shows no growth-rate dependence when the film is prepared at substrate temperatures lower than 300 °C, whereas the defect density monotonically decreases with increasing growth rate when the film is deposited at 400 °C and 450 °C, and a defect density of 10^{14} cm^{-3} has been demonstrated [26.35].

26.3.2 Dangling Bond Defect Density in μc-Si:H

Figure 26.20 shows typical dangling bond defect densities in μc-Si:H as a function of substrate temperature together with those in its a-Si:H counterpart [26.33, 35, 36]. As seen from the figure, the substrate temperature-dependent dangling bond defect density in μc-Si:H in the temperature range above 300 °C shows a similar trend to that for a-Si:H, indicating that the defect determination reaction on the film-growing surface is identical for both μc-Si:H and a-Si:H growth in this temperature range; in other words, the increase in defect density with increasing substrate temperature is controlled by the thermal

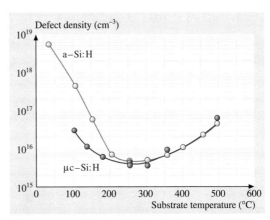

Fig. 26.20 Dangling bond defect density in μc-Si:H films as a function of substrate temperature during film growth in comparison to that in a-Si:H

H removal process on the film-growing surface. On the other hand, the number density of defects in μc-Si:H prepared at low substrate temperatures shows much lower values than those in a-Si:H. Considering that the defect density in the resulting film is largely determined by the defect density on the steady-state film-growing surface, the much lower defect density in μc-Si:H is caused by an increase in the defect annihilation rate on the film-growing surface due to the local heating from H-exchange reactions that occur during the course of μc-Si:H film growth.

It should be noted here that the dangling bond defect density in the resulting μc-Si:H is strongly influenced by the ion bombardment during film growth and by the collisions of crystals growing from different nuclei, which are different aspects to those that are important during a-Si:H growth.

26.4 Device Applications

a-Si:H and μc-Si:H are highly promising materials applicable to electronic or optoelectronic thin-film devices such as thin-film transistors (TFT), position sensors, color sensors, solar cells, etc. [26.1, 2].

A thin-film transistor array with a-Si:H active layer has been developed for switching devices in liquid crystal displays (LCD), and large-area LCDs more than 40 in across have already been made commercially available based on this technology. A laser crystallization technique has been developed in order to increase the carrier mobility of a-Si:H-based TFTs and thus reduce the active area of the transistors in the LCD. LCDs with a-Si:H or laser-crystallized thin-film Si-based transistor arrays are widely used as flat panel displays in televisions and monitors, where they are in competition with plasma display panels (PDP) and other flat panel display systems. If the carrier mobility in as-deposited μc-Si:H is drastically improved through the enhanced control of the film growth process, μc-Si:H will be widely used not only for thin-film transistors but also for signal-scanning devices such as charge-coupled devices monolithically arranged in the periphery of the LCD.

Thin-film Si-based solar cells has also been widely expected to provide low-cost photovoltaics. Actually, a-Si:H-based solar cells have already been widely used in pocket calculators, and now large-area solar cells are being developed for electricity generation. One big advantage of using a-Si:H for solar cell applications is its large optical absorption coefficient compared to single-crystalline or polycrystalline silicon counterparts, resulting in indirect optical transition properties, and so a thickness of less than 1 μm is enough to absorb sufficient sunlight for electricity generation when using a-Si:H-based solar cells. Low-temperature processes using PECVD are also advantageous in terms of reducing the cost of producing a-Si:H-based solar cells. However, there is a well-known phenomenon that occurs in a-Si:H called photo-induced degradation [26.37], where the conversion efficiency of a-Si:H-based solar cells, usually ≈ 10%, is degraded to less than 8% after prolonged exposure to light. Recombinations photogenerated electrons and holes are believed to trigger this photoinduced degradation in a-Si:H; therefore, increasing the field in the a-Si:H-based solar cell has proved an effective way to reduce the degradation. In fact, 0.1 μm-thick a-Si:H-based solar cells do not show any photoinduced degradation, although the conversion efficiency naturally deteriorates due to the reduced absorption of sunlight. The fabrication of a tandem-type stacked solar cell structure consisting of a top cell with a thin a-Si:H layer and a bottom cell containing narrow-gap materials such as a-SiGe : H, μc-Si:H and μc-SiGe:H has been proposed as a promising way to overcome photoinduced degradation and so to achieve high conversion efficiency in thin-film silicon-based solar cells. Initially, a-SiGe:H was adopted for the bottom cell material [26.38]; however, this material also shows severe photoinduced degradation. Recently, μc-Si:H or μc-SiGe:H have been proposed as promising candidates for bottom cell materials, because these materials do not exhibit photoinduced degradation [26.39, 40]. In this proposal, high rates of growth of those materials are crucial to the low-cost fabrication of tandem-type solar cells, since μc-Si:H and μc-SiGe:H undergo largely indirect optical transitions. Therefore, urgent material issues for the realization of low-cost/high-efficiency thin-film silicon-based solar cells include the need to improve the photoinduced stability of high-quality a-Si:H and the need to achieve a high rate of growth of device-grade μc-Si:H or μc-SiGe:H.

26.5 Recent Progress in Material Issues Related to Thin-Film Silicon Solar Cells

26.5.1 Controlling Photoinduced Degradation in a-Si:H

A relationship between the degree of photoinduced degradation and the dihydride bonding (Si−H$_2$) density has been reported in a-Si:H prepared under a variety of deposition conditions, where the substrate temperature, plasma-excitation frequency, gas-flow rate, hydrogen dilution ratio, working pressure, power density, etc. have all been varied [26.41]. Figure 26.21 shows the degree of photoinduced degradation, defined as the difference in the fill factors of photo-I–V characteristics of Ni-a-Si:H Schottky diode before and after light soaking plotted against the Si−H$_2$ density in a-Si:H film, as measured by infrared absorption spectroscopy. Furthermore, it has been suggested from mass spectrometric results that the Si−H$_2$ density in the resulting a-Si:H is strongly in-

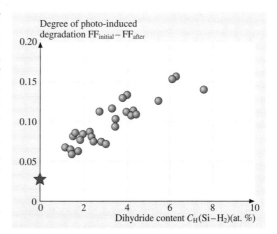

Fig. 26.21 Relationship between the degree of photoinduced degradation and the dihydride (Si−H$_2$) content in a-Si:H prepared under various deposition conditions. The *star* symbol represents for a-Si:H prepared under conditions of a reduced ratio of contributions from higher silane-related species versus the SiH$_3$ contribution at a substrate temperature of 250 °C

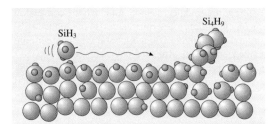

Fig. 26.22 Surface reaction image of the incorporation of Si–H_2 bonds into the resulting a-Si:H due to the contributions from higher silane-related species such as Si_4H_9

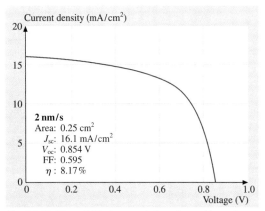

Fig. 26.23 Photo-I–V characteristics of a stable n–i–p a-Si:H-based solar cell fabricated at a high growth rate of 2 nm/s

creased by the contributions from higher silane-related species (HSRS) such as Si_4H_9 when the substrate temperature is kept constant. The contribution of these HSRS to the film precursor (SiH_3) has been theoretically analyzed using a couple of gas-phase reaction-rate equations [26.42]. This analysis predicted that the contribution ratio is a complex function of the electron temperature in the plasma, the electron density in the plasma, the gas temperature, the hydrogen dilution ratio R, and the gas residence time during film growth. Based on our understanding of the network structure (Si–H_2 bonding configuration) responsible for the photoinduced degradation in a-Si:H and the chemical species (HSRS) responsible during film growth (shown schematically in Fig. 26.22), a guiding principle for obtaining highly stabilized a-Si:H has been proposed [26.42, 43].

By following the guiding principle, a-Si:H with minimized Si–H_2 density in the network has been prepared by adjusting the plasma parameters during film growth under high growth rate conditions, and an a-Si:H-based solar cell showing a stable conversion efficiency of 8.2% (Fig. 26.23) has been fabricated at a high growth rate of 2 nm/s [26.41].

Moreover, a-Si:H containing a Si–H_2 density of 0% has been successfully prepared using a triode reactor at substrate temperatures as low as 250 °C by making use of the difference in the gas phase diffusion coefficients of SiH_3 (light) and HSRS (heavy) during film growth.

26.5.2 High Growth Rates of Device-Grade μc-Si:H

In the past, conventional high hydrogen dilution methods performed at relatively low working gas pressures (several tens of mTorr) have been used to obtain device-grade μc-Si:H (with low dangling bond defect densities of $\approx 10^{16}/cm^3$) [26.17]. Recently, a simple concept for preparing device-grade μc-Si:H at a high growth rate has been proposed, known as the high-pressure depletion (HPD) method [26.21, 44]. The production rate of the film precursor SiH_3 in the plasma, which is proportional to the growth rate of μc-Si:H, is determined by the product of the electron density and the number density of SiH_4. To increase the production rate of SiH_3 in the plasma, a high power density (radio frequency, RF, or very high frequency, VHF) and high partial pressures are needed, because the electron density is basically a function of the power density applied

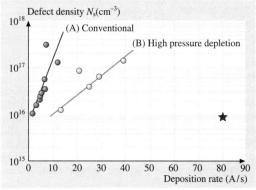

Fig. 26.24 Relationships between the dangling bond defect density in μc-Si:H films and their deposition rates for (A) μc-Si:H prepared using the conventional low-pressure regime and for (B) μc-Si:H prepared using the high-pressure depletion (HPD) method. The *star* symbol represents μc-Si:H prepared under HPD conditions with the novel cathode design

to the plasma and the number density of SiH_4 is proportional to the partial pressure. However, hydrogen atoms (the chemical species responsible during the formation of μc-Si:H) are strongly scavenged by SiH_4 molecules during their transportation from their production site to the film-growing surface. In order to enhance the survival of the hydrogen atoms, it is has been suggested that the SiH_4 molecules should be depleted by applying high power density to the plasma. As conditions of high total pressure are also useful for decreasing the effects of ion bombardment during film growth through the reduction of the electron temperature in the plasma, high working pressures along with SiH_4 depletion conditions (HPD) have recently been popularly adopted for the high-rate growth of device-grade μc-Si:H.

Figure 26.24 shows the number density of dangling bond defects, as measured by electron spin resonance (ESR), in the resulting μc-Si:H as a function of growth rate [26.45]. When the conventional low pressure regime is used for the growth of μc-Si:H, the defect density increases exponentially with increasing growth rate, as seen in the figure (see A). This is caused by both an increase in the ion bombardment (due to the high power density) and an increase in the contributions from short-lifetime chemical species such as SiH_2, SiH and Si due to the reduced ability of those species to react with SiH_4 and H_2 (due to the low pressure and SiH_4 depletion) during film growth. On the other hand, the slope in Fig. 26.24 becomes shallower when HPD conditions are used during the growth of μc-Si:H, as shown in the figure (see B), which illustrates the usefulness of the HPD method for obtaining high-quality μc-Si:H at high growth rates.

The validity of the HPD method has also been demonstrated during the fabrication of μc-Si:H-based solar cells. Figure 26.25 shows the photo-I–V characteristics of a μc-Si:H-based p–i–n single-junction solar cell prepared at a high growth rate of 2 nm/s using the HPD method, which exhibits a reasonably high conversion efficiency of 8.1% [26.46].

However, HPD conditions require that the spacing between the cathode and anode is reduced in the conventional capacitively coupled plasma reactor, giving rise to nonuniform plasma production and nonuniform film growth. To overcome this problem encountered when using HPD conditions, the structure of the cathode surface has been designed to produce uniform plasma

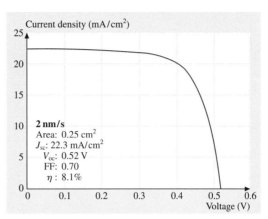

Fig. 26.25 Photo-I–V characteristics of a p–i–n μc-Si:H-based solar cell fabricated at reasonably high growth rate of 2 nm/s under HPD conditions

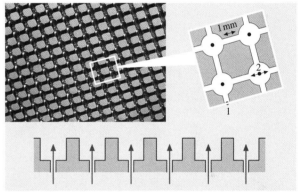

Fig. 26.26 Photograph of the surface structure on the cathode designed for the production of high-density/uniform plasmas

production even in a large-area parallel plate electrode configuration. Figure 26.26 shows the structure of this novel design of cathode. A multitude of holes (hollows) with interconnecting slots are arranged on the cathode surface, which cause strong coupling between the high-density plasmas produced in each hole where source gas injection is performed. Using this newl type of cathode, quite high growth rates (more than 8 nm/s) have been obtained along with reasonably low defect densities in the resulting μc-Si:H, as shown by star symbol in Fig. 26.24 [26.47].

26.6 Summary

In this chapter, the processes involved in the growth of a-Si:H and μc-Si:H from SiH_4 and SiH_4H_2 plasma have been interpreted in detail. The defect density determination reaction that takes place on the film-growing surface was discussed in order to obtain clues that may lead to enhanced optoelectronic properties in those materials. The recent status of work done in the fields of thin-film transistors and solar cells was reviewed, as these are the main device applications of those materials. Recent progress in resolving material issues related to solar cell applications were also described.

Finally, we note here that the concepts used in and our understanding of the film growth process mentioned here are widely applicable to other processes, especially processes where thin films are grown from reactive plasmas.

References

26.1 W. E. Spear, P. G. LeComber: Solid State Commun. **17**, 1193 (1975)
26.2 C. R. Wronski, D. E. Carlson, R. E. Daniel: Appl. Phys. Lett. **29**, 602 (1976)
26.3 T. Moustakas: *Semicond. Semimet.*, Vol. 21A (Academic, New York 1984) p. 55
26.4 T. Saito, S. Muramatsu, T. Shimada, M. Migitaka: Appl. Phys. Lett. **42**, 678 (1983)
26.5 Y. Mishima, M. Hirose, Y. Osaka, K. Nagamine, Y. Ashida, K. Isogaya: Jpn. J. Appl. Phys. **22**, L46 (1983)
26.6 T. Fuyuki, K. Y. Du, S. Okamoto, S. Yasuda, T. Kimoto, M. Yoshimoto, H. Matsunami: J. Appl. Phys. **64**, 2380 (1988)
26.7 A. H. Mahan, B. P. Nelson, S. Salamon, R. S. Crandall: Mater. Res. Soc. Proc. **219**, 673 (1991)
26.8 M. Tsuda, S. Oikawa, K. Saito: J. Chem. Phys. **91**, 6822 (1989)
26.9 J. Perrin, O. Leroy, M. C. Bordage: Contrib. Plasma Phys. **36**, 3 (1996)
26.10 A. Matsuda, K. Nakagawa, K. Tanaka, M. Matsumura, S. Yamasaki, H. Okushi, S. Iizima: J. Non-Cryst. Solids **35-36**, 183 (1980)
26.11 A. Matsuda, K. Tanaka: Thin Solid Films **92**, 171 (1982)
26.12 Y. Matsumi, T. Hayashi, H. Yoshikawa, S. Komiya: J. Vac. Sci. Technol. A **4**, 1786 (1986)
26.13 N. Itabashi, N. Nishiwaki, M. Magane, T. Goto, A. Matsuda, C. Yamada, E. Hirota: Jpn. J. Appl. Phys. **29**, 585 (1990)
26.14 N. Itabashi, N. Nishiwaki, M. Magane, S. Saito, T. Goto, A. Matsuda, C. Yamada, E. Hirota: Jpn. J. Appl. Phys. **29**, L505 (1990)
26.15 K. Tachibana, T. Mukai, H. Harima: Jpn. J. Appl. Phys. **30**, L1208 (1991)
26.16 A. Kono, N. Koike, H. Nomura, T. Goto: Jpn. J. Appl. Phys. **34**, 307 (1995)
26.17 A. Matsuda: J. Non-Cryst. Solids **59-60**, 767 (1983)
26.18 A. Matsuda, T. Goto: Mater. Res. Soc. Proc. **164**, 3 (1990)
26.19 Y. Toyoshima, K. Arai, A. Matsuda, K. Tanaka: J. Non-Cryst. Solids **137-138**, 765 (1991)
26.20 J. L. Guizot, K. Nomoto, A. Matsuda: Surf. Sci. **244**, 22 (1991)
26.21 M. Kondo, M. Fukawa, L. Guo, A. Matsuda: J. Non-Cryst. Solids **266-269**, 84 (2000)
26.22 C. C. Tsai, G. B. Anderson, R. Thompson, B. Wacker: J. Non-Cryst. Solids **114**, 151 (1989)
26.23 K. Nakamura, K. Yoshida, S. Takeoka, I. Shimizu: Jpn. J. Appl. Phys. **34**, 442 (1995)
26.24 A. Matsuda: Thin Solid Films **337**, 1 (1999)
26.25 K. Saito, M. Kondo, M. Fukawa, T. Nishimiya, W. Futako, I. Shimizu, A. Matsuda: Res. Soc. Proc. Mater. **507**, 843 (1998)
26.26 J. Koh, Y. Lee, H. Fujiwara, C. R. Wronski, R. W. Collins: Appl. Phys. Lett. **73**, 1526 (1998)
26.27 H. Fujiwara, M. Kondo, A. Matsuda: Surf. Sci. **497**, 333 (2002)
26.28 H. Fujiwara, M. Toyoshima, M. Kondo, A. Matsuda: J. Non-Cryst. Solids **266-269**, 38 (2000)
26.29 H. Fujiwara, M. Kondo, A. Matsuda: Phys. Rev. B **63**, 115306 (2001)
26.30 S. Suzuki, M. Kondo, A. Matsuda: J. Non-Cryst. Solids **299-302**, 93 (2002)
26.31 H. Fujiwara, M. Kondo, A. Matsuda: Jpn. J. Appl. Phys. **41**, 2821 (2002)
26.32 A. Matsuda, K. Nomoto, Y. Takeuchi, A. Suzuki, A. Yuuki, J. Perrin: Surf. Sci. **227**, 50 (1990)
26.33 G. Ganguly, A. Matsuda: Phys. Rev. B **47**, 3361 (1993)
26.34 G. Ganguly, A. Matsuda: J. Non-Cryst. Solids **164-166**, 31 (1993)
26.35 G. Ganguly, A. Matsuda: Jpn. J. Appl. Phys. **31**, L1269 (1992)
26.36 Y. Nasuno, M. Kondo, A. Matsuda: Tech. Digest of PVSEC-12. Jeju, Korea, (2001) 791
26.37 D. L. Staebler, C. R. Wronski: Appl. Phys. Lett. **28**, 671 (1977)
26.38 Ke. Saito, M. Sano, K. Matsuda, T. Kondo, T. Nishimoto, K. Ogawa, I. Kajita: Proc. WCPEC-2 Vienna, Austria (1998) p.351
26.39 J. Meier, P. Torres, R. Platz, S. Dubail, U. Kroll, J. A. Anna Selvan, N. Pellaton Vaucher, Ch. Hof, D. Fischer, H. Keppner, A. Shah, K.-D. Ufort, P. Giannoules: Mater. Res. Soc. Proc. **420**, 3 (1996)

26.40 K. Yamammoto, M. Yoshimi, T. Suzuki, Yu. Tawada, Y. Okamoto, A. Nakajima: Mater. Res. Soc. Proc. **507**, 131 (1998)
26.41 T. Nishimoto, M. Takai, H. Miyahara, M. Kondo, A. Matsuda: J. Non-Cryst. Solids **299–302**, 1116 (2002)
26.42 M. Takai, T. Nishimoto, T. Takagi, M. Kondo, A. Matsuda: J. Non-Cryst. Solids **266–269**, 90 (2000)
26.43 M. Takai, T. Nishimoto, M. Kondo, A. Matsuda: Appl. Phys. Lett. **77**, 2828 (2000)
26.44 L. Guo, M. Kondo, M. Fukawa, K. Saito, A. Matsuda: Jpn. J. Appl. Phys. **37**, L1116 (1998)
26.45 M. Kondo, T. Nishimoto, M. Takai, S. Suzuki, Y. Nasuno, A. Matsuda: Tech. Digest of PVSEC-12 Jeju, Korea (2001) 41
26.46 T. Matsui, M. Kondo, A. Matsuda: Proc. WCPEC-3 Osaka, Japan (2003) 50-A3-02
26.47 C. Niikura, M. Kondo, A. Matsuda: Proc. WCPEC-3 Osaka, Japan (2003) 5P-D4-03

27. Ferroelectric Materials

Ferroelectric materials offer a wide range of useful properties. These include ferroelectric hysteresis (used in nonvolatile memories), high permittivities (used in capacitors), high piezoelectric effects (used in sensors, actuators and resonant wave devices such as radio-frequency filters), high pyroelectric coefficients (used in infra-red detectors), strong electro-optic effects (used in optical switches) and anomalous temperature coefficients of resistivity (used in electric-motor overload-protection circuits). In addition, ferroelectrics can be made in a wide variety of forms, including ceramics, single crystals, polymers and thin films – increasing their exploitability. This chapter gives an account of the basic theories behind the ferroelectric effect and the main ferroelectric material classes, discussing how their properties are related to their composition and the different ways they are made. Finally, it reviews the major applications for this class of materials, relating the ways in which their key functional properties affect those of the devices in which they are exploited.

	27.0.1	Definitions and Background	597
	27.0.2	Basic Ferroelectric Characteristics and Models	599
27.1	**Ferroelectric Materials**		601
	27.1.1	Ferroelectric Oxides	601
	27.1.2	Triglycine Sulphate (TGS)	607
	27.1.3	Polymeric Ferroelectrics	607
27.2	**Ferroelectric Materials Fabrication Technology**		608
	27.2.1	Single Crystals	608
	27.2.2	Ceramics	609
	27.2.3	Thick Films	613
	27.2.4	Thin Films	613
27.3	**Ferroelectric Applications**		616
	27.3.1	Dielectrics	616
	27.3.2	Computer Memories	616
	27.3.3	Piezoelectrics	617
	27.3.4	Pyroelectrics	620
References			622

27.0.1 Definitions and Background

Ferroelectric materials offer a very wide range of useful properties for the electronic engineer to exploit. As we will see, they are also a class of materials that is hard to define accurately in a single sentence. It is useful to start from the class of insulating materials that form dielectrics; in other words materials that will sustain a dielectric polarisation under the application of an electric field. There exists a set of these materials for which the crystal structure lacks a centre of symmetry. (If a crystal structure has a centre of symmetry, it means that for every atom in the structure there is a point in the unit cell through which inversion will bring one to the same type of atom.) A list of the non-centrosymmetric, or *acentric*, point groups is given in Table 27.1. All of the crystalline materials whose structures possess these point groups (with the exception of group 432) exhibit the phenomenon of piezoelectricity, which means that stress will generate a charge separation on the faces of the crystal (the direct piezoelectric effect) and will undergo mechanical strain when subjected to an electric field (the converse piezoelectric effect). Both effects are widely exploited in electronic devices. A well-known example of a non-centrosymmetric material is the mineral α-quartz, which is used for the piezoelectric resonators employed for frequency filtering and electronic clocks.

Table 27.1 Polar and acentric (non-centrosymmetric) point groups

Crystal system	Polar (acentric)	Nonpolar (acentric)
Triclinic	1	
Monoclinic	2, *m*	
Orthorhombic	*mm*2	222
Trigonal	3, 3*m*	32
Hexagonal	6, 6*mm*	$\bar{6}, \bar{6}m2$
Tetragonal	4, 4*mm*	$\bar{4}, 422, \bar{4}2m$
Cubic	None	23, $\bar{4}3m$, 432

A sub-set of the non-centrosymmetric crystals also possess a unique axis of symmetry. These crystals are said to be *polar*. The polar point groups are also listed in Table 27.1. Polar crystals are piezoelectric (as they are acentric) and also exhibit pyroelectricity, which means that a charge separation will appear on their surfaces when their temperature is changed. Polar structures effectively have a dielectric polarisation *built in* to the unit cell of the crystal structure. This is sometimes called a *spontaneous polarisation*. The application of stress or a change in temperature causes a change in this dipole moment and it is this change that causes the separation of charge on the surfaces of the crystal. The direction and magnitude of the spontaneous polarisation in a polar dielectric can be changed by the application of an electric field, but on removal of the field it will return to its zero-field value. A well-known example of a polar dielectric is ZnO, which possesses the wurtzite crystal structure in which Zn^{2+} ions sit in the tetrahedrally coordinated sites between hexagonal close-packed layers of oxygen ions. Thin films of ZnO are widely used for piezoelectric applications.

A sub-set of the set of polar dielectrics exists for which the application of a field of sufficient magnitude will cause the spontaneous polarisation to switch to a different, stable direction. Upon removal of the field the polarisation will not spontaneously return to its original direction and magnitude. These crystals are called *ferroelectric*. We can view the relationships between the sets of ferroelectric, polar, acentric and centrosymmetric dielectrics as shown in the Venn diagram in Fig. 27.1.

The history of ferroelectrics is long, as has been described in the excellent review by *Busch* [27.1]. Many very distinguished scientists were involved in its early development, including Brewster (who was one of those who studied pyroelectricity), J. & P. Curie (who discovered piezoelectricity), Boltzmann, Pockels and Debye,

to name but a few. Indeed, the term ferroelectricity was first coined by E. Schrodinger. However, credit for the discovery of the effect goes to Joesph Valasek, who found in 1920 that the polarisation of sodium potassium tartrate (Rochelle salt) could be switched by the application of an electric field, thus providing the first demonstration of the process that is the hallmark of ferroelectricity.

A reasonable working definition of ferroelectricity is "a polar dielectric in which the polarisation can be switched between two or more stable states by the application of an electric field". However, as we will see in the discussion which follows, there are exceptions to this definition: some ferroelectrics are semiconducting (and thus are not dielectrics because they cannot sustain an electrical polarisation); the spontaneous polarisation in some ferroelectrics cannot be switched because they cannot sustain an electric field of sufficient magnitude to effect the switching, either because they reach electrical breakdown first, or because they are too conducting.

Since the initial discovery of ferroelectricity in Rochelle salt, the effect has been demonstrated in a wide range of materials, from water-soluble crystals through oxides to polymers, ceramics and even liquid crystals. Many of these will be discussed in this chapter. The range of useful properties exhibited by ferroelectrics covers:

- Ferroelectric hysteresis is used in nonvolatile computer information storage.
- Ferroelectrics can exhibit very high relative permittivities (several thousand) which means that they are widely used in capacitors.
- The direct piezoelectric effect (the generation of charge in response to an applied stress) is widely used in sensors such as accelerometers, microphones, hydrophones etc.
- The converse piezoelectric effect (the generation of strain in response to an applied electric field) is widely used in actuators, ultrasonic generators, resonators, filters etc.
- The pyroelectric effect (the generation of charge in response to a change in material temperature) is widely used in uncooled infra-red detectors.
- The electro-optic effect (a change in birefringence in response to an applied electric field) is used in laser Q-switches, optical shutters and integrated optical (photonic) devices.
- Ferroelectrics exhibit strong nonlinear optical effects that can be used for laser frequency doubling and optical mixing.

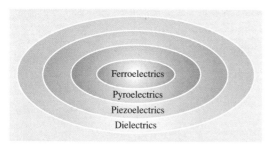

Fig. 27.1 Venn diagram showing how ferroelectrics fit into the different classes of dielectric materials

- Illumination of transparent ferroelectrics with light of sufficient energy causes excitation of carriers into the conduction band. Their movement under the internal bias field caused by the spontaneous polarisation causes a refractive-index modulation that can be used for a variety of optical applications, including four-wave mixing and holographic information storage.
- Ferroelectrics exhibit strong coupling between stress and birefringence, which can be used to couple acoustic waves to optical signals with applications in, for example, radar signal processing.
- Doping certain ferroelectric ceramics with electron donors (e.g. $BaTiO_3$ with La^{3+}) can render them semiconducting. Heating these ceramics through their Curie temperature causes a very large, reversible increase in resistivity (by several orders of magnitude in some cases) over a narrow range of temperature (ca. $10\,°C$). This large positive temperature coefficient of resistance (PTCR) is widely exploited in electric-motor overload-protection devices and self-stabilising ceramic heating elements.

27.0.2 Basic Ferroelectric Characteristics and Models

The switchable spontaneous polarisation in ferroelectrics gives rise to the first characteristic property of the materials: ferroelectric hysteresis. Figure 27.2 shows a schematic plot of polarisation versus electric field as would be observed on a typical ferroelectric. As the field is increased from zero, the overall polarisation in the crystal increases as the polarisations in different dipolar regions are aligned. Eventually, it reaches a saturation point where the only further increase in P is that due to the relative permittivity of the material. (The gradient of the P/E curve for a linear dielectric is equal to its permittivity). Extrapolation of this line back to the abscissa gives the saturation value of the spontaneous polarisation (P_s). Reduction of the field to zero leaves a *remanent polarisation* (P_r), which is usually slightly less than P_s. A negative field will cause the polarisation to reduce, until it reaches zero at the *coercive field* ($-E_c$). A further negative increase in the field will eventually cause a reverse saturation polarisation ($-P_s$) to develop. When the field returns to zero the crystal is left with a negative remanent polarisation ($-P_r$). Increasing the field once more, increases the polarisation from $-P_r$ to zero at E_c, and then to $+P_s$, completing the ferroelectric hysteresis loop. The ability to switch the polarisation between two states gives rise to the first application of ferroelectrics – as a nonvolatile memory storage medium. It also permits polycrystalline ferroelectrics (especially ceramics) to be polarised. The fact that the polarisation can possess different directions within different regions of the same crystal gives rise to the existence of *ferroelectric domains*. These are given names according to the angle between the polarisation between adjacent regions. Hence, two adjacent regions in which the polarisations are orientated at 180° to each other are, naturally, called 180° domains. In many ferroelectrics, the polarisation can choose between many different possible directions so that adjacent regions can have polarisations at, for example, 90° to one another.

For most ferroelectrics, the polar state only exists over a limited range of temperatures. As the temperature is raised, a point is reached at which there is a transition from the polar, ferroelectric phase to a nonpolar, non-ferroelectric phase (called the *paraelectric* phase). In all cases, the paraelectric phase possesses a higher crystal symmetry than the ferroelectric phase into which it transforms. The temperature at which this occurs is called the *Curie temperature* (T_C). This ferroelectric-to-paraelectric phase transition is a characteristic of most ferroelectrics, but again there are exceptions. Some ferroelectric materials melt or decompose before T_C is reached; the polymeric ferroelectric polyvinylidene fluoride (PVDF) is one such example. For most ferroelectrics (called *proper ferroelectrics*), as the Curie temperature is approached from above, the relative permittivity is observed to increase, reaching a peak at T_C and decreasing below T_C. (There is a class of ferroelectrics, known as *improper* or *extrinsic ferroelectrics*, for which there is no peak in permittivity, just an anomaly. These are described further below.) Very

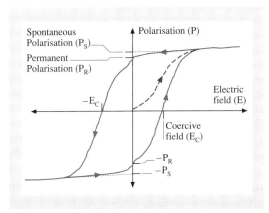

Fig. 27.2 Ferroelectric hysteresis loop

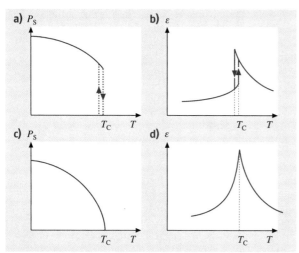

Fig. 27.3a–d The variations of spontaneous polarisation (a and c) and permittivity (b and d) with temperature for typical ferroelectric materials. (**a**) and (**b**) are for a first-order (discontinuous) phase transition while (**c**) and (**d**) are for a second-order (continuous) phase transition

high values of relative permittivity (many thousands) can be reached at T_C, leading to another use of ferroelectric materials – as dielectrics in high-value capacitors. The phase transition at T_C can either be continuous (known as second-order) or discontinuous (known as first-order). In either case, the dependence of permittivity (ε) on temperature above T_C can be described by $\varepsilon = C/(T - T_0)$. T_0 is called the *Curie Weiss temperature*, and is only equal to T_C for a second-order phase transition. C is called the *Curie Weiss constant*. Figure 27.3 shows typical plots of P_s and ε as functions of temperature for ferroelectrics around T_C. Figs. 27.3a and 27.3b show the behaviour for a first-order phase transition. As can be seen, P_s drops discontinuously to zero at the phase-transition temperature. The permittivity rises as the temperature decreases, peaking at the transition, where there is a discontinuous step down. First-order phase transitions tend to show thermal hysteresis in the transitions, so that the transition occurs at a higher temperature when approached from the low-temperature side than from the high-temperature side. Frequently in practical observations there is actually a region of coexistence where both phases exist together in the same sample at the same temperature. In this case, the Curie temperature is defined as the temperature at which the ferroelectric and paraelectric phases have the same Gibbs free energies. In the case of a second-order phase transition (Figs. 27.3c and 27.3d), P_s drops continuously to zero at T_C and the permittivity rises to a sharp peak. There is no hysteresis in the transition.

There are two types of theoretical model that describe the ferroelectric phenomenon. The first is the phenomenological or thermodynamic theory first proposed by *Devonshire* [27.2]. Here, the Gibbs free energy of the ferroelectric system is described in terms of a power series in the spontaneous polarisation. In this case, P_s is termed the order parameter for the phase transition and the theory is similar in form to the generalised phenomenological theory of phase transitions known as Landau theory. It is not appropriate to go into the details of the theory here, which has been described very well by *Lines* and *Glass* [27.3]. The theory successfully predicts many of the behavioural characteristics of ferroelectric materials in terms of the measurable macroscopic properties, including the ferroelectric hysteresis and the behaviour of the permittivity near T_C. However, it tells us nothing about the microscopic origins of ferroelectricity. This is the province of the two types of microscopic models of ferroelectric behaviour. The first of these is the order–disorder model of ferroelectric behaviour. The second is the displacive model.

According to the order–disorder model, the electric dipoles exist within the structure in the paraelectric phase above T_C, but are thermally disordered between two or more states so that the average polarisation is zero. There are several materials for which that is the case, an example being potassium dihydrogen phosphate (KDP), which is paraelectric, tetragonal ($\bar{4}2m$) above 123 K and ferroelectric, orthorhombic ($mm2$) below. The polarisation appears along the tetragonal c-axis. In this crystal structure, the PO_4 groups form tetrahedra that are linked by hydrogen bonds at their corners. The protons in these sit in double potential wells. Above T_C they are delocalised between the the two minima in the wells. Below T_C, they are localised and the PO_4 tetrahedra become distorted. The phosphorous and potassium ions become displaced relative to the oxygen framework, forming the polarisation in the lattice. Replacement of the hydrogen in the structure by deuterium raises T_C to about 220 K because the heavier deuteron delocalises between the two potential minima at a much higher temperature. Another example of an order–disorder ferroelectric is sodium nitrite ($NaNO_2$) in which the polar groups are NO_2^- ions, which become ordered below T_C (163 °C) with their dipoles all pointing along the b-axis of the crystal structure.

In displacive ferroelectrics, there are considered to be no dipoles in the structure above T_C, but the dipoles appear in the structure due to the cooperative motion of ions. Typically this is seen as the *softening* of a zone-centre optical phonon. What this means is that the frequency of an optical phonon mode with zero wavevector goes to zero at T_C. Such a phonon mode involves the displacement in opposite directions of cations and anions in the structure. When the frequency of this mode goes to zero, a dipolar displacement results. Examples of such ferroelectrics include the perovskite-structured oxides such as $BaTiO_3$, which will be discussed further below. One of the successes of soft-mode theory is that it provides a convincing mechanism for the peak in the dielectric permittivity at T_C. According to the Lyddane–Sachs–Teller relationship (see, for example, [27.3] p. 216), the softening of a zone-centre phonon will make the permittivity diverge at T_C.

In practice, the distinction between order–disorder and displacive ferroelectrics becomes rather blurred, as some ferroelectrics which may be considered to be purely displacive may exhibit significant disordered cation displacement well above T_C, while some order–disorder ferroelectrics show soft-mode behaviour. For a more detailed discussion of these theories, the reader is referred to *Lines* and *Glass* [27.3].

27.1 Ferroelectric Materials

27.1.1 Ferroelectric Oxides

This is by far the most technologically important class of ferroelectric materials. They will be discussed according to the most important crystal classes.

Perovskite Ferroelectrics

These materials possess crystal structures isomorphous with the mineral perovskite ($CaTiO_3$). They all have the general chemical formula ABO_3, where A and B are cations. Typically the A cation will be around 1.2–1.6 Å in radius (similar to the oxygen ions) while the B cations will be around 0.6–0.7 Å in radius. The crystal structure is illustrated in Fig. 27.4. It consists of a network of corner-linked BO_6 octahedra, within which is enclosed the large A cation. Another way to look at the structure is as cubic-close-packed AO_3 layers, with the small B cations sitting in the octahedral sites between these close-packed layers. A pair of layers are shown in Fig. 27.5. The structure is a very tolerant one and will accommodate many different ions. Because of this, it is exhibited by a large number of oxides. The basic criteria for the structure to be stable is that the valencies of the ions should balance and that the ionic radii meet the *Goldschmidt criteria* [27.4]. A tolerance factor t is

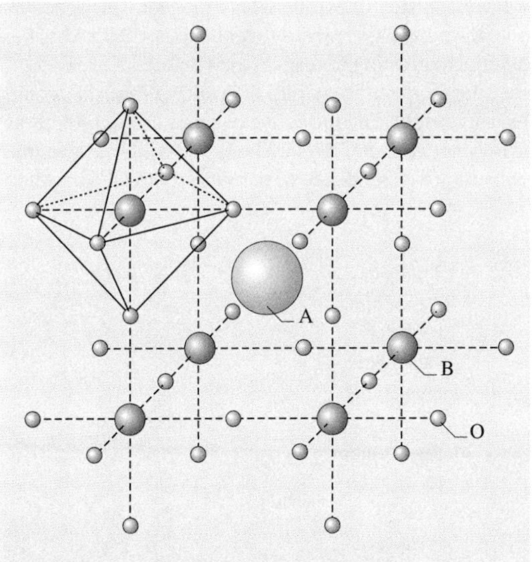

Fig. 27.4 The perovskite crystal structure

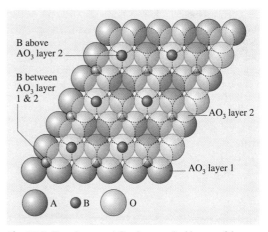

Fig. 27.5 Showing two AO_3 close-packed layers of the perovskite structure, with the B cations sitting in the sixfold coordinated sites between the layers. The second layer has been made semitransparent for clarity

Table 27.2 List of cations frequently found to form perovskite-structured oxides and their ionic radii. Ionic radii taken from [27.5]

A-site cation	Ionic radius when [12][a] by O^{2-} (Å)	B-site cation	Ionic radius when [6][b] by O^{2-} (Å)
Na^+	1.32	Nb^{5+}	0.64
K^+	1.6	Ta^{5+}	0.68
Ba^{2+}	1.6	Zr^{4+}	0.72
Sr^{2+}	1.44	Ti^{4+}	0.605
Pb^{2+}	1.49	Pb^{4+}	0.775
Bi^{3+}	1.11	Sc^{3+}	0.73
Ca^{2+}	1.35	Fe^{3+}	0.645

[a] "12-fold coordinated"; [b] "6-fold coordinated"

defined:

$$t = \frac{r_A + r_O}{\sqrt{2}(r_B + r_O)},$$

where r_X is the ionic radius of the X cation. The ideal cubic perovskite structure, where the ions are just touching each other, will possess $t = 1$. However, the structure will be stable with $0.85 < t < 1.05$. A list of ions that commonly form perovskites, together with their ionic radii, is given in Table 27.2. The closer t is to unity, the more likely the structure will be to be cubic. Conversely, perovskites which have values of $t < 1$ show distorted structures that are frequently ferroelectric. Examples of some perovskites and their tolerance factors are listed in Table 27.3. This table also lists the structures formed by the compounds at room temperature, and whether or not they are ferroelectric. Some of the most interesting perovskites from the point of view of applications are $BaTiO_3$, $PbTiO_3$ and $KNbO_3$. $BaTiO_3$ is cubic above 135 °C, but transforms to a tetragonal ferroelectric structure below this temperature. In this case, the Ba and Ti ions are displaced relative to the anion framework along one of the cubic $\langle 001 \rangle$ directions. This means that the polar axis has six choices for direction in the tetragonal phase. At 5 °C there is a second-phase transition from the tetragonal to an orthorhombic phase, where the polarisation now appears due to cation displacements along one of the cubic $\langle 110 \rangle$ directions, for which there are 12 choices. Finally, at -90 °C there is a transition to a rhombohedral phase with the cations being displaced along one of the cubic $\langle 111 \rangle$ directions, for which there are eight choices. In the case of $PbTiO_3$ there is a single transition to a tetragonal phase at 490 °C, again by cationic displacements along $\langle 100 \rangle$. Some perovskites show phase transitions that are not ferroelectric. For example, $SrTiO_3$ shows a transition to a tetragonal phase at 110 K which involves linked rotations, or tilts, of the TiO_6 octahedra about the cubic [100] direction. Tilting of the octahedra is a common feature of the phase transitions that occur in perovskites and can lead to very complex series of phase transitions, as has been observed for $NaNbO_3$. This type of structural modification and a commonly-used notation for it has been described in detail by *Glazer* [1972].

It is very easy to make solid solutions of the end-member perovskites, such as those listed in Table 27.3, and this has been used to great effect to provide materials with a wide range of properties. For example, Fig. 27.6 shows the temperature dependence of the relative permittivity of a $BaTiO_3$ single crystal [27.6]. There is a peak at each transition where the value perpendicular to the polar axis reaches several thousand, making the material interesting for use in capacitor dielectrics. However, the temperature variation in such a material would make it useless. The formation of ceramic solid solutions of $BaTiO_3$ with $SrTiO_3$, $CaTiO_3$ or $PbTiO_3$ allows the temperature dependence of permittivity to be controlled

Table 27.3 Some end-member perovskites and their properties

Perovskite oxide	Tolerance factor	Structure at 20 °C	Type	T_C (°C)
$BaTiO_3$	1.06	Tetragonal	Ferroelectric	135
$SrTiO_3$	1.00	Cubic	Paraelectric	
$CaTiO_3$	0.97	Tetragonal	Paraelectric	
$PbTiO_3$	1.02	Tetragonal	Ferroelectric	490
$PbZrO_3$	0.96	Orthorhombic	Antiferroelectric	235
$NaNbO_3$	0.94	Monoclinic	Ferroelectric	-200
$KNbO_3$	1.04	Tetragonal	Ferroelectric	412
$KTaO_3$	1.02	Cubic	Ferroelectric	-260
$BiScO_3$	0.83	Rhombohedral	Ferroelectric	370
$BiFeO_3$	0.87	Tetragonal	Ferroelectric	850

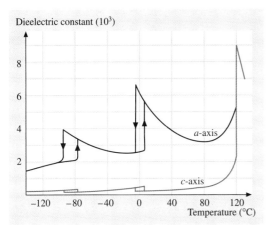

Fig. 27.6 Temperature dependence of the relative permittivity of BaTiO$_3$ measured along [001] and [100] (After [27.6])

so that useful specifications can be met with average permittivities of 2000 or more over a wide range of temperatures. *Herbert* [27.7] has discussed the details of these modifications and their effects on the BaTiO$_3$ transition temperatures. BaTiO$_3$-based ceramics are widely used in ceramic capacitors and form the basis of an industry worth billions of dollars annually. They have also been used for piezoelectric ceramics, but are no longer so important in that field, as they have largely been displaced by ceramics in the PbZrO$_3$–PbTiO$_3$ system. Nevertheless, there is a renewed interest in BaTiO$_3$-based ceramics for lead-free piezoelectrics in response to the legislative drive to reduce lead in the environment.

The perovskite solid solution system between PbZrO$_3$ and PbTiO$_3$ is of great technological importance and is thus worth discussing in some detail. The phase diagram given in Fig. 27.7 shows several phases. Starting at PbTiO$_3$, the ferroelectric tetragonal (F$_T$) phase persists well across the diagram, until the composition Pb(Zr$_{0.53}$Ti$_{0.47}$)O$_3$ is reached, where there is a transition to a ferroelectric, rhombohedral phase. The composition where this occurs is called the morphotropic phase boundary (MPB). The rhombohedral phase region splits into two. There is a high-temperature phase (F$_{R(HT)}$) in which the cations are displaced along the cubic [111] direction. There is also a low-temperature ferroelectric rhombohedral (F$_{R(LT)}$) phase in which the (Zr, Ti)O$_6$

Fig. 27.7 (a) PbZrO$_3$–PbTiO$_3$ phase diagram (after [27.8]) (b) Region close to PbZrO$_3$ (after [27.9, 10]) (c) Region close to the MPB (after [27.11])

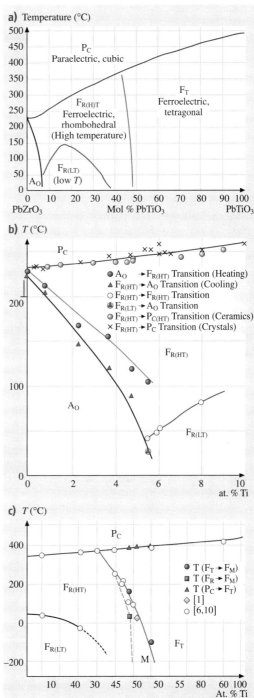

octahedra are rotated about the [111] axis. This doubles the unit cell. Close to $PbZrO_3$ (Fig. 27.7b) the room-temperature structure transforms to the antiferroelectric, orthorhombic (A_O) structure. This is a complex structure in which the cations are displaced in double antiparallel rows along the cubic ⟨110⟩ directions, coupled with octahedral tilts. A higher-temperature A_T phase, which is frequently shown when this phase diagram is cited in the literature, has been shown not to exist in pure solid solutions [27.10]. There is another, more modern, modification to this phase diagram. It has been shown by *Noheda* et al. [27.11] that a monoclinic phase exists at the MPB (Fig. 27.7c). The enormous technological and commercial importance of this system derives from the very high piezoelectric, pyroelectric and electro-optic coefficients that can be obtained from ceramic compositions in different parts of the phase diagram, particularly when the base composition is doped with selected ions. These are discussed in more detail below. This solid solution system is frequently referred to generically as *PZT* although, strictly, PZT was the brand name for a set of piezoelectric ceramic compositions manufactured by the Clevite Corporation.

There is another very important class of oxides, termed complex perovskites, where the A or, more commonly B, sites in the structure are occupied by ions of different valency in a fixed molar ratio. These are not solid solutions as they possess a fixed composition. Examples are $PbMg_{1/3}Nb_{2/3}O_3$ (PMN), $PbSc_{1/2}Ta_{1/2}O_3$ (PST) and $Bi_{1/2}Na_{1/2}TiO_3$. These are also technologically important and can form end members to solid solutions in their own right, $PbMg_{1/3}Nb_{2/3}O_3-PbTiO_3$ (PMN-PT), for example. Many of the complex ferroelectrics (PMN-PT is a classic example) exhibit *ferroelectric relaxor* behaviour. In these materials, the Curie point is no longer a sharp transition but is actually observed over a very wide range of temperatures. A broad permittivity peak is observed and the temperature and height of the peak are strongly dependent on the frequency of measurement. Figure 27.8 shows a typical variation of permittivity with temperature and frequency in PMN [27.12].

The broad tolerance of the perovskite structure to different cations has led to a wide exploration of the inclusion of different iso-valent dopants to obtain different electronic properties. It is also possible to include alio-valent dopants over a wide range, which the structure tolerates through the introduction of cation or anion vacancies, or free charge carriers. For example, it is possible to incorporate Nb^{5+} into the B site of the $BaTiO_3$ system [27.13], and charge balance is main-

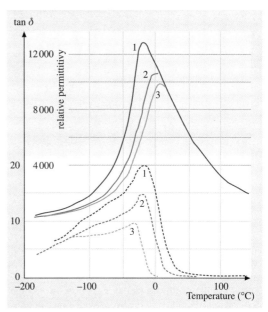

Fig. 27.8 Permittivity (*solid lines*) and dielectric loss (*dashed lines*) vs. temperature in PMN. The measurements were made at: (1) 0.4, (2) 45 and (3) 4500 kHz. (Adapted from [27.12])

tained through the presence of free electrons. Similarly, Fe^{3+} can be introduced into the B site of the system and this substitution creates O vacancies [27.14].

The vast majority of perovskites are employed in ceramic form, but thin- and thick-film materials are becoming of increasing importance and single-crystal materials are starting to emerge which have exceptionally high piezoelectric coefficients.

Ilmenite Ferroelectrics

The ilmenite structure is related to the perovskite structure in that it is exhibited by materials with the general formula ABO_3, where the A cation is too small to fill the [12] coordinated site of the perovskite structure. The structure is made up of hexagonal close-packed layers of oxygen ions, with the A and B ions occupying the octahedrally coordinated sites between the layers. Hence, this structure can also be considered to be related to the perovskite structure in that both are based on oxygen octahedra. The two best known ilmenite ferroelectrics are $LiNbO_3$ and $LiTaO_3$, whose structure is shown in Fig. 27.9. The materials have high T_C values (ca. $1200\,°C$ and $620\,°C$, respectively). In the case of $LiNbO_3$, the T_C is only $50\,°C$ below the melting point.

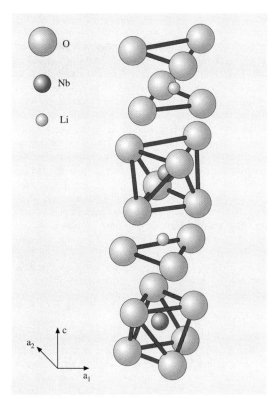

Fig. 27.9 Crystal structure of LiNbO$_3$ (after [27.15])

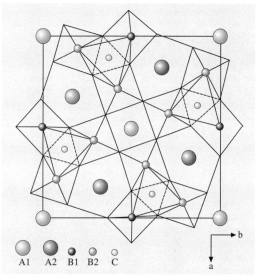

Fig. 27.10 Tungsten bronze crystal structure (after [27.16])

Referring to Fig. 27.9, it can be seen that the cations occupy octahedral sites progressing along the c-axis in the order Nb, vacancy, Li, Nb, vacancy, Li etc. The polarisation occurs through displacement of the cations along the three-fold axis, so the structure transforms from $\bar{3}m$ to $3m$ symmetry at T_C. These are *uniaxial* ferroelectrics, in that the polarisation can only be up or down, and only 180° domains exist. The materials are mainly used in single-crystal form for piezoelectric and electro-optic devices.

Tungsten Bronze Ferroelectrics

This is another very large family of oxygen octahedral ferroelectrics possessing the general formula [A1$_2$A2$_4$C$_4$][B1$_2$B2$_8$]O$_{30}$. The crystal structure is complex, shown schematically in Fig. 27.10. The B1 and B2 sites are octahedrally coordinated by oxygens and have similar sizes and valencies to the B sites in the perovskites. The A1 and A2 sites are surrounded by four and five columns of BO$_6$ octahedra respectively. The three-fold coordinated C sites in the structure are frequently empty, but can be occupied by small uni- or divalent cations (e.g. Li$^+$ or Mg^{2+}). There are a wide range of ferroelectric tungsten bronzes, which frequently show nonstoichiometry. All are tetragonal in their paraelectric phase, and can transform to either a tetragonal ferroelectric phase, in which the polar axis appears along the tetrad axis of the paraelectric phase, or (more commonly) into an orthorhombic phase in which the polar axis appears perpendicular to the original tetrad axis. Examples of tungsten bronzes are PbNb$_2$O$_6$ (lead metaniobate) in which five out of the available six A sites are occupied by Pb^{2+} and the B sites by Nb^{5+}. This crystal and its Ta analogue are metastable below about 1200 °C. They can be stabilised by rapid cooling, or more commonly by doping. Lead metaniobate ceramics are difficult to make, but are commercially available and are occasionally used for piezoelectric devices by virtue of their higher Curie temperatures than members of the PZT system, combined with reduced lateral piezoelectric coupling factors. Substitution of Pb^{2+} by Ba^{2+} can also be used to produce a useful piezoelectric ceramic material Pb$_{1/2}$Ba$_{1/2}$Nb$_2$O$_6$. Other tungsten bronze ferroelectrics such as Sr$_x$Ba$_{1-x}$Nb$_2$O$_6$, Ba$_2$NaNb$_5$O$_{15}$ and K$_3$LiNb$_5$O$_{15}$ have been investigated as single crystals for use in electro-optic devices. However, the growth of highly homogeneous single crystals is difficult because of the formation of optical striations caused by compositional fluctuations.

Aurivillius Compounds

The aurivillius compounds form another important class of ferroelectrics based on oxygen octahedra. They have a good deal in common with the perovskites in that they consist of layers or slabs of perovskite blocks with the general formula $(A_{m-1}B_mO_{3m+1})^{2-}$ separated by $(M_2O_2)^{2+}$ layers in which the M cation is in pyramidal coordination with four oxygens, the M being at the apex of the pyramid. The structure possesses the general formula $M_2(A_{m-1}B_mO_{3m+3})$. In the ferroelectric phases, M is usually Bi and m is usually between one and five. The A and B cations follow the usual ionic radius and valency criteria of the perovskites, and we see ferroelectric compounds in which A = Bi, La, Sr, Ba, Na, K etc. and B = Fe, Ti, Nb, Ta, etc. Part of the crystal structure is shown schematically in Fig. 27.11, illustrating in this case a structure with $m = 3$. This is just over half of the unit cell, showing one slab of perovskite units and two of the $(Bi_2O_2)^{2-}$ layers. The structures are tetragonal in their paraelectric phases, with the tetrad c-axis being perpendicular to the $(Bi_2O_2)^{2+}$ plane. In the majority of cases they become orthorhombic in their ferroelectric phase, with the spontaneous polarisation appearing in the $a-b$ plane, at 45° to the tetragonal a-axis. Bismuth titanate ($Bi_4Ti_3O_{12}$) is an example of an aurivillius ferroelectric in which M = A = Bi, B = Ti and $m = 3$, so that there are three blocks in the perovskite slab. In this case, there is a small component of the spontaneous polarisation (ca. $4\,\mu C/cm^2$) out of the a–b plane, with the majority of the polarisation (ca. $50\,\mu C/cm^2$) being in the plane, so that the ferroelectric structure is monoclinic. The T_C is high (675 °C) and ceramics have been explored for use as high-temperature piezoelectrics. The problem with the compounds when used as ceramics is that the piezoelectric coefficients are rather small. Single crystals have been explored for optical light-valve applications but have not received widespread use because of the problems of growth. Thin films of $SrBi_2Ta_2O_9$ (SBT) are receiving considerable attention for applications in ferroelectric nonvolatile memories.

Other Oxide Ferroelectrics

There is a wide range of other oxide ferroelectric materials, although none have anywhere near the breadth of application of the oxygen octahedral ferroelectrics discussed above. There are two groups of phosphates (in which the phosphorous ions are tetrahedrally coordinated by oxygens) which have found applications in optical systems. The potassium dihydrogen phosphate (KDP) family was referred to above as an example of an order–disorder ferroelectric. KDP, and its deuterated analogue KD*P, can be grown from aqueous solution as large, high-quality single crystals. These can be lapped and polished into plates which are used for longitudinal electro-optic modulators, in which the modulating electric field is applied parallel to the direction of light propagation in the crystal. In this application, the crystals are used at room temperature in the paraelectric phase. One spectacular application for these crystals is as the Q-switches in the large ultra-high-power lasers used for nuclear fusion research, where extremely large (ca. 1 m) crystal slices are needed. Another class of phosphate ferroelectrics is the potassium titanyl phosphate system (KTP). These crystals can be grown relatively easily from the melt and are used for optical frequency doubling applications, particularly using periodically poled crystals [27.17, 18]. This is a large family in which the potassium can be replaced by caesium or rubidium and the phosphorous by arsenic. Lasers using KTP are finding significant uses in surgical procedures [27.19].

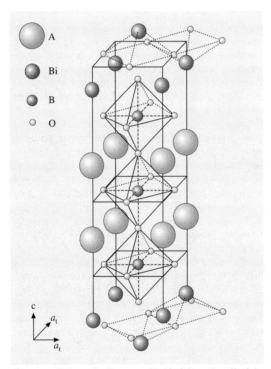

Fig. 27.11 Schematic diagram of half of the unit cell of the crystal structure of an Aurivillius compound with $m = 3$. The cell axes shown are those corresponding to the tetragonal structure

There are other examples of oxygen tetrahedral ferroelectrics. Lead germanate ($Pb_5Ge_3O_{11}$) is an interesting material which has been prepared as a single crystal. It is a hexagonal uniaxial ferroelectric which is also optically active. Ge can be substituted up to 50% by Si, which reduces the Curie temperature from 177 °C to 60 °C. The sense of the optical rotation switches with the sign of the spontaneous polarisation. This has been considered for optical devices and pyroelectric infra-red detectors, but has not reached any commercial applications. It has been prepared in ceramic and thin-film form and is occasionally used as a sintering aid in lead-containing ferroelectric ceramics. Gadolinium molybdate, $Gd_2(MoO_4)_3$ (GMO), possesses a structure consisting of corner-linked MoO_4 tetrahedra. It is an example of an improper ferroelectric material in which the phase transition from the paraelectric phase is controlled by the softening of a zone boundary mode, which would lead to a nonpolar low-temperature phase, which is in fact *ferroelastic*. The high-temperature phase is acentric, and the ferroelastic lattice distortion couples via the piezoelectric coefficients to give a spontaneous polarisation. The lattice distortion can be reversed by electrically switching the spontaneous polarisation and vice versa. GMO has no commercial applications, although it was at one time considered for optical switches.

27.1.2 Triglycine Sulphate (TGS)

The TGS family is a salt of the simplest amino acid, glycine (NH_2CH_2COOH) with sulphuric acid. In the crystal structure, the glycine groups are almost planar and pairs of them are connected by hydrogen bonds. A ferroelectric transition at 49.4 °C between two monoclinic phases is driven by an ordering of the protons in these hydrogen bonds. The replacement of these protons by deuterons (which can be accomplished by repeated crystallisation of TGS from heavy water) to give DTGS results in a 10 °C increase in T_C. There has been a great deal of academic interest in TGS because it possesses a well-behaved second-order order–disorder ferroelectric phase transition. In addition it has been of considerable technological interest for uncooled pyroelectric infra-red detectors and thermal imaging devices [27.20]. As a consequence, there has been considerable work on the replacement of various components with the objective of either improving performance or increasing T_C, such as can be achieved by deuteration. The replacement of glycine by l-alanine has been demonstrated to result in an internal bias field in the crystal, which means that the crystals will retain their single-domain structure, even if heated well above T_C.

27.1.3 Polymeric Ferroelectrics

The first polymeric ferroelectric to be discovered was polyvinylidene fluoride (PVDF). The monomer for this possesses the formula CH_2CF_2. It was first shown to be piezoelectric by *Kawai* [27.21] and later *Bergman* et al. [27.22] speculated about the possibility of it being ferroelectric, although no Curie temperature could be shown to exist, as the polymer melts first. The structure of the polymer is complex, as the polymer backbone can adopt a number of configurations depending upon whether the neighbouring carbon-carbon linkages adopt trans or gauche configurations. In a trans (T) configuration, the groups bonded to the carbon atoms sit on opposite sides of the carbon-carbon bond. In a gauche (G) configuration, they sit on the same side. In an all-trans configuration, the carbon backbone of the polymer forms a simple zigzag. In the case of PVDF, there are four different phases with different T and G sequences, and with different ways the polymers stack together in crystal structures. The polymer crystallises from the melt into *form II*, also called the α-phase, in which the bonds arrange themselves into the sequence TGTḠ, see Fig. 27.12. The fluorine atoms are strongly electronegative which makes the C−F bond polar so

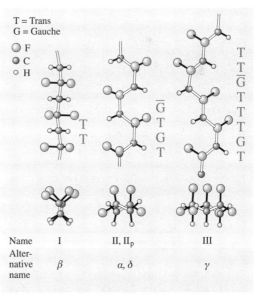

Fig. 27.12 PVDF bond structures

that the molecule has a net dipole moment perpendicular to its length. However, the molecules of polymer arrange themselves in the unit cell so that the dipoles cancel each other out. Form II is neither ferroelectric nor piezoelectric, but application of an electric field will convert it into form II$_P$, also called the δ-phase, in which the polymer molecules are arranged so that the unit cell has a net dipole moment. High-temperature annealing of either of these forms will produce form III (the γ-phase), which has a new TGTTT$\overline{\text{G}}$TT configuration which also has a net dipole moment perpendicular to the long axis of the molecule and these arrange in a crystal structure which is also polar. Subjecting forms II or III to stretching or drawing will produce form I (the β-phase), which is an all-trans configuration (Fig. 27.12). It can be seen from Fig. 27.12 that of the three molecular configurations, the polar bonds all point most-nearly in the same direction and this is retained in the unit cell. Electrical poling makes this the most-strongly piezoelectric phase of PVDF. The stretching can be in a single direction, called uniaxial and usually achieved by drawing through rollers, or in two perpendicular directions, called biaxial and usually achieved by inflating a tube of polymer. Strong electric fields are needed for poling and these are frequently applied by placing the polymer under a corona discharge. The formation of a copolymer of vinylidene fluoride with between 10% and 46% trifluoroethylene (TrFE) leads to a polymer which will crystallise directly into form I (the β-phase) from either melt or solution. This can be poled to produce a material which is as active as pure PVDF. The copolymers also show clear signs of a ferroelectric-to-paraelectric transition, such as exhibiting peaks in permittivity at T_C, with a T_C that depends on the amount of TrFE in the copolymer. Extrapolation to 0% TrFE implies a T_C in PVDF of 196 °C. PVDF is readily available and has achieved moderately widespread use as a piezoelectric sensor material. It is particularly useful where light weight and flexibility are important, or where very large areas or long lengths are needed. The P(VDF-TrFE) copolymers are not as readily available and are not widely used. Other polymers that have been shown to exhibit ferroelectric behaviour include the odd-numbered nylons, but these are only weakly piezoelectric and have not achieved any technological uses.

27.2 Ferroelectric Materials Fabrication Technology

Ferroelectrics are used commercially in a very wide range of forms, from single crystals through polycrystalline ceramics and thin films to polymers. Hence, only a summary of the fabrication techniques can be presented here.

27.2.1 Single Crystals

As has been shown above, ferroelectrics often have a wide range of complex compositions. This renders the problems of single-crystal growth much more difficult than for single-element crystals such as silicon. In many cases, the most technologically useful compositions are themselves solid solutions of complex ferroelectric end members. Frequently these do not melt congruently and sometimes one or more components of the melt will be volatile. Some of the growth techniques which are used are:

Czochralski Growth

Certain ferroelectric crystals melt congruently and can be pulled from the melt using the Czochralski technique. Examples include LiNbO$_3$ and LiTaO$_3$. Both materials are widely used technologically. LiNbO$_3$ is used in surface acoustic wave (SAW) and electro-optic devices and LiTaO$_3$ in pyroelectric infrared detectors, piezoelectric resonators, SAW and electro-optic devices. LiNbO$_3$ melts congruently at 1240 °C, but the congruently melting composition is not at the stoichiometric composition (Li : Ta = 1), but lies at around 49% Li$_2$O. Crystals pulled from such melts in platinum crucibles are perfectly adequate for piezoelectric and most optical applications. However, it is known that the stoichiometric composition possesses rather better optical properties than the congruent composition and there is now a premium-grade material available commercially. LiTaO$_3$ melts congruently at 1650 °C. Again, the composition is not at Li : Ta = 1, but is slightly Ta deficient. The growth temperature is above the melting point for platinum. The solution is to use crucibles made from iridium, which is very expensive. Also, IrO$_2$ is volatile so the growth atmosphere must be N$_2$, which means that the crystals come out of the melt oxygen-deficient and must be embedded in LiTaO$_3$ powder and annealed under O$_2$ after growth to make them clear. Pt/Rh crucibles have been used to grow LiTaO$_3$ crystals, but the Rh absorbed into the crystals makes them brown. Such crystals are acceptable for piezoelectric and pyroelectric, but not for optical, applications. Virtually all ferroelectric materials need to be electrically poled before use. In the case of LiTaO$_3$, silver electrodes are applied to faces perpendicular to the polar (c) axis and a pulsed field of about $10\,\text{V/cm}$ (1 ms with a 50/50 duty cycle) applied while the crystal is cooled slowly through T_C (620 °C). A similar process can be applied to pole LiNbO$_3$ crystals, or as the T_C is very close to the growth temperature, the field can be applied between the growing crystal and the crucible. This process has also been used to grow periodically poled crystals in which the field is reversed periodically during the growth of the crystal. Such crystals have some advantages in optical frequency-doubling in a process known as quasi-phase matching [27.23].

Solution and Flux Growth

The growth of crystals from aqueous solution is a technology very readily applied to water-soluble materials such as KDP and TGS. There are two processes that have been used. The first consists of rotating crystal seeds in a bath of saturated solution, which is slowly cooled so that material that comes out of solution is deposited on the seeds. The second consists of recirculating the bath solution over feed material, which is held at a slightly higher temperature and then back over the slightly cooler seeds, where the feed material is deposited. In both techniques, very careful control of temperature is essential (better than 0.002 °C) and the growth process can take many days.

The process of growing oxide single crystals from solution or flux is closely related to the above process, in that the crystals are grown from a saturated solution of the required oxide in a molten flux as it is slowly cooled. In the simplest version of this process, the oxides are sealed in a platinum crucible, heated to a *soak* temperature and slowly cooled to room temperature. Small crystals (usually up to a few mm in size) can then be recovered from the solidified melt. This has been applied to growth of a wide variety of small single crystals, such as PZT and PMN-PT solid solutions. PZT, for example, can be grown from PbO−B$_2$O$_3$ flux mixtures, or even pure PbO [27.25]. BaTiO$_3$ crystals can be grown from KF or TiO$_2$ fluxes. A major issue with the growth of Pb-containing ferroelectrics from PbO rich fluxes is the evaporation of PbO from the melt at the crystallisation temperature over the long time periods required to grow large crystals. This can be avoided to some degree by the use of sealed crucibles. A variation on the flux growth process is top-seeded solution growth, whereby a seed crystal is immersed from above into a supersaturated solution of the material to be grown. The crystal then grows

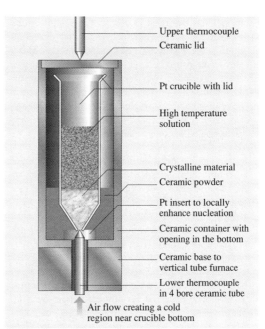

Fig. 27.13 A modified Bridgman apparatus used for growing piezoelectric single crystals (after [27.24])

as the solution is cooled further. The process is difficult to control and it has the major problem, especially when growing from a PbO flux, that the melt is open and therefore very prone to flux evaporation. It has been successfully applied to the growth of crystals of PMN-PT and PZN-PT solid solutions. The Bridgman process seeks to get over this problem by placing a dense compact of the flux/solute mixture inside a sealed crucible with a conical end, as shown in Fig. 27.13. A seed crystal is placed at the base of the cone which is held cooler than the rest of the crucible by a cooled rod. A molten zone is then allowed to travel up the crucible and the crystal solidifies inside it. Good-quality crystals have been grown this way and it is the preferred technique for the manufacture of PMN-PT and PZN-PT solid solutions. One problem with all flux-growth techniques is that the flux composition is constantly changing during the growth process and therefore the crystals tend to be nonuniform in composition. This is a particular problem for larger crystals, especially for solid solutions.

27.2.2 Ceramics

The vast majority of ferroelectric materials that are used commercially are used in the form of polycrystalline ceramics. These are used in many of the applications listed above, including as dielectrics and in piezoelectric, pyroelectric, PTCR and electro-optic devices. Many different ferroelectric ceramic fabrication technologies have been developed over the years, but they are all based around a similar fabrication sequence. The key points are discussed below. Note that as the vast majority of commercially exploited ferroelectrics are oxides, the discussion will centre around these materials.

- Raw-material selection: the usual starting point for a ceramic fabrication sequence is the selection of the raw materials. For oxide ferroelectrics, the raw materials are usually oxides or carbonates, but occasionally other compounds (e.g nitrates, citrates etc.) are used which will decompose when heated to form the oxides required. High purity (usually $> 99.9\%$ with respect to unwanted cations) is important for good reproducibility. Small quantities of dopants can have a major effect on the final electrical properties, as would be expected for any electronic material, but can also seriously affect the sintering characteristics and grain size of the final ceramic body. Note that high purity does not usually imply the kind of purity that would be needed for a semiconductor material. Indeed, ultra-high purity ($> 99.99\%$) is often obtained by raw-material manufacturers by applying processes which seriously affect the reactivity of the oxide powders. A high degree of solid-state reactivity in the powders is important for the processes which follow. This is usually determined by a combination of raw-material particle size and specific surface area, although selecting the right crystallographic phase can be important. When using TiO_2, for example, it is usually found that the anatase phase is more reactive than rutile. These factors can be analysed for quality control purposes by using laser particle size analysis (taking care to break up loosely bound agglomerates by ultrasonic dispersion in water, with a dispersing agent prior to measurement), BET (Brunauer, Emmett and Teller)-specific gas absorption and X-ray powder diffraction respectively. The majority of ceramics are made by mixing together powdered raw materials, but occasionally solution mixing techniques are used and these will be discussed further below.
- Raw-material mixing: this is usually done in a ball mill consisting of a cylinder containing a mixture of small balls or cylinders made of steel, steatite, zircon or yttria-stabilised zirconia (YSZ). The oxide raw materials are accurately weighed into the ball mill, together with a predetermined amount of a milling fluid (usually deionised water, but occasionally an organic fluid such as acetone). It is usual to add a dispersant to aid the breakup of agglomerates. The ball mill is then sealed and rotated to mix the ingredients. Precise milling conditions are determined by the materials being used, but the mixing time is usually of the order of a few hours. It is advantageous to use milling balls that are made of a material whose wear products will be reasonably innocuous in the final ceramic. For this reason, YSZ balls are preferred, as small amounts of iron and silica contaminants can have unwanted or deleterious effects on the properties of many ferroelectric ceramics. For similar reasons, the ball mills are frequently rubber lined. Many variations on this process have been explored. High-energy ball milling is receiving considerable attention. In this process, the milling balls are given very high energy by either aggressively vibrating the ball mill or by stirring them at high speed with a paddle. Raw materials trapped between the balls are both comminuted and, if the energy is sufficiently high, can be forced to react together. Crystalline raw materials can be made amorphous in this process. A significant amount of energy can be stored in the powders after this process so that subsequent sintering can be undertaken at lower temperatures. In an adaptation of this process, it can be done in a continuous flow-through mill, whereby the slurry is pumped through the high-energy mill, making it less of a batch process.
- Drying: the slurry that results from the mixing process is dried after the milling balls are removed. On a small scale, this can be done in an oven, but on a commercial scale, this is usually achieved by spray drying.
- Calcination: the purpose of this process is to react the raw materials into the required crystallographic phase. The dried powders are placed into a sealed crucible (usually high-purity alumina or zirconia), which is baked in a furnace at a temperature high enough to decompose any non-oxide precursors and cause a solid-state reaction between the raw materials, but not so high as to sinter the particles and form hard agglomerates that will be difficult to break up in subsequent processing. A simple example is the reaction between $BaCO_3$ and TiO_2 to form $BaTiO_3$:

$$BaCO_3 + TiO_2 \rightarrow BaTiO_3 + CO_2 \uparrow .$$

This reaction will go to completion at about $600\,°C$. Note that a much higher temperature ($> 1000\,°C$) is required for the decomposition of $BaCO_3$ to BaO in the absence of TiO_2.

- Milling: this is usually done in a similar manner to the mixing process described above, but is done for longer. The objective is to break up hard agglomerates and reduce the powder to its primary particle size. Again, high-energy milling techniques can be used. Dispersants are usually added to aid the process, and pH is controlled so that the powder does not flocculate. Organic binders can be added at this stage. The slurry from the milling process is usually spray-dried or freeze-dried as a free-flowing aggregated powder is required. The objective is to form soft aggregates that will break up on subsequent die pressing. Alternatively, the slurry (or *slip*) can be taken straight to a slip- or tape-casting process.
- Shape forming: after drying, the powders from the above process can be uniaxially pressed into the *green* shape required, typically using a steel punch and die-set. It is important in this process to make sure that the force is applied equally to top and bottom punches. Usually, the die is designed with a slight taper to ease removal of the workpiece after pressing. Faults that can occur at this stage of the process include *capping* or radial cracks in the work piece, which usually result from an uneven distribution of pressure. One way to avoid this is to use cold isostatic pressing. In this process the powder, or more frequently a lightly uniaxially cold-pressed green block, is vacuum-sealed into a rubber mould, which is then immersed in a bath of oil. The oil is taken up to a high pressure, which isostatically compresses the block. The advantage of using this process is that it avoids the nonuniform distribution of pressure that is frequently a problem with uniaxially cold pressing. The high pressure allows the agglomerates to be broken up and the process usually results in a much higher green density, which facilitates the sintering process. Anther process that can be used to form the green item is slip-casting, which involves pouring the ceramic slip into a plaster mould, which is usually gently rotated. The plaster absorbs moisture from the slip and the ceramic particles are deposited as a layer. Surplus slip is poured off and the deposited layer can be separated from the mould after it is dry. This process has been used to make large cylinders of piezoelectric ceramics. A third process that is frequently used to make thin sheets of ceramic is tape-casting, by which the slip is passed under a set of *doctor blades* in the apparatus shown schemati-

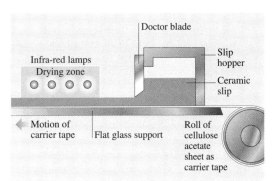

Fig. 27.14 Tape-casting apparatus

cally in Fig. 27.14. In this case, the slip is cast onto a continuous roll of plastic tape and a plastic binder is added to the slip which, when dry, holds the powder together and makes a flexible green tape. Sometimes the slip is cast directly onto a glass sheet. The green tapes can be screen-printed with patterns of metallic electrode inks and then laminated by warm-pressing to make multilayer structures with interleaved electrodes. This type of fabrication is widely used for multilayer ceramic (MLC) capacitors and multilayer ceramic piezoelectric actuators.

- Sintering: the green ceramic bodies are sintered in furnaces at temperatures of 1150–1300 °C, depending on the ceramic being manufactured. There is a good deal of process know-how in the sintering of the ceramic bodies. Key issues are:
  - The heating profile, which must be carefully controlled, especially over the lower range of temperatures up to about 500 °C where the organic matter (dispersants and binder) is burned off. Too rapid heating will cause large pores, or even cracks, to form.
  - For lead-containing ceramics, it is important that air be allowed free access to the vaporising organic material, or it will be carbonised and this can lead to reduction of the PbO in the ceramic to free Pb.
  - Loss of PbO by volatilisation at higher temperatures ($> \approx 800$ °C) means that the ceramic body needs to be sintered in a PbO-rich environment, which is often produced by packing the ceramic body in a *spacer powder* of $PbZrO_3$, or PZT ceramic chips within well-sealed ceramic crucibles or saggers. This tends to restrict the access of air to the sintering ceramic body, which conflicts with the previous requirement.

- The ceramic body must be free to move during the sintering process, as there is significant linear shrinkage (about 15 to 18%). If the body is unable to slide over the surface on which it sits, this can lead to cracking, especially for large components. One solution to this problem is to sit the body on zirconia sand.

The sintering process has been very well described elsewhere. It is possible to obtain very high densities (> 98%) by careful control of the sintering conditions. For Pb-containing ferroelectric ceramics, a small excess of PbO is usually added to compensate for PbO loss by evaporation. This also tends to act as a liquid-phase-sintering aid, lubricating the grains of ceramic as they slide over each other in the sintering process, and providing a surface-tension force that pulls the ceramic grains together. The PbO also acts as a solvent, aiding the movement of the ceramic components during sintering and further densifying the ceramic. It is possible to sinter such materials to transparency, which means that there is virtually zero porosity. Very high densities can also be obtained by hot-pressing, in which the green body is placed inside a ceramic punch and die-set and raised to the sintering temperature under a pressure of about 35 MPa. Alternatively, hot isostatic pressing (HIP) can be used. In this process the pressure transmitting medium is a high-pressure gas (usually argon, but if the ceramic is Pb-containing it is important to include a few percent of oxygen to prevent reduction of the ceramic). In this case pressures of 100 MPa or more can be used. If the ceramic is pre-sintered so that there is no open porosity, then this can be achieved without the need for any container, but if there is open porosity, the body must be encapsulated in a suitable metal container, which will usually need to be a noble metal such as Pt.

- Electroding: good-quality electroding is essential for all ferroelectric devices for a variety of reasons. Any low-permittivity layer between the electrode and the ferroelectric material will manifest itself in a fall-off in capacitance and an increase in loss as the frequency of measurement is increased. Poor-quality electrodes, or even the wrong type of conductive material can lead to problems with device ageing, or even an inability for the device to function as intended. Most piezoelectric materials are supplied with a *fired-on* silver electrode, which is a mixture of silver powder, a finely divided glass frit and a fluxing agent. Obtaining the correct balance between metal and glass contents is important, as too high a proportion of metal will lead to poor electrode adhesion while too high a proportion of glass will lead to poor electrode conductivity. Such electrodes are solderable with the use of appropriate fluxes. Sputtered or evaporated metal electrodes such as Ni or Cr/Au can also be used, as can metal electrodes deposited by electroless processes (e.g. Ni). The ohmic nature of the electrode contact to the ceramic is not usually important for highly insulating materials, such as piezoelectrics, but is important for semiconducting ceramics such as PTCR $BaTiO_3$. In this case it is usual to use Ni electrodes, which need to be annealed after deposition to develop the ohmic contact. In some cases it is necessary to fire on the electrode at the same time as ceramic sintering (cofired electrodes). This is particularly important where the electrodes are buried in the structure, as with MLC capacitors and actuators. Clearly, there are problems to solve here in terms of the potential oxidation or melting of the electrode material. The conventional materials to use in such electrodes have been noble metals such as palladium (sometimes alloyed with silver) or platinum, which are very expensive. There have been serious efforts to develop ferroelectric compositions that can be fired with base metal electrodes such as Ni. In the case of $BaTiO_3$-based dielectrics, this has entailed the development of heavily acceptor-doped compositions that can stand being fired in a neutral or slightly reducing atmosphere. In the case of MLC actuators, there has recently been some success in developing piezoelectric PZT compositions that can stand being fired in such atmospheres, using Cu as the electrode material.

- Poling: many devices made from ferroelectric ceramics (all piezoelectric and pyroelectric devices) require poling before they will develop useful properties. This entails applying a field that is significantly in excess of the coercive field (typically 3×), usually at an elevated temperature. It is not usually necessary to exceed the Curie temperature. For example, PZT ceramics with T_C in the range 230–350 °C can be poled by applying $35-kV/cm$ (depending on the type of PZT – see below – soft PZTs need lower poling fields than hard PZTs) at about 150 °C, with the field kept applied while the workpiece is cooled to room temperature. It is usual to immerse the ceramic in a heated bath of oil (mineral or silicone) during the process. One disadvantage of this is that the ceramic then needs to be carefully cleaned after poling. Silicone oil can be very hard to remove completely and its presence as a residue will compromise electrode solderability. For this reason, some workers have developed a process whereby the ceramic is poled under SF_6 gas. There is a rapid decay of properties after poling. This decay stabilises after a few hours, so it usual to wait at least 24 hours before electrical properties are measured.

There are many variations on the above basic process route which have been researched. One of these is the use of solution techniques to prepare the oxide powders. The basic principle here is that if the cations are mixed in solution, then they will be mixed on the interatomic scale without the need for milling processes that can introduce impurities. (This is a matter of discussion, as the act of precipitation and decomposition can lead to separation of the components.) Many routes have been explored, including the use of inorganic precursors such as nitrates and metalorganic precursors such as oxalates, citrates or alkoxides and acetates. Some of these have achieved a degree of commercial success, although the use of solution routes is more complex and the raw materials much more expensive than the mixed oxide routes. The use of mixed barium titanium oxalates has been very successful in producing high-quality fine-grained barium titanate powder for use in the capacitor industry. The use of metal citrates (frequently called the Pechini process) has been successfully used on a research scale to prepare many different types of ferroelectric oxide, but this type of process has not been applied on a commercial scale. Metal alkoxides, such as titanium isopropoxide, are readily soluble in alcohols and will react quickly with water to precipitate a hydroxide gel. Workers have used mixtures of titanium and zirconium alkoxides with lead and lanthanum acetates to coprecipitate a mixed hydroxide gel that could be calcined and sintered to make transparent lead lanthanum zirconate titanate ceramics for electro-optical applications.

27.2.3 Thick Films

There has been considerable interest in the integration of thick (10–50 μm thick) films of ferroelectric materials with alumina and other types of substrates such as silicon, to complement the wide range of other thick-film processes that are available, covering conductors, dielectrics, magnetic materials etc. There are many potential advantages to thick film processing for making certain types of sensor, especially the ability to use screen-printing for depositing the patterns of the materials required. Screen-printing involves using a sheet of wire mesh (the *screen*) that is coated with a photosensitive polymer. Exposure and development of the polymer allows selected areas to be removed, opening regions through which a paste of the required material can be pushed using a rubber blade or squeegee. The principle is simple, but there is a considerable amount of know-how in the formulation of the paste, which consists of the active material (in this case a ferroelectric powder such as PZT), an organic vehicle (a mixture of a solvent and polymer) and a glass frit. The screen is stretched over a former, and held close to, but not in contact with, the surface onto which the print is required. The paste is placed on the screen, and then spread over the screen with the squeegee, which prints the paste onto the substrate. Successive layers of different materials can be printed and cofired, provided there is good compatibility between them. The process has been well developed for piezoelectric films and adequate properties have been obtained from the films, although they are still well below the values that could be expected from a bulk ceramic material. (See review by *Dorey* and *Whatmore* [27.26] for further details.)

27.2.4 Thin Films

The integration of high-quality thin films ($< 0.1–5$ μm thick) of ferroelectric materials onto substrates such as silicon has excited considerable interest for potential applications ranging from nonvolatile information storage to their use as active materials in microelectro-mechanical systems (MEMS), where they can potentially be used for microsensors and actuators. Almost all the interest has been in the use of oxide ferroelectrics, but there has been some interest in the use of P(VDF-TrFE) copolymer films. These can be spun onto electroded substrates from methyl ethyl ketone solution. They are dried at relatively low temperatures ($< 100\,°C$) and crystallised by annealing at $180\,°C$ for several hours [27.27]. Such films have been applied to pyroelectric devices. However, the activity coefficients which can be obtained from such films are much lower than those that can be obtained from oxide materials. A range of deposition techniques have been developed for growing ferroelectric oxide films, which are summarised below:

- Chemical solution deposition (CSD): this term is applied to a wide range of processes that involve taking the metal ions into metalorganic solution, which is then deposited on the substrate by spinning, followed by drying and annealing processes to remove the volatile and organic components and convert the layer into a crystalline oxide. There are two broad classes of CSD process: metalorganic deposition (MOD) and sol gel. The MOD processes usually involve dissolving metal complexes with long-chain carboxylic acids in relatively heavy solvents such as toluene. The carbon content of the precursors is quite high, so there is a good deal of thickness shrinkage during firing. MOD solutions tend to be quite stable with time and resistant to hydrolysis. Sol-gel processes use precursors such as metal alkoxides, acetates and β-diketonates in alcohol solution. (For example, a set of precursors to deposit PZT would be Ti isopropoxide, Zr n-propoxide and lead acetate). Alkoxide precursors are very susceptible to hydrolysis, and so careful control of moisture content during the sol synthesis is essential and stabilisers such as ethylene glycol are usually added to the solutions to extend the useable lifetimes of the sols. Whereas MOD solutions are true solutions, the sols are actually stable dispersions of metal oxide/organic ligand particles with a size of about 4–6 nm. Sols possess lower viscosities and tend to produce the oxide layer at a somewhat lower temperature than the MOD processes, but the individual layer thicknesses produced by a single spin tend to be lower. Single crack-free layers tend to be in the range 100–200 nm thick. CSD processes have the advantages that they are low cost, the composition can be easily changed and they produce very smooth layers. The processes are planarising and will not follow underlying surface topology, which can be a disadvantage. Also, the processes are not industry standard in that they are wet and tend to have many variations.

- Metalorganic chemical vapour deposition (MOCVD): this is a variation on the process that has been very successfully applied to the growth of group III–V semiconductor layers. The principle is simple: volatile metalorganic compounds are passed over a heated substrate, where they decompose to form a layer of the desired compound. The problem with the growth of ferroelectric oxides is that most of the available metal organic precursors are relatively nonvolatile at room temperature. There has, therefore, been a great deal of research into the available precursors for the compounds that are required. Metal alkyls (such as lead tetraethyl) are very volatile, but only available for relatively few of the metal ions of interest (Pb being the main one). They are also pyrophoric and highly toxic.

Some metal alkoxides, such as Ti isopropoxide, are suitable MOCVD precursors. Metal β-diketonates and related compounds such as tetramethyl heptane dionates (THDs) have received considerable attention as Ba and Sr precursors. All of these precursors need to be heated to give them suitable volatility and this means that the lines connecting the precursor source to the growth chamber need to be heated as well. Some of the precursors (especially THDs) are solids, which means that they are not really suitable for use in conventional bubbler-type sources. There has been considerable success in using solutions of these compounds in tetrahydrofuran (THF). The solutions are sprayed into a vaporiser that consists of a cylinder, containing wire wool or ball bearings, heated to a temperature at which the solution will flash-evaporate. A carrier gas is passed through the cylinder and this carries the precursor vapour into the growth chamber. The growth chamber is usually held at reduced pressure and a certain amount of oxygen is introduced to aid the oxide deposition. Frequently a radio-frequency (RF) or microwave plasma is also introduced to aid the growth of a high density film and reduce the required substrate temperature. The major problem with the MOCVD process for complex ferroelectric oxides which have many cation components is finding the right combination of precursors that will all decompose at the same substrate temperature (usually ca. 550–650 °C) at a rate that will give the desired composition in the film. The process has been very successful in growing thin films of materials such as (Ba, Sr)TiO$_3$, with potential applications in the dynamic random-access memory (DRAM). The major advantage of MOCVD is that it is a truly conformal growth technique, with major advantages for semiconductor devices with complex topologies, but is a very expensive technique to set up because of the complex growth and control systems needed. Also, precursor availability is still a problem for many systems.

- Sputtering: a range of sputtering processes have been applied to the growth of ferroelectric thin films, including RF magnetron sputtering, direct-current (DC) sputtering and dual ion-beam sputtering. The RF magnetron process is probably the most popular. With all the processes, the major problem is one of obtaining the correct balance of cations in the growing film. Many different solutions have been found to this problem. In reactive sputtering, a composite metal target can be used. This can be made of segments of the metals to be sputtered (for example, Pb, Zr and Ti for PZT) and their relative areas changed to obtain the right composition in the film. Alternatively, multiple targets can be used and the substrate exposed to each one for different lengths of time, or the power applied to each one can be varied. In reactive sputtering, it is necessary to have an amount of oxygen in the sputtering gas (usually Ar). It is possible to sputter ferroelectric thin films from ceramic or mixed powder targets, but it is necessary to adjust the target composition to allow for different yields for each element. In any sputtering process there are many variables to adjust to optimise the process, including the sputtering power, and RF or DC substrate bias, which will affect the ion bombardment of the growing film, the sputtering atmosphere pressure and gas mixture and the substrate temperature. All of these can affect the film growth rate, composition, crystallite size and crystalline phases that are deposited, and the stress in the growing film. For this reason, the development of a sputtering process for a complex ferroelectric oxide can be a time-consuming business, and once a set of conditions has been arrived at for one particular composition, it cannot quickly be changed to accommodate new compositions. Dual ion-beam sputtering differs from the RF and DC processes in that a much lower background pressure is used, the material is sputtered from the target using an ion beam and a second lower-energy ion beam is used to stimulate and densify the growing film. The sputtering processes have the advantage of being well accepted industrially, as they are dry and can readily coat large-area substrates.

- Laser ablation: this process involves bombarding a ceramic target with a pulsed, focussed laser beam, usually from an ArF excimer source. The target is held under vacuum. A plasma plume is produced and the products ablated from the target are allowed to fall on a heated substrate. The advantages of the process are that there is usually good correspondence between the composition of the target and the growing film. Relatively small ceramic targets are acceptable for the process, and it is thus a good method for getting a rapid assessment of the properties of thin films of a given material. The disadvantages of the process are that the plasma plume will only coat a relatively small area of substrate, although there are now systems which use substrate translation to coat large areas, and particles can be ablated from the target, causing defects in the growing film.

In all the techniques used for the growth of ferroelectric oxide thin films, the key issues are control of composition and the formation of the desired crystalline phase (usually perovskite) with the desired crystallinity (crystallite size, morphology and orientation). All of the ferroelectric perovskites have a tendency to crystallise into a non-ferroelectric fluorite-like pyrochlore phase at low temperatures. In the case of the CSD processes, this means that as the film is heated from room temperature, after it loses the organic components, it first forms an amorphous oxide which then crystallises into a nanocrystalline pyrochlore phase, finally forming the desired perovskite phase. The temperatures at which this will occur depend very much on the ferroelectric oxide that is being grown and the precise composition. In the case of PZT, the pyrochlore phase will form above about 300–350 °C. The perovskite phase will start to form above about 420 °C, depending upon the ratio of Zr : Ti in the solid solution. The compositions close to $PbTiO_3$ will crystallise into perovskite much more readily than those close to $PbZrO_3$. In the case of a complex perovskite, such as $Pb(Mg_{1/3}Nb_{2/3})O_3$ or $PbSc_{1/2}Ta_{1/2}O_3$, the pyrochlore phase is much more stable and much higher temperatures (> 550 °C) are needed to convert it to perovskite. Excesses of PbO will tend to favour the formation of perovskite, and deficiencies favour pyrochlore. Higher annealing temperatures will promote PbO loss and it is possible to get into a position, through PbO loss, where the pyrochlore becomes the most stable phase, even at high annealing temperatures. Residual pyrochlore phase invariably compromises the electrical properties of the films through reduced permittivity and piezoelectric/pyroelectric coefficients. The other growth techniques have the advantage that the films can be deposited onto heated substrates, at temperatures where they will grow directly into the perovskite phase (at least in principle), although there are many examples in the literature of films being deposited (e.g. by sputtering) at low substrate temperatures and converted to the desired perovskite phase by post-deposition annealing, in which case the same problems of pyrochlore formation apply. The control of film crystallinity (crystallite orientation and size) is important as it has a direct effect on the electrical properties. This is usually achieved through control of the crystallite nucleation. Sputtered Pt is frequently used as a substrate onto which ferroelectric thin films are grown. Like many metals, this will naturally grow with a ⟨111⟩ preferred orientation. It is face-centred cubic (FCC), with a lattice parameter of about 3.92 Å, which matches the lattice parameters of many of the ferroelectric perovskites, which are about

4 Å. This means that, with appropriate process control, it is quite possible to get a highly orientated (111) ferroelectric film on Pt, with a crystallite size of about 100 nm. Changing the underlying nucleation layer can allow other orientations to be grown. For example, the use of thin films of TiO_2 or PbO on top of the Pt electrode can induce a (100) orientation. It is important to control the nucleation density of the perovskite phase. If this is allowed to become too low, than large circular grains several microns in diameter (called *rosettes*) can form, which tend to induce defects at their boundaries. (Further details on thin-film ferroelectric growth techniques can be found in the book by *Pas de Araujo* et al. [27.28]).

27.3 Ferroelectric Applications

27.3.1 Dielectrics

As noted above, $BaTiO_3$ possesses a very high permittivity close to T_C. The inclusion of selected dopants can reduce T_C and optimise the properties of the material for capacitor applications. The substitution of Sr^{2+} for Ba^{2+}, for example, will reduce it, so that at about 15% substitution it will occur at about 20 °C. Substitution of Zr^{4+} for Ti^{4+} has a similar effect. The use of off-valent substitutions will have an effect upon resistivity and degradation characteristics, so that substituting La^{3+} on the A site will reduce resistivity at low concentrations. Substituting Nb^{5+} on the B site at the 5% level has been shown to confer resistance to degradation. The addition of Mn to the lattice has been shown to have a positive effect on dielectric loss. Control of grain size has also been shown to have a marked effect on dielectric properties. Reducing the grain size has the effect of increasing the concentration of domain walls per unit volume of ceramic, and thus increasing the domain-wall contribution to the permittivity. A reduction in grain size also brings about an increase in the unrelieved stress on the grains, which further increases the permittivity. Once the grain size falls below about 0.5 μm, the stress on the grains reduces their tetragonality and the permittivity falls. It is also possible to use heterogeneity in the ceramic grains to flatten the curve of dielectric constant versus temperature. A range of Electronic Industries Alliance (EIA) codes have been introduced to define the variation of capacitance with temperature. The X7 temperature range, for example, is $-55-+125$ °C, while following this with the letter R would specify a capacitance change of no more than ±15%. Other dielectrics have been developed based on PMN that have higher peak permittivities, but worse temperature characteristics, capable of meeting a Z5U specification (+22 to −56% capacitance variation over the range 10–85 °C) but not much better. There has been huge progress in MLC technology, with capacitors now available with > 100 layers of submicron thickness and capacitances ranging from a few hundred pF to 100 μF. The majority of these now have base-metal (Ni) electrodes. The fabrication of the ceramic dielectrics for these is a very complex area and a great deal of technology has been developed to service a very large MLC capacitor market (615 Bn units in 2002, worth an estimated $ 8.4 Bn).

27.3.2 Computer Memories

The storage of digital information is of great interest and growing technological importance. The ability of ferroelectrics to store information via the sense of the spontaneous polarisation has made them candidates for this application ever since the advent of the electronic computer. However, it is only with the development of the techniques for the growth of ferroelectric thin films onto silicon at relatively low temperatures (see above) that their use has become a reality. There are two ways in which ferroelectrics can be used in computer memories. The first is the replacement of the dielectric layer in dynamic random-access memories (DRAMs). Here, the motivation is to exploit the high permittivities exhibited by ferroelectric oxides to reduce the area of silicon required to store a single bit of information. The favoured materials here are based on (Ba, Sr)TiO_3, while the favoured technique for thin-film growth is plasma-enhanced MOCVD because of its ability to deposit conformal coatings. No DRAM devices using these thin films have yet reached the market place. The use of the switchable polarisation for the nonvolatile storage of information has received a great deal of research. A good deal of this work has been based upon the use of PZT thin films, grown by a variety of techniques, including CSD, RF sputtering and MOCVD. Initially, there were many problems to solve, including the decay of the switchable polarisation (ca. 35 μC/cm^2) as the number of switching cycles increased (fatigue), the tendency for the polarisation in a bit to become less easily switched with time and the number of times it is switched in the same sense (imprint) and the tendency for the switchable polarisation in any one bit to decay with time (retention). The fatigue and imprint problems in PZT have been ascribed to the motion of oxygen vacancies addressed by the use of oxide, rather than metallic, electrodes and doping to reduce the vacancy concentration. It has also been demonstrated that aurivillius compounds such as strontium bismuth tantalate (SBT) and La-doped $Bi_4Ti_3O_{12}$ not show fatigue behaviour, although they do require rather higher growth temperatures than the PZT family and have a significantly smaller switchable polarisation (ca. $7\,\mu C/cm^2$). A number of companies are now making chips with nonvolatile on-board embedded ferroelectric random-access memory (FRAM). These are going into applications such as smart cards and computer games.

27.3.3 Piezoelectrics

This represents the most diverse range of applications for ferroelectric materials, as noted in the introductory section, covering a wide range of sensors, actuators and acoustic wave components for frequency control and filtering applications. The key properties determining the performance of a piezoelectric material in a particular application are the piezoelectric coefficients, elastic and dielectric constants. These are all tensor properties (see Nye [27.29]) and are combined through the constitutive equations:

$$D_i = d_{ijk}X_{jk} + \varepsilon_{il}E_l,$$
$$x_{ij} = s^E_{ijkl}X_{kl} + d_{mij}E_m,$$

where D_i = electric displacement, E_i = electric field, X_{ij} = stress, x_{ij} = strain, d_{ijk} = piezoelectric coefficients, ε^X_{ij} = dielectric permittivity coefficients at constant stress, and s^E_{ijkl} = elastic compliance coefficients at constant electric field.

The above equations use the Einstein repeated-suffix tensor notation. Note that most piezoelectric equations use the reduced-suffix notation for the piezoelectric and elastic constants [27.29].

There are two basic modes of operation for piezoelectric sensors and actuators. The first is well away from any mechanical resonance – usually at low frequency or quasi-DC. Here, the electric signals generated by an applied stress, or the strain generated by an applied electric field can be easily calculated using the constitutive equations of the type quoted above. The commonly used modes for operating piezoelectric ceramic actuators are longitudinal, where the electric field is applied

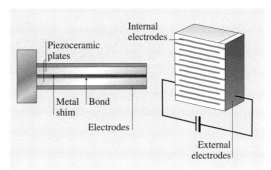

Fig. 27.15 (a) Piezoelectric bimorph and (b) MLC actuator structures

parallel to the direction in which strain is required and the coupling is via the piezoelectric d_{33} coefficient, and transverse, where a field is applied along the polar axis, but the exploited strain is perpendicular to this direction and coupling is via the d_{31} and d_{32} coefficients. Corresponding modes are frequently used for piezoelectric sensors. The two most-commonly used actuator structures are illustrated in Fig. 27.15. Figure 27.15a shows a piezoelectric bimorph, in which two pieces of piezoelectric ceramic are bonded to a central metal shim. Fields are applied parallel to the polar axes of the ceramic elements so that their transverse extensions are equal and opposite. Usually one end of the bimorph is clamped so that the other end bends. In a variation on this structure, a single piece of ceramic is bonded to a piece of metal, making a unimorph. Bimorph and unimorph devices can provide high displacements (several tens to thousands of microns at up to 200-V drive) but relatively low forces (typically up to 1 N), depending on the size of the bimorph. The second widely used actuator structure is the multilayer ceramic actuator (Fig. 27.15b). This device operates in longitudinal mode, with applied field and extension parallel to the polar axis. The division of the structure into many thin layers reduces the voltage required to obtain a given extension. Typical devices can produce a few to a few tens of microns extension at a few hundred Volts (depending on dimensions) but high forces (in the kiloNewtons range). It is also possible to use a shear mode of operation for both sensing and actuation. In this actuation mode, the field is applied perpendicular to the polar axis and a shear strain is generated in the plane containing the vectors parallel to the applied field and the polar axis.

Many piezoelectric devices are operated under an alternating electric field whose frequency is tuned to match a mechanical resonance. A piezoelectric element oper-

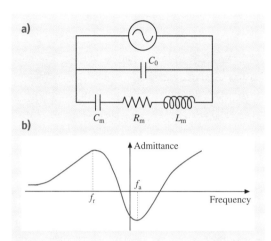

Fig. 27.16 (a) Piezoelectric element equivalent circuit (b) Admittance versus frequency for a typical piezoelectric resonator

ated well away from resonance will exhibit a capacitive impedance characteristic, As a mechanical resonance is approached, the electrical behaviour of the element can be characterised using an equivalent circuit as shown in Fig. 27.16a. C_0 is the static capacitance. The L_m, R_m, and C_m form the motional arm of the circuit. If the admittance of such a circuit is plotted as a function of frequency, a behaviour is obtained as shown in Fig. 27.16b. The maximum in the admittance is very close to the mechanical resonant frequency (f_r), while the minimum in the admittance is close to antiresonance (f_a); f_r is determined by the elastic properties of the piezoelectric, its density and the dimensions relevant to the particular resonant mode. It is useful to define an effective electromechanical coupling coefficient k_{eff} which is defined by:

$$k_{eff}^2 = \frac{\text{mechanical energy converted to electrical energy}}{\text{input mechanical energy}}$$

or

$$k_{eff}^2 = \frac{\text{electrical energy converted to mechanical energy}}{\text{input electrical energy}},$$

k_{eff} can be determined from f_r and f_a:

$$k_{eff}^2 \approx \frac{f_a^2 - f_r^2}{f_a^2}.$$

It can be shown that, for a simple resonant mode of a piezoelectric element, k_{eff} takes the form:

$$k_{eff}^2 = \frac{d^2}{s^E \varepsilon^X}$$

where the d, s and ε coefficients in this equation are combinations of the piezoelectric, elastic and dielectric permittivity coefficients. The precise combination of the coefficients will depend on the shape of the piezoelectric element and the mode being excited into resonance. A list of the more commonly used modes and the dependence of k_{eff} on the relevant materials constants is given in Table 27.4. For more complex structures, such as those which involve a piezoelectric element (or elements) bonded to a metal component, it is still possible to define and measure a k_{eff}, but its value will be a much more complex combination of the materials constants, and will also involve the properties of the metal parts and the relative dimensions of the piezoelectric and metal components. (For more details refer to the IEEE standards on piezoelectricity).

By far the most widely exploited piezoelectric materials system is based on the $PbZr_xTi_{1-x}O_3$ ceramic solid solution series. Most of the compositions of-interest are centred on the morphotropic phase boundary (MPB) region (Fig. 27.7) with $x = 0.52$. In the undoped series, the piezoelectric coefficients and dielectric constants peak at this composition. However, their properties can be greatly enhanced by the inclusion of selected cation dopants. These can be classified as follows:

- Isovalent: these are divalent cations that will substitute for Pb^{2+} on the A site (e.g. Sr^{2+}) or tetravalent ions (e.g. Sn^{4+}) substituting for Ti/Zr^{4+} on the B site.
- Donors: these are cations with a valency greater than the site onto which they substitute (e.g. La^{3+} for Pb^{2+} or Nb^{5+} for Ti/Zr^{4+}).
- Acceptors: these are cations with a valency lower than the site onto which they substitute (e.g. K^+ for Pb^{2+} or Ni^{3+} for Ti/Zr^{4+}).
- Multivalent: these are ions that take multiple valency (e.g. Mn substituted on the B site).

The broad effects of each class of dopants are listed in Table 27.5.

The resulting PZT ceramics are broadly classified into *hard* or *soft* according to whether they are doped with acceptors or donors, respectively. Each manufacturer has its own nomenclature scheme for its piezoceramic products. However, the US Navy produced a set of specifications under MIL-STD-1376B

Table 27.4 A list of the most commonly used piezoelectric resonances and the dependence of the k_{eff} for each on the piezoelectric, elastic and dielectric constants [27.30]

Mode type	Mode shape	Coupling factor
Length extensional mode of a bar using transverse excitation $L > 5w, L > 10t$		$k_{31}^l = d_{31} / \sqrt{\varepsilon_{33}^T s_{11}^E}$
Length extensional mode of a cylindrical rod using logitudinal excitation $L > 10 \times$ diameter		$k_{33}^l = d_{33} / \sqrt{\varepsilon_{33}^T s_{33}^E}$
Radial mode of a round plate using longitudinal excitation $D > 20 \times$ thickness		$k_p = k_{31} / \sqrt{2/(1-\sigma^p)}$

Grey areas denote the electroded surfaces

Table 27.5 The effects of substituents in the PZT system (after [27.8])

Isovalent substituents e.g. Sr^{2+} for Pb^{2+}, Sn^{4+} for Zr^{4+}	Donor substituents e.g. La^{3+} for Pb^{2+}, Nb^{5+} for Zr/Ti^{4+}
• Lower Curie point • Increase permittivity • Small improvement in linearity at high drive • No change in coupling factor, aging, volume resistivity or low amplitude mechanical or dielectric loss	• No change in Curie point • Higher permittivity • Higher electromechanical coupling • Poorer high drive linearity • Lower aging • Increase resistivity • Reduced Q_E and Q_M
Acceptor substituents e.g. K^+ for Pb^{2+}; Fe^{3+} for Zr/Ti^{4+}	**Variable valence substituents e.g. Mn – can be 2+, 3+, 4+ – for Zr^{4+}**
• No change in Curie point • Lower permittivity • Lower electromechanical coupling • Improved drive linearity • Little change in aging • Volume resistivity reduced • Increased Q_E and Q_M	• No change in Curie point • Range of permittivity • Range of electromechanical coupling – usually lower • High drive/linearity not changed • Reduced aging

which are universally recognised. Hard ceramics (e.g. US Navy type III) have relative permittivities in the range of 1000 with high Qs (> 500) and d_{33} of about 215 pC/N. Soft ceramics (e.g. US Navy type VI) have much higher permittivities (3250) and piezoelectric coefficients ($d_{33} = 575$ pC/N) but lower Q's (65). Hard ceramics have much better drive-voltage stability and higher Curie temperatures than soft ceramics. The soft compositions tend to be used for actuator applications, where the maximum piezoelectric displacement for a given drive voltage is sought, while soft ceramics tend to be used for resonant and sound-generation applications. There are, however, exceptions to this. Medical ultrasound transducer arrays tend to use soft, high-permittivity ceramics, because they need to have the maximum charge sensitivity in the reception of the sound waves reflected from tissues within the body, and

Table 27.6 Typical values for the piezoelectric, dielectric and elastic properties of some selected piezoelectric ceramic materials (taken from [27.31])

Material type	I	III	V	VI	Units
Stress-free relative permittivity (ε_{33})	1275	1025	2500	3250	
Dielectric loss tangent ($\tan \delta$)	≤ 0.006	≤ 0.004	≤ 0.025	≤ 0.025	
d_{33}	290	215	495	575	pC/N
k_p	0.58	0.50	0.63	0.64	
Mechanical quality factor Q_m	≥ 500	≥ 800	≥ 70	≥ 65	
Curie temperature (T_C)	325	325	240	180	°C

high permittivities because the small elements usually need to drive transmission lines with significant capacitance.

There are other types of piezoelectric ceramics that are used for specialist applications. Modified PbTiO$_3$ ceramics possess very low values of d_{31} and tend to be used for ultrasound array applications where small cross-coupling between adjacent elements is important. They also possess much higher response to hydrostatic stress than MPB PZT ceramics. Because of this, arrays of small blocks of this type of ceramic are used in flank array sensors in submarines. As noted above, lead metaniobate ceramics are used in high-temperature applications. Table 27.6 lists the piezoelectric and other properties of some selected piezoelectric ceramics.

Single crystals of PZN-PT and PMN-PT have received considerable attention because of the very high piezoelectric d_{33} coefficients and k_{33} coupling factors that can be obtained and single-crystal plates of these are now commercially available. Table 27.7 lists some of the properties that have been measured from these materials, and compares them with a commercial soft PZT ceramic.

Single crystals of LiNbO$_3$ are widely used in surface acoustic wave devices. In these, interdigitated electrodes (IDE) are applied to one polished face of a crystal. Excitation of these with an RF field will cause the generation of piezoelectrically excited Rayleigh waves, which will propagate with very little attenuation. These can be detected by further sets of IDE placed on the same face of the crystal. Such devices are widely used in signal processing and filtering applications in mobile telecommunication applications.

Thin films of piezoelectric materials, excited into thickness-mode resonance, are being explored for very-high-frequency (> 1 GHz) resonant filter applications.

27.3.4 Pyroelectrics

The pyroelectric effect is now widely used in uncooled detectors of long-wavelength infrared radiation (IR). The principle behind the operation of these devices is very simple. The radiation to be detected is allowed to fall upon a thin chip of the pyroelectric material. The energy absorbed causes a change in temperature and the generation of a pyroelectric charge, which will cause the flow of current in an external circuit. This can be amplified and used, for example, to switch an alarm. The basic circuit of a pyroelectric detector is shown schematically in Fig. 27.17. The field-effect transistor (FET) amplifier needs to be close to the pyroelectric element, ideally in the same package. This is because the latter has a very high output impedance and thus is very noise sensitive. The pyroelectric signal is represented in this circuit as a current source, i_p. i_p is proportional to the rate of change of the element temperature with time. Pyroelectric devices do not have a DC response. They only see changes in the intensity of the radiation with time. This is a major advantage in many applications, where usually there is a requirement to detect changes in the radiation coming from a scene, rather than the average or unchanging background intensity. An example of this is the requirement to detect the radiation from a person coming into the field of view of an IR detector.

Pyroelectric devices have several advantages over other radiation detectors. Their response is independent of the wavelength of the incident radiation, provided there is some means to absorb the radiation. Hence, pyroelectric radiation detectors have been used across the full range of the electromagnetic spectrum, from microwaves to X-rays. Basically a single pyroelectric detector design can be used for different wavelengths of radiation, simply by equipping the package with different windows coated with filters according to

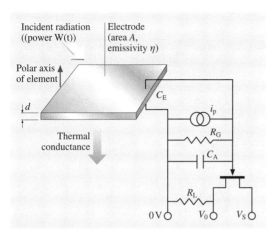

Fig. 27.17 Schematic diagram of a pyroelectric IR sensor

Table 27.7 Piezoelectric properties of PMN-PT and PZN-PT single-crystal materials [27.33] compared with those of a commercial soft-piezoelectric ceramic HD3203 [27.34]

Property	Composition		
	PZNT9307	PZNT9109	HD3203
$\varepsilon_{33}^T/\varepsilon_o$	3500	3000	3370
T_{RT} (°C)	100	70	NA
T_C (°C)	165	175	195
k_t (%)	58	56	0.536
k_{33} (%)	84	82	0.763
$N_{33}.a$	1230	1380	2000
d_{33} (pC/N)	1700	1900	564

the radiation it is desired to sense. Such devices are widely used in spectroscopic sensors for, e.g. automotive exhaust-gas analysis. However, by far the widest application of pyroelectric detectors is in the sensing of IR in the wavelength range 8–14 μm. The reasons for this are that there is an atmospheric window of low absorption and that objects with temperatures in the range of room temperature (300 K) emit most of their radiative energy in that waveband. However, all of the semiconductor detectors of -8–14 μm IR need to be cooled to about 77 K to work efficiently. On the other hand, pyroelectric detectors work perfectly well uncooled. Cheap detectors are readily available for under $ 1, and have thus found their way into a host of applications, notably detectors of people in such things as intruder alarms and remote light switches. Pyroelectric arrays have been developed in which a plane of pyroelectric material is interfaced to a silicon chip bearing a two-dimensional array of FET amplifiers and a set of switches which can be used to multiplex the pyroelectric signals onto a single output. These arrays have been used for uncooled thermal imagers with performances comparable to those that can be achieved with cooled semiconductor devices [27.32].

The requirements for the active materials in a pyroelectric device has been described by, for example, *Whatmore* [27.20], *Whatmore* and *Watton* [27.32]. There are two time constants that determine the frequency response of a pyroelectric device: the electrical time constant determined by $R_G(C_E + C_A)$ (see Fig. 27.17) and the thermal time constant determined by H/G where H is the heat capacity of the pyroelectric element and G the thermal conductance from the pyroelectric element to the environment. The capacitance of the element (C_E) is determined by the permittivity (ε) of the pyroelectric material while the heat capacity is determined by its specific heat per unit volume (c'). The pyroelectric current is proportional to the material's pyroelectric coefficient ($p = dP_s/dT$). A full treatment of the physics of device operation shows that detector performance (as determined by specific detectivity) is proportional to one of three basic figures-of-merit (FOMs), which combine the pyroelectric, dielectric and thermal properties of the pyroelectric material together. If the pyroelectric element capacitance is small in comparison with the amplifier capacitance (C_A), then the appropriate FOM is $F_I = p/c'$. If $C_E \gg C_A$ and the noise in the device is dominated by the amplifier noise sources, then the appropriate FOM is $F_V = p/c'\varepsilon$. If the noise is dominated by alternating-current (AC) Johnson noise in the pyroelectric element then the appropriate FOM is $F_D = p/c'\sqrt{\varepsilon \tan \delta}$, where $\tan \delta$ is the dielectric loss tangent of the pyroelectric material. Note that it is very important to measure the dielectric constant and loss at a frequency relevant to the device use. As most pyroelectric devices are used in the frequency range 0.1–100 Hz, and dielectric loss is usually much greater at < 100 Hz than at 1 kHz, a low-frequency measurement is essential. Table 27.7 lists the pyroelectric properties of several ferroelectric materials. Note that many of the literature papers on the subject only quote 1-kHz dielectric measurements. The differences between the FOM for the materials where high- and low-frequency data are available are clear. It can be seen that that the TGS family gives the largest figures of merit. However, these materials are water soluble and have relatively low Curie temperatures and so they are only still used in the high-performance devices for instruments such as in Fourier-transform infra-red (FTIR) spectrometers. LiTaO₃ (LT) is a good pyroelectric material with relatively low permittivity.

It is inert and relatively easy to handle. It is used in many single-element detectors. Ceramic materials such as those based on modified PbZrO$_3$ (Mod PZ) or modified PbTiO$_3$ (Mod PT) are widely used in low-cost detectors. Note that, although their FOM are worse than LT, they are low cost and their performance is perfectly adequate for many applications. The PVDF family (represented here by a P(VDF/TrFE70/30 copolymer) have good F_V, but relatively low F_D values compared with the other materials listed. They tend to be used in large-area detectors because of their low permittivities, low cost and the fact that they are readily made in very thin films with low thermal mass, which is an advantage in some circumstances. They have been demonstrated in linear arrays. It is advantageous for very small-area detectors (say $< 100\,\mu$m square), such as those used in arrays, for the pyroelectric material to have a relatively high permittivity (a few hundred) so that the detector can have a capacitance that matches the input capacitance of the FET amplifier (usually ca. 1 pF). The ceramic materials are well suited to this application for that reason. There has been considerable research into the use in thermal imaging arrays of ferroelectrics with T_C close to room temperature under an applied bias field, which will provide an induced pyroelectric effect well above the normal T_C. This has been called dielectric bolometer mode of operation and the best materials researched for this have been PST and (Ba$_x$Sr$_{1-x}$)TiO$_3$ solid solutions with $x \approx 0.35$. These materials have relative permittivities of > 1000 under the operational conditions and very high pyroelectric coefficients which can give an effective F_D some three times greater than can be achieved using conventional pyroelectric ceramics [27.35].

References

27.1 G. Busch: Ferroelectrics **74**, 267 (1987)
27.2 A. F. Devonshire: Phil. Mag. **40**, 1040 (1949)
27.3 M. E. Lines, A. M. Glass: *Principles and Applications of Ferroelectric Materials* (Clarendon, Oxford 1977)
27.4 V. M. Goldschmidt: *Geochemistry* (Oxford Univ. Press, Oxford 1958)
27.5 R. D. Shannon, C. T. Prewitt: Acta Cryst. B **25**, 925 (1969)
27.6 W. J. Merz: Phys. Rev. **76**, 1221 (1949)
27.7 J. M. Herbert: *Ceramic Dielectrics and Capacitors* (Gordon Breach, Philadelphia 1985)
27.8 B. Jaffe, W. R. Cook Jr., H. Jaffe: *Piezoelectric Ceramics* (Academic, New York 1971)
27.9 R. W. Whatmore: Ph.D. Thesis, Cambridge University, UK (1977)
27.10 R. W. Whatmore, A. M. Glazer: J. Phys. C.: Solid State Phys. **12**, 1505 (1979)
27.11 B. Noheda, J. A. Gonzalo, A. C. Caballero, C. Moure, D. E. Cox, G. Sirane: Ferroelectrics **237**, 541 (2000)
27.12 G. A. Smolenskii, V. A. Isupov, A. A. Agranovskaya, S. N. Popov: Fiz. Tverd. Tela **2**, 2906 (1960)
27.13 N.-H. Chan, D. M. Smyth: J. Am. Ceram. Soc. **67**, 285 (1984)
27.14 N.-H. Chan, R. K. Sharma, D. M. Smyth: J. Am. Ceram. Soc. **65**, 168 (1981)
27.15 S. C. Abrahams, E. Buehler, W. C. Hamilton, S. J. Laplaca: J. Phys. Chem. Solids **34**, 521 (1973)
27.16 P. B. Jamieson, S. C. Abrahams, J. L. Bernstein: J. Chem. Phys. **48**, 5048 (1968)
27.17 M. E. Hagerman, K. R. Poeppelmeier: Chem. Mater. **7**, 602 (1995)
27.18 Y. J. Ding, X. D. Mu, X. H. Gu: J. Non-Lin. Opt. Phys. Mater. **9**, 21 (2000)
27.19 D. Madgy, S. F. Ahsan, D. Kest, I. Stein: Arch. Otolaryng. Head Neck Surg. **127**, 47 (2001)
27.20 R. W. Whatmore: Rep. Prog. Phys. **49**, 1335 (1986)
27.21 H. Kawai: Jpn. J. Appl. Phys. **8**, 967 (1969)
27.22 J. G. Bergman, J. H. McFee, G. R. Crane: Appl. Phys. Lett. **18**, 203 (1971)
27.23 J. A. Ghambaryan, R. Guo, R. K. Hovsepyan, A. R. Poghosyan, E. S. Vardanyan, V. G. Lazaryan: J. Optoelectron. Adv. Mater. **5**, 61 (2003)
27.24 A. Dabkowski, H. A. Dabkowska, J. E. Greedan, W. Ren, B. K. Mukherjee: J. Cryst. Growth **265**, 204–213 (2004)
27.25 R. Clarke, R. W. Whatmore: J. Cryst. Growth **33**, 29–38 (1976)
27.26 R. A. Dorey, R. W. Whatmore: J. Electroceram. **12**, 19 (2004)
27.27 N. Neuman, R. Köhler: Proc. SPIE **2021**, 35 (1993)
27.28 C. Pas de Araujo, J. F. Scott, G. W. Taylor: *Ferroelectric Thin Films: Synthesis and Basic Properties* (Gordon Breach, Princeton 1996)
27.29 J. F. Nye: *Physical Properties of Crystals, Their Representation by Tensors and Matrices*, (Oxford Univ. Press, Oxford 1957)
27.30 ANSI/IEEE Standard on Piezoelectricity: IEEE Trans. **UFFC 43**, 717 (1996)
27.31 MIL-STD-1376B(SH) (1995) US Navy Military Standard for Piezoelectric Ceramics
27.32 R. W. Whatmore, R. Watton: Pyroelectric Materials and Devices. In: *InInfrared Detectors and Emitters: Materials and Devices*, ed. by P. Capper, C. T. Elliott (Chapman Hall, New York 2000) pp. 99–148
27.33 Y. Hosono, K. Harada, T. Kobayashi, K. Itsumi, M. Izumi, Y. Yamashita, N. Ichinose: Jpn. J. Appl. Phys. **41**, 7084–8 (2002)

27.34 S. Sherrit, H. D. Wiederick, B. K. Mukherjee: Proc. SPIE **3037**, 158 (1997)

27.35 R. W. Whatmore, P. C. Osbond, N. M. Shorrocks: Ferroelectrics **76**, 351 (1987)

28. Dielectric Materials for Microelectronics

Dielectrics are an important class of thin-film electronic materials for microelectronics. Applications include a wide swathe of device applications, including active devices such as transistors and their electrical isolation, as well as passive devices, such as capacitors. In a world dominated by Si-based device technologies, the properties of thin-film dielectric materials span several areas. Most recently, these include high-permittivity applications, such as transistor gate and capacitor dielectrics, as well as low-permittivity materials, such as inter-level metal dielectrics, operating at switching frequencies in the gigahertz regime for the most demanding applications.

This chapter provides a survey of the various dielectric material systems employed to address the very substantial challenge associated with the scaling Si-based integrated circuit technology. A synopsis of the challenge of device scaling is followed by an examination of the dielectric materials employed for transistors, device isolation, memory and interconnect technologies. This is presented in view of the industry roadmap which captures the consensus for device scaling (and the underlying economics) – the International Technology Roadmap for Semiconductors. Portions of the survey presented here are selected from work previously published by the author [28.1–3].

28.0.1	The Scaling of Integrated Circuits.	625
28.0.2	Role of Dielectrics for ICs	629
28.1	**Gate Dielectrics**	**630**
28.1.1	Transistor Structure	630
28.1.2	Transistor Dielectric Requirements in View of Scaling	630
28.1.3	Silicon Dioxide	635
28.1.4	Silicon Oxynitride: SiO_xN_y	641
28.1.5	High-κ Dielectrics	643
28.2	**Isolation Dielectrics**	**647**
28.3	**Capacitor Dielectrics**	**647**
28.3.1	Types of IC Memory	647
28.3.2	Capacitor Dielectric Requirements in View of Scaling	648
28.3.3	Dielectrics for Volatile Memory Capacitors	648
28.3.4	Dielectrics for Nonvolatile Memory	649
28.4	**Interconnect Dielectrics**	**651**
28.4.1	Tetraethoxysilane (TEOS)	651
28.4.2	Low-κ Dielectrics	651
28.5	**Summary**	**653**
References		**653**

This chapter considers the role of dielectric materials in microelectronic devices and circuits and provides a survey of the various materials employed in their fabrication. We will examine the impact of scaling on these materials, and the various materials utilized for their dielectric behavior. Extensive reviews are available on the device characteristics for the reader to consult [28.4–7]. We will primarily confine the discussion here to Si-based microelectronic circuits.

Dielectric materials are an integral element of all microelectronic circuits. In addition to their primary function of electrical isolation of circuit and device components, these materials also provide useful chemical and interfacial properties. The material (and resulting electrical) properties of dielectrics must also be considered in the context of the thin films used in semiconductor microelectronics, as compared to bulk properties. The dimensions of these dielectric thin films are determined by the device design of the associated integrated circuit technology, and these dimensions decrease due to a calculated design process called *scaling*.

28.0.1 The Scaling of Integrated Circuits

The ability to reduce the size of the components of integrated circuits (ICs), and therefore the circuits themselves, has resulted in substantial improvements in device and circuit speeds over the last 30 years. Equally important, this calculated size reduction permits the fabrication of a higher density of circuits per unit area on semiconductor substrates. The economic implication of this scaling was captured by G. Moore more than 40 years ago [28.8].

Moore's Law

Moore observed that the minimum cost of manufacturing integrated circuits per component actually decreases with increasing number of IC components, and thus with greater circuit functionality and computing power. This is obviously an important economic driving force, as the ability to increase the number of circuits per unit area would lead to a lower minimum cost, and thus higher market demand and more potential profit. Moreover, Moore noted that the rate of increase in the number of components for a given circuit function roughly doubled each year in the early 1960s, and predicted that it would continue to do so through 1975. In 1975, *Moore* revised this estimate of doubling time to 24 months due to the anticipated complexity of circuits [28.9].

The semiconductor industry has generally confirmed (and aligned to meet) these extrapolations, often referred to as Moore's law, over the last 30 years. The extrapolation is often analyzed in the semiconductor industry, and the doubling period, which has varied between 17 and 32 months over the life of the industry, is now roughly 23 months [28.10]. Indeed the cost per transistor has decreased from ≈ 5 \$/transistor in 1965 to less than 10^{-6} \$/transistor today [28.11]. Current advanced Si IC production technology results in the fabrication of well over 500 000 000 transistors on a microprocessor chip.

Technology Roadmap

The contemporary industry analysis encompassing this observation is presented in the International Technology Roadmap for Semiconductors (ITRS) where the extrapolations of the future technological (and economic) requirements for the industry are annually updated [28.12]. The current scaling trends indicate that the compound annual reduction rate (CARR) in device dimensions is currently consistent with the following equation:

$$\mathrm{CARR}(T) = \frac{1}{2}^{\left(\frac{1}{2T}\right)} - 1 , \qquad (28.1)$$

where T is the technology node cycle time measured in years. Thus in two years, the rate of reduction is -15.9%. This corresponds to a scaling factor of $\approx 0.7\times$ from a given technology node to the next, or a factor of $\approx 0.5\times$ over the time of two technology nodes. Recent reviews and predictions for the limits of scaled integrated circuit technology continue to be available [28.13, 14].

Table 28.1 provides selected scaling targets from the ITRS through 2010 [28.12]. As may be seen, the industry roadmap now segregates scaling targets among three categories: microprocessor (MPU)/application-specific IC (ASIC) applications, low-operational-power applications, and low-stand-by-power applications. Higher-performance technologies, such as MPU/ASIC applications, require aggressive scaling, while low-stand-by-power applications require less-aggressive scaling. A key criteria to enable these technologies is the associated power management in the *on* and *off* state, hence leakage current remains an important distinction amongst the various roadmap applications.

Performance and Scaling

The concomitant reduction in device dimensions, such as transistor channel length, associated with increasing the number of components per unit area has resulted in a significant increase in processing performance – the speed at which computations can be done. For example, as scaling reduces the distance that carriers must travel in a transistor channel, the response time of the transistor as a digital switch also decreases (as long as sufficient mobility is maintained). Scaling has resulted in the speeds of microprocessors increasing from 25–50 MHz clock frequencies in the early 1990s ($\approx 10^6$ transistors) to 2.2 GHz in 2003 ($\approx 10^8$ transistors) [28.15]. Current predictions suggest that 20-GHz frequencies are possible for complementary metal oxide semiconductor (CMOS)-based microprocessors incorporating 10^9 transistors [28.11].

The various dielectric materials associated with the components impacts the overall performance of the corresponding IC technology. In the case of transistors, the gate dielectric is integral to the performance of transistor electrical characteristics such as the drive current I_d. The interconnection performance of circuit elements is influenced by the dielectric material that isolates the various metal interconnection lines through the line-to-line capacitance. Memory elements incorporate very-low-leakage dielectric materials for charge-storage purposes. The behavior of these materials with component size reduction is thus an important design consideration in IC technology.

Figure 28.1 shows the dependence of the delay time as a function of technology node (scaling) associated with conventional CMOS transistor gates interconnected with metal lines that are isolated with different di-

Dielectric Materials for Microelectronics 627

Table 28.1 Selected scaling targets from the ITRS roadmap [28.12]

Year of production	2003	2004	2005	2006	2007	2008	2009	2010
Technology node	100 nm	90 nm	80 nm	70 nm	65 nm	57 nm	50 nm	45 nm
MPU/ASIC applications								
Physical gate length, MPU/ASIC (nm)	45	37	32	28	25	22	20	18
Equivalent oxide thickness, T_{ox} (nm)	1.3	1.2	1.1	1.0	0.9	0.8	0.8	0.7
Drive voltage V_{dd} (V)	1.2	1.2	1.1	1.1	1.1	1	1	1
Gate dielectric leakage ($\mu A/\mu m$)	0.1	0.2	0.2	0.2	0.2	0.2	0.2	0.3
Gate dielectric leakage (A/cm^2)	222	450	521	595	933	1061	1167	1852
Low operating power								
Physical gate length, low operating power (nm)	65	53	45	37	32	28	25	22
T_{ox} (nm)	1.6	1.5	1.4	1.3	1.2	1.1	1.0	0.9
V_{dd} (V)	1.00	0.90	0.90	0.90	0.80	0.80	0.80	0.70
Gate dielectric leakage ($nA/\mu m$)	0.33	1.00	1.00	1.00	1.67	1.67	1.67	2.33
Gate dielectric leakage (A/cm^2)	0.51	1.89	2.22	2.70	5.21	5.95	6.67	10.61
Low stand-by power								
Physical gate length, low stand-by power (nm)	75	65	53	45	37	32	28	25
T_{ox} (nm)	2.2	2.1	2.1	1.9	1.6	1.5	1.4	1.3
V_{dd} (V)	1.2	1.2	1.2	1.2	1.1	1.1	1.1	1.0
Gate dielectric leakage ($pA/\mu m$)	3	3	5	7	8	10	13	20
Gate dielectric leakage (A/cm^2)	0.004	0.005	0.009	0.015	0.023	0.031	0.048	0.080

electrics [28.16]. The total delay time associated with the circuit response for a given technology node is a combination of the temporal response of the transistor gate (influenced by the gate capacitance) as well as the propagation through the interconnecting wire lines associated with the circuit [influenced by the wire resistance and the line-to-line capacitance, i. e. a resistive–capacitive (RC) delay] [28.16, 17]. Changing both the interconnect wire material (from Al to Cu) to decrease the line resistivity as well as decreasing the dielectric constant in the material isolating these lines (from SiO_2 to a lower dielectric constant material) to decrease line-to-line parasitic capacitance minimizes the overall RC delay time response. Of course, the challenge in adopting this solution was the successful integration of these material changes into a high-volume fabrication line, which took the investment of considerable time and effort.

In order to compare the performance of different CMOS technologies, one can consider a performance metric that captures the dynamic response (i. e. charging and discharging) of the transistors associated with a specific circuit element to the supply voltage provided to the element at a representative (clock) frequency [28.1, 18, 19]. A common element employed to examine such switching-time effects in a new transistor design technology node is a CMOS inverter. This circuit element is shown in Fig. 28.2 where the input signal is attached to the gates and the output signal is connected to both the n-type MOS (NMOS) and p-type MOS (PMOS) transistors associated with the CMOS stage. The switching time is limited by both the fall time required to discharge

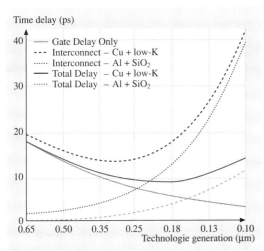

Fig. 28.1 Time delay contributions for integrated circuits. The resultant delay is dependent upon the circuit architecture as well as the material constituents. After [28.16] (© 2004 IEEE, with permission)

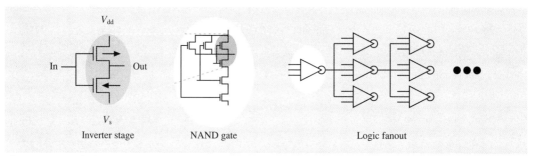

Fig. 28.2 Example circuits employed to evaluate the performance of a transistor technology. For example, elements are combined to simulate processor performance adequately. After [28.1] (AIP, with permission© 2001)

the load capacitance by the n-type field-effect transistor (n-FET) drive current and the rise time required to charge the load capacitance by the p-FET drive current. That is, the switching response times τ are given by:

$$\tau = \frac{C_{LOAD} V_{dd}}{I_d}, \quad \text{where}$$

$$C_{LOAD} = FC_{GATE} + C_j + C_i, \quad (28.2)$$

and C_j and C_i are the parasitic junction and local interconnection capacitances, respectively. Of course, as one considers the combination of transistors to produce various circuit logic elements to enable computations (viz. a *processor*), it is also clear that the manner in which the transistors are interconnected also plays an important role on the overall performance of the circuit. The *fan out* for interconnected devices is given by the factor F. Ignoring delay in the gate electrode response, as $\tau_{GATE} \ll \tau_{n,p}$, the average switching time is therefore:

$$\bar{\tau} = \frac{\tau_p + \tau_n}{2} = C_{LOAD} V_{dd} \left(\frac{1}{I_d^n + I_d^p} \right). \quad (28.3)$$

The load capacitance in the case of a single CMOS inverter is simply the gate capacitance if one ignores parasitic contributions such as junction and interconnect capacitance. Hence, an increase in I_d is desirable to reduce switching speeds. For more realistic estimates of microprocessor performance, the load capacitance is connected (*fanned out*) to other inverter elements in a predetermined fashion. When coupled with other NMOS/PMOS transistor pairs in the configuration shown in Fig. 28.2, one can create a logic NAND gate, which can be used to investigate the dynamic response of the transistors and thus examine their performance under such configurations. For example, a fan out $F = 3$ can be employed in microprocessor performance estimates, as shown in Fig. 28.2.

One can then characterize the performance of a circuit technology (based on a particular transistor structure) through this switching time. To do this, various *figures of merit* (FOM) have been proposed that incorporate parasitic capacitance as well as the influence of gate sheet resistance on the switching time [28.18–20]. For example, a common FOM employed is related to (28.3) simply by

$$\text{FOM} \cong \frac{1}{\bar{\tau}} = \frac{2}{\tau_p + \tau_n}. \quad (28.4)$$

In the case where parasitic capacitance and resistance effects are ignored, it is easily seen then that an increase in the device drive current I_d results in a decrease in the switching time and thus an increase in the FOM value (performance). However, the incorporation of parasitic effects results in the limitation of FOM improvement, despite an increase in the gate dielectric capacitance. All of these issues depend critically upon the materials constituents of the integrated circuit, and dielectrics are an important component [28.1].

Impact of Scaling on Dielectrics

The scaling associated with the device dimensions that constitute an integrated circuit impacts on virtually all film dimensions such as thickness, including the dielectrics employed in the circuit. Various scaling methodologies have been applied since the pioneering work by *Dennard* and coworkers [28.21, 22]. These various methods have been recently summarized [28.5, 23]. For example, following the work by *Dennard* et al., the scaling of the transistor from one node N to the next $N + 1$ under a constant-electric-field condition would result in a relationship between the *new* transistor gate (or channel) length $L_{N+1} = L_N/\alpha$, where $\alpha > 1$ is the constant scaling factor (Fig. 28.3). Other dimensions, such as the gate dielectric thickness and the gate width,

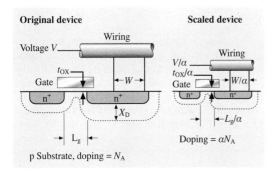

Fig. 28.3 Scaling methodology for integrated circuit technology. After [28.23] (© 2004 IEEE, with permission)

scale similarly. The voltage would also scale according to $V_{N+1} = V_N/\alpha$ in this scenario as well. Table 28.1 provides the drive voltage scaling (V_{dd}) anticipated by the ITRS roadmap, depending upon the application.

This scaling approach has several shortcomings, including the reduction of power-supply voltages with each node, raising problems with circuit power-supply compatibility. So, for earlier technology nodes ($L \geq 0.35\,\mu\text{m}$), the power supply was kept constant while the device dimensions were scaled as described above. Eventually, the resultant higher electric fields associated with this scaling approach led to reliability concerns for the gate dielectric, mobility degradation and hot-carrier effects. For further scaling, the power supply was scaled at a different factor (such as $\alpha^{1/2}$) to compensate for such effects. For modern devices with *deep-submicron* gate length, other techniques are used to scale from node to node. Nevertheless, for CMOS technology over the last ≈ 25 years, it has been noted that the gate dielectric thickness can be related to the channel length as shown in Fig. 28.4 for a wide variety of device technologies [28.24].

The critical point in the context of dielectric films is that the dimensions of the devices, and therefore the dimensions (thicknesses) of the dielectric layers, are significantly reduced as scaling proceeds (Table 28.1). At some point, the materials properties associated with the dielectric in a component will be predicted to no longer provide the desirable electrical behavior attained in previous nodes, thus stimulating research on new dielectric materials. In the case of gate dielectrics, scaling results in thinner layers to a point where quantum-mechanical tunneling becomes an issue. For memory capacitor dielectrics, a specific capacitance density is required, regardless of capacitor dimensions, for a reliable memory element. Indeed, scaling impacts on the selection and integration of numerous constituents of electronic devices including gate electrode materials, source/drain junction regions, contacts, interconnect dielectrics.

28.0.2 Role of Dielectrics for ICs

Dielectrics are pervasive throughout the structure of an integrated circuit. Applications include gate dielectrics, tunneling oxides in memory devices, capacitor dielectrics, interconnect dielectrics and isolation dielectrics, as well as sacrificial or masking applications during the circuit fabrication process.

The basic properties utilized in dielectric materials include their structure and the resultant polarizability behavior. As noted in Fig. 28.5, there are several mechanisms to consider in the polarization of materials in the presence of an electric field. For nonmetallic (insulating) solids in CMOS applications, there are two main contributions of interest to the dielectric constant which give rise to the polarizability: electronic and ionic dipoles. Figure 28.5 illustrates the frequency ranges where each contribution to the real part of the complex relative dielectric permittivity ε_r dominates, as well as the imaginary part of the complex permittivity (dielectric losses), ε_r'' [28.25]. The region of interest to CMOS applications includes the shaded area where switching frequencies are in the GHz range at the present time.

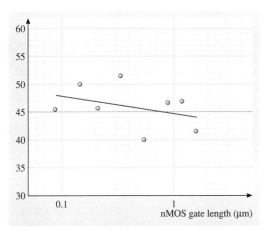

Fig. 28.4 Scaling trends of the transistor gate length and the gate insulator thickness for various generations of integrated technologies. The *solid line* is a linear fit to the data. The *dashed line* represents the approximate scaling trend between the gate dielectric thickness and gate length observed over several technology generations: $L_G = 45 \times t_{ox}$. After [28.24], with permission

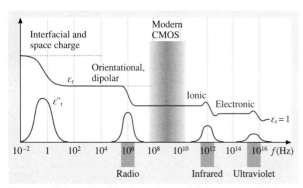

Fig. 28.5 Schematic of the dependence of dielectric permittivity on frequency. After [28.25], with permission

In general, atoms with a large ionic radius (i. e. high atomic number, Z) exhibit more electron dipole response to an external electric field, because there are more electrons to respond to the field (electron screening effects also play a role in this response). This electronic contribution tends to increase the permittivity of materials with higher-Z atoms.

The ionic contribution to the permittivity can be much larger than the electronic portion in cases such as perovskite crystals of $(Ba,Sr)TiO_3$ (BST) and $(Pb,Zr)TiO_3$ (PZT), which exhibit ferroelectric behavior below the Curie point. In these cases, Ti ions in unit cells throughout the crystal are uniformly displaced in response to an applied electric field (for the case of ferroelectric materials, the Ti ions reside in one of two stable, non-isosymmetric positions about the center of the Ti−O octahedra). This displacement of Ti ions causes an enormous polarization in the material, and thus can give rise to very large dielectric constants in bulk films of 2000–3000, and has therefore been considered for dynamic random-access memory (DRAM) capacitor applications [28.26, 27]. Since ions respond more slowly than electrons to an applied field, the ionic contribution begins to decrease at very high frequencies, in the infrared range of $\approx 10^{12}$ Hz, as shown in Fig. 28.5.

28.1 Gate Dielectrics

28.1.1 Transistor Structure

A cross-sectional schematic of the structure of a metal–oxide–semiconductor (MOS) field-effect transistor (FET) and a modern CMOS transistor, consisting of the n-FET and p-FET pair, is shown in Fig. 28.6. The source- and drain-region dopant profiles reflect modern planar CMOS technologies where an extended, doped region is produced under the gate region in the channel. Additionally, so-called *halo* or *pocket* dopant implantation regions are also shown. These dopant profile approaches have been incorporated in recent years in an effort to permit transistor channel scaling and yet maintain performance [28.28].

As can be seen, there exist several regions of the device which require dielectric materials to enable useful transistor operation. Both n-MOS and p-MOS transistors are isolated with a dielectric, typically deposited SiO_2 in most modern device technologies, which is deposited in trench structures. This isolation technique is called *shallow* trench isolation (STI) as the associated trench depth is $\leq 0.5\,\mu m$. Earlier generations of CMOS (and often devices under research) also utilized the so-called local oxidation of silicon (LOCOS) isolation approach [28.29, 30]. So-called spacer dielectrics (typically SiO_2, SiO_xN_y, or SiN_x) are also used around the transistor *gate stack* for isolation and implantation-profile control. The gate stack is defined here as the films and interfaces comprising the gate electrode, the underlying gate dielectric, and the channel region. The gate dielectric, typically SiO_2 or SiO_xN_y for transistors currently in production, electrically isolates the gate electrode from the underlying Si channel region while allowing the modulation of the carrier flow in the channel. As we shall see, the interface between the gate dielectric and the channel regions is particularly important in regard to device performance.

28.1.2 Transistor Dielectric Requirements in View of Scaling

In addition to the thickness reduction associated with scaling, the gate dielectric must adhere to several requirements that are dependent upon the specific product application, including adequate drive current, suitable capacitance, minimal leakage current, and reliable performance. Table 28.1 provides some selected values describing the ITRS 2003 roadmap.

Drive Current
The improved performance associated with the scaling of logic device dimensions can be seen by consider-

ing a simple model for the drive current associated with a FET [28.1, 3, 18, 20]. In the gradual channel approximation, the drive current can be written as:

$$I_d = \frac{W}{L}\mu C_{inv}\left(V_G - V_T - \frac{V_D}{2}\right)V_D, \quad (28.5)$$

where W is the width of the transistor channel, L is the channel length, μ is the channel carrier mobility (assumed to be constant in this analysis), C_{inv} is the capacitance density associated with the gate dielectric when the underlying channel is in the inverted state, V_G and V_D are the voltages applied to the transistor gate and drain, respectively, and the threshold voltage is given by V_T. It can be seen that in this approximation the drain current is proportional to the average charge across the channel (with a potential $V_D/2$) and the average electric field (V_D/L) along the channel direction. Initially, I_D increases linearly with V_D and then eventually saturates to a maximum when $V_{D,sat} = V_G - V_T$ to yield:

$$I_{D,sat} = \frac{W}{L}\mu C_{inv}\frac{(V_G - V_T)^2}{2}. \quad (28.6)$$

A goal here is to increase the saturation drive current as much as possible for a given transistor technology design. The term $(V_G - V_T)$ is limited in range due to reliability concerns as too large a V_G would create an undesirable, high electric field across the gate dielectric. Moreover, practical operation of the device at room temperature (or above) requires that V_T cannot easily be reduced below about $8 \times k_B T \cong 200$ meV ($k_B T \approx 25$ meV at room temperature). Elevated operating temperatures could therefore cause statistical fluctuations in thermal energy, which would adversely affect the desired V_T value. Thus, even in this simplified approximation, one is left to pursue a reduction in the channel length, increasing the width, increasing the gate dielectric capacitance, or a carefully engineered combination of all of these in order to increase $I_{D,sat}$. Note that none of these dimensions are, in reality, independent and so substantial resources are utilized to simulate device performance in order to maximize the performance prior to the fabrication of a representative transistor [28.5, 30, 32].

Gate Capacitance

In the case of increasing the inversion capacitance to improve I_D, consider the gate stack structure idealized as a parallel-plate capacitor (ignoring quantum-mechanical and depletion effects from a Si substrate and

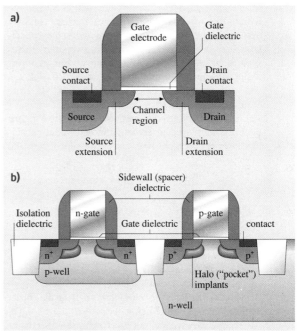

Fig. 28.6a,b Important regions of (a) a MIS field-effect transistor, and (b) a planar CMOS transistor structure. After [28.31] (© 2004 IEEE, with permission)

gate [28.33]),

$$C = \frac{\kappa \varepsilon_0 A}{t}, \quad (28.7)$$

where κ is the dielectric constant of the material, ε_0 is the permittivity of free space ($= 8.85 \times 10^{-3}$ fF/μm), A is the area of the capacitor, and t is the thickness of the dielectric between the capacitor electrode plates. (Note that the relative permittivity of a material is often given by ε or ε_r, such as with the expression $C = \varepsilon \varepsilon_0 A/t$. Note that the relation between κ and ε varies depending on the choice of units (e.g. when $\varepsilon_0 = 1$), but since it is always the case that $\kappa \approx \varepsilon$, we shall assume here that $\kappa = \varepsilon$.) It is clear that transistor designs utilizing planar CMOS scaling approaches discourage increasing the area of the capacitor, and so one is left to considering the reduction of the dielectric thickness in order to increase the capacitance.

In order to compare alternative dielectric materials which exhibit a dielectric constant higher than the standard for the industry, SiO_2, this expression for C can also be rewritten in terms of t_{eq} (i.e. equivalent oxide thickness) and th eκ_{ox} of the capacitor ($\kappa_{ox} = 3.9$

for the low-frequency dielectric constant of SiO_2). The term t_{eq} then represents the theoretical thickness of SiO_2 that would be required to achieve the same capacitance density as the dielectric (ignoring important issues such as leakage current and reliability). For example, if the capacitor dielectric is SiO_2, $t_{eq} = 3.9\,\varepsilon_0 (A/C)$, and so a capacitance density of $C/A = 34.5\,\text{fF}/\mu\text{m}^2$ corresponds to $t_{eq} = 10\,\text{Å}$.

Thus, the physical thickness of an alternative high-κ dielectric employed to achieve the equivalent capacitance density of $t_{eq} = 10\,\text{Å}$ can be obtained from the expression

$$\frac{t_{eq}}{\kappa_{ox}} = \frac{t_{high\text{-}\kappa}}{\kappa_{high\text{-}\kappa}} \text{ or simply,}$$

$$t_{high\text{-}\kappa} = \frac{\kappa_{high\text{-}\kappa}}{\kappa_{ox}} t_{eq} = \frac{\kappa_{high\text{-}\kappa}}{3.9} t_{eq}. \qquad (28.8)$$

From this expression, a dielectric with a relative permittivity of 16 therefore results in a physical thickness of $\approx 40\,\text{Å}$, to obtain $t_{eq} = 10\,\text{Å}$. The increase in the physical thickness of the dielectric impacts on properties such as the tunneling (leakage) current through the dielectric, and is discussed further below.

The industry scaling process clearly presents a major challenge for the core transistor gate dielectric as predictions call for a much thinner effective thickness for future alternative gate dielectrics: $t_{eq} \leq 1\,\text{nm}$ [28.12]. The interfacial regions between the gate electrode, dielectric and channel [in totality termed the metal–insulator–semiconductor (MIS) *gate stack* as shown schematically in Fig. 28.7] require careful attention, as they are particularly important in regard to transistor performance. These regions, less than approximately 0.5 nm thick, serve as a transition between the atoms associated with the materials in the gate electrode, gate dielectric and Si channel, and can alter the overall capacitance of the gate stack, particularly if they have a thickness that is substantial relative to the gate dielectric. Additionally, these interfacial regions can be exploited to obtain desirable properties. The upper interface, for example, can be engineered in order to block boron out-diffusion from the poly-Si gate. The lower interface, which is in direct contact with the CMOS channel region, must be engineered to provide low interface trap densities (e.g. dangling bonds) and minimize carrier scattering (maximize channel carrier mobility) in order to obtain reliable, high-performance device. Mobility degradation, relative to that obtained using SiO_2 (or SiON) gate dielectrics, associated with the incorporation of high-κ gate dielectrics is an important issue currently under investigation and discussed further below.

Reactions at either of these interfaces during the device fabrication process can result in the formation of a significant interfacial layer that will likely reduce the desired gate stack capacitance. Additionally, any suitable interfacial layer near the channel must result in a low density of electrically active defects ($\lesssim 10^{11}/\text{cm}^2\text{eV}$ is often obtained for SiO_2) and avoid degradation of carrier mobility in the region near the surface channel.

The reduced capacitance can be seen by noting that the dielectric film that has undergone interfacial reactions results in a structure that essentially consists of several dielectric layers in series. If we suppose that the dielectric consists of two layers, then from electrostatics the total capacitance of two dielectrics in series is given by $1/C_{tot} = 1/C_1 + 1/C_2$, where C_1 and C_2 are the capacitances of the two layers. Thus, the lowest-capacitance layer will dominate the overall capacitance and also set a limit on the minimum achievable t_{eq} value. If we consider a dielectric stack structure such that the bottom layer (layer 1) of the stack is SiO_2, and the top layer (layer 2) is the high-κ alternative gate dielectric, (28.8) is expanded (assuming equal areas) to:

$$t_{eq} = t_{SiO_2} + \frac{\kappa_{SiO_2}}{\kappa_{high\text{-}\kappa}} t_{high\text{-}\kappa}. \qquad (28.9)$$

From (28.9), it is clear that the layer with the lower dielectric constant (SiO_2 in this case) limits the ultimate capacitance of the MIS gate stack.

Of course, the actual capacitance of a CMOS gate stack for ultra-large-scale integration (ULSI) devices does not scale simply with $1/t$ due to parasitics, quantum-mechanical confinement of carriers, and depletion effects. Indeed, consideration of these important effects can result in much confusion on the defini-

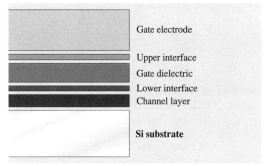

Fig. 28.7 Schematic of the important regions of the transistor gate stack. After [28.1] (AIP, with permission © 2001)

tion of dielectric thickness as extracted from electrical measurements.

Parasitic resistances and capacitances associated with various portions of the transistor structure can result in an overall degradation of performance as defined by a delay time figure of merit [28.1, 19]. There exist several materials issues associated with the control of such parasitics including source/drain dopant-profile control, gate/contact sheet resistance minimization, etc. which are beyond the scope of this chapter.

Quantum-mechanical confinement effects on carriers in the channel region occur as a result of the large electrical fields in the vicinity of the Si substrate surface. These fields quantize the available energy levels resulting in the displacement of the charge centroid from the interface into the Si and at energies above the Si conduction-band edge. The extent of the penetration of the charge centroid is dependent upon the biasing conditions employed for the metal–insulator–semiconductor (MIS) structure and, for accurate estimates of transistor drain-current performance, the inversion capacitance measurement provides accurate determination of the equivalent oxide thickness [28.33, 36, 37].

Current MIS gate-stack structures employ heavily doped polycrystalline Si (poly-Si) as the *metal* gate electrode for CMOS transistors. Poly-Si gates were introduced as a replacement for Al metal gates in the 1970s for CMOS integrated circuits and have the very desirable property of a tunable work function for both n-MOS and p-MOS transistors through the implantation of the appropriate dopant into the poly-Si gate electrode.

However, in deep-submicron scaled devices, the *poly-Si depletion effect* impacts the overall gate-stack capacitance significantly, as the scaling results in an increased sensitivity to the effective *electrical* thickness resulting from all of the gate-stack component films, including the gate electrode. The depletion effect (which occurs in any doped semiconductor) is a result of the decrease in the density of majority carriers near the poly-Si/dielectric interface upon biasing, resulting in an increased depletion width in the poly-Si [28.28, 38]. This results in a voltage drop across the depletion region, rendering a smaller voltage drop across the remaining portions of the gate stack (i.e. the dielectric and the substrate). As a result, the (low-frequency) capacitance under strong inversion biasing conditions is significantly lower than that obtained under accumulation (Fig. 28.8). Moreover, the inversion-layer charge in the substrate is reduced from that possible in the ideal case where no depletion layer exists in the gate (i.e. an ideal metal conductor for the gate electrode). This reduction in the

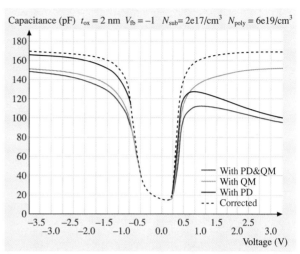

Fig. 28.8 Capacitance–voltage curves demonstrating the effects of poly-Si depletion and quantum-mechanical behavior for scaled transistors. N_{sub} and N_{poly} denote the dopant density for the substrate and poly-Si gate, respectively. After [28.34]

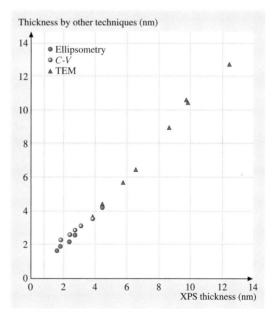

Fig. 28.9 Metrology results for various measurement techniques of the thickness of thin SiO_2 films. X-ray photoelectron spectroscopy (XPS) measured thicknesses are compared to thicknesses derived from high-resolution transmission electron microscopy (HRTEM), C–V and ellipsometry. After [28.35] (AIP, with permission © 1997)

inversion-layer charge in turn reduces the ideal transistor drive current. As the dielectric thickness is reduced in the scaling process, this effect becomes a serious limitation for transistor drive current.

To solve this problem, a return to metal gate electrode materials is now under development to eliminate any depletion region in the gate, and thus the interaction of any dielectric with various metal gate candidates is of interest. Compatibility of dielectric materials with metal gate materials is briefly discussed later in this chapter.

When a capacitance–voltage curve is properly corrected for these quantum-mechanical and dopant depletion effects, as seen in Fig. 28.8, one can accurately extract the equivalent oxide thickness (EOT), t_{eq} [28.34]. This thickness definition is in contrast to that derived directly from the raw data in a capacitance–voltage (C–V) measurement, which is termed the *capacitance equivalent thickness* (CET > EOT), or the *physical thickness*, which can be determined by non-electrical characterization methods such as ellipsometry or high-resolution transmission electron microscopy, as seen in Fig. 28.9 [28.34, 35]. Please note that thicknesses derived from quantum-mechanical-corrected accumulation capacitance data are also often abbreviated as t_{qm}.

Leakage Current

Of course, physically thinning the dielectric layer to the nanometer regime raises the prospect of quantum-mechanical tunneling through the dielectric – often referred to as *leakage*. To minimize gate leakage (tunneling) currents, quantum mechanics indicates that the gate dielectric must be sufficiently thick and have a sufficient energy barrier (band offset) to minimize the resultant tunneling current. Consider the band diagram shown in Fig. 28.10, where the electron affinity (χ) and gate electrode work function (Φ_M) are defined (q is the charge). For electrons traveling from the Si substrate to the gate, this barrier is the conduction-band offset, $\Delta E_C \cong q[\chi - (\Phi_M - \Phi_B)]$; for electrons traveling from the gate to the Si substrate, this barrier is Φ_B. The effect of the height of the energy barrier and the film thickness on tunneling current can be seen by considering the expression for the direct tunneling current where electron transport occurs through a trapezoidal energy barrier [28.18]:

$$J_{dt} = \frac{A}{t_{diel}^2} \exp(-Bt_{diel})$$
$$\times \left\{ \left(\Phi_B - \frac{V_{diel}}{2}\right) \exp\left[\sqrt{\left(\Phi_B - \frac{V_{diel}}{2}\right)}\right] \right.$$
$$\left. - \left(\Phi_B + \frac{V_{diel}}{2}\right) \exp\sqrt{\left(\Phi_B + \frac{V_{diel}}{2}\right)} \right\}.$$
(28.10)

Here A is a constant, $B = 4\pi/h(2m^*q)^{1/2}$, Φ_B is the potential-energy barrier associated with the tunneling process, t_{diel} is the physical thickness of the dielectric, V_{diel} is the voltage drop across the dielectric, and m^* is the electron effective mass in the dielectric. From (28.10), one observes that the tunneling (leakage) current increases exponentially with decreasing barrier height and thickness for electron direct-tunneling transport. In the context of charge transport through the dielectric, electrically active defects (electron or hole *traps*) can result in charging of the dielectric, which in turn deleteriously affects the electric field in the channel region and therefore mobility. Conduction mechanisms through such fixed charge traps can also be evaluated through electrical characterization techniques [28.18]. In the case of dielectrics layers, tunneling transport has also been previously examined [28.39].

The introduction of a dielectric material that exhibits a suitable energy barrier and thickness, while performing electrically as a thin SiO_2 layer, is now a key research and development goal for the industry to continue the scaling trend. It is also clear that the gate dielectric must exhibit adequate thickness uniformity and integrity over the surface of the wafer. The reduction of this leakage current impacts important circuit properties such as stand-by power consumption, as shown in Fig. 28.11. With the minimization of leakage current as a key driving force, alternate gate dielectric materials appear to be required for low-power CMOS device technologies in the near future and are therefore the subject of intense research.

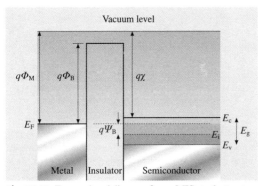

Fig. 28.10 Energy-band diagram for an MIS stack structure

Dielectric Materials for Microelectronics | 28.1 Gate Dielectrics

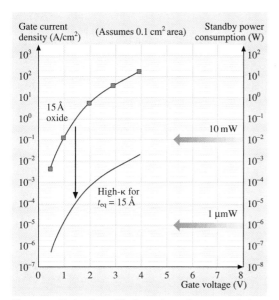

Fig. 28.11 Reduction of leakage (tunneling current) from the incorporation of a higher permittivity gate dielectric. After [28.1] (AIP, with permission © 2001)

Table 28.2 Properties of SiO_2

Geometrical parameters	
Si−Si bond length	3.12 Å
Si−O bond length	1.62 Å
O−O bond length	2.27 Å
Mean bond angle	144° (tetrahedral bonding)
Bond angle range	Si−O−Si: 110° to 180°
Density (g/cm³)	
Thermally grown (fused silica)	2.20
Quartz	2.65
Index of refraction (optical frequencies)	
Thermally grown (fused silica)	1.460
Quartz	1.544
Quasi-static dielectric constant (≤ 1 kHz)	
Thermally grown (fused silica)	3.84
Quartz	3.85
Band gap	
Thermally grown (fused silica)	8.9 eV
Quartz	≈ 9.0 eV
Electrical breakdown strength	
Thermally grown (fused silica)	10 MV/cm
Quartz	≈ 10 MV/cm

Reliability

Any gate dielectric material must possess the ability to enable a specified, stable operational lifetime for the transistor. Key areas of concern in this regard includes defect generation/mitigation resulting in dielectric breakdown phenomenon, charge formation within the dielectric and at the associated interfaces, and resistance to energetic ("hot") carrier interactions. This is often evaluated in the context of accelerated lifetime testing through electrical stressing in combination with thermal stressing of devices. Reviews on these topics can be found in this handbook as well as other sources [28.6, 28, 40, 41].

28.1.3 Silicon Dioxide

As articulated by *Hummel*, silicon has an amazing number of desirable materials (and therefore electronic) properties that no other electronic material can rival [28.42]. Silicon constitutes 28% of the Earth's crust, it is nontoxic, exhibits a useful band gap ($E_G = 1.12$ eV) for applications at room temperature, and can be grown in crystalline form (from abundant sand, i.e. SiO_2). However, it can be argued that the ability of Si to form the stable, high-quality insulator, SiO_2, may be the primary reason that Si-based transistor technology has dominated the industry since the 1960s [28.8, 43]. This compound has a number of material properties that result in outstanding electrical performance. Some of these properties are summarized in Table 28.2. Over the last 25 years, abundant compendia of the properties of SiO_2 with particular emphasis on materials and electrical properties have been published, as well as conferences which have focused almost exclusively on SiO_2 [28.44–53].

It is noteworthy that the early development of transistor technology utilized the semiconductor Ge, which has a rather unstable oxide, GeO_2, that is water soluble. The utilization of crystalline Si and its stable oxide presented a superior solution for the practical manufacturing of the planar transistor and the associated integrated circuit of the era [28.43]. It was the utilization of the properties of SiO_2 that resulted in the dominance of Si-based device technology over Ge-based technology in the 1960's.

Bonding Arrangements for Thin SiO_2

As can be seen in Fig. 28.12, the phase diagram for SiO_2 has a number of crystalline phases and associated polymorphisms [28.52, 54]. In the context of accessible phase space appropriate for typical CMOS processes (pressure ≈ 1 bar and $T \leq 1100\,°C$), the quartz and tridymite

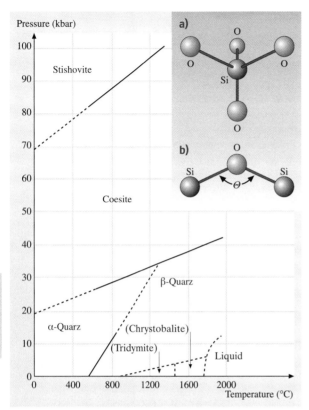

Fig. 28.12a,b Phase diagram for SiO$_2$. *Inset*: bonding configurations for the O–Si–O system. (**a**) Tetrahedral unit for the Si–O bonding arrangement. (**b**) Bond angle θ defined for O–Si–O bonds. After [28.54] (IOP Publishing © 1994, with permission)

phases are stable [28.55]. High-pressure phases are also well studied [28.56].

The bulk SiO$_2$ tetrahedral bonding coordination (inset, Fig. 28.12) results in an average Si–O–Si bond angle of $\theta \approx 145°$ [28.57] comparable to that observed for vitreous silica [28.54, 58]. As noted originally in the classic work by *Zachariasen*, the bonding interconnections among the various SiO$_4$ tetrahedra can result in an amorphous SiO$_2$ film consisting of locally ordered ring structures commonly observed in silica glasses [28.50, 52, 59]. The numerous available variations of the ring structure provide substantial flexibility and minimize bond strain and the density of defects at interfaces [28.60].

For the purposes of microelectronic applications, we first consider so-called *thermally grown* SiO$_2$. As the name implies, a SiO$_2$ layer is formed on Si from a thermally activated oxidation process (often referred to as a *thermal oxide*) rather than simply a (low-temperature) deposition process. Thermal oxide films are stable enough to endure high-temperature ($\approx 1000\,°C$) post-implantation/dopant activation anneals. These films have useful electrical properties including a stable (thermal and electrical) interface with minimal electrically-active-defect densities ($\approx 10^{10}/\text{cm}^2$) as well as high breakdown strength ($\approx 10\,\text{MV/cm}$) over a large area ($\approx 1\,\text{cm}^2$) [28.18]. The breakdown strength is important for high reliability of devices fabricated with a dielectric, as the scaling of devices typically results in somewhat higher electric fields across the oxide.

Recent experimental results on thin thermal SiO$_2$ [28.34, 40, 61] combined with modern computational modeling [28.57, 62] of the defects associated with the SiO$_2$/Si(001) interface, which is the dominant microelectronic materials system, and *bulk* SiO$_2$ [28.63] has provided a reasonably consistent picture of the known defects and interface structure.

High-resolution X-ray photoelectron spectroscopy (XPS) evidence, coupled with other experimental and theoretical modeling results, appears to support the presence of SiO$_x$ ($x = 4$) species (often called *sub-oxides*) detected as a chemical shift in the associated Si oxidation states (Fig. 28.13) [28.64–67].

More recent high-resolution angle-resolved XPS analysis of ultrathin (6-Å) SiO$_2$ has resulted in a further understanding of the average oxidation state observed with depth along the interfacial transition region [28.68, 69]. From these experimental results, models of the interface have been constructed from first-principles methods that reproduce the experimental results, as shown in Fig. 28.14 [28.57, 62]. As seen in Fig. 28.14, the transition region between the Si substrate and the SiO$_2$ consists of interfacial suboxide species of varying abundance (and therefore density). These results appear to be in general agreement with recent high-resolution transmission electron microscopy as well [28.70].

Oxidation Kinetics of Si

The oxidation kinetics of Si have also been exhaustively studied and reviewed [28.28, 40, 71]. Two regimes for oxidation are generally characterized as *passive* and *active* in nature. The passive oxidation regime refers to the low-temperature/high-pressure region of the pressure–temperature (p–T) phase space shown in Fig. 28.15 [28.40, 72–77]. In this portion of the p–T phase space, the reaction between Si and O$_2$ proceeds

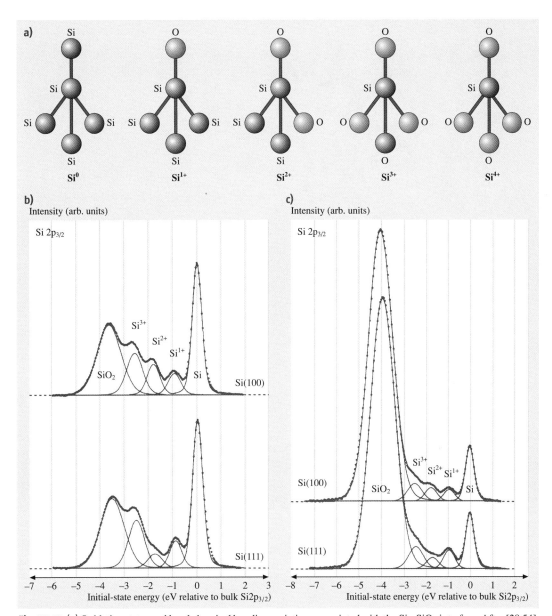

Fig. 28.13 (a) Oxidation states and local chemical bonding variations associated with the Si–SiO$_2$ interface. After [28.54]. High-resolution XPS results demonstrating the existence of such bonding variations at the interface for substrate orientation and with **(b)** 5-Å and **(c)** 14-Å SiO$_2$ film thickness. Photon energy $h\nu = 130$ eV. After [28.66] (APS, with permission © 1988)

as a surface reaction resulting in a thin (≤ 2 − nm) SiO$_x$ layer, where $0.5 \leq x \leq 2$. Decreasing pressure and increasing temperature during the reaction results in both Si etching and oxidation – the so-called *active* regime – and the formation of thicker SiO$_2$ films.

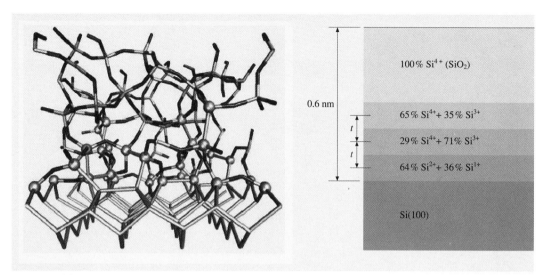

Fig. 28.14 (a) First-principles simulations of the SiO$_2$/Si interface. After [28.62] © 2002 Elsevier. (b) Angle-resolved XPS results for the SiO$_2$/Si interface showing the detected O−Si−O bonding environments. The single layer thickness $t = 0.137$ nm associated with the bonding distance in Si in this evaluation. After [28.69] (© 2001 by APS, with permission)

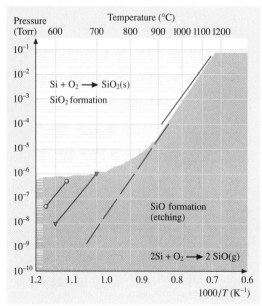

Fig. 28.15 Pressure–temperature phase diagram for thin SiO$_2$/Si. After [28.77] (© 1997 AIP, with permission)

The clever utilization of isotopic labeling techniques coupled with depth profiling from ion scattering [28.78] and nuclear reaction analysis [28.71] has generally confirmed that molecular oxygen (O$_2$) is the mobile species that interstitially diffuses through a growing oxide and only reacts at the Si interface to form the SiO$_2$ on the Si surface, as first empirically proposed by *Deal* and *Grove* in 1965 for SiO$_2$ thicker than about 4 nm [28.79, 80]. The details of the oxidation process for thin films deviates from the Deal–Grove model, and is discussed thoroughly in the review by *Green* et al., for example [28.40].

Defects for Thin SiO$_2$

In addition to the electrical breakdown strength associated with SiO$_2$, the physicochemical properties of the interface between the substrate Si and SiO$_2$ plays a major role in establishing the observed electrical properties [28.61, 67]. Like all interfaces of solids, the interface between bulk Si and bulk SiO$_2$ naturally results in the presence of point defects (Fig. 28.16). Examples include strained bonds, atoms that are not fully bonded to neighboring atoms (so-called "dangling"), bonds and hydrogenic species bonded among the Si−O bonding network. These defects can result in energy levels in the band structure associated with an MIS device, and serve as *traps* for carriers (electrons and holes).

The dominant orientation for contemporary Si-based integrated circuit technology substrates is Si(100). This is, of course, no accident and the choice of the substrate

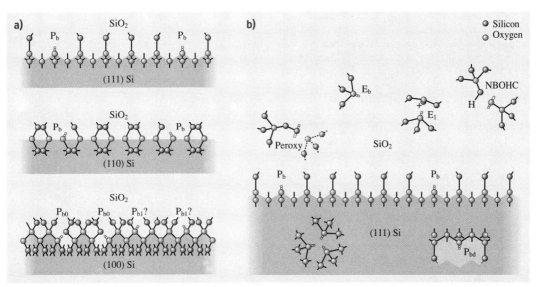

Fig. 28.16a,b Schematic representations of (**a**) the various point defects for the SiO$_2$/Si interface and (**b**) bulk SiO$_2$. After [28.54] (IOP, with permission © 1994)

orientation has much to do with the resultant SiO$_2$/Si interface. Early in the development of Si-based transistors, it was observed that the density of interface states for SiO$_2$ grown on Si(100) was significantly lower than that observed on Si(111) [28.81–85]. This difference in interface state behavior is attributed to the density of unsatisfied (*dangling*) bonds at the SiO$_2$/Si interface.

Techniques sensitive to the presence of such point defects, such as electron (paramagnetic) spin resonance (EPR or ESR) [28.88] have provided valuable insight on the atomic nature of the various defects in the bulk and interfacial regions of SiO$_2$ [28.89–94]. For such defects, although the electrical manifestation of significant densities ($\geq 10^{10}$/cm^{2}eV) of such *trap states* is easily detected by conventional electrical characterization methods, the unambiguous identification of the moiety responsible for the observed electrical behavior is very difficult using only electrical characterization. From techniques like ESR employed over the last 30 years, point-defect structural models have been established in the bulk regions of SiO$_2$ and Si, as well as at the interface (including various substrate orientations). A schematic illustration of various known point defects for the Si/SiO$_2$ system is shown in Fig. 28.16. The major classes of defects that have been examined over the last 35 years by ESR have included interfacial dangling bonds (P$_b$ centers) and bulk defects in SiO$_2$ (E$'$ centers) [28.91, 94].

The early discovery and study of P$_b$ centers [28.95–97] eventually led to the interpretation of their ESR signals as due to dangling bonds and the realization of their importance to MOS device physics [28.86, 90,

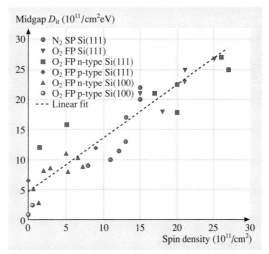

Fig. 28.17 Correlation of mid-gap interface state density with detected spin density from ESR measurements for SiO$_2$/Si(111) and SiO$_2$/Si(100). FP and SP denote *fast* and *slow* pullout conditions for the wafers after thermal treatments. After [28.86, 87], with permission

91, 98]. It has also been noted that P_b-center defects are generated for unannealed, thermally stressed and radiation-damaged MOS structures, and moreover can account for roughly 50% of the density of interface defect states (D_{it}) for Si/SiO$_2$ interfaces [28.54, 99, 100]. It has been demonstrated that the interface state density can be directly proportional to the density of P_b centers (dangling bonds) for the SiO$_2$/Si(111) and SiO$_2$/Si(100) interfaces, as seen in Fig. 28.17. Recently, further ESR work has been performed to establish the detailed structure of the defect on Si(100) and Si(111) [28.101–108].

Mitigation of Defects for Thin SiO$_2$

Synopses of the early MOS transistor work [28.109–111] examining the importance of the dielectric–semiconductor interface are available [28.6, 43, 112]. Indeed, the close connection between interfacial chemical behavior and electrical device performance was investigated and realized in pioneering surface science work by *Law* and coworkers on the reaction of gaseous species with atomically clean Ge [28.113] and Si surfaces [28.114–116]. In particular, the relative interface state (dangling-bond) density, as measured by the areal density of surface reactions with technologically important species such as H_2O, H_2, O_2, CO, CO_2, provided important clues on the control of the Si/SiO$_2$ interface and the resultant electrical properties reported 5–10 years later.

The control of the density of interface defects through the chemical reaction of species, introduced mainly through gaseous exposure at elevated temperatures, has proven to be fruitful. As noted above, early surface science work [28.114] indicated the rapid reaction of H_2 with the atomically clean Si surface. By definition, the atomically clean Si surface is saturated with dangling bonds. The reaction of these bonds with H_2 results in the chemical passivation of the surface – that is, the reaction of the surface to eliminate the dangling bonds and produce a H-terminated Si surface. In the context of the MOS structure [28.47, 117], it was realized early that annealing the structure in ambient H_2 resulted in beneficial transconductance performance [28.118–121]. The use of anneals in forming gas (N_2 : H_2 of various mixtures, typically 90–95% N_2 : 10–5% H_2) were originally developed to improve electrical contacts for the gate and source/drain regions of the MOSFET. *Balk* and *Kooi* established the effect of hydrogen ambients on the reduction of fixed charge in the Al/SiO$_2$/Si MOS structure [28.120, 122].

Subsequent studies demonstrated that anneals of the MOS structure in hydrogenic environments, typically at 400–500 °C for 30–60 min, results in the passivation of dangling bonds at the interface [28.86, 101–106, 123–126].

Hydrogen incorporation into the bulk of SiO$_2$ can, however, also be detrimental to dielectric performance in MOS capacitors and FETs [28.54, 127]. For example, silicon bonded to hydroxyl (silanol) species have been identified with fixed charge in the oxide, resulting in undesirable, irreversible voltage shifts. This charge induced shift is shown in Fig. 28.18 for n-type and p-type MIS diode structures from their associated capacitance–voltage response [28.1, 6, 7]. (An analogous threshold-voltage shift would be observed in a transistor turn-on characteristic.) More complicated effects such as negative-bias temperature instability (NBTI), where an increase of the density of fixed charge (Q_f) and interface trap (Q_{it}) density is noted with time upon thermal stress and/or under negative bias, has been attributed to H_2O-induced depassivation of Si dangling bonds (i.e., generation of P_b centers) at the Si/SiO$_2$ interface [28.54, 128]. Defects generated by radiation damage have also been extensively investigated [28.54, 129] as well as interactions with annealing ambients such as vacuum [28.130] or SiO [28.124].

Dielectric Breakdown and Reliability of SiO$_2$

As noted in Sect. 28.1.2, the reliability of dielectrics is obviously an important phenomenon to control. The scaling of microelectronic devices necessarily results in increased stress on the dielectric due to the higher electric fields placed across the dielectric film. As a result, power-supply voltages are also scaled to minimize the likelihood of catastrophic (*hard*) breakdown, which would generally result in complete failure in the associated integrated circuit. As noted by *Hori*, such a catastrophic breakdown phenomenon is dependent upon the presence of defects in the dielectric, and thus requires a statistical analysis of many devices (and therefore films) to enable a reliability prediction for the dielectric layer [28.18]. Evaluation of the breakdown is often performed under the conditions of constant-field stress until a time at which breakdown is observed [called time-dependent dielectric breakdown (TDDB)]. Breakdown fields for thermally grown SiO$_2$ with thicknesses larger than ≈ 10 nm often exceed 10 MV/cm, providing outstanding insulator prosperities for microelectronic applications. Various models have also been developed to accelerate such testing to perform reliability predictions. A concise review of these is offered in the literature as well [28.18, 28]. Defects such as film nonuniformity, bond stress, surface asperities, contam-

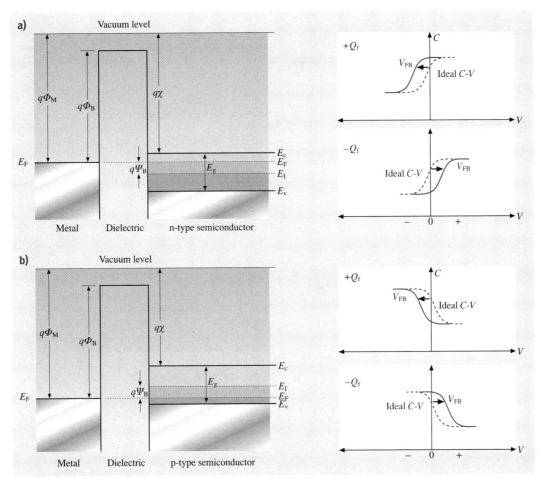

Fig. 28.18a,b The effect of fixed charging of the capacitor dielectric, demonstrating the shift of the associated flat-band voltage for MIS structures with (**a**) n-type and (**b**) p-type substrates. After [28.1] (© 2001 AIP, with permission)

ination, and particulates embedded in the film or the Si substrate are a few examples of potential causes of catastrophic breakdown phenomenon.

With the scaling of the SiO_2 gate dielectric layer to thicknesses well below 10 nm, extensive evaluations have been made of the various models that attempt to predict the reliability of SiO_2 in regard to breakdown. Often, a *soft* breakdown can be observed, where the film exhibits a sudden increase in conductivity, but not to the same degree as that observed in hard breakdown phenomenon. In spite of the many years of research on SiO_2 thin films, the nuances of the dependence of voltage acceleration extrapolation on dielectric thickness and the improvement of reliability projection arising from improved oxide thickness uniformity, have only recently become understood, despite decades of research on SiO_2 [28.131].

28.1.4 Silicon Oxynitride: SiO_xN_y

The introduction of nitrogen in SiO_2, often described as *silicon-oxynitride* (or more simply SiON) has provided the opportunity to maintain the scaling expectations for integrated circuits while minimizing the process variations associated with the modification of the gate dielectric material [28.18, 40, 71, 132]. (It should be noted that engineers and technologists often loosely refer to SiON films simply as oxide or nitride – frequently lead-

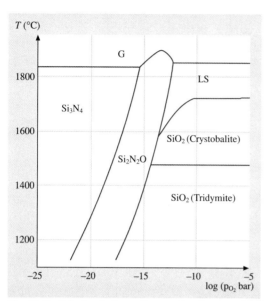

Fig. 28.19 Calculated phase diagram for the growth of Si–O–N for various oxygen partial pressures. After [28.134], with permission

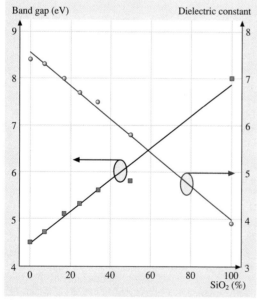

Fig. 28.20 Dependence of permittivity and insulator band gap of SiO_xN_y with SiO_2 content. After [28.135], with permission

ing to confusion on the detailed chemical composition of the film.) Introduction of N into SiO_2 has been accomplished by thermal treatments (in N_2, N_2O, NO and NH_3) as well as plasma treatments [28.18, 133]. The bulk phase diagram for the SiON system is presented in Fig. 28.19, which indicates that, under equilibrium conditions, Si_2N_2O is the only stable SiON species and SiO_2 and Si_3N_4 would not coexist [28.134]. As noted by *Green* et al., N would not be expected to incorporate into SiO_2 whenever an even very small partial pressure of oxygen $P_{O_2} > 10^{-20}$ atm is present. However, the nonequilibrium surface reaction kinetics and/or a reduction in interfacial bond strain likely plays a key role in the tendency for interfacial N incorporation for SiON films [28.40].

The dielectric constant and band gap as a function of the SiO_2 content are presented in Fig. 28.20 [28.135]. As noted in Sect. 28.1.2, it is the increase in the dielectric constant that increases the capacitance of an associated MOS structure. Thus, for a SiON film with $\kappa = 7$, a 2-nm SiON film would ideally exhibit electrical behavior (such as capacitance and leakage-current behavior) equivalent to a 1.1-nm SiO_2 film, according to (28.8). The data in Fig. 28.20 implies that the index of refraction (at 635 nm) varies between $n(SiO_2) = 1.46$ and $n(Si_3N_4) = 2$, indicating that optical techniques such as ellipsometry can provide very useful information for such films.

Like the SiO_2/Si system, point defects have also been studied for SiON films on silicon [28.136]. *Lenahan* and coworkers examined the role of strain at the interface and the impact on the associated dangling bonds [28.137]. More recently, work on the effects of hydrogen annealing has also examined the SiON system [28.138].

It is also noted that the SiON material system also exhibits useful diffusion-barrier properties. This property is utilized to inhibit uncontrolled dopant diffusion from the polysilicon gate electrode through the dielectric layer and into the channel region of transistors, for example. The formation of SiON films from the reaction of NO or N_2O with Si, as well as nitridation of Si and SiO_2 from N_2 and NH_3 exposure, has been examined in detail and has been summarized in recent reviews [28.40, 71].

For transistor gate dielectric applications, a controlled variation in the depth concentration profile for N is often required throughout the dielectric to maintain the improved device reliability and mobility associated with SiON thin films [28.28]. An example of such N profiles is shown in Fig. 28.21, where a somewhat reduced N content is noted in the vicinity of the Si channel/dielectric interface relative to that near the dielectric/poly-Si gate

28.1.5 High-κ Dielectrics

Recent research has focused on high-κ dielectrics to further enable scaling of transistor (and memory capacitor) technology. In the context of the industry roadmap, the term high-κ dielectric more generally refers to materials which exhibit dielectric constants higher than the SiON films ($\kappa \approx 7$) described above.

Desirable Properties for High-κ Dielectrics

Many materials systems are currently under consideration as potential replacements for SiO_2 and SiO_xN_y as the gate dielectric material for sub-100-nm (CMOS) technology. It should be stated at the outset that this field of materials research remains very active, and a conclusive summary on the topic of high-κ materials, particularly for conventional MOSFET gate dielectric applications, is not yet possible. The field continues to evolve with interesting research being reported almost weekly in peer-review journals as well as industry newsletters and magazines. Nevertheless, a systematic consideration of the required properties of gate dielectrics indicates that the key desirable properties for selecting an alternative gate dielectric include: permittivity and the associated band gap/alignment to silicon as well as mobility, thermodynamic stability, film morphology, interface quality, compatibility with the current or expected materials to be used in processing for CMOS devices, process compatibility, and reliability. The desirable properties are summarized in Table 28.3. Reviews of the developing field and recent research on high-κ dielectric materials candidates are available [28.1, 2, 144].

Many gate dielectric materials appear favorable in some of these areas, but very few materials are promising with respect to all of these properties. Indeed, many of these desirable properties are interrelated. Many of these materials have been examined over the last 20 years for capacitor applications, for example. Materials currently under extensive investigation include oxides, silicates and aluminates of Hf, Zr, La, Y, and their mixtures, which results in dielectric constants in the range 10–80. Table 28.4 summarizes some of the measured or calculated properties associated with high-κ dielectric material constituents. There are also some recent studies of oxides associated with the rare-earth lanthanide series of elements in the periodic table [28.145]. One notes the tradeoff between the band gap (and therefore the conduction/valence-band offsets to those of Si) for these dielectrics and the associated dielectric constant, as seen in Fig. 28.22 [28.146].

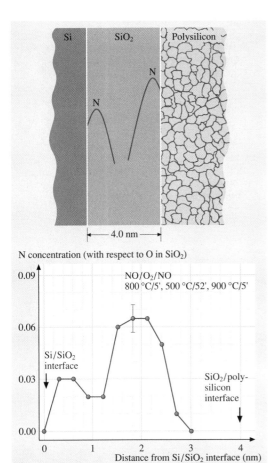

Fig. 28.21 N concentration depth profiles for tailoring the overall dielectric constant and materials properties of a SiON layer. After [28.34] (© 1999 Elsevier, with permission)

interface [28.34]. Such profiles are thought to control effects such as interfacial strain and roughness (which can degrade mobility) and inhibit dopant out-diffusion from the gate electrode. Incorporation of N in thin dielectric films using plasma processes has been studied extensively [28.133, 139–143].

Given the various leakage-current constraints afforded by power limitations and reliability, the scaling of SiON appears to be limited to $t_{eq} \approx 1$ nm, corresponding to the 65-nm node for high-performance microprocessor products set to begin in ≈ 2007 [28.12]. As a result, investigations of alternative high-κ dielectric materials have been initiated over the last several years.

Table 28.3 Desirable properties for high-κ gate dielectrics (65-nm node)

Physical property	Value/Criteria	Electrical property	Value/Criteria
Permittivity	15–25	Equivalent oxide thickness	< 1 nm
Band gap	> 5 eV	Gate leakage current (low power)	< 0.03 A/cm^2
Band offset	> 1.5 eV	Gate leakage current (high performance)	< 10^3 A/cm^2
Thermodynamic stability to 1000 °C	When in direct contact with Si channel	CV dispersion	Minimal (meV)
Compatibility with metal electrodes	Mimimized extrinsic (pinning) defects	CV hysteresis	Minimal (meV)
Morphology control	Resistance to interdiffusion of constituents, dopants, and capping metals	V_T (V_{FB}) shift (fixed charge, defects, trapping, etc.)	Minimal
Deposition process	Suitable for high-volume production	Channel mobility	Near SiO$_2$ universal curve
Etching	Suitable control for patterning after processing/annealing	Interface quality	Near SiO$_2$; $D_{IT} \approx 5 \times 10^{10}$ /cm^{2}eV

Table 28.4 Comparison of relevant properties for selected high-κ candidates. Key: mono. = monoclinic; tetrag. = tetragonal

Material	Dielectric constant (κ)	Band gap E_G (eV)	ΔE_C (eV) to Si	Crystal structure(s) (400–1050 °C)
SiO$_2$	3.9	8.9–9.0	3.2–3.5[b]	amorphous
Si$_3$N$_4$	7	4.8[a]–5.3	2.4[b]	amorphous
Al$_2$O$_3$	9	6.7[h]–8.7	2.1[a]–2.8[b]	amorphous*
Y$_2$O$_3$	11[d]–15	5.6–6.1[d]	2.3[b]	cubic
Sc$_2$O$_3$	13[d]	6.0[d]		cubic
ZrO$_2$	22[d]	5.5[a]–5.8[d]	1.2[a]–1.4[b]	mono., tetrag., cubic
HfO$_2$	22[d]	5.5[d]–6.0	1.5[b]–1.9[c]	mono., tetrag., cubic
La$_2$O$_3$	30	6.0	2.3[b]	hexagonal, cubic
Ta$_2$O$_5$	26	4.6[a]	0.3[a,b]	orthorhombic
TiO$_2$	80	3.05–3.3	≈ 0.05[b]	tetrag. (rutile, anatase)
ZrSiO$_4$	12[d]	6[d]–6.5	1.5[b]	tetrag.
HfSiO$_4$	12	6.5	1.5[b]	tetrag.
YAlO$_3$	16–17[d]	7.5[d]		**
HfAlO$_3$	10[e]–18[g]	5.5–6.4[f]	2–2.3[f]	**
LaAlO$_3$	25[d]	5.7[d]		**
SrZrO$_3$	30[d]	5.5[d]		**
HfSiON	12–17[i,j]	6.9[k]	2.9[k]	amorphous

* (γ-Al$_2$O$_3$ phase) has been recently reported [28.148], ** Onset of crystallization depends upon Al content, [a] [28.149], [b] [28.150–152], [c] [28.153], [d] [28.146], [e] [28.154], [f] [28.153], [g] [28.155], [h] [28.156], [i] [28.157], [j] [28.158], [k] [28.159]

The compatibility of alternative dielectrics with metal gate electrodes is also an important consideration [28.1, 3, 147]. The interfacial reactions between some gate electrode metals and gate dielectric is thought to lead to extrinsic states due to Fermi-level pinning, which shifts the threshold voltages for transistors to a fixed value. Another clear challenge is compatibility with other process steps that entail substantial thermal budgets (≈ 1000 °C, ≤ 10 s) for conventional CMOS. Future scaling may well require a modification (reduction) of such thermal budget desires to incorporate metal gate electrodes in CMOS, and thus may open

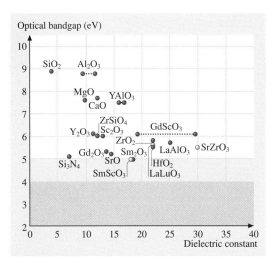

Fig. 28.22 Band gap versus dielectric constant for a number of dielectric materials. After [28.146] MRS Bulletin, with permission

the door to the consideration of other gate dielectric materials that exhibit stability at somewhat lower temperatures. However, key issues that must be addressed under such a scenario include alternative source/drain engineering (e.g. dopant activation at lower thermal budgets) and/or gate electrode formation in the device fabrication flow (e.g. gate electrode insertion after high-temperature anneals). Previous research on such CMOS process modifications has indicated that a variety of challenges exist to address such scenarios with adequate process margin and yield.

Mobility Degradation for Transistors with High-κ Dielectrics

In the case of many of the alternate gate dielectric materials (mainly metal oxides) currently under consideration, the polarizability of the metal–oxygen (*ionic*) bond is responsible for the observed low-frequency permittivity enhancement. Such highly polarizable bonds are described to be *soft* relative to the less polarizable *stiff* Si–O bonds associated with SiO_2. Unlike the relatively stiff Si–O bond, the polarization frequency dependence of the M–O bond is predicted to result in an enhanced scattering coupling strength for electrons with the associated low-energy and surface optical phonons. This scattering mechanism can therefore degrade the electron mobility in the inversion layer associated with MOSFET devices. A theoretical examination of this effect is provided by *Fischetti* et al. where calculations

of the magnitude of the effect indicate that pure metal-oxide systems, such as ZrO_2 and HfO_2, suffer the worst degradation, whereas materials which incorporate Si–O bonds, such as silicates, fare better [28.160]. In that work, it is also noted that the presence of a thin SiO_2 interfacial layer between the Si substrate and the high-κ dielectric can help boost the resultant mobility by screening this effect, although the maximum attainable effective mobility is still below that for the ideal SiO_2/Si system. These researchers further suggest that the effect is also minimized by the incorporation of Si–O in the dielectric, as in the case of pseudo-binary systems such as silicates.

Comparisons of this model with experimental evidence have recently been reported explicitly through the comparison of HfO_2 and Hf–silicate mobility studies. The results indicate that n-MOS and p-MOS poly-Si gated devices with Hf–silicate dielectrics exhibit better mobilities (approaching those of SiO_2) over those obtained using HfO_2 due to a diminished soft-phonon scattering mechanism [28.161]. Incorporation of metal gate electrodes results in further improvement [28.162].

Current High-κ Dielectric Research and Development

At this time (end of 2004), work on Hf-based dielectric materials, mainly Hf–silicates [28.163] and HfO_2 [28.1], dominates the recent engineering literature (Fig. 28.23) and they appear to exhibit useful properties for integrated-circuit scaling down to the 20-nm node. These properties include a relative stability for interfacial reactions that prefer silicate formation thus can avoid a lower overall dielectric constant of the high-κ dielectric stack, due to a pure SiO_2 interfacial layer, as described by equation (28.9). The papers on these materials systems are too numerous to list individually here and so the reader is referred to the reviews on the topic [28.1, 2, 144].

Of particular importance for gate dielectrics have been investigations of the stability (in particular, changes in morphology and interdiffusion) of these films upon thermal processing in view of the required integration constraints for CMOS applications [28.3]. A significant hurdle for the integration of all dielectrics currently under consideration for conventional CMOS is the stability of all gate-stack constituents (stable film morphology, minimal interdiffusion, etc.) under a thermal budget of $\approx 1000\,°C$ for 10 s for dopant activation and adequate process integration margin. Under such thermal treatments, gate dielectric constituents such as Zr [28.164] or Al [28.165] have been reported to penetrate the Si chan-

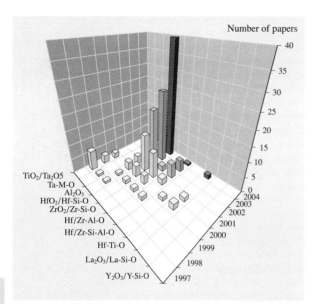

Fig. 28.23 Distribution of papers presented at recent IEEE international electron devices meeting (IEDM) and very-large-scale integration (VLSI) conferences indicate the current emphasis of materials systems examined for gate dielectric research

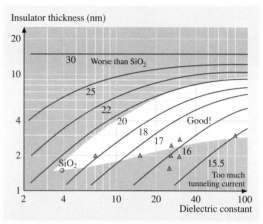

Fig. 28.24 Contours of constant scale length, which is proportional to the transistor gate length, versus dielectric constant and insulator thickness, showing the useful design space for high-κ gate dielectrics. Data points are rough estimates of the tunneling constraints for various high-κ insulators. The depletion depth is assumed to be 15 nm. The useful design space will shrink with decreasing depletion depth. After [28.23] (© 2004 IEEE, with permission)

nel region and thus present a potential source of impurity scattering for mobility degradation. In contrast, Hf does not appear to exhibit such penetration within detectible limits [28.166].

It is also important to note that most of the high-κ metal oxides listed in Table 28.4 crystallize at relatively low temperatures ($T \approx 500-600\,°C$). An exception is Al_2O_3 where crystallization in thin films is observed at $T \approx 800\,°C$ [28.167]. In the case of Hf–silicates, film morphology (viz. suppression of crystallization) and diffusion-barrier properties can be controlled through the incorporation of N, as in the case of SiO_2 [28.157, 168–172]. Currently, HfSiON dielectrics appear to provide desirable properties in this regard.

Suppression of crystallization is also observed for aluminates (depending upon the Al content) [28.173], but there are fewer studies of interdiffusion with the Si substrate available at this time. As noted above however for the case of Al_2O_3 thin films, Al penetration into the Si substrate has been reported upon thermal annealing budgets appropriate for dopant activation [28.165]. Such penetration could be a concern for any high-κ dielectric incorporating Al as well.

The morphology of the dielectric film may also impact on the propensity for capping layer constituents (from metal gate material layers) to diffuse through the stack as well. Metal gate materials currently under investigation include well-known minority-lifetime killers such as Ni (in NiSi) as well as tunable-work-function alloys and layers [28.3, 147, 174, 175]. The potential for the interdiffusion of metals into the channel region is clearly undesirable.

The limits of a useful high-κ dielectric constant value have also been examined. For example, the study by *Frank* et al. (Fig. 28.24) suggests that there exist a range of useful permittivity values as planar CMOS transistor structures are scaled [28.23]. It is noted that the physical thickness of the gate insulator should be less than the Si depletion length under the channel region, and that this results in a limited design space that enables further device scaling. Again, one notes that the dielectric constant cannot be arbitrarily increased without careful consideration of the complete transistor design.

28.2 Isolation Dielectrics

Dielectrics are also employed to electrically isolate various regions of an integrated-circuit technology. The dominant material for these applications is SiO_2. A method commonly employed to obtain such isolation in larger-scale IC technologies (generally with gate lengths larger than 0.25 μm) is the so-called local oxidation of silicon (LOCOS) technique, where regions of Si between various components are preferentially (thermally) oxidized [28.18, 28]. Such isolation oxides are typically several hundred nanometers thick and the tapered shape of the edge of the LOCOS isolation oxide near the transistor gate dielectric and source/drain region (often referred to as the *bird's beak*) is an important region to control during the device fabrication at these gate lengths. Thinning of the dielectrics in this region results in breakdown reliability concerns, as the electric fields in this region can be quite high. A *reoxidation process* is often performed to improve the thickness and reliability of the SiO_2 in this region as well.

Scaling ICs beyond this gate-length regime however has required the placement of isolated regions utilizing deposited SiO_2 trench structures, as seen in Fig. 28.6. As summarized by *Wolf*, extensive work has been done to control the shape of the trench walls and the SiO_2 layer initially formed on these walls. Chemical vapor deposition methods are normally utilized for the dielectric deposition as this approach provides superior conformality in the trench structure. Filling the trench without void formation, controlling film stress, and chemical mechanical polishing properties are also important aspects that must be addressed in the fabrication process [28.12, 28]. Scaling of integrate circuits will result in a decrease of the area available for these isolation trenches, and so the increasing aspect ratio of the trench depth to width requires considerable attention in regard to trench filling. The control of the shape of the top corner regions of the trench structures is also an area of concern due to high-field reliability as well as the formation of essentially a parasitic *edge transistor*. Initially, thermal oxidation methods were employed to round off the shape of the corner, but further scaling will require etch processing methods.

The fabrication of transistor source/drain regions utilizing implantation processes also utilizes a dielectric spacer layer that surrounds the transistor gate region (Fig. 28.6). This spacer provides isolation between the gate and source/drain regions, and also permits the control of the depth (and therefore profile) of the implanted dopant species – a so-called self-aligned dopant implantation process. Dielectrics used for spacer technologies include deposited SiO_2 and Si_3N_4, and the extent of the spacer dimensions from the gate is an important device parameter to control [28.18]. Further scaling of transistor structures will likely result in the need for elevated source/drain regions, and so process compatibility of the spacer material with the source/drain formation processes and high-κ dielectrics will become an important consideration [28.12].

28.3 Capacitor Dielectrics

Capacitors are employed in a variety of integrated circuits including storage (memory) circuit elements and input/output coupling circuitry. Clearly, dielectrics are critically important in this application. In contrast to the MIS structure associated with transistors, capacitors are passive devices incorporating a metal–insulator–metal (MIM) structure. Early electrodes were composed of degenerately doped Si while more recent work focuses on integrating metals for modern devices.

28.3.1 Types of IC Memory

A dominant memory technology for the IC industry includes dynamic random-access memory (DRAM) in which capacitors play the essential role of storing charge, and thereby useful information. This type of memory element requires refreshing in order to maintain useful information, and the rate of refresh is fundamentally related to the dielectric associated with the capacitor structure and leakage of charge from that capacitor. As a result, this class of memory devices is called volatile. In contrast, a static random-access memory (SRAM) nonvolatile memory element required a capacitor which stores charge without the refresh requirement, and can preserve charge for many years. Again capacitor design must include the consideration of the dielectric [28.27].

In addition to DRAM capacitors, other types of capacitors are used in the back end of the transistor flow. These devices are MIM capacitors, and they typically reside between levels of metal interconnects, and serve as decoupling or radio-frequency (RF) capacitors for input/output functions. For these applications, the capacitance density is still of critical importance, but other factors must be considered, such as the linearity of the capacitance as a function of voltage and temperature.

28.3.2 Capacitor Dielectric Requirements in View of Scaling

In brief, scaling the capacitor dimensions requires a tradeoff between the amount of stored charge required for a reliable memory element and the area occupied by the capacitor (and the associated transistors for the memory cell). From generation to generation of devices, the amount of stored charge is kept roughly constant (≈ 30 fF/cell), even though dimensional scaling occurs, for bit detection (sense amplifier), retention and reliability reasons [28.28, 176].

As noted in (28.6) for a simple parallel-plate capacitor, the area associated with the capacitor as well as the dielectric permittivity play an important role. Maintaining the required capacitance with scaling was (and will be) accomplished by increasing the capacitor area, for a suitable dielectric thickness where leakage currents are severely limited; this can be accomplished by utilizing three-dimensional capacitor structures produced in the Si bulk, such as deep-trench capacitors [28.176, 177]. Clever and intricate processing methods to increase the capacitor area were adopted for several generations for well-established SiO_2 and Si–oxynitride dielectric capacitor materials, such as the implementation of hemispherical-grain (HSG) poly-Si and *crown* structures (Fig. 28.25). Eventually, practical concerns about cost and device yield make manufacturing such structures problematic. Thus, substantially increasing the dielectric constant of the insulating material was considered, and less-complicated structures can be fabricated. At this point, both stacked and trench capacitor structures are dominant in the industry. Indeed, the ability to fabricate deep-trench structures will help guide the need for the incorporation of alternative dielectrics. These structures often require the use of chemical vapor deposition (CVD) methods to ensure conformality over the stacked structure topography or deep within trenches that have large aspect ratios. More recently, atomic-layer CVD processes have been examined for this purpose as well. Although capacitors do not have requirements regarding lateral transport of carriers, as transistors do, capacitors have much more rigorous charge-storage-capacity requirements when they serve as memory elements. This additional constraint results in projected limitations.

28.3.3 Dielectrics for Volatile Memory Capacitors

Capacitor dielectric properties for dynamic random-access memory (DRAM) applications have been important since these devices were introduced commercially in the early 1970s for volatile memory [28.178]. *Fazan* provides some rules of thumb that provide guidance for the selection of desirable materials (and electrical) properties for DRAM capacitor dielectrics [28.176]. In addition to the desire to maintain the storage capacitance around 30 fF/cell, leakage of charge from the storage capacitor must be kept to a cur-

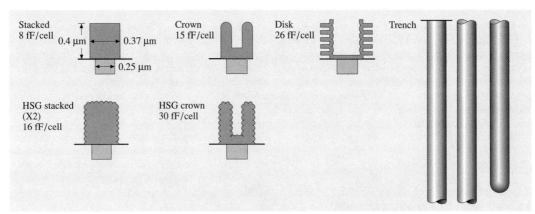

Fig. 28.25 Assorted capacitor structures employed in IC technology. Courtesy of S. Summerfelt

rent density of less than $0.1\,\mu\text{A}/\text{cm}^2$ so that less than $\approx 10\%$ of the capacitor charge is lost during the associated refresh (recharge) cycle. Scaling the DRAM cell size has resulted in the consideration of a variety of dielectric materials [28.12, 27, 28].

SiO_2 and SiO_xN_y

The earliest DRAM planar capacitors utilized SiO_2 as the dielectric. Subsequent scaling into the Mbit regime required an increased capacitance density, and thus a higher dielectric constant. So-called oxide/nitride (ON) or oxide/nitride/oxide (ONO) dielectric stacks were incorporated [28.18]. As the names imply, the dielectric consisted of a stack of SiO_2 and Si_3N_4 layers. The incorporation of the Si nitride layer, typically by a deposition method such as chemical vapor deposition, results in an overall increase in the dielectric constant of the stack. Scaling the stacked capacitors has required the development of alternative dielectrics (Al_2O_3 and Ta_2O_5) at the 130-nm node. Nevertheless, according to the 2003 ITRS, Si_3N_4 will be utilized for DRAM capacitor structures to the 45-nm node.

Al_2O_3

Aluminium oxide exhibits a dielectric constant of $\kappa \approx 9$ and a significant band gap of 8.7 eV, as noted in Table 28.4. However, the rate of capacitor scaling and the charge storage per cell appears to require a dielectric constant significantly larger than this value. As a result, substantial research and development was also conducted to incorporate tantalum pentoxide into DRAM capacitor dielectrics. According to the 2003 ITRS, Al_2O_3 will be utilized for DRAM capacitor structures to the 45-nm node.

Ta_2O_5

Amorphous tantalum pentoxide provides a dielectric constant of $\kappa \approx 25$ with a concomitantly smaller band gap ($\approx 4.4\,\text{eV}$). Ta_2O_5 films are often deposited by CVD processes for conformality in stacked or trench capacitor structures. Metal–insulator–metal structures are required to preserve the maximum capacitance density, as reactions with polysilicon results in the formation of a thin SiO_2 layer as discussed above in connection with gate dielectric materials. Moreover, the use of MIM structures permits the possibility of a Ta_2O_5 crystalline microstructure, which enables a $\kappa \approx 50$. *Chaneliere* et al. has summarized the research and development associated with Ta_2O_5 films [28.179]. According to the 2003 ITRS, Ta_2O_5 will also be utilized for DRAM capacitor structures to the 45-nm node.

Barium Strontium Titanate (BST)

Considerable effort has also been exerted in the search for CMOS-compatible "ultra high-κ" dielectrics. Such materials are envisioned to be required for scaling at and beyond the 45-nm node [28.12]. CVD BST films have been a focus of these efforts in the recent past, with a dielectric constant of $k \approx 250$. The movement of the Ti atom (ion) in the BST lattice structure results in a substantial contribution to the polarization of this materials system. Utilization of this material has required the use of noble-metal electrodes including Pt, Ru (RuO_2) and Ir to control interfacial reactions. As a result, considerable process complexity is introduced into the manufacturing process.

Alternative Dielectric Materials

We also note that recent research on alternative gate dielectric materials, such as HfO_2 and HfSiO, has also rekindled interest in these materials for capacitor applications. The prospect of better interfacial oxide formation control has been one motivating factor, although substantial further work is still required.

28.3.4 Dielectrics for Nonvolatile Memory

As the name implies, nonvolatile memory devices retain their state whether power is applied to the device or not. A thorough description of such devices is provided by *Hori* [28.18]. For example, the electrically erasable programmable read-only memory (EEPROM) device requires an erase operation prior to programming (*writing*) new data to the device. To accomplish this, a stacked-gate MOS structure is utilized where the intermediate gate is embedded in a dielectric – a so-called *floating gate*. The placement of a higher electric field to permit Fowler–Nordheim tunneling through such dielectrics to the floating gate is utilized to program arrays of these elements – called *flash memory*. Scaling and the reliability required for such nonvolatile memory devices has required the evolution from using SiO_2 to SiO_xN_y.

SiO_2 and SiO_xN_y

For flash memory elements, the formation of nitrided SiO_2 enables a suitable dielectric for reliable operation. The nitridation process, as described earlier, entails the exposure of an SiO_2 dielectric to anneals with N_2, NH_3, or N_2O (or combinations of these), often under rapid thermal annealing conditions. Details of this process have been summarized by *Hori* [28.18]. Subsequent reoxidation of the dielectric is also employed to improve reliability properties.

So-called silicon oxide nitride oxide silicon (SONOS) structures are under examination to improve the scaling limitations of the dielectric thickness and tunneling properties [28.180].

Further scaling of transistors presents significant challenges for flash memory. The thickness of the dielectric is again a key constraint given the need for reliable operation under high-electric-field conditions. Memories based on Si nanocrystals embedded within SiO_2 are currently under development to address this challenge [28.182].

Ferroelectrics

Ferroelectric materials, such as $PbZr_{1-x}Ti_xO_3$ (PZT) or $SrBi_2Ta_2O_9$ (SBT) have been examined for some time as potential dielectric candidates for memory elements based upon the ferroelectric effect [28.26]. In storage capacitors incorporating such materials, the polarization state of the ferroelectric is preserved (once poled) without the presence of an electric field [28.25]. This state is then sensed by associated circuitry and can therefore be used as a memory device.

Principles. Ferroelectricity describes the spontaneous alignment of dipoles in a dielectric as the result of an externally applied electric field. This alignment behavior has thermal constraints in that heating the dielectric above the Curie temperature results in a phase transformation to a paraelectric state. Therefore, capacitors and circuits which utilize the ferroelectric effect must contain materials that have a relatively high Curie temperature compared to that experienced during operation and reliability testing (generally $> 200\,°C$). As shown in Fig. 28.26, the polarization response of these materials to an externally applied electric field exhibits a hysteresis behavior [28.181]. (This behavior is analogous to that observed in ferromagnetic materials, hence the name. Note however that ferroelectric materials contain no iron.) Upon increasing the voltage across the ferroelectric above a coercive value (V_c), the polarization of the material is limited to the spontaneous value (P_s), corresponding to maximum domain alignment. Removal of the electric field results in a decrease of the polarization to the remnant value (P_r), which can be sensed accordingly. The time required for the polarization to be switched is on the order of nanoseconds for these devices. Moreover, this can be accomplished in low-power circuits, making these materials attractive, in principle, for scaled CMOS.

Materials. In addition to the perovskite PZT and SBT materials mentioned above, other materials investigated for nonvolatile memory applications include $BaTiO_3$, $PbTiO_3$, $Pb_{1-x}La_xZr_{1-y}Ti_yO_3$ (PLZT), $PbMg_{1-x}Nb_xO_3$ (PMN), $SrBi_2Nb_2O_9$ (SBN), and $SrBi_2(Ta_{1-x}Nb_x)_2O_9$ (SBTN). These materials generally have the cubic structure ABO_3 shown in Fig. 28.27, with the larger (A) cations in the corner of the cubic unit cell, and the smaller (B) cations (e.g. Ti, Zr, Mg, Nb, Ta) located in the center of the unit cell. The oxygen anions are located at the face-centered-cube positions. The distortion of the unit cell in response to the externally applied electric field results in the observed polarization. In particular, the movement of the B cation (e.g. Ti) in the associated lattice between the equilibrium

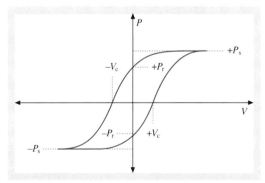

Fig. 28.26 Polarization–voltage curve for a ferroelectric material. After [28.181] (© 2004 IEEE, with permission)

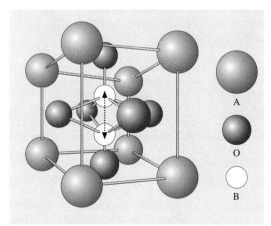

Fig. 28.27 Perovskite ABO unit cell showing motion of the B ion in the cell among the equilibrium positions, resulting in ferroelectric polarization behavior. After [28.181] (© 2004 IEEE, with permission)

positions results in substantial distortion and therefore polarization [28.26, 181].

Issues for Ferroelectric Materials. Substantial research is underway to understand the reliability issues associated with ferroelectric memory devices. The retention of the polarization state is one area of investigation. It is observed that the polarization state decreases slowly over time (*log-time* decay behavior), even in the absence of an electric field, and the reasons are still poorly understood. Another area of concern is *imprint*, where a polarization state, if repeatedly poled to the same state, becomes preferred. Subsequent switching to the opposite state can result in relatively poor retention times. Again, the physical mechanism associated with this phenomenon is poorly understood. The role of hydrogen exposure is another area of investigation, where exposure of ferroelectric random-access memory (FeRAM) elements to hydrogen, commonly from forming gas during CMOS back-end processing, can result in the suppression of the remnant polarization [28.183–185]. Barriers for hydrogen permeation into the ferroelectric are an area of investigation as well.

Phase-Change Memory

An alternative to charge-storage devices, based upon a controlled phase change, is also now under consideration for scaled integrated circuits [28.186]. Chalcogenides, such as GeSbTe, have been utilized in compact-disk memory storage technology. In that technology, a laser heats a small volume of the material, resulting in a phase change between crystalline and amorphous states, which obviously changes the reflectivity of the exposed region. For the IC application, an electric current is passed through these materials to accomplish the phase-change effect, dramatically altering the resistance of the region. This utilization of alternative (non-dielectric) material is another example of new directions under consideration for IC scaling.

28.4 Interconnect Dielectrics

As noted in Fig. 28.1, the performance of the integrated circuit, as measured by the time delay for signal propagation, also depends upon the interconnections between circuit elements. The scaling of CMOS has resulted in a substantial increase in interconnect metal lines throughout the IC chip, which make a major contribution to the delay time. The industry segments these levels into *local* (interconnection between neighboring devices), *intermediate* (metal 1 interconnection between neighboring circuits), and *global* (interconnection across the chip), as shown in Fig. 28.28. A cross section of a contemporary 65-nm-node IC is shown in Fig. 28.29 where eight layers of metallization and the associated dielectric isolation are evident.

As discussed in Sect. 28.0.1, the *RC* time delay for the interconnect contribution to performance can be attributed to the metal lines and their isolation dielectrics (see Fig. 28.1). The resistivity of the lines has been reduced in the industry by recently adopting copper metallization processes in lieu of aluminium metallization in 1998. Further reductions in the delay time then require the consideration of the dielectric between the lines, as these essentially form a parasitic capacitor structure, and therefore low-κ dielectrics are required. It is noted that for the global interconnect level, new concepts such as RF or optical communication, will likely be needed for continued CMOS scaling.

28.4.1 Tetraethoxysilane (TEOS)

For many years, CVD-deposited SiO_2 provided adequate isolation for interconnection of ICs. This was frequently accomplished through the deposition of tetraethoxysilane (TEOS) and subsequent densification thermal treatments to render a dielectric constant of $\kappa \approx 4$. Films produced in this manner were relatively easy to process and provided good mechanical strength. The incorporation of fluorine into these films [fluorinated silicate glass or (FSG)] succeeded in a reduction of the dielectric constant to $\kappa \approx 3.5$–3.7 after considerable efforts. As F is among the most electronegative elements, the incorporation of F into the silica matrix renders the film less polarizable due to Si−F bond formation, and therefore results in a lower permittivity. Scaling CMOS however has driven the industry to consider interconnect dielectric materials with $\kappa \ll 4$.

28.4.2 Low-κ Dielectrics

As seen in Figures 28.28 and 28.29, several levels of dielectrics must be incorporated with the metallization schemes associated with ICs. As can be seen, dielectrics utilized in this back end of (fabrication) line (BEOL) portion of the IC fabrication process are also segmented into pre-, inter- and intra-metal dielectrics. Addition-

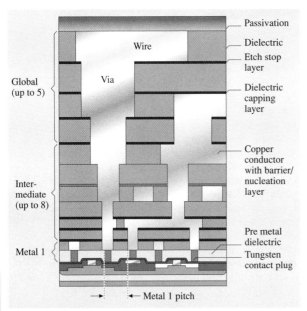

Fig. 28.28 Schematic of a typical chip cross section showing key interconnect metallization and dielectric layers. From [28.12]

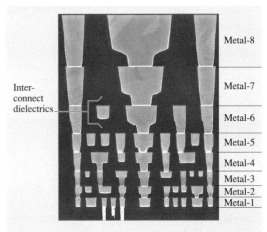

Fig. 28.29 Cross section of a modern IC chip for the 65-nm node to be in production in 2005 showing eight levels of metal interconnects isolated by low-κ dielectrics. (Courtesy of Intel)

ally, dielectric layers are employed to facilitate etching control for Cu metal patterning of the various interconnection lines. This requires materials integration with barrier layers (that control Cu diffusion), etch stop layers, CMP (chemical mechanical polishing) stop layers as well as overall mechanical stability. Additionally, for the local and metal 1 layers, the future incorporation of metal gate and/or NiSi materials may require extensive limitations on processing temperatures ($< 500\,°C$).

Several materials are under investigation for the low-κ application, and some of these are summarized in Table 28.5 [28.12, 28]. Most consist of polymeric (low-Z) materials with varying porosity to enable a sufficiently low dielectric constant for interconnect applications. Recent technology announcements for the 90-nm and 65-nm nodes often refer to these materials as carbon-doped oxides [28.187, 188]. At this point, films are deposited by either CVD or spin-on methods. Desirable properties for low-κ dielectrics include thermal stability (to $\approx 500\,°C$), mechanical stability (to withstand packaging), electrical isolation stability/reliability (similar to SiO_2), chemical stability (minimal moisture absorption, resistance to process chemicals, resist compatibility), compatibility with BEOL materials (e.g., diffusion barriers) and processes (etching, cleaning, post-metallization and forming-gas anneals, chemical mechanical polishing, etc.) and of course low cost. As noted by Wolf, thermal properties of low-κ materials are important

Table 28.5 Interconnect dielectric materials (after [28.12, 28])

Dielectric constant	Material	Pre-production year
3.5–3.7	FSG	Current
3.0–3.6	Polyimides	Current
2.7–3.1	Spin-on glass	Current
2.6–2.9	Organo-silicate glasses (OSG)	Current – 2011
2.6–2.8	Parylene-based polymers	Current
2.5–3.2	Methyl/hydrogen silsesquioxane (MSQ/HSQ)	2005
1.8–2.4	Porous MSQ, parylene-based polymers	2008
1.1–2.2	Silica aerogels	2011
1.5–2.2	Silica xerogels	2011

for effective power dissipation in high-performance applications.

According to the ITRS roadmap, deposited silicon oxides are envisioned to be employed for pre-metal dielectrics through the 45-nm node ($\approx$ 2012). The introduction of Ni–silicide contacts, metal gate electrodes and high-κ dielectrics will certainly have an impact on the development of alternatives to these deposited oxides. Methyl/hydrogen silsesquioxane (MSQ/HSQ) appear to be under consideration for further development in this regard.

For inter/intra-metal dielectrics, a wide array of materials are envisioned, as seen in Table 28.5. Both FSG and organo-silicate glasses (OSG) are envisioned to address the requirements of these interconnect dielectrics to the 45-nm node. Thereafter, alternative materials ($\kappa <$ 2.8) mentioned in Table 28.5 will need to be developed.

Patterning (hard mask) and etch-stop dielectrics will continue to employ Si-oxides, Si-nitrides, Si-oxycarbides, and Si-carbonitrides to the 45-nm node. Thereafter, alternative materials for patterning may be required.

28.5 Summary

It should be evident that IC technology is critically dependent upon suitable dielectrics throughout the entire chip. Materials properties, and their resultant electrical properties, must be carefully evaluated throughout the research and development process associated with integrated circuit technology.

Recent years and roadmap predictions clearly place an emphasis on the development of new materials for the various dielectrics employed to achieve scaling. Researchers and technologists engaged in this endeavor must be able to span several disciplines to enable the successful integration of these new dielectric materials.

References

28.1   G. Wilk, R. M. Wallace, J. M. Anthony: J. Appl. Phys. **89**(10), 5243 (2001)
28.2   R. M. Wallace, G. Wilk: Critical Rev. Solid State Mater. Sci. **28**, 231 (2003)
28.3   R. M. Wallace: Appl. Surf. Sci. **231-232**, 543 (2004)
28.4   H.-S. P. Wong: *ULSI Devices*, ed. by C. Y. Chang, S. M. Sze (Wiley, New York 2000) Chap. 3
28.5   S. Wolf: *Silicon Processing for the VLSI Era*, Vol. 3 (Lattice, Sunset Beach 1995)
28.6   E. Nicollian, J. Brews: *MOS Physics and Technology* (Wiley, New York 1982)
28.7   S. M. Sze: *Physics of Semiconductor Devices*, 2nd edn. (Wiley, New York 1981)
28.8   G. Moore: Electronics **38**, 8 (1965). Also see: http://www.intel.com/labs/index.htm
28.9   G. Moore: *Tech. Dig. Int. Electron. Dev. Meet* (IEEE, Washington, D.C. 1975) p. 11
28.10 P. E. Ross: IEEE Spectrum **40**(12), 30 (2003)
28.11 Intel: *Expanding Moore's Law*; see: http://www.intel.com/labs/eml.htm (2002)
28.12 ITRS, see: http://public.itrs.net/ (2003)
28.13 H. Iwai: Microelec. Eng. **48**, 7 (1999)
28.14 H. Iwai, H. S. Monose, S.-I. Ohmi: *The Physics and Chemistry of SiO$_2$ and the Si–SiO$_2$ Interface*, Vol. 4, ed. by H. Z. Massoud, I. J. R. Baumvol, M. Hirose, E. H. Poindexter (The Electrochemical Society, Pennington 2000) p. 1
28.15 G. Moore: *No Exponential is Forever … But We Can Delay "Forever"*, keynote address at Int. Solids State Circuits Conference. See: http://www.intel.com/labs/eml/doc.htm (2003)
28.16 S. C. Sun: IEEE Tech. Dig. Int. Electron. Dev. Meet. Washington, DC, 765 (1997)
28.17 M. T. Bohr: IEEE Tech. Dig. Int. Electron. Dev. Meet. Washington, DC, 241 (1995)
28.18 T. Hori: *Gate Dielectrics and MOS ULSI's*, Series in Electronics and Photonics, Vol. 34 (Springer, Berlin 1997)
28.19 A. Chatterjee, M. Rodder, I-C. Chen: IEEE Trans. Electron. Dev. **45**, 1246 (1998)
28.20 I-C. Chen, W. Liu: *ULSI Devices*, ed. by C. Y. Chang, S. M. Sze (Wiley, New York 2000) Chap. 10
28.21 R. Dennard, F. Gaensslen, H-N. Yu, V. L. Rideout, E. Bassous, A. R. LeBlanc: J. Solid State Circuits **SC-9**, 256 (1974)
28.22 R. Dennard, F. Gaensslen, E. Walker, P. Cook: J. Solid State Circuits **SC-14**, 247 (1979)
28.23 D. Frank, R. H. Dennard, E. Nowak, P. M. Solomon, Y. Taur, H.-S. P. Wong: Proc. IEEE **89**, 259 (2001)
28.24 S. Thompson, P. Packan, M. Bohr: Intel Technol. J. Q **3**, 223–225 (1998)
28.25 S. O. Kasap: *Principles of Electrical Engineering Materials and Devices* (McGraw–Hill, New York 2002)
28.26 R. Ramesh: *Thin Film Ferroelectric Materials and Devices* (Kluwer, Boston 1997)

28.27 D.-S. Yoon, J. S. Roh, H. K. Baik, S.-M. Lee: Crit. Rev. Solid State Mater. Sci. **27**, 143 (2002)

28.28 S. Wolf: *Silicon Processing for the VLSI Era*, Vol. 4 (Lattice, Sunset Beach 2002)

28.29 S. Wolf: *Silicon Processing for the VLSI Era*, Vol. 1 (Lattice, Sunset Beach 1986)

28.30 S. Wolf: *Silicon Processing for the VLSI Era*, Vol. 2 (Lattice, Sunset Beach 1990)

28.31 B. Bivari: IEEE Tech. Dig. Int. Electron. Dev. Meet., 555 (1996)

28.32 S. Banerjee, B. Streetman: *ULSI Devices*, ed. by C. Y. Chang, S. M. Sze (Wiley, New York 2000) Chap. 4

28.33 R. Rios, N. D. Arora: IEEE Tech. Dig. Int. Electron. Dev. Meet. San Francisco, 613 (1994)

28.34 A. C. Diebold, D. Venables, Y. Chabal, D. Muller, M. Weldon, E. Garfunkel: Mater. Sci. Semicond. Proc. **2**, 104 (1999)

28.35 Z. H. Lu, J. P. McCaffrey, B. Brar, G. D. Wilk, R. M. Wallace, L. C. Feldman, S. P. Tay: Appl. Phys. Lett. **71**, 2764 (1997)

28.36 Y.-C. King, C. Hu, H. Fujioka, S. Kamohara: Appl. Phys. Lett. **72**, 3476 (1998)

28.37 K. Yang, Y.-C. King, C. Hu: Symp. VLSI Tech. Tech. Dig. Papers, Kyoto, Japan, 77 (1999)

28.38 C. Y. Wong, J. Y. Sun, Y. Taur, C. S. Oh, R. Angelucci, B. Davari: IEEE Tech. Dig. Int. Electron. Dev. Meet. San Francisco, 238 (1988)

28.39 E. M. Vogel, K. Z. Ahmed, B. Hornung, W. K. Henson, P. K. McLarty, G. Lucovsky, J. R. Hauser, J. J. Wortman: IEEE Trans. Electron. Dev. **45**, 1350 (1998)

28.40 M. L. Green, E. P. Gusev, R. Degraeve, E. L. Garfunkel: J. Appl. Phys. **90**, 2057 (2001)

28.41 D. K. Schroder: *Semiconductor Material and Device Characterization*, 2nd edn. (Wiley, New York 1998)

28.42 R. Hummel: *Electronic Properties of Materials*, 2nd edn. (Springer, New York 1993)

28.43 W. R. Runyan, K. E. Bean: *Semiconductor Integrated Circuit Processing Technology* (Addison-Wesley, New York 1990)

28.44 S. T. Pantilides: *The Physics of SiO and its Interfaces* (Pergamon, New York 1978)

28.45 G. Lucovsky, S. T. Pantilides, F. L. Galeener: *The Physics of MOS Insulators* (Pergamon, New York 1980)

28.46 C. R. Helms, B. E. Deal: *The Physics and Chemistry of SiO and the Si−SiO Interface* (Plenum, New York 1988)

28.47 P. Balk (ed): *The Si−SiO$_2$ System*, Mater. Sci. Monogr. (Elsevier, New York 1988) p. 32

28.48 C. R. Helms, B. E. Deal: *The Physics and Chemistry of SiO and the Si−SiO Interface*, 2 (Plenum, New York 1993)

28.49 H. Z. Massoud, E. H. Poindexter, C. R. Helms: *The Physics of SiO and its Interfaces − 3*, Vol. 96-1 (Electrochemical Society, Pennington 1996)

28.50 R. A. B. Devine: *The Physics and Technology of Amorphous SiO* (Plenum, New York 1988) p. 2

28.51 E. Garfunkel, E. Gusev, A. Vul': *Fundamental Aspects of Ultrathin Dielectrics on Si-based Devices*, NATO Science Series, Vol. 3/47 (Kluwer, Nowell 1998)

28.52 R. A. B. Devine, J.-P. Duraud, E. Dooryhee: *Structure and Imperfections in Amorphous and Crystalline Silicon Dioxide* (Wiley, New York 2000)

28.53 H. Z. Massoud, I. J. R. Baumvol, M. Hirose, E. H. Poindexter: *The Physics of SiO and its Interfaces − 4*, Vol. PV2000-2 (Electrochemical Society, Pennington 2000)

28.54 C. R. Helms, E. H. Poindexter: Rep. Prog. Phys. **57**, 791 (1994)

28.55 G. Dolino: *Structure and Imperfections in Amorphous and Crystalline Silicon Dioxide*, ed. by R. A. B. Devine, J.-P. Duraud, E. Dooryhee (Wiley, New York 2000) Chap. 2

28.56 L. W. Hobbs, C. E. Jesurum, B. Berger: *Structure and Imperfections in Amorphous and Crystalline Silicon Dioxide*, ed. by R. A. B. Devine, J.-P. Duraud, E. Dooryhee (Wiley, New York 2000) Chap. 1

28.57 A. Bongiorno, A. Pasquarello: Appl. Phys. Lett. **83**, 1417 (2003)

28.58 F. Mauri, A. Pasquarello, B. G. Pfrommer, Y.-G. Yoon, S. G. Louie: Phys. Rev. B **62**, R4786 (2000)

28.59 W. H. Zachariasen: J. Am. Chem. Soc. **54**, 3841 (1932)

28.60 P. Balk: J. Nanocryst. Sol. **187**, 1−9 (1995)

28.61 F. J. Grunthaner, P. J. Grunthaner: Mater. Sci. Rep. **1**, 65 (1986)

28.62 A. Bongiorno, A. Pasquarello: Mater. Sci. Eng. B **96**, 102 (2002)

28.63 A. Stirling, A. Pasquerello: Phys. Rev. B **66**, 24521 (2002)

28.64 G. Hollinger, F. R. Himpsel: Phys. Rev. B **28**, 3651 (1983)

28.65 G. Hollinger, F. R. Himpsel: Appl. Phys. Lett. **44**, 93 (1984)

28.66 F. J. Himpsel, F. R. McFeely, A. Taleb-Ibrahimi, J. A. Yarmoff, G. Hollinger: Phys. Rev. B **38**, 6084 (1988)

28.67 T. Hattori: Crit. Rev. Solid State Mater. Sci. **20**, 339 (1995)

28.68 F. Rochet, Ch. Poncey, G. Dufour, H. Roulet, C. Guillot, F. Sirotti: J. Non-Cryst. Solids **216**, 148 (1997)

28.69 J. H. Oh, H. W. Yeom, Y. Hagimoto, K. Ono, M. Oshima, N. Hirashita, M. Nywa, A. Toriumi, A. Kakizaki: Phys. Rev. B **63**, 205310 (2001)

28.70 D. A. Muller, T. Sorsch, S. Moccio, F. H. Baumann, K. Evans-Lutterodt, G. Timp: Nature **399**, 758 (1999)

28.71 I. J. R. Baumvol: Surf. Sci. Rep. **36**, 1 (1999)

28.72 J. J. Lander, J. Morrison: J. Appl. Phys. **33**, 2089 (1962)

28.73 F. W. Smith, G. Ghidini: J. Electrochem. Soc. **129**, 1300 (1982)

28.74 K. Wurm, R. Kliese, Y. Hong, B. Röttger, Y. Wei, H. Neddermeyer, I. S. T. Tsong: Phys. Rev. B **50**, 1567 (1994)

28.75 J. Seiple, J. P. Pelz: Phys. Rev. Lett. **73**, 999 (1994)

28.76 J. Seiple, J. P. Pelz: J. Vac. Sci. Technol. A **13**, 772 (1995)
28.77 Y. Wei, R. M. Wallace, A. C. Seabaugh: J. Appl. Phys. **81**, 6415 (1997)
28.78 E. P. Gusev, H. C. Lu, T. Gustafsson, E. Garfunkel: Phys. Rev. B **52**, 1759 (1995)
28.79 B. E. Deal, A. S. Grove: J. Appl. Phys. **36**, 3770 (1965)
28.80 J. D. Plummer: Silicon oxidation kinetics–from Deal-Grove to VLSI process models. In: *The Physics of SiO and its Interfaces – 3*, Vol. 96-1, ed. by H. Z. Massoud, E. H. Poindexter, C. R. Helms (Electrochemical Society, Pennington 1996) p. 129
28.81 P. Balk: Trans. IEEE **53**, 2133 (1965)
28.82 G. Abowitz, E. Arnold, J. Ladell: Phys. Rev. Lett. **18**, 543 (1967)
28.83 B. E. Deal, M. Sklar, A. S. Grove, E. H. Snow: J. Electrochem. Soc. **114**, 266 (1967)
28.84 E. Arnold, J. Ladell, G. Abowitz: Appl. Phys. Lett. **13**, 413 (1968)
28.85 R. R. Razouk, B. E. Deal: J. Electrochem. Soc. **126**(9), 1573–1581 (Sept. 1979)
28.86 P. J. Caplan, E. H. Poindexter, B. E. Deal, R. R. Razouk: J. Appl. Phys. **50**, 5847 (1979)
28.87 P. J. Caplan, E. H. Poindexter, B. E. Deal, R. R. Razouk: *The Physics of MOS Insulators*, ed. by G. Lucovsky, S. T. Pantilides, F. L. Galeener (Pergamon, New York 1980) p. 306
28.88 J. H. Weil, J. R. Bolton, J. E. Wertz: *Electron Paramagnetic Resonance: Elementary Theory and Practical Applications* (Wiley, New York 1994)
28.89 P. J. Caplan, J. N. Helbert, B. E. Wagner, E. H. Poindexter: Surf. Sci. **54**, 33 (1976)
28.90 E. H. Poindexter, P. J. Caplan: Prog. Surf. Sci. **14**, 201 (1983)
28.91 E. H. Poindexter, P. J. Caplan: J. Vac. Sci. Technol. A **6**, 390 (1988)
28.92 J. F. Conley: Mater. Res. Soc. Symp. Proc. **428**, 293 (1996)
28.93 J. F. Conley, P. M. Lenahan: A review of electron spin resonance spectroscopy of defects in thin film SiO_2 on Si. In: *The Physics of SiO_2 and its Interfaces – 3*, Vol. 96-1, ed. by H. Z. Massoud, E. H. Poindexter, C. R. Helms (Electrochemical Society, Pennington 1996) p. 214
28.94 P. M. Lenahan, J. F. Conley: J. Vac Sci. Technol. B **16**, 2134 (1998)
28.95 A. G. Revesz, B. Goldstein: Surf. Sci. **14**, 361 (1969)
28.96 Y. Nishi: J. Appl. Phys. **10**, 52 (1971)
28.97 I. Shiota, N. Miyamoto, J-I. Nishizawa: J. Appl. Phys. **48**, 2556 (1977)
28.98 E. H. Poindexter, E. R. Ahlstrom, P. J. Caplan: *The Physics of SiO_2 and its Interfaces*, ed. by S. T. Pantilides (Pergamon, New York 1978) p. 227
28.99 G. J. Gerardi, E. H. Poindexter, P. J. Caplan, N. M. Johnson: Appl. Phys. Lett. **49**, 348 (1986)
28.100 D. Sands, K. M. Brunson, M. H. Tayarani-Najaran: Semicond. Sci. Technol. **7**, 1091 (1992)
28.101 K. L. Brower: Phys. Rev. B **38**, 9657 (1988)
28.102 K. L. Brower: Phys. Rev. B **42**, 3444 (1990)
28.103 A. Stesmans, V. V. Afanas'ev: J. Appl. Phys. **83**, 2449 (1998)
28.104 A. Stesmans, V. V. Afanas'ev: J. Phys. Condens. Matter **10**, L19 (1998)
28.105 A. Stesmans, V. V. Afanas'ev: Micro. Eng. **48**, 116 (1999)
28.106 A. Stesmans, B. Nouwen, V. Afanas'ev: Phys. Rev. B **58**, 15801 (1998)
28.107 E. H. Poindexter, P. J. Caplan: *Insulating Films on Semiconductors*, ed. by M. Schulz, G. Pensl (Springer, Berlin, Heidelberg 1981) p. 150
28.108 M. Shulz, G. Pensl: *Insulating Films on Semiconductors* (Springer, New York 1981)
28.109 W. H. Brattain, J. Bardeen: Bell Syst. Tech. J. **13**, 1 (1953)
28.110 W. L. Brown: Phys. Rev. **91**, 518–527 (1953)
28.111 E. N. Clarke: Phys. Rev. **91**, 756 (1953)
28.112 H. R. Huff: J. Electrochem. Soc. **149**, S35 (2002)
28.113 J. T. Law: J. Phys. Chem. **59**, 67 (1955)
28.114 J. T. Law, E. E. Francois: J. Phys. Chem. **60**, 353 (1956)
28.115 J. T. Law: J. Phys. Chem. **61**, 1200 (1957)
28.116 J. T. Law: J. Appl. Phys. **32**, 600 (1961)
28.117 P. Balk: Microelectron. Eng. **48**, 3 (1999)
28.118 E. Kooi: Philips Res. Rep. **20**, 578 (1965)
28.119 P. Balk: Electrochem. Soc. Ext. Abstracts **14**(109), 237 (1965)
28.120 P. Balk: Electrochem. Soc. Ext. Abs. **14**(111), 29 (1965)
28.121 P. Balk: J. Electrochem. Soc. **112**, 185C (1965d)
28.122 E. Kooi: Philips Res. Rep. **21**, 477 (1966)
28.123 A. Stesmans: Appl. Phys. Lett. **68**, 2076 (1996)
28.124 A. Stesmans, V. Afanas'ev: Micro. Eng. **36**, 201 (1997)
28.125 K. L. Brower, S. M. Myers: Appl. Phys. Lett. **57**, 162 (1999)
28.126 A. Stesmans: Phys. Rev. B **48**, 2418 (1993)
28.127 A. G. Revesz: J. Electrochem. Soc. **126**, 122 (1979)
28.128 G. J. Gerardi, E. H. Poindexter: J. Electronchem. Soc. **136**, 588 (1989)
28.129 T. R. Oldham, F. B. McLean, H. E. Jr. Boesch, J. M. McGarrity: Semicond. Sci. Technol. **4**, 986 (1989)
28.130 A. Stesmans, V. Afanas'ev: Phys. Rev. B **54**, 11129 (1996)
28.131 E. Wu, B. Linder, J. Stathis, W. Lai: IEEE Tech. Dig. Int. Electron. Dev. Meet. Washington, DC, 919 (2003)
28.132 D. A. Buchanan: IBM J. Res. Devel. **43**, 245 (1999)
28.133 Y. Wu, G. Lucovsky, Y.-M. Lee: IEEE Trans. Electron. Dev. **47**, 1361 (2000)
28.134 M. Hillert, S. Jonsson, B. Sundman: Z. Metallkd. **83**, 648 (1992)
28.135 D. M. Brown, P. V. Gray, F. K. Heumann, H. R. Philipp, E. A. Taft: J. Electrochem. Soc. **115**, 311 (1968)
28.136 E. H. Poindexter, W. L. Warren: J. Electrochem. Soc. **142**, 2508 (1995)
28.137 J. T. Yount, P. M. Lenahan, P. W. Wyatt: J. Appl. Phys. **74**, 5867 (1993)

28.138 K. Kushida-Abdelghafar, K. Watanabe, T. Kikawa, Y. Kamigaki, J. Ushio: J. Appl. Phys. **92**, 2475 (2002)

28.139 G. Lucovsky, T. Yasuda, Y. Ma, S. Hattangady, V. Misra, X.-L. Xu, B. Hornung, J.J. Wortman: J. Non-Cryst. Solids **179**, 354 (1994)

28.140 S. V. Hattangady, H. Niimi, G. Lucovsky: Appl. Phys. Lett. **66**, 3495 (1995)

28.141 S. V. Hattangady, R. Kraft, D. T. Grider, M. A. Douglas, G. A. Brown, P. A. Tiner, J. W. Kuehne, P. E. Nicollian, M. F. Pas: IEEE Tech. Dig. Int. Electron. Dev. Meet. San Francisco, 495 (1996)

28.142 H. Yang, G. Lucovsky: Tech. Dig. Int. Electron. Dev. Meet. Washington, DC, 245 (1999)

28.143 J. P. Chang, M. L. Green, V. M. Donnelly, R. L. Opila, J. Eng Jr., J. Sapjeta, P. J. Silverman, B. Weir, H. C. Lu, T. Gustafsson, E. Garfunkel: J. Appl. Phys. **87**, 4449 (2000)

28.144 R. M. Wallace, G. Wilk: Mater. Res. Soc. Bull.,, 192 (March 2002) also see this focus issue for reviews of other aspects on gate dielectric issues

28.145 H. Iwai, S. Ohmi, S. Akama, C. Ohshima, A. Kikuchi, I. Kashiwagi, J. Taguchi, H. Yamamoto, J. Tonotani, Y. Kim, I. Ueda, A. Kuriyama, Y. Yoshihara: IEEE Tech. Dig. Int. Electron. Dev. Meet. San Francisco, 625 (2002)

28.146 D. G. Schlom, J. H. Haeni: Mater. Res. Soc. Bull. **27**(3), 198 (2002) and refs. therein

28.147 Y-C. Yeo: Thin Solids Films **462-3**, 34 (2004) and references therein

28.148 S. Guha, E. Cartier, N. A. Bojarczuk, J. Bruley, L. Gignac, J. Karasinski: J. Appl. Phys. **90**, 512 (2001)

28.149 S. Miyazaki: J. Vac. Sci. Technol. B **19**, 2212 (2001)

28.150 J. Robertson, C. W. Chen: Appl. Phys. Lett. **74**, 1168 (1999)

28.151 J. Robertson: J. Vac. Sci. Technol. B **18**, 1785 (2000)

28.152 J. Robertson: J. Non-Cryst. Solids **303**, 94 (2002)

28.153 H. Y. Yu, M. F. Li, B. J. Cho, C. C. Yeo, M. S. Joo, D.-L. Kwong, J. S. Pan, C. H. Ang, J. Z. Zheng, S. Ramanathan: Appl. Phys. Lett. **81**, 376 (2002)

28.154 E. Zhu, T. P. Ma, T. Tamagawa, Y. Di, J. Kim, R. Carruthers, M. Gibson, T. Furukawa: *IEEE Tech. Dig. Int. Electron. Dev. Meet.* (IEEE, Washington, D.C. 2001) p. 20.4.1.

28.155 G. D. Wilk, M. L. Green, M.-Y. Ho, B. W. Busch, T. W. Sorsch, F. P. Klemens, B. Brijs, R. B. van Dover, A. Kornblit, T. Gustafsson, E. Garfunkel, S. Hillenius, D. Monroe, P. Kalavade, J. M. Hergenrother: IEEE Tech. Dig. VLSI Symp. Honolulu, 88 (2002)

28.156 H. Nohira, W. Tsai, W. Besling, E. Young, J. Petry, T. Conard, W. Vandervorst, S. De Gendt, M. Heyns, J. Maes, M. Tuominen: J. Non-Cryst. Solids **303**, 83 (2002)

28.157 M. R. Visokay, J. J. Chambers, A. L. P. Rotondaro, A. Shanware, L. Colombo: Appl. Phys. Lett. **80**, 3183 (2002)

28.158 M. S. Akbar, S. Gopalan, H.-J. Cho, K. Onishi, R. Choi, R. Nieh, C. S. Kang, Y. H. Kim, J. Han, S. Krishnan, J. C. Lee: Appl. Phys. Lett. **82**, 1757 (2003)

28.159 K. Torii, T. Aoyama, S. Kamiyama, Y. Tamura, S. Miyazaki, H. Kitajima, T. Arikado: Tech. Dig. VLSI Symp. Honolulu, 112 (2004)

28.160 M. V. Fischetti, D. A. Nuemayer, E. A. Cartier: J. Appl. Phys. **90**, 4587 (2001)

28.161 Z. Ren, M. V. Fischetti, E. P. Gusev, E. A. Cartier, M. Chudzik: IEEE Tech. Dig. Int. Electron. Dev. Meet. Washington, DC, 793 (2003)

28.162 R. Chau, S. Datta, M. Doczy, B. Doyle, J. Kavalieros, M. Metz: IEEE Elecron. Dev. Lett. **25**, 408 (2004)

28.163 G. D. Wilk, R. M. Wallace: Appl. Phys. Lett. **74**, 2854 (1999)

28.164 M. Quevedo-Lopez, M. El-Bouanani, S. Addepalli, J. L. Duggan, B. E. Gnade, M. R. Visokay, M. J. Bevan, L. Colombo, R. M. Wallace: Appl. Phys. Lett. **79**, 2958 (2001)

28.165 S. Guha, E. P. Gusev, H. Okorn-Schmidt, M. Copel, L. Å. Ragnarsson, N. A. Bojarczuk: Appl. Phys. Lett. **81**, 2956 (2002)

28.166 M. Quevedo-Lopez, M. El-Bouanani, S. Addepalli, J. L. Duggan, B. E. Gnade, M. R. Visokay, M. Douglas, L. Colombo, R. M. Wallace: Appl. Phys. Lett. **79**, 4192 (2001)

28.167 R. M. C. de Almeida, I. J. R. Baumvol: Surf. Sci. Rep. **49**, 1 (2003)

28.168 R. M. Wallace, R. A. Stolz, G. D. Wilk: Zirconium and/or hafnium silicon-oxynitride gate, US Patent 6 013 553; 6 020 243; 6 291 866; 6 291 867 (2000)

28.169 G. D. Wilk, R. M. Wallace, J. M. Anthony: J. Appl. Phys. **87**, 484 (2000)

28.170 A. L. P. Rotondaro, M. R. Visokay, J. J. Chambers, A. Shanware, R. Khamankar, H. Bu, R. T. Laaksonen, L. Tsung, M. Douglas, R. Kuan, M. J. Bevan, T. Grider, J. McPherson, L. Colombo: Symp. VLSI Technol. Tech. Dig. Papers, Honolulu, 148 (2002)

28.171 M. Quevedo-Lopez, M. El-Bouanani, M. J. Kim, B. E. Gnade, M. R. Visokay, A. LiFatou, M. J. Bevan, L. Colombo, R. M. Wallace: Appl. Phys. Lett. **81**, 1609 (2002)

28.172 M. Quevedo-Lopez, M. El-Bouanani, M. J. Kim, B. E. Gnade, M. R. Visokay, A. LiFatou, M. J. Bevan, L. Colombo, R. M. Wallace: Appl. Phys. Lett. **82**, 4669 (2003)

28.173 M.-Y. Ho, H. Gong, G. D. Wilk, B. W. Busch, M. L. Green, W. H. Lin, A. See, S. K. Lahiri, M. E. Loomans, P. I. Räisänen, T. Gustafsson: Appl. Phys. Lett. **81**, 4218 (2002)

28.174 Y-C. Yeo, T-J. King, C. Hu: J. Appl. Phys. **92**, 7266 (2002)

28.175 I. S. Jeon, J. Lee, P. Zhao, P. Sivasubramani, T. Oh, H. J. Kim, D. Cha, J. Huang, M. J. Kim, B. E. Gnade, J. Kim, R. M. Wallace: *IEEE Tech. Dig. Int. Electron. Dev. Meet.* (IEEE, San Francisco 2004)

28.176 P. C. Fazan: Integr. Ferroelectr. **4**, 247 (1994)

28.177 H. Schichijo: *ULSI Devices*, ed. by C.Y. Chang, S.M. Sze (Wiley, New York 2000) Chap. 7

28.178 P.J. Harrop, D.S. Campbell: Thin Solid Films **2**, 273 (1968)

28.179 C. Chaneliere, J.L. Autran, R.A.B. Devine, B. Balland: Mater. Sci. Eng. R **22**, 269 (1998)

28.180 S.S. Chung, P.-Y. Chiang, G. Chou, C.-T. Huang, P. Chen, C.-H. Chu, C.C.-H. Hsu: *Tech. Dig. Int. Electron. Dev. Meet.* (IEEE, Washington, D.C. 2003) p. 26.6.1.

28.181 G.F. Derbenwick, A.F. Isaacson: IEEE Circuits Dev., 20 (2001)

28.182 B. De Salvo: *Tech. Dig. Int. Electron. Dev. Meet.* (IEEE, Washington, D.C. 2003) p. 26.1.1.

28.183 A.R. Krauss, A. Dhote, O. Auciello, J. Im, R. Ramesh, A. Aggarwal: Integr. Ferroelectr. **27**, 147 (1999)

28.184 J. Im, O. Auciello, A.R. Krauss, D.M. Gruen, R.P.H. Chang, S.H. Kim, A.I. Kingon: Appl. Phys. Lett. **74**, 1162 (1999)

28.185 N. Poonawala, V.P. Dravid, O. Auciello, J. Im, A.R. Krauss: J. Appl. Phys. **87**, 2227 (2000)

28.186 S. Lai: *Tech. Dig. Int. Electron. Dev. Meet.* (IEEE, Washington, D.C. 2003) p. 10.1.1

28.187 S. Thompson: *IEEE Tech. Dig. Int. Electron. Dev. Meet.* (IEEE, San Francisco 2002) p. 765

28.188 M. Bohr: 65 nm Press Release, August 2004, www.intel.com/research/silicon

29. Thin Films

This chapter provides an extended introduction to the basic principles of thin-film technology, including deposition processes, structure, and some optical and electrical properties relevant to this volume. The material is accessible to scientists and engineers with no previous experience in this field, and contains extensive references to both the primary literature and earlier review articles. Although it is impossible to provide full coverage of all areas or of the most recent developments in this survey, references are included to enable the reader to access the information elsewhere, while the coverage of fundamentals will allow this to be appreciated.

Deposition of thin films by the main physical deposition methods of vacuum evaporation, molecular-beam epitaxy and sputtering are described in some detail, as are those by the chemical deposition methods of electrodeposition, chemical vapour deposition and the Langmuir–Blodgett technique. Examples of structural features of some thin films are given, including their crystallography, larger-scale structure and film morphology. The dependence of these features on the deposition conditions are stressed, including those required for the growth of epitaxial films and the use of zone models in the classification of the morphological characteristics. The main optical properties of thin films are reviewed, including the use of Fresnel coefficients at

29.1	Deposition Methods	661
	29.1.1 Physical Deposition Methods	661
	29.1.2 Chemical Deposition Methods	677
29.2	Structure	682
	29.2.1 Crystallography	682
	29.2.2 Film Structure	683
	29.2.3 Morphology	688
29.3	Properties	692
	29.3.1 Optical Properties	692
	29.3.2 Electrical Properties	696
29.4	Concluding Remarks	708
References		711

media boundaries, reflectance and transmittance, matrix methods and the application of these techniques to the design of antireflection coatings, mirrors and filters. The dependence of electrical conductivity (or resistivity) and the temperature coefficient of resistivity in metallic thin films is discussed, in particular the models of Thomson, Fuchs–Sondheimer and the grain-boundary model of Mayadas–Shatzkes. For insulating and semiconducting thin films the origin and effects of several high-field conduction processes are examined, including space-charge-limited conductivity, the Poole–Frenkel effect, hopping, tunnelling and the Schottky effect. Finally, some speculations regarding future developments are made.

The earliest use of thin films by mankind is probably in the application of glazes to brickwork and pottery. Tin glazing was discovered by the Assyrians and was used to coat decorative brickwork. Glazed brick panels have been recovered from various archeological sites, with a fragment from Nimrud being dated to about 890 B.C. Various alternative types of glaze have since been developed, and are used both to overcome the effects of porosity in pottery and for decorative purposes. The importance of thin films in optics was first recognised in the 17th century by Hooke, Newton and others. In 1675

Newton described observations of colours in thin films of transparent material and from geometrical considerations was able to calculate effective film thickness and correlate this with the periodicity of observed colours. The films in question were typically of air or water in the space between two glass surfaces having different radii of curvature. The phenomena described are seen most conveniently as *Newton's rings*, where an optically flat glass surface is in contact with a convex glass surface having a large radius of curvature. Under monochromatic illumination in reflection, a series of concentric coloured rings were observed, the radii being proportional to $\sqrt{1}$, $\sqrt{2}$, $\sqrt{3}$, etc. Although the wave nature of light and interference phenomena were not universally accepted at that time (especially by Newton) these observations, and others where the radii of the rings were seen to be larger for red than for violet light, established one of the most important properties of thin films. Since these early observations of thin film properties, they have been found to display not only characteristic and distinctive optical behaviour, but also mechanical, electrical and magnetic properties. These properties underpin many of their present-day industrial applications, such as antireflection coatings and optical filters, surface acoustic wave devices, electronic components (both discrete and integrated) such as resistors, capacitors, thin-film transistors and other active devices, magnetic data storage and superconductors.

Notwithstanding this wide range of properties and applications, it should be stressed that there is no watertight definition of the thickness below which a film becomes thin. The reason for this is primarily because different properties scale differently with thickness. *Eckertová* [29.1] has pointed out that it is permissible to say in general that the physical limit is determined by the thickness below which certain described anomalies appear, but that this differs for different physical phenomena. For instance a film which is optically thin may be of the order of the wavelength of light (≈ 500 nm) although some electrical properties are determined by the mean free path of conduction electrons (≈ 50 nm for metals) or by the thickness at which a given applied voltage produces an electric field which leads to high-field (nonlinear) behaviour. Generally speaking a thin film has a thickness of less than 1 μm, and is deposited on the surface of a substrate by one of several distinct deposition methods [29.2]. In keeping with the remit of this Handbook, the emphasis in this chapter is on the electrical and optical properties of thin films, as these are clearly the most important in the field of electronic and optoelectronic materials. It is not the intention to dismiss thermal and mechanical properties as irrelevant in this sphere, but to stress that, although they may both also have an influence on the electrical and optical properties, they are both mainly dependent on the material itself. Mechanical properties of thin films have been reviewed by *Hoffman* [29.3] and also by *Campbell* [29.4]. However, the *structure* of thin films often has a crucial effect in determining other thin-film properties and, unlike the thermal and mechanical properties, is very largely determined by the film deposition method and conditions. For these reasons sections on both the structure and deposition methods of thin films are included in this review.

The basic structure of the chapter, following this brief introduction, covers the various aspects of thin films starting from their deposition methods, through their structures to their optical and electrical properties. There is no attempt to cover all of these aspects comprehensively; in a chapter such as this there is sufficient space to cover only the most important techniques and properties. In Sect. 29.1 the major deposition methods are described, such as physical processes like evaporation and sputtering, and chemical methods including chemical vapour deposition (CVD) and the Langmuir–Blodgett technique for molecular films. In Sect. 29.2 some examples of the crystalline structure of various types of films are examined. These can be relatively simple in the case of evaporated metal films, whereas compound films (such as compound semiconductors) can exhibit a variety of different structural forms depending on the deposition conditions and the thickness. Films prepared from larger molecules, such as the organic phthalocyanines having molecular weights of the order of 500, can also show several quite complex structures. In addition to the crystal structure per se, which is normally determined using X-ray, electron or neutron diffraction techniques, the morphologies of thin films often show interesting features, which are observed using electron microscopy or one of the newer scanning-probe imaging techniques such as scanning–tunnelling microscopy (STM) or atomic force microscopy (AFM). Section 29.3 covers the basic optical and electrical properties of thin films. The optical properties covered in Sect. 29.3.1 are largely the result of the electromagnetic wave nature of light, leading to the interference phenomena which were first observed in the 17th century. Films of suitable thickness and refractive index may be used in simple antireflection coatings, and quite complex multilayer film structures are used as optical filters. In Sect. 29.3.2 electrical properties are described. These include electrical conductivity (or resistivity) in relatively high-conductivity materials such as metals, where the conductivity depends not only on the carrier concentration and bulk mean free path, but is also modified by the effects of scattering at one or both of the film surfaces and internal grain boundaries. Lateral conduction in discontinuous (island) films where there are significant potential barriers between highly conducting regions has been investigated by *Neugebauer* and *Webb* [29.5], although this not discussed in the present work. This mechanism is reviewed elsewhere in the literature [29.1, 6]. Various nonlinear conduction processes arise as the result of the high electric fields that can be applied across thin films; for example a film of thickness 100 nm having a potential difference of only 1 V applied between its surfaces, would experience a field of 1×10^7 V/m, which approaches the dielectric breakdown strength of many materials. Finally in Sect. 29.4 the main points or the discussion are summarised and some speculations concerning future trends are given.

Much of the technology associated with thin-film deposition and growth is explicitly excluded from this discussion. The measurement of film thickness, both during deposition (monitoring) or post-deposition, is not covered, although good accounts are given in the literature [29.7, 8]. Furthermore the condensation and growth mechanisms of thin films are also omitted [29.9, 10], although the structure and other properties are largely influenced by these. Many of the up-to-date materials properties and device applications are covered elsewhere in this volume. The reader is referred to Chapt. 14 for details of epitaxial growth techniques and to Chapt. 17 for structural characterisation. Optical properties and characterisation are discussed in Chapt. 3, while the corresponding materials and devices are addressed in Chapt. 37. Sensors and transducers, which are frequently based on various thin-film techniques (both inorganic and organic) are described in Chapts. 39 and 54. In the area of electronics, much current effort is in the areas of molecular electronics as described in Chapt. 53. Indeed the present author has predicted that many of the high-field conduction effects commonly observed in thin films will, in due course, also be detected in nanostructures in a suitably modified form [29.11].

Many of the topics coved in this chapter are explained in more detail in various books. The *Handbook of Thin Film Technology* [29.12], although published over three decades ago, has an enormous amount of detail concerning basic thin-film phenomena, most of which is still of interest today. Another very useful resource is the series of volumes originally entitled *Physics of Thin Films* and later simply *Thin Films* (recent volumes retitled *Thin Films and Nanotechnology*) published by Academic Press. Volume 25 of this series [29.13] contains an index for all the earlier volumes covering the years 1963–1998. Detailed references to several articles from both these sources are given in the references. Also of particular interest in the materials field is the *Handbook of Thin Film Materials* [29.14]. The reader is also referred to several other books covering the general field of thin films and technology [29.1, 15–19]. Other books covering more restricted areas of thin-film technology (e.g. optical properties, electronics) are referred to in the appropriate sections.

The SI system of units is used throughout, and where original work was presented in non-SI units, these have normally been converted. Other customary units, such as the electronvolt, are used where appropriate. The main emphasis in this chapter is on the physics of the deposition processes and the optical and electrical properties, and on their relationships with film structure, while keeping mathematical details to a minimum. As mentioned earlier, only the most important methods and properties are covered, although references to techniques omitted and fundamental factors are included to aid the acquisition of a deeper understanding, where this is required, and to appreciate the applications of thin films in those areas not covered elsewhere in this volume.

29.1 Deposition Methods

The deposition methods described in this section are those most commonly used. The physical methods of vacuum evaporation and sputtering are perhaps the most controllable, giving the best-quality films. In particular, molecular-beam epitaxy (MBE) is capable of providing very high-quality films. Examples of chemical methods, which are somewhat less flexible and limited to particular types of films, are also given. Perhaps the most important of these is CVD, but as will be seen the particular route chosen depends on the chemical properties of the material to be deposited.

29.1.1 Physical Deposition Methods

These could perhaps be described as physical vapour deposition (PVD) methods, since they involve direct deposition from the vapour phase, normally under reduced pressure. There are two main classes which are considered. The first of these is vacuum evaporation, which takes place at a low background pressure, i.e. high vacuum (HV) of the order of 10^{-4} Pa, although MBE systems normally operate at pressures significantly below this level. Secondly we have sputtering, which relies on the ejection of atoms or molecules from the surface of a material target by a mechanism of momentum transfer involving an impact from (usually) ionised inert gas atoms such as argon; in order for this to occur a sufficient inert gas pressure is required to sustain a discharge, and thus the typical operating pressure is of the order of 1 Pa. Thus evaporated films are liable to have a higher intrinsic purity than sputtered films, which are prone to the incorporation of sputtering gas atoms. However, as these are generally inert there is often no deleterious effect on the film properties.

Vacuum Evaporation

During evaporation the evaporant is heated in vacuum by, for example, employing a high current through a refractory boat in which the evaporant is placed, or by using a high-current electron beam focussed on the evaporant. The purpose of this heating is to convert the condensed phase of the evaporant into the vapour phase, which then condenses elsewhere in the system in the form of a thin film. At a given absolute temperature T, there will in general be some of the evaporant in a condensed phase and some in the vapour phase. In thermodynamic equilibrium the Clausius–Clapeyron equation, relating the equilibrium vapour pressure p^* to the temperature T may be applied:

$$\frac{dp^*}{dT} = \frac{L}{T(v_g - v_c)}, \qquad (29.1)$$

where L is the latent heat of vaporisation, representing the amount of energy required to convert a quantity of the condensed phase (liquid or solid) into the vapour phase at the given temperature. If L is expressed in J/mol, then v_g and v_c represent the molar volumes of the gaseous and condensed phases respectively. Although evaporation is not an equilibrium process, the importance of the equilibrium vapour pressure to evaporation is that maximum evaporation rates are dependent on the vapour pressure. Expressions for p^* may be obtained from (29.1) provided the temperature dependence of L is known. For the simplest assumption, where L is independent of T, we assume that the vapour obeys the ideal gas law. For one mole of the vapour

$$p^* v_g = RT, \qquad (29.2)$$

where R is the universal gas constant. Further assuming that the volume of the vapour phase is very much greater than that of the condensed phase ($v_g \gg v_c$), combining the above two equations and integrating, an expression for p^* is obtained. This can be expressed in logarithmic form as

$$\log_{10} p^* = AT^{-1} + B, \qquad (29.3)$$

where $A = -(L/R)\log_{10} e$ and B is a constant of integration. The constants A and B are tabulated by *Dushman* and *Lafferty* [29.21] for various metals using non-SI units, although in a slightly different form from the definitions of (29.3). In many cases such an expression for determining equilibrium vapour pressure is inadequate. The main reason for this is that the assumption of a constant latent heat L is invalid, particularly over a wide temperature range. When temperature variations in L are taken into account, (29.3) is modified by the addition of several additional terms, which may be determined from standard thermodynamic enthalpy and entropy data [29.22]. Tabulated values of coefficients for a logarithmic $p^* - T$ expression involving four coefficients are given by *Kubaschewski* et al. [29.23] for various elements and compounds over defined temperature ranges in both the liquid and solid phases. Even more helpful are direct tabulated and graphical data of the p^*-T relationship in the case of various solid and liquid elements [29.20, 22, 24]. Examples of the p^*-T dependence are given in Fig. 29.1 for several elements using the data of *Honig* and *Kramer* [29.20]. From this figure it is clear that p^* increases very rapidly with temperature, and that most metals require temperatures of up to about 1500 K to attain significant vapour pressures; however, the refractory metals (Pt, Ta, W, etc.) require far higher temperatures for evaporation and thus they are not normally deposited by this method, although they are very useful materials for evaporation boat ma-

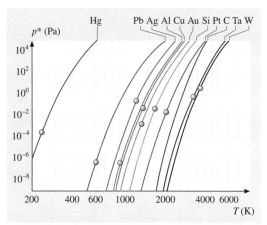

Fig. 29.1 Experimental dependence of the equilibrium vapour pressure p^* on absolute temperature T for several elements, derived from the compilation of *Honig* and *Kramer* [29.20]. The *point* (•) on each *curve* indicates the melting point, except for that of carbon, which sublimes

terials, which require very low vapour pressures and high melting points.

It is a well-known result from the kinetic theory of gases, that the molecules of a gas at a pressure p impinge on surfaces at a rate

$$N_i = \frac{p}{(2\pi m k_B T)^{1/2}} \qquad (29.4)$$

where $N_i (\mathrm{m^{-2} s^{-1}})$ is termed the impingement rate, m is the mass of a gas molecule, k is Boltzmann's constant and T is the absolute temperature. In principle this equation applies equally well to residual gas molecules (N_2 and O_2) in an evaporation chamber, as well as to evaporant molecules in the vapour phase. It was noted previously that the equilibrium vapour pressure of a material largely determines the maximum evaporation rate. Hertz [29.25] observed for the case of mercury that evaporation rates were proportional to the difference between the equilibrium vapour pressure p^* and the reverse hydrostatic pressure p_h exerted at the surface of the evaporant. The *maximum* evaporation rate occurs when $p_h = 0$, and is equal to the impingement rate a vapour of the evaporant would exert at its equilibrium vapour pressure. This is given by an expression similar to (29.4), with pressure p^*, evaporation temperature T_e and mass of evaporant molecule m_e. This rate is rarely achieved, not only due to the hydrostatic pressure p_h, but also to a molecular reflection phenomenon proposed by *Knudsen* [29.26], which takes place at the evaporant surface. It was proposed that only a certain fraction α_v (the evaporation coefficient) of molecules make the transition from the condensed to the vapour phase. The net *molecular evaporation rate*, N_e, is then given by the Hertz–Knudsen equation

$$N_e = \frac{\alpha_v (p^* - p_h)}{(2\pi m_e k_B T_e)^{1/2}}. \qquad (29.5)$$

Under equilibrium conditions $p_h = p^*$ and there is no net evaporation; however, under nonequilibrium conditions, especially when $p_h \ll p^*$, the evaporation rate can be considerable. Values of the coefficient α_v have been tabulated in the literature for a selection of different elements and compounds [29.27].

Equipped with vapour pressure and evaporation coefficient data, it is straightforward to estimate the evaporation rate using (29.5). The molecular deposition rate N_R represents the number of evaporated molecules deposited on the substrate per unit area per second. It is clearly proportional to the evaporation rate N_e, depending on the geometry of the evaporation/deposition system. N_R is simply related to the thickness deposition rate R(m/s) by

$$N_R = \frac{\rho R}{m_e}, \qquad (29.6)$$

where ρ is the density of the depositing film. An indication of the film quality is given by the impingement ratio K, representing the ratio of the rate at which ambient gas molecules impinge on the substrate to the rate of evaporated molecules depositing on the substrate. Hence

$$K = \frac{N_i}{N_R} \qquad (29.7)$$

From (29.4) and (29.6) it is clear that $K \propto p/R$ for a given material. In order to reduce contamination by gas incorporation into the film, it is therefore necessary to operate at a low background gas pressure with a relatively high deposition rate. A simple calculation for copper shows that for $R = 1$ nm/s, $N_R \approx 8.5 \times 10^{19}\,\mathrm{m^{-2}\,s^{-1}}$. If we require $K < 10^{-3}$ for impinging oxygen molecules then the maximum value of N_i allowed is $8.5 \times 10^{16}\,\mathrm{m^{-2}\,s^{-1}}$, which from (29.4) corresponds at room temperature to a pressure of about 3×10^{-6} Pa. Such a pressure is in the range of ultra-high vacuum (UHV), and is not attainable in normal vacuum deposition systems. Increasing the deposition rate is the normal method of lowering K without moving into UHV, as is used in MBE. This may in principle be achieved either by increasing the evaporation rate by increasing T_e to raise p^*, or by increasing the proportionality constant between N_e and N_R by shortening the distance between the evaporant and the substrate. A useful table of values of K for various combinations of chamber pressure and deposition rate is given by *Eckertová* [29.1].

There is in principle a second limitation on the level of the background pressure in a vacuum system which is also derived from kinetic theory. By using a Maxwell–Boltzmann distribution function for the molecular velocities in a gas, it can be shown that the mean free path between collisions in the gas λ is given by [29.22]

$$\lambda = \frac{k_B T}{p \pi \delta^2 \sqrt{2}}, \qquad (29.8)$$

where δ is the effective molecular diameter of a gas molecule. Values of δ are tabulated in the literature [29.21], and range from about 0.22 nm (He) to 0.49 nm (Xe). From (29.8) λ is inversely proportional to the pressure p, and this general relationship is shown for He, O_2 and Xe in Fig. 29.2. Since the values of δ do not vary greatly for different gases, the λ values for all common gases fall

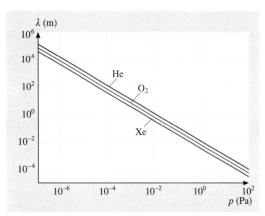

Fig. 29.2 The dependence of mean free path λ on pressure p for helium, oxygen and xenon at 300 K. Mean free paths are calculated from (29.8) using values of molecular diameter δ from *Dushman* and *Lafferty* [29.21]

within a narrow range, as illustrated in the figure. The proportion of molecules $F(x)$ which have not suffered an intermolecular collision as a function of distance x since their last collision is given by

$$F(x) = \exp(-x/\lambda) \,. \tag{29.9}$$

From this expression it is evident that for a deposition chamber size of about $\lambda/10$, less than 10% of molecules suffer a collision before hitting the walls of the chamber or the substrate, e.g. in a large vacuum chamber of dimensions about 1 m where the mean free path is about 10 m. From Fig. 29.2 a pressure of 10^{-4} Pa is sufficiently low to ensure that $\lambda > 10$ m. In evaporation processes we require that the evaporant molecules are not significantly dispersed by collisions with gas molecules en route to the substrate, and since the mean free path is relatively insensitive to the molecular diameter, λ is a suitable criterion to determine this. Thus smaller deposition chambers will not require as low a residual gas pressure as larger ones.

The geometry of the evaporation source and its positioning with respect to the substrate is an important consideration in determining the uniformity of a thin-film deposit. Expressions for the film thickness variation over the substrate for various types of source have been derived by *Holland* and *Steckelmacher* [29.28]. Traditionally the *point source* and the small *surface source* are used. It is assumed that the evaporation source, of either type, is located a distance h below the centre of a flat substrate, as shown in the inset to Fig. 29.3. It is obvious that the film thickness deposited directly above

the source at the point O will be thicker than elsewhere, since this is the closest point of the source to the substrate. For the case of the point source, molecules are emitted uniformly in all directions, and the ratio of the film thickness d deposited at the point P a distance l from O to the thickness d_0 deposited at O is given by

$$\frac{d}{d_0} = \frac{1}{[1+(l/h)^2]^{3/2}} \,. \tag{29.10}$$

For a small surface source, the emission follows a cosine law, and the thickness distribution is given by

$$\frac{d}{d_0} = \frac{1}{[1+(l/h)^2]^2} \,. \tag{29.11}$$

Plots of these two functions are shown on Fig. 29.3 for the point source P and the small surface source S. For both sources, there is a rapid diminution in the thickness ratio d/d_0, which falls off to 0.985 and 0.98 respectively for the point and surface sources at a value of $l/h = 0.1$. When the substrate is relatively small, the thickness uniformity is thus better than 2%. Uniformity can be enhanced, for these two small sources, by increasing the source–substrate distance h, but at the expense of consuming more evaporant material. It has also been shown that a substantially uniform deposit will be obtained for a point source if the source is placed at the centre of

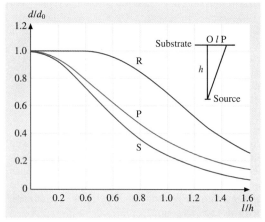

Fig. 29.3 Film thickness profiles calculated for evaporation from various sources onto a flat parallel substrate. Thickness ratio d/d_0 is given as a function of the ratio l/h, as shown in the *inset*. Sources include the small surface source (S), the point source (P) and a thin ring source (R) of radius b, where $b/h = 0.75$. Profiles are calculated, respectively, from (29.11), (29.10) and from an expression given by *Glang* [29.22]

a sphere and the substrates placed on the surface. For a surface source a uniform thickness distribution is obtained if both the source and the substrates are placed on the surface of a sphere [29.28].

Other systems for obtaining uniform thickness are based on the use of either extended sources, or substrates rotating around a central axis with respect to an off-centre source, such that variations in the thickness average out. *Holland* and *Steckelmacher* [29.28] have addressed ring sources in addition to strip and cylindrical sources evaporating onto a plane-parallel substrate. The ring source has also been treated by von *Hippel* [29.29] and the small surface source evaporating onto a rotating plane parallel substrate has been discussed by *Macleod* [29.30]. Expressions for ring and large surface sources are given by *Glang* [29.22]. A ring source can be considered to be a narrow ring of radius b, which is usually placed parallel to a flat substrate. This introduces an additional controllable variable which influences the thickness distribution. Thickness distributions may be obtained by calculating the distribution arising from an element of the ring source and then summing over all elements in the ring. For small b/h the thickness distribution reduces to that of the small surface source. For b/h larger than about unity the uniformity is also poor as the maximum in the distribution moves away from the central axis towards the source radius. However, for intermediate ranges, in particular b/h of about 0.7–0.8 there is considerable uniformity in thickness up to l/h of about 0.6. This is also illustrated in Fig. 29.3, where the curve R shows the thickness distribution for a narrow ring source with $b/h = 0.75$, calculated from the formula given by *Glang* [29.22]. Extended disc source thickness distributions may be obtained by integrating the thickness distribution derived for a narrow ring source from radius zero to radius b, but in general the thickness uniformity is not as good as for ring sources. Expressions for ring source of finite width can be determined by subtracting the disc source distribution for the inner radius from that of the outer radius. Diagrams illustrating the thickness distributions for several sets of ring and disc sources are given by *Glang* [29.22]. *Behrndt* [29.7] gives further details and references for such sources, and a discussion of the results of using combinations of several rod-like sources. Further attempts to minimise thickness variations have resulted in the evolution of various *planetary* deposition systems, so called because the rotating substrate plane is itself rotated around a vertical axis. A typical example of such a system and its performance is given in the literature [29.31].

Before describing the practical details of the evaporation process, it should be mentioned that, whatever type of evaporation process is used, some consideration needs to be given to the substrate. There are three basic types of substrate commonly used: glasses, single crystals and polycrystalline ceramics. Ceramics are particularly useful for high-temperature deposition processes. Although the substrate material may be predetermined (e.g. silicon wafers for microelectronics, optical glass for coatings on lenses, etc.), there are frequently factors which can be selected to increase the probability of obtaining good-quality films. The substrate planarity is important, in order to reduce the effect of defects. Electrical resistivity should in general be high in order to minimise surface leakage currents. Thermal properties include a consideration of the coefficient of expansion, which should be comparable to that of the deposited film to reduce the probability of thermally induced stresses in the films. Thermal conductivity should be high for films whose applications will entail significant power dissipation. A full description of thin-film substrates and the associated problems of substrate cleaning are discussed in detail by *Brown* [29.32].

Resistive evaporation is the most commonly used evaporation technique, although it is unfortunately not suitable for certain applications. It entails the resistive heating of an evaporation source, by passing a high current either through the source itself or through an adjacent heater. There are a large number of practical evaporation sources for various applications which are described in the literature [29.1, 22, 28]. The main requirement for any evaporation source is that the source itself will not emit its own vapour at the operating temperature. If metallic sources are used, the equilibrium vapour pressure of the metal must be negligible in comparison to that of the evaporant. A glance at Fig. 29.1 shows that the refractory metals (such as Pt, Ta and W) have equilibrium vapour pressures very low compared to about 1 Pa at temperatures up to about 2000 K, and are therefore suitable for evaporation of many materials. Carbon and oxides of aluminium or magnesium may also be used in evaporation, although owing to their low electrical conductivity refractory metal heaters are often used in conjunction with these. However, at elevated temperatures the oxides themselves may dissociate, with the liberated oxygen reacting with the evaporant to form its own oxide. For this reason the refractory metals cannot themselves be evaporated from oxide crucibles. Furthermore carbon may react with metals to form carbides well below 2000 K. Having chosen a suitable evaporation source, the vacuum deposition chamber is pumped down to the required background pressure and deposition may commence. The temperature of the source is increased such that the vapour pressure increases to a value consistent with a reasonable evaporation and hence deposition rate. The substrate is normally shielded using a metallic shutter until the required deposition rate is established, when the shutter is removed. The deposition rate is usually monitored by one of several methods [29.7, 22]. The quartz crystal method entails depositing the evaporant material on a vibrating quartz crystal, whose frequency varies linearly with the thickness of material deposited [29.33, 34]; the deposition rate is obtained by electronically differentiating the thickness signal with respect to time. Optical thickness monitoring methods are also used, particularly in applications where films are required for optical purposes, and require films having an optical thickness of multiples of $\lambda/4$, where λ is the wavelength for which the optical film is designed [29.7, 22]. These thickness multiples correspond to maxima and minima in reflected light intensity, which can easily be monitored during deposition. When the required film thickness is attained, a shutter is again used to curtail further film deposition and the evaporation process is stopped. Sequences of several film layers may be deposited by using a suitable sequential masking system [29.35].

Apart from simple vacuum evaporation there are also several other types of evaporation, which entail using alternative methods to heat the evaporant, or cater for various problems encountered in the deposition of alloys and compounds. In electron-beam evaporation the evaporant material is placed in a ceramic hearth, which is bombarded by high-energy electrons The main advantage of this method is that considerably higher temperatures of above 3000 K may be obtained than with resistive evaporation methods. This allows evaporation of the refractory metals and also elements such as boron and carbon (which sublimes). Other advantages are the evaporation of reactive metals such as Al, which are prone to contamination due to alloying, and situations where high purity is required.

For non-elemental (multicomponent) films there are several methods which may be employed. The main problem with such films is that, because of differences between the vapour pressures of the evaporant components, the Hertz–Knudsen expression of (29.5) predicts that the compound will evaporate noncongruently, i.e. with different evaporation rates for each element, and the deposited film will not in general replicate the composition of the evaporant. The problem is compounded because the proportion of each component present will therefore vary throughout the evaporation process, and the overall composition of the deposited films will vary throughout its thickness. In the evaporation of alloys, each component establishes its own vapour pressure at a given temperature, which may be predicted by *Raoult's law* for an ideal solution. This states that the vapour pressure established by each individual component is proportional to the mole fraction of the component present. Under these circumstances it is relatively simple to adjust the proportions of the alloy components to achieve the required film composition. However, complications occur in nonideal solutions, where the interactions between different types of atoms vary and an empirical parameter known as the activity coefficient, a, is defined. Knowing this it is possible to determine the vapour pressure for the component and hence the evaporation and deposition rates. Activity coefficients for various important systems, such as nichrome (a Ni–Cr alloy), are given in the literature [29.36]. However, as noted previously the mole fractions present in the evaporant are time-dependent and the composition of the films will vary during deposition. A parameter K has been determined, which if maintained close to unity should result in congruent evaporation throughout the entire deposition process. Further details are given by *Zinsmeister* [29.37]. Additional comments on the evaporation of both nichrome and permalloy alloys are given by *Glang* [29.22].

On evaporation of compounds, a wealth of chemical reactions can occur, which leads to a variety of different evaporation mechanisms. Frequently more than one reaction may operate simultaneously. Four simple generic reactions have been listed by *Glang* [29.22], which include many of the materials which can be successfully evaporated. We consider only compounds AB composed of the elements A and B, and denote the phase of the material by s (solid), l (liquid) or g (gas). Some compounds (solids or liquids) do not dissociate on evaporation and the reaction can be expressed simply as

$$AB(\text{s, l}) \rightarrow AB(\text{g}) \, . \tag{29.12}$$

There is then only the single vapour phase $AB(\text{g})$ produced. Other compounds dissociate on evaporation into two different vapour species. One reaction which is typical of the evaporation of chalcogenides and some simple oxides is given by

$$AB(\text{s}) \rightarrow A(\text{g}) + \frac{1}{2} B_2(\text{g}) \tag{29.13}$$

whereas the group IV dioxides tend to dissociate according to the reaction

$$AB_2(s) \rightarrow AB(g) + \frac{1}{2}B_2(g) \,. \tag{29.14}$$

In compounds of elements whose volatilities are significantly different, frequently one element enters the vapour phase while the other remains in the solid (or possibly liquid) phase. The compound decomposes into its different elements, with the simplest reaction of this type being

$$AB(s) \rightarrow A(s) + \frac{1}{2}B_2(g) \,. \tag{29.15}$$

Examples of materials evaporating without dissociation following (29.12) are the simple oxides GeO, SiO, SnO, the fluorides CaF_2 and MgF_2, and PbS. In general the evaporation of these materials can therefore be considered as analogous to those of the elements, and where available simple p^*–T data give the vapour pressure. Materials which follow (29.13) are primarily the II–VI chalcogenides, in particular the cadmium compounds CdS, CdSe and CdTe, and several simple oxides such as BaO, BeO, CaO, MgO, NiO and SrO. A full review of the electrical and structural properties of the cadmium chalcogenides is given by *Gould* [29.38]. Equation (29.14) is followed by many of the group IV dioxides, such as SiO_2, SnO_2, TiO_2 and ZrO_2, which dissociate into their lower oxides and molecular oxygen. Decomposition, as described by (29.15) is followed by most of the metallic borides, carbides and nitrides [29.22]. Other modes of decomposition, such as those for the technologically important III–V compounds, are also described in this reference.

Provided the relevant reaction for the evaporation process is known [e.g. as given by Eqs. (29.12–29.15)] it is possible to determine the equilibrium vapour pressure using standard thermodynamic tables. The law of mass action may be used to determine the equilibrium state of the chemical reaction, and for evaporation, the pressures of the various phases present contribute to a temperature-dependent equilibrium constant K_p, which is given by

$$K_p = \prod_i (P_i)^{\nu_i} \tag{29.16}$$

where P_i represents the vapour pressure for each of the various components of the vapour expressed in standard atmospheres and the ν_i are the molecular coefficients. K_p is related to the thermodynamic parameter ΔG^0, the standard change in Gibbs free energy per mole for the reaction, by the expression

$$\Delta G^0 = -RT \ln K_p \,, \tag{29.17}$$

where R is the universal gas constant. Free-energy data is tabulated for many materials in the literature [29.23, 39] and the use of (29.16) and (29.17) allows realistic estimates of the equilibrium vapour pressures to be determined and the evaporation rates estimated. Further discussions are given by *Glang* [29.22]. Nevertheless, although the theoretical framework is in place for the controlled evaporation of compounds, the exact type of dissociation and decomposition are not known in all cases, and for compounds consisting of three or more elements this problem is compounded. Some materials are therefore not amenable to direct evaporation from a single source.

For these cases various special evaporation techniques have been developed, all of which aim to replicate the composition of the evaporant in the deposited film, by ensuring that the different molecules arrive at the substrate in the desired proportions and at a constant rate. Reactive evaporation is a process whereby one component of the desired film is evaporated resistively in the normal manner, while the other component is present in the evaporation chamber in the form of a gas. This method is most commonly used where oxide films are required and oxygen gas present in the chamber combines with a metal at the substrate to form the oxide. The impingement ratio K, given by (29.7) is optimised so that evaporant and gas molecules reach the substrate in a predetermined ratio. Oxides of the common metals have been deposited by this method, including those of Al, Cr, Cu and Fe; additionally tantalum and titanium oxides have been prepared for dielectric applications. Reactive evaporation has also been used for compounds other than oxides. For instance, stoichiometric CdS films have been obtained by the evaporation of CdS in a sulphur vapour [29.40]. The sulphur vapour has the effect of discouraging the appearance of nonstoichiometric Cd-rich films which occur when CdS is evaporated alone. Nitrides may also be deposited in some cases by the use of a nitrogen atmosphere. The utility of this technique has been demonstrated in the reactive evaporation of silicon oxide, SiO_x, by *Timson* and *Hogarth* [29.41]. The evaporant is a mixture of the monoxide, the dioxide and possibly also free silicon. A sequence of films, varying in composition between SiO and SiO_2 as determined by electron spin resonance studies were obtained, depending on the ratio of the deposition rate to the oxygen pressure R/p. A useful nomogram for use in reactive evaporation is given by *Glang* [29.22].

Co-evaporation, sometimes known as the three-temperature technique, involves the simultaneous evaporation of two or more materials from separate evap-

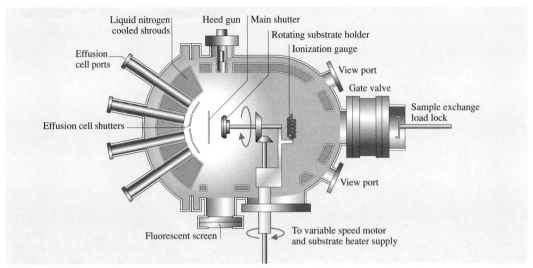

Fig. 29.4 Schematic diagram of a molecular-beam epitaxy (MBE) system for the deposition of III–V semiconductor compounds. After [29.42] with permission from Elsevier Science

oration sources. The essential feature of this process is that the rate of molecular arrival at the substrate may be controlled by individually controlling the temperatures of the evaporation sources. A major advantage of this method is that materials with widely different vapour pressures at a given temperature may be co-deposited by increasing the temperature of the less-volatile material. This method has found applications in the deposition of II–VI compounds such as CdS and CdSe, and also the III–V compounds such as AlSb, GaAs, InAs and InSb. It is also used for the deposition of alloy films.

Flash evaporation is another technique used for deposition of layers whose components have widely different vapour pressures. The evaporation is constrained so that only very small amounts of the compound completely vaporise almost instantaneously. The various components normally evaporate noncongruently, but because only very small quantities are evaporated rapidly, any inhomogeneity in the deposited films is limited to a few atomic layers. The process is repeated by evaporating further small quantities of evaporant. Several alternative arrangements have been developed to evaporate small particles, by continuously feeding powdered evaporant from a hopper into a heated evaporation source. The rate of delivery needs to be controlled such that at any given time there will be a number of evaporant particles in the source at various stages of the evaporation process. This ensures that the vapour emerging from the source contains contributions from many individual particles, resulting in an overall vapour composition which approximates to that of the evaporant powder. The method has been used particularly for III–V compound semiconductors such as GaAs.

Finally, the method of molecular-beam epitaxy (MBE) should be included here as a particularly sophisticated example of evaporation (or co-evaporation). In MBE the background pressure is considerably lower than in the case of the simpler evaporation techniques. i. e. well into the UHV range. This effectively eliminates the effects of background gas impingement, drastically improving the purity of the films. Several sources (effusion ovens) may be used simultaneously, each of which is separately controlled. These permit the deposition of compounds and precisely controlled doping of semiconductors. Sophisticated monitoring and isolation techniques are employed, and deposition sequences are usually under computer control. MBE is used particularly for the preparation of III–V compounds [29.42, 43] and silicon [29.44]. Figure 29.4 illustrates a typical MBE system for use in the deposition of III–V compounds [29.42]. A full description of MBE, together with other epitaxial growth techniques is given elsewhere in this volume in Chapt. 14.

Sputtering

The technique of sputtering in its various forms is of secondary importance only to evaporation for the deposition of thin films, offering certain advantages for the deposition of high-melting-point and dielectric materials. Sputtering is essentially the removal of particles from the surface of a *target* of the deposition material, by the action of incident energetic particles (normally positive ions). It is generally considered that the sputtering mechanism is the result of the transfer of momentum from the incident particles to the target atoms, some of which become dislodged. Energy acquired by other atoms interacting with the dislodged atoms can be sufficient to overcome the surface binding energy potential barrier. The net effect of a large number of atomic interactions is for a certain proportion of the target to be released (or sputtered) from the surface. Sputtered atoms typically have energies much higher than their evaporated counterparts. For low-energy sputtering with incident energy 1 keV, an ejected particle may have an energy of 10 eV, since typically 1% of the incident energy is transferred to each sputtered atom. In contrast, the thermal (kT) energy of an atom evaporated from a source at 2000 K is less than 0.2 eV. Such considerations have implications regarding the sticking coefficients for deposited materials (i.e. the probability that an atom reaching the substrate condenses on it), which is generally higher for higher-energy atoms.

Sputtering is limited by a threshold energy, below which it will not occur for a given ion/target combination. Threshold energies for a wide range of incident ions and target materials are of the order of 20–30 eV; experimental values for some common ion/target combinations are given by *Wehner* and *Anderson* [29.45]. These workers have pointed out that the threshold energy appears to be approximately four times the latent heat of sublimation per atom, equivalent to the surface binding energy. However, for practical sputtering considerations, incident energies are normally at least a few hundred electronvolts, well above the threshold energy. The efficiency of sputtering, measured by the sputtering yield, is also very low at low incident ion energies.

The sputtering yield is the most important sputtering parameter, and its variation with the target material and the incident ion energies is of major concern in both theories of sputtering and in practical sputtering deposition techniques. Sputtering yield $S(E)$ is defined as the ratio of the number of sputtered atoms to the number of incident ions. Hence

$$S(E) = \frac{N_s}{N_{ion}}, \qquad (29.18)$$

where N_{ion} is the impingement rate of incident ions and N_s is the sputtering rate of the target. Its primary variation is with the incident ion energy E. However, it also depends on the atomic properties of the incident ions and the target atoms. In the following discussion Z_1 and m_1 refer to the atomic number and mass of the incident ions, and Z_2 and m_2 refer to the corresponding quantities for the target material. The dependence of sputtering yield on the atomic number of the incident ions is quite striking, as shown in Fig. 29.5 [29.46]. This shows the variations of $S(E)$ for 45-keV incident ions onto targets of silver, copper and tantalum. Maxima in $S(E)$ occur corresponding to the noble gases Ne, Ar, Kr, and Xe. This dependence is one reason why the noble gases are most commonly used as the incident ion species, the second being that their inclusion into deposited films is unlikely to cause significant problems, due to their chemically inactive nature.

A number of theories have been developed to predict the dependence of sputtering yield $S(E)$ on the incident beam energy E. *Keywell* [29.47] assumed that the incident ions interacted with the sputtering target as if they were hard spheres, with the incident ion energy decreasing exponentially with the number of collisions. A discussion of this and other early models is given in the literature [29.45, 48]. More recent models [29.49, 50] have taken account of the scattering of incident ions by the atomic nucleus, and screening effects due to the electron cloud are also included, based on the Thomas–Fermi potential. This predicts a characteristic screening length a, which depends on the atomic numbers of the incident and target atoms, Z_1 and Z_2, respectively. The screening length is given by

$$a = \frac{a_0}{(Z_1^{2/3} + Z_2^{2/3})^{1/2}}, \qquad (29.19)$$

where $a_0 = 5.29 \times 10^{-11}$ m is the Bohr radius of the ground state of the hydrogen atom. Another parameter which is often invoked in sputtering models is the energy transfer coefficient, γ. For an incident ion of mass m_1 and energy E, impacting on a stationary target atom of mass m_2, the maximum energy E_{max} the target atom will attain is given by

$$E_{max} = \gamma E, \qquad (29.20)$$

where

$$\gamma = \frac{4 m_1 m_2}{(m_1 + m_2)^2}. \qquad (29.21)$$

In most collisions the incident ion will lose some energy to the target lattice, resulting in the acquisition of an energy less than E_{max} by the target atom. γ has a maximum value of unity for $m_1 = m_2$, but

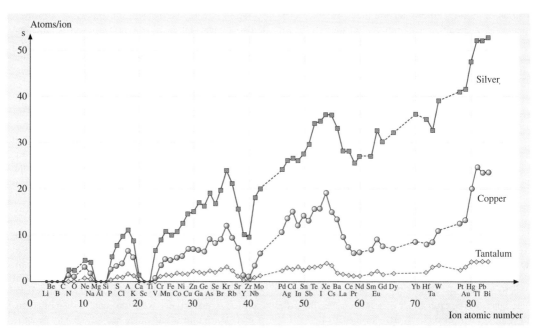

Fig. 29.5 Sputtering ratios of silver, copper and tantalum targets for different ions of energy 45 keV. After [29.46] with permission from Elsevier Science

this is neither a sufficient nor a practical condition for obtaining high sputtering yields owing to the impracticability of having identical sputtering ions and target material as well as the striking periodic dependence of $S(E)$ on Z_1 noted earlier. Two well-known sputtering models are now outlined, both of which have been reasonably successful in predicting sputtering yields.

In the model of *Pease* [29.49], which is also outlined elsewhere in the literature [29.51], ions are considered to interact with the target atoms near to the surface, which if displaced from their lattice sites are termed *primary knock-ons*. The mean energy of primary knock-ons before further interactions is denoted by $\bar{E}$. The threshold energy for displacement of a target atom, either by an incident ion or by a primary knock-on is denoted by E_d. Collisions with stationary atoms result in slowing down of displaced atoms, and in some cases a reversal of momentum and escape from the target surface, providing they still retain energy greater than E_s, the surface binding energy per atom. The number of atomic layers which contribute to the sputtering process is established by considering the average number of collisions n that a primary knock-on of energy $\bar{E}$ will make before its kinetic energy is reduced to less than that required to leave the target surface, i.e E_s. This is given by $2^n = \bar{E}/E_s$ or

$$n = \frac{\ln(\bar{E}/E_s)}{\ln 2} \ . \tag{29.22}$$

While engaged in n collisions the primary knock-ons diffuse on average $n^{1/2}$ atomic layers toward the surface, and thus $(1+n^{1/2})$ layers are involved in sputtering of material liberated by primary knock-ons only. The total number of atomic layers involved in sputtering is also equal to this expression. We also have to consider the probability of interaction in a given atomic layer, and denote the number of atoms per unit volume in the target by N. If these atoms are considered to be arranged in uniform crystallographic layers in three dimensions, the atoms in each layer will have an area density of $N^{2/3}$. If σ_p denotes the cross section for interaction between incident ions and target atoms, the probability of interaction in a given atomic layer is given by the product of the number of atoms per unit area and their interaction cross sections, or $\sigma_p N^{2/3}$. The total probability of producing a primary knock-on which can contribute to the sputtering process is then given by the product of the probability of interaction in a given atomic layer and the number of layers involved in sputtering, i.e.

$\sigma_{\rm p} N^{2/3}(1+n^{1/2})$. An estimate of the number of atoms displaced per primary knock-on is $\bar{E}/2E_{\rm d}$ [29.52] of which half, or $\bar{E}/4E_{\rm d}$, will be directed towards the surface. The total sputtering yield $S(E)$ is then given by the product of the probability of producing primary knock-ons and the number of displaced atoms produced by each. Hence

$$S(E) = \sigma_{\rm p} \bar{E} N^{2/3} \frac{1}{4E_{\rm d}} \left[1 + \left(\frac{\ln(E/E_{\rm s})}{\ln 2} \right)^{2/3} \right] \quad (29.23)$$

where the expression for n in (29.22) has been substituted.

In (29.23) for the sputtering yield, N, $E_{\rm d}$ and $E_{\rm s}$ are constants depending on the target material and $E_{\rm d}$ is normally determined from radiation damage experiments. However, $\sigma_{\rm p}$ and $\bar{E}$ are energy-dependent variables, which depend on the atomic numbers Z_1 and Z_2, the Thomas–Fermi screening length a, and the maximum target-atom energy $E_{\rm max}$. It is not appropriate to give these expressions here, but it is worth mentioning that different expressions are applicable in different incident ion energy ranges, defined by $E < L_{\rm A}$, $L_{\rm A} < E < L_{\rm B}$, and $L_{\rm B} < E$. Further details concerning the threshold energies $L_{\rm A}$ and $L_{\rm B}$ are given in [29.49, 51]. Calculations of sputtering yield predicted by this model are reasonably consistent with experiment, showing a sharp increase above a certain threshold energy which is determined by $E_{\rm s}$, a broad maximum and a relatively slow decrease at higher energies [29.49].

The model of *Sigmund* [29.50] represents a comprehensive theory of sputtering, particularly for single-element solids, although this has led to a certain degree of complexity in its general form. The model has been revised and extended [29.48] and tested against a considerable amount of data produced by many different workers [29.53]. The emphasis in this section is to briefly outline the model, to provide a basis for the calculation of $S(E)$ at normal incidence for low- and high-energy sputtering, and to give useful references to the required data. Various approximations made and some of the conditions of validity are omitted. Being a multiple-collision process, sputtering may be treated by using the formalism of transport theory, and in the Sigmund model the Boltzmann transport equation is solved. The sputtering yield at normal incidence is given by

$$S(E) = \Lambda \alpha N S_{\rm n}(E), \quad (29.24)$$

where Λ is termed the material factor, and depends only on the target material. α is a dimensionless factor, effectively depending only on the ratio of the target atom to incident ion masses m_2/m_1. The factor α is given for various different conditions in the literature [29.48, 50]. As in the model of *Pease* [29.49], N represents the number of atoms per unit volume in the target. $S_{\rm n}(E)$ is the nuclear scattering cross section, depending on the interaction probability between the incident ions and the target atoms at a given energy.

The material factor Λ arises in the sputtering yield expression as a result of integrating the number of sputtered atoms emitted in all directions with all possible energies. According to Sigmund, the material factor

$$\Lambda = \frac{3}{4\pi^2} \frac{1}{NC_0 E_{\rm s}}, \quad (29.25)$$

where $C_0 = \frac{1}{2}\pi\lambda_0 a_{\rm BM}^2$, λ_0 is a constant approximately equal to 24, and $a_{\rm BM} = 2.19 \times 10^{-11}$ m is the Born–Meyer potential characteristic screening radius. $E_{\rm s}$ represents the surface barrier energy, as in the model of Pease. Making these substitutions into (29.24) gives a more useful formula for the sputtering yield [29.45] as

$$S(E) = \frac{1}{16} \frac{\alpha S_{\rm n}(E)}{\pi^3 a_{\rm BM}^2 E_{\rm s}} \quad (29.26)$$

which applies both for low- and high-energy sputtering. An expression for E^*, the boundary between low-energy and high-energy sputtering, is given in the literature [29.50], and is typically a few hundred electronvolts, depending on the atomic masses and atomic numbers of the incident ions and the target atoms. For low-energy sputtering ($E < E^*$) the nuclear scattering cross section can be approximated by

$$S_{\rm n}(E) = C_0 E_{\rm max} = \tfrac{1}{2}\pi\lambda_0 a_{\rm BM}^2 E_{\rm max}, \quad (29.27)$$

where $E_{\rm max} = \gamma E$ from (29.20). Substituting into (29.26) yields

$$S(E) = \frac{3}{4} \frac{\alpha \gamma E}{\pi^2 E_{\rm s}}. \quad (29.28)$$

Hence for low-energy sputtering $S(E)$ is directly proportional to the sputtering energy, where α and γ effectively depend only on the ratio m_2/m_1 and $E_{\rm s}$ is a constant for the target material.

For higher incident ion energies ($E > E^*$) the Thomas–Fermi, rather than the Born–Meyer interaction potential is applicable, and a simple approximation for $S_{\rm n}(E)$ is not available. It is therefore necessary to determine its value from tables or graphical data, calculating $S(E)$ directly from (29.26). Incident ion energies E and nuclear scattering cross sections $S_{\rm n}(E)$ may be expressed in terms of reduced or Thomas–Fermi variables ε and $s_{\rm n}(\varepsilon)$ [29.48, 50]. $s_{\rm n}(\varepsilon)$ is a universal function of ε, depending on the details of the screened Coulomb function used [29.48]. The detailed dependence of $s_n(\varepsilon)$ on ε utilised in the Sigmund model is that given by *Lindhard* et al. [29.54]. For ease of computation E_{TF}, the Thomas–Fermi energy unit has been calculated for various incident ion/target combinations [29.48]. E_{TF} represents the ratio E/ε, and thus ε and hence $s_n(\varepsilon)$ may be obtained. $S_n(E)$ may then be determined from the reduced value.

Agreement between predicted and measured sputtering yields as a function of incident ion energy are good. *Andersen* and *Bay* [29.53] have reviewed sputtering yield measurements at normal incidence for a considerable number of incident ion and target species, and generally concluded that sputtering yields for high-yield materials (e.g. Ag, Au, Cu, Zn) are predicted well by the theory, while for low-yield materials (e.g. Nb, Ta, Ti, W) it was over-estimated by a factor of up to three. A fuller discussion, to which the reader is referred, shows that the model breaks down for certain combinations of Z_1, Z_2 and E due to the existence of inelastic collisions, non-isotropic and nonlinear effects (spikes) in the collision cascade and unallowed-for surface effects. The regions of validity of the model are neatly presented in terms of a three-dimensional (Z_1, Z_2, E) space, and illustrate the fact that (29.26) is an amazingly good prediction for a large number of (Z_1, Z_2, E) combinations of practical interest.

Sputtering yields may be determined experimentally by measuring the decrease in mass of a sputtering target. Normally such experiments are performed using ion-beam sputtering, in a fairly high-vacuum environment. The target area bombarded by the beam is small, and fairly low sputtering rates are achieved, but it has the advantage that ion energies are accurately given by the accelerating potential. The methods of sputtering normally used for the preparation of thin films are based on the establishment of a population of positive ions in a low-pressure gaseous environment. Such techniques depend on the establishment of a glow discharge, and require a considerably higher operating pressure, whereas ion-beam sputtering does not require the presence of gas molecules in the deposition chamber, and can therefore be used under UHV conditions, where gaseous contamination effects are minimised. A complicating factor, which often has to be taken into account, is that some incident ions may become embedded in the target, thus tending to increase its mass. The net loss of material from the target Δm is then the difference between the mass of material sputtered and the mass gain of the target. If N_{ion} and N_s are the incident ion impingement rate and the atomic sputtering rate of the target respectively (29.18), and m_1 and m_2 are the incident ion and target atom masses, it is simple to calculate the mass loss and increase of a given target with bombarded area A sputtered for a time t. The loss of mass is given by $N_s m_2 A t$, and the increase in mass is given by $N_{ion} \Gamma m_1 A t$, where Γ is the probability that an incident ion becomes embedded in the target. Hence

$$\Delta m = N_s m_2 A t - N_{ion} \Gamma m_1 A t \qquad (29.29)$$

or

$$S(E) = \frac{N_s}{N_{ion}} = \frac{\Delta m}{m_2 N_{ion} A t} + \Gamma \frac{m_1}{m_2} . \qquad (29.30)$$

Frequently the second term on the right-hand side of this equation is negligible, and the sputtering yield is then given by the first term only. This is particularly applicable for large ion doses, while for intermediate doses a detailed empirical knowledge of the value of Γ is required. For low ion doses the approximation $\Gamma = 1 - R_0$, where R_0 is an ion reflection coefficient, may be used [29.53]. It is often useful to express the sputtering yield in terms of the ion current at the cathode $I_c = e N_{ion} A$, where e is the electronic charge and the sputtering yield is then given by [29.1]

$$S(E) \approx \frac{e \Delta m}{m_2 I_c t} . \qquad (29.31)$$

Measurements of the ion current, target mass loss and the sputtering time are then all that are necessary to calculate $S(E)$. Although strictly speaking the above expressions are only applicable for sputtering performed with ions of a single energy, they may also be applied to a first approximation for practical sputtering methods using the glow discharge.

Although ion-beam sputtering is a very useful technique for measuring sputtering yields and comparing with theoretical predictions, it is not normally used for routine film deposition, owing to the need for UHV conditions, the low sputtering rate obtained and the associated expense. Virtually all sputtering methods used for the deposition of thin films utilise the glow discharge phenomenon. When a gas at reduced pressure is subjected to an applied voltage, usually in the range of a few hundred to several kilovolts, any free electrons will be accelerated in the field, acquiring energy and ionising the gas molecules to produce a plasma. Under the influence of the field the positively ionised gas molecules are accelerated towards the cathode and free electrons towards the anode. The gas pressure required for the establishment of such a discharge is of the order of a few pascals, which in turn is determined by the molecular concentration required to maintain the mean free path within suitable limits to initiate and sustain the discharge. The basic principles of the glow discharge are well known and documented [29.55, 56] and need not concern us here. However, it should be noted that the positive ions tend to accumulate in front of the cathode (the cathode-fall region), across which most of the applied voltage is dropped. Thus the electric field between the cathode and the anode is distinctly nonlinear, and most of the energy acquired by the ions in their journey to the cathode is obtained in the cathode-fall region.

It is relatively simple to set up a self-sustained glow discharge, although the behaviour of the discharge depends in a fairly complex manner on the pressure, the applied voltage and the geometry of the system. Inert gases such as argon are generally used, unless the sputtered film relies on a reaction between the sputtering gas and the sputtered material to produce the required compound film. Once the discharge is established, the cathode is continuously bombarded by energetic ions, and provided the energy of these ions is sufficient, sputtering of the cathode will occur. Thus if the target is sufficiently conductive, all that is necessary is to make the cathode of the target material. Substrates on which deposition will occur are normally placed parallel to the target at the anode so that the sputtered atoms need only cover a short distance before deposition. The energy of the sputtered atoms when they reach the substrate is considerably higher than in the case of evaporation, leading to better adhesion. Furthermore, sputtered compounds are normally deposited without dissociation, and therefore the stoichiometry of the target material is usually preserved in the deposited film. This basic form of sputtering is normally termed *diode sputtering*, since only two electrical connections are required.

Basic diode sputtering operates reasonably satisfactorily, but has a major disadvantage in that the sputtering rate N_s depends directly on the impingement rate of the incident ions N_{ion} and thus also on the discharge current. At low gas pressures the supply of ions is limited and thus the discharge current and the sputtering rate are low. Moreover, at sufficiently low pressures the electrons do not ionise a sufficient number of gas atoms to sustain the glow discharge. This low-pressure limit is in fact a major drawback in diode sputtering, since enforced operation at higher pressures above about 3 Pa results in significant film contamination. For this reason sputtering was for many years rejected in favour of evaporation for thin-film deposition. A major improvement in sputtering technology, which enabled sputtering to be performed at lower pressures was the introduction of *triode sputtering*. In this technique, the discharge is maintained by using an additional source of electrons, so that it is not totally dependent on secondary electron emission from the cathode. Electrons are supplied by an additional thermionic emitter, consisting of a filament through which a current is passed. The electron concentration in the plasma can be readily controlled by varying the anode potential or the filament current, while the ion current is directly controlled by the target potential. A further innovation usually employed in triode sputtering systems is the use of a coil to set up a magnetic field, which forces electrons from the emitter to travel in spiral paths, thus greatly increasing the distance travelled from the emitter before they are collected by the anode. This enhances the ionisation probability α, and thus the supply of positive ions.

It should be emphasised that both diode and triode sputtering may not be used for the sputtering of insulators – a very important class of materials. Since positive ions are responsible for the sputtering process, these need to be neutralised by electrons from the external circuit. In the case of insulators a positive charge builds up at the surface of the target, and this charge cannot leak away through the insulator. Sputtering ceases when the positive charge causes the potential at the target surface to approach that of the plasma. Various methods, such as bombarding the target surface with electrons, have been attempted to solve this problem, but in general with little success. Glow discharge sputtering of insulators is generally therefore performed reactively, or with alternating radio-frequency (RF) fields, as described later.

As with evaporated films, there are a number of problems associated with the deposition of alloys and compounds. The best theories of sputtering yields are typically inaccurate by a factor of two or so. Empirical determinations are normally required for films of controlled composition to be deposited. Nevertheless, sputtering of multicomponent or compound targets is frequently all that is necessary to obtain films varying only marginally from the target composition. *Maissel* [29.57] suggests that during the initial sputtering run an *altered region* is formed on the surface of the target, which is deficient in the target component having the highest sputtering yield. Thus although stoichiometric films will not normally be deposited during the initial sputtering run, such films may generally be deposited during subsequent runs, since the deficiency in the higher sputtering rate material compensates for its greater removal rate [29.51, 57]. Minor differences between the compositions of the film and target do still occur, but are normally the result of oxidation or of the evaporation of one component at high temperatures. When compounds are sputtered the incident ion energies are often sufficient to break the chemical bonds, again leading to a deficiency in one component. *Reactive sputtering* and *co-sputtering* are specifically designed to give greater flexibility during the deposition of multicomponent films.

Reactive sputtering involves the sputtering of a target using a sputtering gas which is amenable to chemical combination with the target material. The gas used may be either solely the reactive species, or a mixture of the reactive species and an inert sputtering gas, such as argon. Clearly the process is suitable for preparation of certain insulators which cannot be prepared by the previously mentioned sputtering techniques. Deposition rate depends primarily on the target sputtering yield, through the energy of the incident sputtering ions, the gas pressure and the distance between the target and the substrate. For example, *Pernay* et al. [29.58] have concluded that the deposition rate is primarily determined by a reduced electric field E^*, given by E/p, where E is the mean electric field between the cathode and the anode, and p is the gas pressure. When a copper target was sputtered in a mixture of oxygen and argon, either Cu_2O, CuO or Cu were deposited depending on the value of E^* and the oxygen concentration [29.59]. Reactive sputtering has been used mainly to deposit oxide films which have insulating or semiconducting properties. Among the common oxides that have been studied are those of aluminium, cadmium, copper, iron, silicon, tantalum, tin and titanium. Oxides of niobium, thorium, vanadium and zirconium have also been investigated, as have the rare-earth oxides of hafnium, lanthanum and yttrium. The other major non-oxide compounds which have been investigated are the nitrides, with silicon nitride, Si_3N_4, receiving most attention due to its utility both as an insulator and in variants of the silicon planar process for integrated-circuit fabrication. This material has been prepared using a Si_3N_4 target and a mixture of argon and nitrogen as the sputtering gas [29.60]. Carbides and sulphides have also been sputtered reactively. In all cases it is imperative that the stoichiometry of the deposited films is determined using the normal chemical techniques, and the deposition parameters adjusted until the required composition is attained. A very useful review of reactive sputtering, covering both the mechanisms and the primary reactive sputtering techniques, has been given by *Westwood* [29.61].

Co-sputtering has certain advantages from the point of view of the preparation of films whose components have different sputtering yields. Segregation of components on cooling can lead to a nonstoichiometric composition of material in the target, and differences in sticking coefficients between components can lead to the deposited composition differing from that arriving at the substrate. Co-sputtering has the advantage that the proportion of each sputtered components may be controlled during deposition, and may readily be adjusted in response to varying conditions to ensure stoichiometry of the deposited film. Co-sputtering may be performed with two or more conventional sputtering targets operating simultaneously for each component, each directed towards a common substrate. Another method of co-sputtering is again to use conventional targets, but with the substrate alternately subjected to the sputtered material from each target. This may be achieved by mounting the substrate on a disc which rotates in front of the targets. Alternatively a specially designed target assembly may be used such that the components arrive at the substrate in their stoichiometric proportions; the potentials of the components may be independently controlled to compensate for differing sputtering yields and sticking coefficients. Generally a plane-parallel substrate arrangement, with the target large in comparison to the target–substrate distance is employed, which maximises the uniformity of the film thickness. *Sinclair* and *Peters* [29.62] have described some target-electrode assemblies which have proven suitable for co-sputtering, including a concentric ring and disc assembly and an interdigitated design. Each of these allows independent control of the voltage applied to the two parts of the electrode assemblies and thus control over the composition of the sputtered films. Although the spatial dependence of the electric field in such assemblies is complex, and the deposition rate of one component may be influenced by the potential applied to the electrode for the other component, it is relatively straightforward to calibrate such a source for different materials, and to compensate for such effects by varying the acceleration potentials accordingly.

It is clear that, in general, impurities present in the gaseous phase may be incorporated into the deposited films. The fraction f_i of the impurity species i trapped within a film is given by [29.57]

$$f_i = \frac{s_i N_i}{s_i N_i + N_R} \tag{29.32}$$

where s_i and N_i ($m^{-2}\,s^{-1}$) are respectively the sticking coefficient and impingement rate of the impurity species and N_R is the molecular deposition rate of the required film material; f_i can be reduced by increasing N_R, or by decreasing either or both of s_i or N_i. In *bias sputtering* the phenomenon of re-sputtering is employed, i.e. components of the sputtered film are re-sputtered by ions from the plasma. This is achieved by applying an additional small negative potential to the substrate. Preferential sputtering of the impurities takes place, providing the binding energy of the impurities is lower than that of the host film material. This leads to an effective reduction in f_i, below that predicted by (29.32) [29.63]. The value of s_i in the expression is effectively reduced by re-sputtering of impurities from the film surface. *Asymmetric alternating current (AC) sputtering* [29.64] achieves similar objectives to bias sputtering, i.e. bombardment of the cathode target by ions to cause sputtering and bombardment of the films by lower-energy ions to preferentially re-sputter the impurities. These are accomplished by establishing an alternating current between the substrate and cathode and the use of a simple diode-resistance network. During the negative half-cycle the full negative potential is applied to the cathode, allowing sputtering to take place. In the following positive half-cycle, a relatively small positive potential (limited by the resistance) is applied to the cathode. This allows low-voltage sputtering of the films, which is sufficient to remove only the gaseous impurities. In this technique, again the value of s_i is effectively reduced, leading to a decrease in f_i. In *getter sputtering*, the gettering action of a freshly deposited thin film is utilised to purge the chamber of unwanted active gases. The concentration of impurities in thin films has been observed to decrease as the film thickness increases, and is attributed to the trapping (or gettering) action of the film material. In getter sputtering systems this action is exploited so that the majority of impurities are trapped in a film region deposited between the gas entry point and the substrates, while the required film is deposited elsewhere in the chamber where the gas impurity concentration will already have been depleted by the gettering action. The design of a typical getter sputtering system is described by *Theuerer* and *Hauser* [29.65]. In the case of getter sputtering, N_i is effectively reduced in (29.32), lowering the value of f_i.

Trapping of inert gas molecules in the films is also a possibility, and their concentration should be given by (29.32). However, at thermal energies the sticking coefficient is effectively zero, and thus trapping ought not to be of great concern. Nevertheless, for sputtering systems using argon, concentrations of 0.1% have been observed [29.67], generally for ion energies greater than 100 eV. Although not a major problem for many appli-

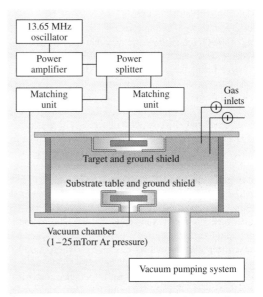

Fig. 29.6 A typical radio-frequency (RF) sputtering system. After [29.66] with permission from Oxford Univ. Press

cations the possibility of inert-gas contamination should not therefore be discounted.

It has been mentioned earlier that simple direct current (DC) discharges as used in the diode and triode sputtering techniques described above, are unable to deposit insulating films. Although reactive sputtering may be used, the deposition of films by this method requires careful control of sputtering rates and gas pressure, and often calibrations of individual systems and materials. The use of *radio-frequency (RF) sputtering*, allows insulators to be sputtered directly [29.68]. When a high-frequency (typically several MHz) alternating signal is applied to a metal plate, on which an insulating (dielectric) target is mounted and is placed within an auxiliary low-pressure discharge, the positive charge which accumulates on the target is neutralised during half of each cycle. Owing to the much higher mobility of the electrons than the ions in the plasma, the plasma potential becomes positive with respect to the target. RF sputtering was first employed by *Davidse* and *Maissel* [29.69], who found that a self-sustained discharge could be started and maintained by the application of RF power to the electrode only. The mechanism by which sputtering takes place requires the frequency to be in the radio-frequency range [29.16]. The electron mobility is several orders of magnitude greater than that of the ions which have difficulty in keeping up with the field variations, and at high frequencies the ions accumulate in front of the target, in the same way that they do in a DC discharge. Because the number of electrons arriving at the target during the positive half-cycle are not matched by the number of ions arriving during the negative half cycle, the front of the target acquires a negative charge, which repels most of the electrons that would arrive at the surface during the positive half-cycle. The plasma itself is said to be positively *self-biassed* with respect to the target. It is therefore possible to both feed energy into the plasma, while also giving the ions sufficient energy to cause sputtering. The RF discharge is self-sustained, and Davidse and Maissel found that an externally generated plasma is not required. A schematic diagram of the RF sputtering equipment is shown in Fig. 29.6 [29.66] while more detailed discussion of the RF sputtering process is given in previous reviews [29.16, 57]. These authors point out that a major difference between diode and RF sputtering is a requirement for an impedance matching network or *match box* [29.70] in the RF case. The matching network is inserted between the power supply and the discharge chamber, its purpose being to maximise the power that can be delivered to the plasma. Typically a network containing an inductor and a capacitor is used. An example is shown in the literature [29.71]. A blocking capacitor is normally inserted in the network to ensure that the build up of the self-bias voltage proceeds satisfactorily. Although the target itself will normally perform this function in the case of insulator sputtering, the inclusion of a blocking capacitor allows RF sputtering to be used for the deposition of conductors as well as insulators, so semiconductors and metals are also easily deposited by this technique. Mainly for this reason RF sputtering has become a versatile technique which is used near-universally for semiconductor production as well as for insulators and metals [29.70]. A significant increase in the sputtering rate is also observed in RF systems.

Finally, a technique used to increase the sputtering rate still further is that of *magnetron sputtering*. In this technique, a magnetic field is applied between the target and the substrate. Electrons in the plasma then experience the Lorentz force from the magnetic field in addition to that derived from the electric field, so that the total force on the electron is given by

$$F = m\frac{dv}{dt} = eE + ev \times B, \quad (29.33)$$

where F is the total force experienced by the electron, m and e are the electron mass and charge respectively, v is the electron velocity, E is the electric field and B is the magnetic field [29.19, 72]. The effect of the two fields is to cause the electrons to move in spiral rather than linear paths, and therefore to greatly increase the distance travelled by the electrons in the plasma. This substantially increases the time the electrons are within the plasma, and thus the probability of their causing secondary ionisation is also significantly enhanced. In turn this leads to increased current flow and crucially increased deposition rates. The magnetron sputtering source is constructed by mounting a powerful magnet below the target, with one pole concentric with the other as shown in Fig. 29.7 [29.66]. This establishes a magnetic field across the target surface; the direction of the magnetic field is perpendicular to the electric field at the furthest extent of the magnetic force lines from the target. The result of the two interacting fields is to confine the electrons for a significant period within a region termed the *racetrack* at the surface of the target, where the electrons effectively hop across the surface. Ionisation of the sputtering gas occurs mainly above the racetrack, and erosion of the target takes place preferentially in this region. In addition to the enhanced deposition rate, magnetron sputtering also has several other advantages. Since the ionisation region is effec-

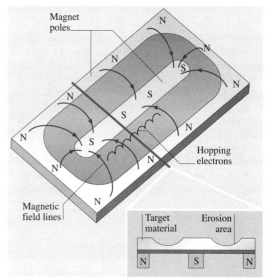

Fig. 29.7 The technique of magnetron sputtering, showing the hopping electrons of the target confined to a *racetrack* region by the magnetic fields from the magnets under the target. After [29.66] with permission from Oxford Univ. Press

tively confined to the racetrack, unwanted sputtering from the substrate and the walls of the chamber are reduced, as is the substrate temperature. The gas pressure required to sustain the discharge is also reduced, owing to the greater ionisation efficiency.

A further improvement in the technique of magnetron sputtering may be obtained by the use of an unbalanced magnetron, as described by *Window* and *Savvides* [29.73]. In spite of its excellent high deposition rate, magnetron sputtering generally produces films which have a columnar structure. This leads to the presence of voids, both within grains and at grain boundaries. Low-energy ion bombardment has been shown to improve adatom mobility, allowing atoms to migrate on the depositing surface and filling some of the voids [29.72]. Various computer simulations described in this reference have illustrated the importance of ion beams in depositing dense, void-free films. It is possible to generate an ion current in the plasma by using an unbalanced magnetron where the inner and outer magnets are deliberately unbalanced from the ideal magnetron configuration. Further details concerning the field characteristics of unbalanced magnetrons and of the unbalanced magnetron technique are given in the literature [29.72, 73]. The method has been used to successfully deposit various types of film, including those intended for electrical and optical applications, corrosion protection, and wear and abrasion resistance. A fuller discussion of materials prepared using this method is given by *Rohde* [29.72].

Magnetron sputtering may also be used in conjunction with other sputtering techniques. A combination of RF and magnetron sputtering is frequently used, for example for Co:M alloys where M represents Al, Si, Ti, Cr, Fe, Ni, Cu, Zr, Nb, Mo, Ag, Ta or W [29.74], and also for the deposition of silicon nitride films [29.60].

Thus sputtering has developed from a relatively simple DC technique capable only of the deposition of conductors at a relatively low rate, to encompass both conductors and insulators (RF sputtering) and with sputtering rates suitable for high-throughput industrial processes (magnetron sputtering). Improvements in film morphology have been accomplished by the use of unbalanced magnetron sputtering.

29.1.2 Chemical Deposition Methods

There are many different chemical methods for preparing thin films, but by their nature many of these are restricted to one or perhaps a few different materials, related to a specific chemical reaction or series of reactions. It is therefore only possible to categorise some general techniques and to give examples for various particular materials. *Campbell* [29.75] divided chemical methods into two classes: those depending on an electrical source of ions, such as electroplating and anodisation, and those requiring a chemical reaction, usually, but not exclusively involving a chemical vapour as in the CVD method. A further technique, which has found favour in recent years, is the deposition of molecular films using the Langmuir–Blodgett technique. Brief examples of these are given below.

Electrodeposition

By electrodeposition, we include all methods which involve passing an electric current through a solution such that a film of a material, usually a metal or an oxide, is deposited on one of the electrodes immersed in the solution. Electrolytic deposition, or *electroplating*, is the deposition of a metal by this method. According to *Campbell* [29.75, 76], of the 70 metallic elements, 33 can be electroplated successfully, although of these only 14 are deposited commercially. The metallic film is deposited on the cathode, and its thickness and the time in which this occurs are determined by the two laws of electrolysis:

1. The mass of the deposit is proportional to the quantity of electricity passed.
2. The mass of material deposited by the same quantity of electricity is proportional to the electrochemical equivalent, E.

This can be expressed as

$$\frac{W}{A} = JtE\alpha , \tag{29.34}$$

where W is the mass deposited on a cathode of surface area A, J is the current density, t is the time of deposition and α is the *current efficiency*, or ratio of the expected to the theoretical mass deposited. Its value is generally in the range 0.5–1.0. The previous expression can be rewritten in terms of the film deposition rate, R (or thickness d per unit time t), by noting that the film density $\rho = W/Ad$. Combining this with (29.34) gives

$$R = \frac{d}{t} = \frac{JE\alpha}{\rho} . \tag{29.35}$$

When a high current is passed through a suitable electrolyte, metallic ions move towards the cathode under the influence of the applied electric field. The mass deposited, and the deposition rate are given by (29.34) and (29.35), and may easily be estimated if E and ρ are known. These are tabulated for the 14 metals commonly deposited by this method [29.75]. Deposition rates can be very high, depending on the current drawn through the electrolyte. For example, silver will deposit at a rate of about 1 nm/s for a current density of $10\,\text{A/m}^2$, rising to 1 μm/s at $10\,000\,\text{A/m}^2$ [29.75]. Further details of the deposition of alloys are given in a standard text [29.77], while the growth and structures of electrodeposits are described by *Lawless* [29.78].

A second technique which is used for the deposition of oxides (or less commonly for other compounds such as nitrides), is that of *anodisation*. As its name implies, films are deposited on the anode of the parent metal, following a chemical reaction. The chemical equations describing the anodisation process are given below [29.75]:

$$M + nH_2O \rightarrow MO_n + 2nH^+ + 2ne^- \text{ (at the anode)},$$
$$2ne^- + 2nH_2O \rightarrow nH_2\uparrow + 2nOH^- \text{ (at the cathode)},$$
(29.36)

where M represents the metal and n is an integer specifying the oxide deposited. The oxide grows in an amorphous form at the metallic anode surface, while hydrogen is evolved at the cathode. The electrolyte is typically an aqueous solution of water containing a dilute acid.

In general a constant current is passed through the cell and film thickness is proportional to the time elapsed. Films grown under constant-voltage conditions, show a decreasing deposition rate with time, as the current falls to zero when the voltage dropped across the film increases to the total applied voltage. Thus the maximum thickness of the film is determined by the voltage used. Growth rates for anodic films under constant-current conditions are of the same order as those for the PVD methods of evaporation and sputtering. Aluminium and tantalum anodisation at a current density of $20\,\text{A/m}^2$ gives growth rates of the order of 1 nm/s [29.76]. Considerably more detailed discussions of anodic oxide films in particular are given in the literature [29.79, 80]. These cover considerably more details of the types of films that may be successfully prepared, their structures, characterisation methods and copious references to related work.

Chemical Vapour Deposition (CVD)

Chemical vapour deposition is a method whereby a volatile compound reacts, with other gas species, to produce the required compound on the substrate. CVD does not require operation under vacuum, as with the PVD processes of evaporation, MBE and sputtering. Moreover, it is capable of producing both silicon and gallium arsenide (GaAs) semiconductor films, and to grow these on substrates of the same material as high-quality epitaxial layers (homoepitaxy). For these reasons it has found wide acceptance in semiconductor processing. High deposition rates relative to PVD deposition methods are possible, and controlled doping performed during the deposition process is relatively straightforward. Objects having quite complex shapes can be coated relatively easily, since the film deposition does not depend on line-of-sight geometry as in the case of evaporation and to a lesser extent sputtering. There are however some formidable disadvantages with the technique. Generally much higher substrate temperatures are required in CVD than for the other methods. The reactive gases used in the deposition process and/or the reaction products can be extremely toxic, explosive or corrosive, requiring considerable investment in safety features. This may well be justified in high-throughput semiconductor production processes, but uneconomic for smaller-scale development work, where PVD methods are preferable. Corrosion and effects of unwanted diffusion, alloying and chemical reactions at the substrate surface may occur under the high operating temperatures, and masking of substrates is particularly difficult [29.81].

In principle CVD may utilise any chemical reaction which produces the required material, and therefore the equipment used for the production of a given material is likely to differ from that required for another material. Nevertheless, there are several different classes of reaction which are used in CVD processes which are described in the literature [29.16, 18, 19, 75, 76]. Some of these are outlined below.

Pyrolysis or thermal decomposition of a gaseous compound entails passing a vapour over a heated substrate, which causes decomposition and the condensation of a stable solid. This has been used primarily in the production of silicon from SiH_4 and nickel from nickel carbonyl. *Oxidation* is frequently employed to provide SiO_2 in silicon processing. Halides of the required metal oxide are reacted with steam to give the oxide and a hydrogen halide. In *nitriding* the steam is replaced with ammonia (NH_3), and nitrides such as Si_3N_4 may be generated. *Reduction* reactions use hydrogen gas instead of steam or ammonia in the preparation of silicon and refractory metals such as tantalum or molybdenum from their halides. In *disproportionation,* a reversible reaction of the type

$$A + AB_2 \rightleftarrows 2AB \qquad (29.37)$$

is usually employed [29.76], where A represents a metal, B is usually a halide and both AB and AB_2 are gases. The higher valence state is more stable at lower temperatures so, if the hot gas AB is passed into a colder region, AB_2 will be formed and A deposited. In *transfer reactions* for the preparation of compounds, two different strategies may be employed. Either a compound can be used as the source material, or the source comprises a volatile compound of one of the elements of the required compound. Gallium arsenide (GaAs) can be prepared by both methods [29.16, 83]. In Fig. 29.8 [29.82] a typical CVD system is shown which uses a transfer reactor of the second type to produce GaAs. Highly purified arsenous chloride ($AsCl_3$) and Ga are the source materials. Purified hydrogen is bubbled through the $AsCl_3$ and passed through the furnace. In zone 1 of the furnace, the following reaction takes place:

$$2AsCl_3 + 3H_2 \rightarrow 6HCl + \frac{1}{2}As_4 \,. \qquad (29.38)$$

The arsenic vapour is completely absorbed by the Ga source in zone 2 until saturation occurs at 2.25 at. % As. No free arsenic condenses beyond the furnace, but gallium is transported as a lower chloride. Deposition of GaAs occurs in zone 3, when hydrogen is diverted to bypass the bubbler.

Tables of materials which can be deposited by CVD and the associated reactions are given in the literature [29.19, 75]. However, several modifications to the basic CVD process are also in current use. These seek to overcome some of the deficiencies associated with CVD, such as the operation at atmospheric pressure and the use of very high temperatures often involved. Systems employing organic precursors also have certain advantages for the deposition of various materials.

We first discuss low-pressure CVD or LPCVD. Typical pressures of the order of 100 Pa are used, approximately a thousandfold reduction from atmospheric pressure. Contamination is normally lower than for atmospheric CVD and film stoichiometry is often improved. To compensate for the reduced pressure, the input reactant gas concentrations are increased, the gas flow velocity is increased, but the gas density is reduced. Because the mean free path of the gas molecules increases due to the reduced pressure, substrates can be placed closer together, giving a higher throughput in commercial applications. Silicon and dielectric films are generally prepared by this method. The operating temperatures required for LPCVD remain high, however.

In plasma-enhanced CVD (PECVD) a glow discharge is initiated inside the chamber where the reaction occurs. A comprehensive review of the PECVD technique is given by *Ohja* [29.84]. Dissociation of the gas in the plasma takes place, owing to the nonequilibrium nature of the plasma, which has a high electron temperature but a low gas temperature. The main advantage of this technique is that the plasma decomposition allows film deposition to take place at much lower temperatures than those employed in atmospheric CVD using decomposition reactions and also in LPCVD. In PECVD, as in LPCVD, the pressure is reduced considerably below atmospheric pressure, although the range of pressures employed varies over a wider range, and can be as high as a few hundred Pa. The discharge is usually excited by an RF field, although both DC and microwave frequency fields have also been used. Average electron energies are of the order of several eV, which is sufficient to cause ionisation and decomposition of the gas molecules. A major application of PECVD is in the deposition of silicon nitride films, an important dielectric used in semiconductor processing for microelectronic

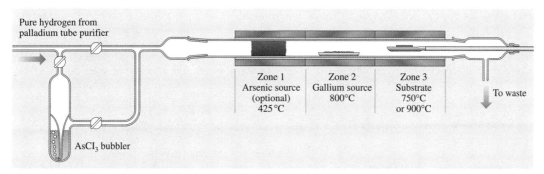

Fig. 29.8 Epitaxial growth apparatus for the production of GaAs using chemical vapour deposition (CVD). After [29.82] with permission from Elsevier Science

passivation. Here the temperatures may not exceed about 300 °C, otherwise damage to the circuitry can result, and so atmospheric CVD and indeed LPCVD are not suitable techniques. For details of the chemical reactions involved, and a comparison of the properties of silicon nitride prepared by LPCVD and by PECVD the reader is referred to the literature [29.19, 85]. The former reference also includes a table giving details for the deposition of many other thin-film materials, including silicon, germanium and other semiconductors, as well as various oxides, nitrides and carbides. PECVD systems operating at microwave frequencies utilise the phenomenon of electron cyclotron resonance (ECR); this deposition method is known as electron cyclotron PECVD (or ECR-PECVD). Frequencies are of the order of 1–10 GHz and the ECR plasma may be generated at pressures of about $10^{-3} - 10^{-1}$ Pa, leading to a degree of ionisation up to about 1000 times that achieved in an RF plasma [29.19]. High deposition rates, low-pressure operation and absence of contamination have made ECRPECVD a very attractive technique for film deposition. A full review of this technique has been given by *Popov* [29.86].

Finally we briefly mention the technique of metalorganic CVD (MOCVD), which is becoming an increasingly important method of film deposition. The technique is essentially the same as CVD, except that at least one of the precursors is a volatile metalorganic compound, such as a metal alkyl. The formation of the required compound takes place by pyrolysis of the metalorganic compound with another (inorganic or organic) gaseous precursor. The main advantage of MOCVD is the relatively high volatility of the organic compounds at moderate temperatures, and the fact that, since all these are in the vapour phase, precise control of gas flow rates and partial pressures are possible [29.19]. The compound semiconductors such as GaAs, GaP and InP may be prepared using, for example, trimethyl-Ga [(CH$_3$)$_3$Ga] or trimethyl-In [(CH$_3$)$_3$In] as precursors. The basic reaction is with a Group V hydride, e.g. AsH$_3$ or PH$_3$ for the compounds listed above. A short table of some film materials which have been successfully generated by MOCVD is given in the literature [29.19]. In addition to its applications in preparing compound semiconductor films MOCVD has also been applied to the deposition of high-temperature superconducting thin films, with quite complex structures [29.88].

Langmuir–Blodgett Film Deposition

The Langmuir–Blodgett technique of film deposition entails the coating of solid substrates with molecular

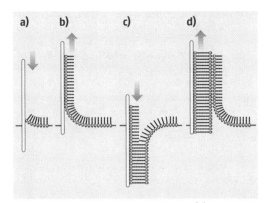

Fig. 29.9a–d Langmuir film deposition: (**a**) immersion of a hydrophilic substrate without film deposition; (**b**) withdrawal of the hydrophilic substrate; (**c**) repeated immersion with film deposition; (**d**) continuation of the process. After [29.87] with permission from Elsevier Science

layers. *Langmuir* [29.89] first described the transfer of fatty-acid organic monolayers from the surface of water to a rigid substrate, while *Blodgett* [29.90] extended the work to show that the process could be repeated so that monolayers were sequentially transferred to form multilayer films. The early history of Langmuir–Blodgett films has been briefly described in the preface to the Proceedings of the First International Conference on Langmuir–Blodgett Films [29.91]. In the basic deposition process, a fatty-acid film is spread over the surface of a subphase (usually purified water). Most molecular-film materials have a molecular structure where one end of the molecule is hydrophilic, which seeks out the water subphase, while the other end of the molecule is hydrophobic and attempts to point away from the water subphase. We therefore have the situation where the hydrophilic end of the molecule is oriented downwards towards the water, with the hydrophobic end oriented upwards. In Langmuir–Blodgett deposition, the required molecules are dispersed in a volatile solution, which eventually evaporates, leaving the molecules with their hydrophilic ends immersed in the water, but with their axes randomly oriented [29.87]. A compressive surface-barrier system is employed to compress the molecules so that they are close-packed and oriented perpendicular to the surface of the subphase, with their hydrophilic ends pointing vertically downwards. The molecules can then be transferred to either a hydrophilic or hydrophobic substrate by slowly dipping into the subphase. The basis process is shown in Fig. 29.9 [29.87], for the case of a hydrophilic substrate. On the initial introduction of

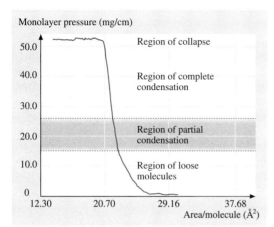

Fig. 29.10 The pressure–area relationship for a cadmium stearate molecular layer obtained at a temperature of 22 °C at pH 5.6. Film deposition should be carried out in the linear region of the curve, i.e. the region of complete condensation without collapse. After [29.87] with permission from Elsevier Science

the substrate into the subphase no coating takes place (a), but on removing it a film attaches to the substrate by polar-bond formation (b). The substrate is then effective hydrophobic, and on dipping for a second time, the hydrophobic ends of the molecules on the subphase form van der Waals bonds with the hydrophobic substrate surface, thus again making the surface hydrophilic (c). On withdrawing the substrate for a second time a further molecular layer is deposited, with the hydrophobic ends of the molecules at the surface (d). Repeated dipping cycles then establish multilayer films with an odd number of layers. The use of an initially hydrophobic substrate gives an even number of molecular layers for each completed dipping cycle.

It is evident that we have a mechanism whereby a predicted number of molecular layers may be deposited, and knowing the molecule dimensions (i.e. the length of the molecule between its hydrophilic and hydrophobic ends) together with the number of dipping cycles, it is easy to accurately calculate the film thickness. It is important, however, that the surface pressure is maintained within a fixed range of values during deposition. If it is too high the surface film collapses in localised regions, giving a transferred layer thickness greater than one monolayer; conversely, if the pressure is too low, there may be voids in the film surface, which transfer on to the substrate. This is illustrated in Fig. 29.10 [29.87], which shows a pressure–area isotherm for a cadmium stearate molecular layer. It is normal to use a feedback system during deposition to control a moveable barrier on the film surface to maintain the surface pressure within a narrow range while the dipping process takes place. Here the molecular area remains effectively constant for a large change in pressure (region of complete condensation in the figure), ensuring that small changes in pressure do not cause the deposited film to fold over or break up.

There are, however, some limitations to the use of the Langmuir–Blodgett deposition method. The material must satisfy the following requirements [29.87]. The molecular chain structure needs to be hydrophilic at one end and hydrophobic at the other. Furthermore, it should be insoluble in water, but soluble in a suitable solvent in order to give a spreading solution. Some materials which are insoluble can, however, be modified by chemical substitution of molecular end groups. For example, metal-free phthalocyanine is normally insoluble, but the chemical addition of $C(CH_3)_3$ end groups to the molecule renders it soluble in some aromatic solutions such as toluene. This type of chemical modification allows the deposition of materials which cannot normally be deposited to be deposited in a substituted form [29.92]. Many other substituted organic compounds may also be deposited, including dyes, porphyrins, fullerenes, charge-transfer complexes and polymers [29.93]. It is also possible to prepare Langmuir–Blodgett film layers with alternating monolayer structures, by dividing the Langmuir–Blodgett trough into two compartments, one for each type of monolayer, separated by a surface barrier [29.94, 95]. Deposition then takes place by alternately removing the substrate from the surfaces of the two compartments, by passing the substrate through the subphase under the surface barrier between deposition of each of the materials.

The advantages of the Langmuir–Blodgett technique over other thin-film deposition methods lie primarily in the low temperature at which deposition takes place (close to room temperature), and on the feasibility of depositing very thin mono-moleculer layers of known thickness. Several types of electronic devices incorporating Langmuir–Blodgett films have been discussed [29.96] as has the writing of very fine lines of width less than 10 nm separated by 20 nm in manganese stearate films [29.97]. Further work on the deposition of other materials by this method may well lead to novel device structures and to improved processing methods. An excellent discussion on the preparation and major properties of Langmuir–Blodgett films is given by *Petty* [29.93].

29.2 Structure

In this section some examples of the structures of various types of thin films are described. By the *structure*, we mean mainly the crystallographic form in which the film is deposited, and also factors associated with this, such as the mean size of crystallites and their orientations. However, the gross morphology of films is also included, and this covers features generally larger than the grain size. The basic type of film, i.e. epitaxial, polycrystalline or amorphous, is usually deduced by diffraction methods (X-ray, electron or neutron), although we will mainly be concerned with the first of these only. Thus a brief description of the method of X-ray crystallography is first given, followed by examples of different types of films that have been observed. Finally some examples of the larger-scale morphology are described.

Several reviews of film structure have been given previously [29.98–100], as well as more generalised discussions in most of the texts covering thin films, to which the reader is referred. However, it should be noted that the large number of variables encountered in thin-film deposition processes (e.g. substrate temperature, deposition rate, angle between vapour stream and substrate) permit films of differing structures to be deposited under nominally similar conditions. It is always wise, therefore, before interpreting measurements of any thin-film properties (such as the electrical or optical properties) to first fully determine the structure of the films, and to take this into account in any interpretation.

29.2.1 Crystallography

In general films tend to deposit from the vapour phase in one or more of their documented bulk-crystal structures. In crystals, the spatial arrangement of the atoms exhibits a symmetry, which can be replicated throughout three-dimensional space by repetition of a basic unit, called the primitive unit cell. The unit cell of any crystal structure may be specified by the magnitudes of three vectors a, b and c representing the unit cell lengths, and the angles α, β and γ between the three vectors. Symmetry considerations allow all crystals to be classified into seven crystallographic structures, namely the cubic, tetragonal, orthorhombic, hexagonal, trigonal, monoclinic and triclinic forms, of which only the first three are orthogonal. Conditions on the equality or otherwise of the lengths a, b and c and the angles between them determine into which type the structure of a given material falls. In three dimensions the atoms arrange themselves in planes, with gaps between the planes known as the planar spacing, d_{hkl}. In any perfect crystal there are many sets of crystal planes that may be identified. The possible planar spacings are given by different expressions, depending on the crystal structures. For example, in the cubic structure ($a = b = c$, $\alpha = \beta = \gamma = 90°$)

$$\frac{1}{d_{hkl}^2} = \frac{h^2 + k^2 + l^2}{a^2} \,. \tag{29.39}$$

In the tetragonal structure ($a = b \neq c$, $\alpha = \beta = \gamma = 90°$)

$$\frac{1}{d_{hkl}^2} = \frac{h^2 + k^2}{a^2} + \frac{l^2}{c^2} \tag{29.40}$$

and in the hexagonal structure ($a = b \neq c$, $c \perp a$, $\gamma = 120°$)

$$\frac{1}{d_{hkl}^2} = \frac{4}{3}\left(\frac{h^2 + hk + k^2}{a^2}\right) + \frac{l^2}{c^2} \,. \tag{29.41}$$

In these equations a, b and c represent the lengths of the sides of the primitive unit cell, and h, k and l are integers. Similar, but more complex, equations exist for the other crystal structures which have less symmetry. For bulk crystals having unit cell sides of the order of nanometres, the crystal planes act as three-dimensional diffraction gratings, which reflect an incident beam of radiation according to Bragg's law,

$$\lambda = 2d_{hkl}\sin\theta \,, \tag{29.42}$$

where λ is the wavelength of the radiation, and θ is the angle between the incident beam and the reflecting planes. The values of λ for X-rays, electrons and neutrons may be chosen such that they are of the same order as d_{hkl}, and therefore diffraction will occur at suitable angles. These angles are those which satisfy Bragg's law, and are therefore determined by the radiation wavelength λ and by the integer values (h, k and l) applying to the sets of reflecting planes present. (hkl) then specifies the set of reflecting planes, and also the orientation.

For perfect crystals, reflections only occur at certain well-defined angles. If the crystal is rotated with respect to the incident beam of radiation, the deviation angle between the incident and reflected beam is 2θ. If the material, instead of being perfect, is made up of a large number of differently oriented microcrystallites, i.e. it is polycrystalline, the diffraction method will pick up reflections from a large number of individual crystallites, and therefore provide data on all the reflecting planes present. A particular set of planes will only reflect at a specific value of 2θ, and by observing the reflecting angles the planar spacings d_{hkl} can be determined. By reference to standard tabulated data (such as the *International Tables of Crystallography*), which give details of the reflections and intensities expected for most crystalline materials, a considerable amount of information can be determined. This includes the nature of the phases present, their preferred orientations and a measure of the microcrystallite size.

X-ray diffraction is the most commonly used technique. The penetration depth of X-rays can be quite significant, and if the film is too thin they can penetrate down to the substrate. Thus X-ray diffraction often gives information on the substrate as well as the film. If this is to be avoided, the film thickness needs to be increased or the X-rays directed at a low glancing angle with respect to the substrate surface. Electrons have a considerably lower penetration ability, and therefore give information only on the immediate surface region. Neutron diffraction is used particularly in the investigation of magnetic films.

29.2.2 Film Structure

Apart from the detailed type of crystal structures observed, it is useful to divide the structures of thin films into three main categories, according to their crystallographic perfection (or lack of it). Epitaxial films are essentially perfect films grown on a substrate surface with all the atoms arranged predictably in their correct crystallographic positions. There may be certain defects, such as vacancies, interstitial atoms or impurities, but the deviation from perfection should be relatively small. Furthermore, there may be some slight deviations in the unit cell dimensions from those of the corresponding bulk material, owing to idiosyncrasies of the deposition method or the substrate surface, and in particular to stresses built up in the film during deposition. Various methods have been devised to deposit epitaxial films, such as molecular-beam epitaxy (MBE) and vapour-phase epitaxy, both of which are used in the semiconductor industry for preparation of high-quality films. Non-epitaxial films are frequently polycrystalline, having a large number of grains, oriented in different directions, and separated by grain boundaries. The grain size is heavily influenced by the deposition conditions, and can have a profound effect on secondary properties, such as the electrical conductivity, in which grain boundaries influence the electron mean free path. Finally, amorphous films have no long-range order, being noncrystalline, with only short-range order apparent.

Substrate temperature during deposition and subsequent thermal history after deposition can have a profound effect in determining the initial structure, orientation and subsequent phase changes. Some examples of these effects are included in the following discussion.

In the case of epitaxial films, a large number of factors determine whether the film is indeed epitaxial, and if so the crystalline orientation that will result. The main concepts that underlay the growth of epitaxial films are the equilibrium thermodynamics of the growing nuclei, the evaporation process as described by the Hertz–Knudsen expression (29.5), surface kinetics and mobility on the substrate and the growth of defects [29.101]. The interdependence and interaction of these various processes, which are generally all temperature-dependent, lead in many cases to a particular value of temperature (usually called the epitaxial temperature) only above which epitaxy can take place. Depending on the relative difference between the unit cell dimensions of the substrate and the films (often quantified as the *misfit*) there is frequently a considerable stress between them, which is released by a network of imperfections (dislocations). Details of the epitaxial growth are strongly related to various theories of nucleation. There are several of these, based on the characteristics of microscopic nuclei on the substrate. These grow during the deposition process, and if the conditions are suitable, can result in the deposition of epitaxial films. In the capillarity model [29.9, 27] the nuclei become stable when they reach a particular size (the critical nucleus). Above this size the nucleus is stable and continues to grow. However, in this case the size of the critical nucleus is of the order of 100 atoms or more. An alternative theory of nucleation has been proposed for critical nuclei consisting of only a few atoms [29.102, 103]. In this model certain geometries have been postulated for the critical nucleus under different conditions of supersaturation and substrate temperature. The rate of nucleation is given by

$$I = Ra_0^2 N_0 \left(\frac{R}{\gamma N_0}\right)^{n^*}$$
$$\times \exp\left(\frac{(n^*+1)Q_{ad} - Q_D + E_{n^*}}{k_B T}\right) . \quad (29.43)$$

In this expression the symbols have the following meanings: R is the impingement rate of atoms on the substrate, a_0 is a characteristic jump distance of the atoms across the substrate surface, N_0 is the concentration of adsorption sites, $\gamma = h/k_B T$ is the vibrational frequency, where T is the substrate temperature, and n^* represents the number of atoms in the critical nucleus. The remaining terms in the numerator of the exponential represent energies: Q_{ad} is the binding energy of a single atom to the substrate surface, Q_D is the activation energy for surface diffusion of adsorbed atoms and E_{n^*} is the dissociation energy of a critical nucleus containing n^* atoms. By substituting $n^* = 1, 2, 3$, etc., into (29.43), expressions can be obtained for critical nuclei of size 1, 2 or 3 atoms. These correspond to stable nuclei of size $(n^* + 1) = 2, 3$ or 4. By simultaneously solving pairs of these equations, it is possible to calculate the temperature at which the size of the critical nucleus changes from one value of n^* to another. For example, the temperature at which the critical nucleus size changes from one to two atoms, T_{12}, is given by

$$T_{12} = -\frac{(Q_{ad} + E_2)}{k \ln(R/\gamma N_0)}, \tag{29.44}$$

where E_2 is the dissociation energy of a two-atom nucleus into single atoms. The model has been tested, notably in the case of a face-centred cubic metal (silver) deposited on a high-purity (100) NaCl substrate [29.103]. At low substrate temperatures and high supersaturations, it was found that the nucleation rate I is proportional to R^2, which from (29.43) implies that $n^* = 1$ (the critical nucleus is just one atom). At higher temperatures, critical nuclei were found to have $n^* = 2$ or 3, with stable nuclei adopting the (111) or (100) orientations. Further discussion is given in several later reviews [29.1, 10, 16, 19]. More-sophisticated models of nucleation are based on the kinetic behaviour of nuclei, and involve a set of coupled differential equations, relating growth and dissociation of nuclei. More details are given in the literature [29.105, 106].

Another example of where the occurrence of epitaxy is defined by well-defined temperature ranges, is given in the work of *Kalinkin* et al. [29.107] *Muravjeva* et al. [29.104]. This important work has also been discussed in other reviews [29.108, 109]. These workers made extensive measurements on the evaporation of both cadmium and zinc chalcogenides (sulphides, selenides and tellurides), which were evaporated from Knudsen cells at a pressure of about 10^{-2}–10^{-3} Pa. The substrates used were mica in all cases. It was found that for all three cadmium compounds, and all three zinc compounds, the conditions for epitaxial growth were dependent on both the temperature of the substrate (the epitaxial temperature T_{Ep}) and the evaporator temperature T_{ev}. The latter influenced the evaporation rate N_e (29.5) and also the supersaturation ratio of the growing films. For each of the six compounds, different combinations of T_{Ep} and T_{ev} were investigated. We discuss in detail only the cadmium compounds, although the behaviour of the zinc compounds is entirely similar, as can be seen from Fig. 29.11 [29.104]. Below about $T_{Ep} \approx 310$–320 °C, the evaporator temperature for epitaxy T_{ev} increased approximately linearly with T_{Ep} according to $T_{ev} = A_1 + T_{Ep}$, where A_1 is a constant. Above $T_{Ep} \approx 310$–320 °C, T_{ev} decreased linearly with T_{Ep} according to $T_{ev} = A_2 - 2T_{Ep}$, where A_2

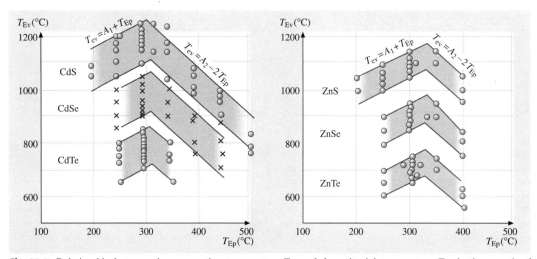

Fig. 29.11 Relationship between the evaporation temperature T_{ev} and the epitaxial temperature T_{Ep} in the growth of cadmium and zinc chalcogenide films on mica. After [29.104] with permission from Elsevier Science

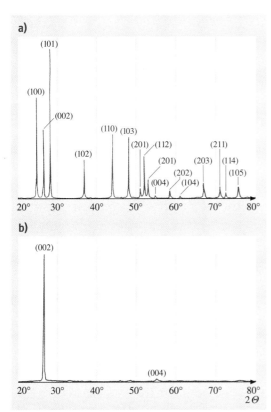

Fig. 29.12a,b X-ray diffraction traces obtained from (**a**) CdS powder, and (**b**) a CdS thin film. After [29.110] with permission from Wiley–VCH

is a second constant. These relationships between the epitaxial temperature and the evaporator temperature are illustrated in Fig. 29.11 [29.104]. For the cadmium chalcogenides, values of the range of the constants A_1 and A_2 for which epitaxial behaviour was observed are $A_1 = 800-950\,°\mathrm{C}$ and $A_2 = 1750-1900\,°\mathrm{C}$ for CdS, $A_1 = 600-750\,°\mathrm{C}$ and $A_2 = 1550-1700\,°\mathrm{C}$ for CdSe, and $A_1 = 400-550\,°\mathrm{C}$ and $A_2 = 1350-1500\,°\mathrm{C}$ for CdTe [29.104]. It is evident from these values that for each substance the evaporator temperature T_ev lies within a range of about $150\,°\mathrm{C}$ for values of T_Ep at which epitaxy is possible. In addition, for a given substrate temperature the evaporator temperature required for epitaxial CdS is highest, becoming progressively lower for CdSe and for CdTe. For example for $T_\mathrm{Ep} = 320\,°\mathrm{C}$, the evaporator temperatures for epitaxy are $1120-1270\,°\mathrm{C}$ for CdS, $920-1070\,°\mathrm{C}$ for CdSe and $720-870\,°\mathrm{C}$ for CdTe. The same sequence is also followed for the zinc chalcogenides. At combinations of temperatures outside the limits given by the values of A_1 and A_2, the films deposited were not epitaxial, being polycrystalline and only partially ordered.

We now consider some examples of polycrystalline thin films, and the factors which account for their various crystallographic characteristics. X-ray diffractometry is particularly useful in determining the basic crystal structure adopted by a particular film, particularly when two or more phases are possible. The cadmium chalcogenides have, on average, four valence electrons per atom, which leads to tetrahedral bonding. However, there are two different ways in which tetrahedra formed from each of the constituent atoms can penetrate into each other. This leads to two allowed crystal structures: the hexagonal wurtzite structure and the cubic zinc-blende structure. The planar spacings for these are described by (29.41) and (29.39), respectively. In the evaporant material, which is usually composed of a polycrystalline powder, a large number of diffraction peaks normally occur, which can be indexed using the appropriate planar spacing expression. Figure 29.12a [29.110] shows an example of a diffraction trace obtained from CdS evaporant powder. The diffraction planes are indexed in terms of the hexagonal structure, and rule out the possibility of the material being of the cubic form. More interestingly, Fig. 29.12b shows a trace obtained from a thin film prepared from this material. Although the two peaks shown correspond to those observed for the evaporant powder, it is clear that reflections from most of the crystal planes are missing. If the intensities of the peaks are different from those for a random powder (or more specifically from those tabulated in the *International Tables of Crystallography*), the crystallites are said to be preferentially oriented. In this case it can be seen that virtually all the crystallites must be aligned in the same direction, with the c-axis directed perpendicular to the substrate plane, along a preferential orientation in the [002] direction.

Apart from the simple determination of film crystal structures and preferential orientations, it is also possible to monitor changes from one crystal structure to another. For example, the phthalocyanines are organic crystals, which are known to possess several basic crystal structures, of which the α-phase and the β-phase are the most common. There is some controversy over whether the α-phase is orthorhombic or monoclinic, and it was originally identified as tetragonal [29.111]. The β-phase is monoclinic. However, X-ray diffraction has identified most films prepared at lower substrate temperatures as being of the α-structure, while those deposited at higher temperatures are of the β-structure. Both copper phthalocyanine (CuPc) and cobalt phthalocyanine (CoPc) films deposited on to substrates held at room temperature were of the α-structure, but on annealing at 240 °C for an extended period, the CuPc films underwent a phase transition to the β-structure [29.112]; the CoPc films also showed a transition to the β-phase on annealing at temperatures up to 325 °C [29.113]. The α-form films of both materials were preferentially aligned in the [001] direction assuming a tetragonal structure, or the [002] direction assuming an orthorhombic or monoclinic structure. The annealed β-form films were also preferentially oriented, in the [20$\bar{1}$] direction for CuPc and the [001] direction for CoPc. In the latter case it was possible to monitor the dominant peak over the temperature range 200–325 °C where the phase change occurred. d_{hkl} increased slightly with annealing temperature up to 200 °C, then decreased significantly at 250–300 °C. The value of 2θ decreased slightly with T_{ann} up to 200 °C, again followed by a significant *increase* at approximately 250–300 °C. For T_{ann} up to 250 °C d_{hkl} lay in the range 1.271–1.277 nm and 2θ in the range 6.92–6.95°, which is close to those expected for α form CoPc; at higher annealing temperatures d_{hkl} was in the range 1.241–1.246 nm and 2θ in the range 7.09–7.12°, which are in better agreement with those expected for β form CoPc.

X-ray diffraction also allows an estimate of the mean grain size L in polycrystalline films to be obtained, by detailed observation of the peak profiles. The well-known Scherrer equation [29.114]

$$L = \frac{K\lambda}{\eta \cos\theta} \quad (29.45)$$

allows an estimate of L to be made from simple details of the peak profiles. In (29.45), λ represents the X-ray wavelength, η the width of a strong peak in radians at half-maximum intensity, and θ is the Bragg angle, where K is a constant which is approximately unity. An example of the use of this method is in the determination of L from the first Bragg peak of α-form CoPc films before annealing [29.113]. The films were subsequently heated to a higher temperature T_{ann} in oxygen-free nitrogen for two hours and the value of L again measured. Measurements were repeated at even higher temperatures until the dependence of L on T_{ann} over the entire temperature range of the α-to-β phase change was determined. These results are illustrated in Fig. 29.13 [29.113]. It is clear that at temperatures up to 200 °C there was an increase in the mean microcrystallite size from about 30 nm to over 40 nm, although there was no indication of a phase

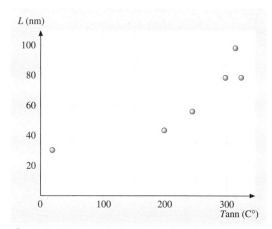

Fig. 29.13 Mean crystallite grain size L as a function of maximum cumulative annealing temperature T_{ann} for a cobalt phthalocyanine film of thickness 0.1 μm. After [29.113] with permission from Wiley–VCH

change. *Ashida* et al. [29.115] ascribed this effect in CuPc to preliminary crystallite growth at the early stages of the phase transformation process. Above annealing temperatures of about 200 °C there was a rapid increase in crystallite size to about 100 nm at 315 °C; similar behaviour has also been observed in which α-form CuPc films were heated to 300 °C when the diffraction patterns became sharper and the crystallites larger [29.115].

Some general conclusions concerning the dependence of grain size in vapour-deposited polycrystalline films on various film attributes have been summarised by *Chopra* [29.18]. The major feature determining mean grain size is the surface mobility of adsorbed atoms and clusters during the deposition process. A high surface mobility will allow adsorbed atoms to move large distances before being incorporated into growing nuclei, resulting in a relatively low concentration of nuclei and thus larger crystallites. The substrate temperature is the most obvious factor contributing to high mobility, and therefore grain size tends to increase with increasing substrate temperature. We have already seen that epitaxy requires a minimum substrate temperature; this too is a consequence of the surface mobility requirement. The total film thickness obviously sets an upper limit on the crystallite size, and therefore the grain size increases in general with film thickness. Subsequent annealing at temperatures greater than the substrate temperature during deposition can also cause an increase in the mean crystallite size as was shown in Fig. 29.13, as surface atoms augment the crystallites and grains coalesce.

The dependence of grain size on deposition rate is not consistent between different materials and substrates. Frequently there is an increase with increasing deposition rate [29.18], although a decrease has been observed in some cases, such as in CdTe films where the establishment of a higher concentration of adsorbed nuclei led to the growth of finer grains [29.38].

The interplay between the various deposition parameters can have a profound effect on the structure of the films. This is illustrated particularly well in Fig. 29.14 [29.116], for cadmium arsenide (Cd_3As_2) films evaporated onto mica substrates. Both the substrate temperature T_s and the growth rate v were varied, with the latter covering a wide range. Diffraction patterns indicated that the crystalline films were of the lower temperature tetragonal α-phase of Cd_3As_2; the structure of this phase is documented in the literature [29.117] and the planar spacings are given by (29.40). Amorphous films recrystallised following interaction with the electron beam in the electron microscope and also reverted to a tetragonal structure. The figure shows that, over the substrate temperature and deposition rate ranges investigated, the films could have an amorphous, polycrystalline or well-oriented crystal structure. Below approximately 125–145 °C the films were amorphous, with polycrystalline films resulting when this temperature was exceeded. The amorphous-to-crystalline transition was only weakly dependent on the deposition rate. However, the epitaxial temperature, which delineates polycrystalline and epitaxial films, increased with increasing deposition rate. For deposition rates less than about 4 nm/s (40 Å/s) there was a direct transition from the amorphous to the well-oriented crystalline structure, whereas at higher rates there was a transitional polycrystalline region. The preferred orientation of the films was [001] with the c-axis of the tetragonal structure perpendicular to the plane of the substrate. Also shown on the figure are examples of X-ray diffraction patterns for the crystalline, polycrystalline and amorphous regions. These are characterised, respectively, by discrete reflection points, *spotty* rings and diffuse rings.

Finally, a brief discussion on amorphous films is appropriate. These are films which show no evidence of a regularly repeated crystalline structure, i.e. they do not exhibit long-range order. We have already seen that in Cd_3As_2 films (Fig. 29.14), a relatively low substrate temperature during deposition leads to the growth of amorphous films. This is a general rule, applicable to most materials, and is basically the opposite to the condition for epitaxy – the deposition of amorphous films is encouraged if the surface mobility of atoms on the substrate is low. This causes the adsorbed atoms to be incorporated into the films almost at the point of impact, rather than after diffusion across the substrate surface and joining a growing nucleus. High deposition rates also predispose atoms to remain at their impact position

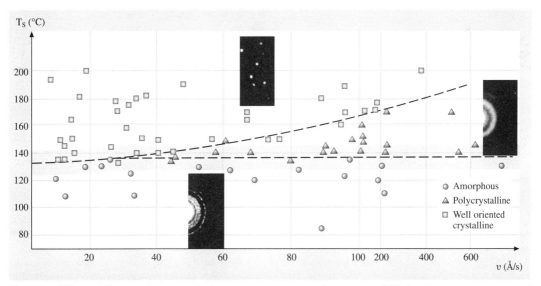

Fig. 29.14 Effects of film growth rate v and substrate temperature t_s on the structure of Cd_3As_2 films vacuum-evaporated onto mica substrates. After [29.116] with permission from Elsevier Science

within the growing film – they have only a very limited time to diffuse before they are locked into the growing film structure. These considerations are pertinent to the deposition of amorphous metallic films, which require substrate temperatures to be lowered to liquid-nitrogen temperature or below, and also to elemental chalcogenide and semiconductor films (Ge and Si) and carbon, all of which have a sufficiently low surface mobility at room temperature. A table giving examples of amorphous thin film materials, their preparation methods, and the transition from the amorphous to a crystalline phase is given by *Maissel* and *Francombe* [29.16]. The amorphous state is metastable, requiring some thermal energy to transfer it to a crystalline state.

Another method of producing amorphous films is the incorporation of residual gases during deposition. A relatively high pressure of oxygen ($\approx 10^{-3}$ Pa) can cause oxidation of the evaporant. Large-scale crystallites of metals are then unable to form. The admission of nitrogen during sputtering of refractory metals is also sufficient to result in the deposition of an amorphous phase [29.1].

Since the structure of amorphous materials do not possess any long-range order, it is necessary to describe the short-range order using some function which can be determined directly from diffraction experiments. The radial distribution function $J(r)$ is defined as the number of atoms lying at a distance between r and $r + dr$ from the centre of an arbitrary atom at the origin [29.119]. It is given by

$$J(r) = 4\pi r^2 \rho(r). \quad (29.46)$$

Here, $\rho(r)$ is an atomic pair correlation function, which is equal to zero at values of $r < r_1$, where r_1 is the average nearest neighbour interatomic separation, $\rho(r)$ is equal to the average value of density ρ^0 at large values of r where the material becomes homogeneous. Between these values $\rho(r)$ is oscillatory, with the peaks in the distribution corresponding to the average interatomic separations [29.119]. $J(r)$ also shows oscillations superimposed on the average density parabola $4\pi r^2 \rho^0$, and the area under a particular peak gives the coordination number for the given shell of atoms. The coordination number represents the number of nearest neighbours in the structure. An example of a radial distribution function, derived from X-ray data, for thin films of the compound Ge_xTe_{1-x} where $x = 0.11$, is shown in Fig. 29.15 [29.118]. For materials having covalent bonds the coordination number is four, whereas for amorphous metals, which generally crystallise in close-packed structures, the coordination number is higher.

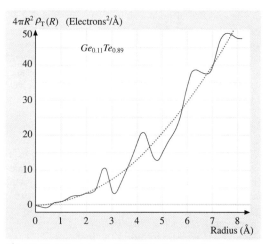

Fig. 29.15 Calculated radial distribution function for a sample of composition $Ge_{0.11}Te_{0.89}$. The *dashed curve* represents the contribution of the average electron density. After [29.118] with permission from Elsevier Science

For covalently bonded materials, the atoms tend to arrange themselves in tetrahedra, which in the amorphous state interconnect at their apexes. However, the apexes themselves interconnect within the material in a random manner, whereas the tetrahedra dimensions are effectively determined by the size of the atoms. Thus there is no long-range order because of the random interconnections, but the short-range order survives, and is similar to that for the corresponding crystalline material. *Bosnell* [29.120] has described the important amorphous semiconducting thin films, and classified them into three major categories: covalent noncrystalline solids, semiconducting oxide glasses and dielectric films. The importance of these materials, in particular their electrical properties, cannot be overestimated. The standard text on this subject [29.121] gives copious information on their electrical properties, as well as their structures and radial distribution functions. Local atomic order in amorphous films is discussed in the literature by *Dove* [29.122], with the primary emphasis being on data derived from electron diffraction. This work also gives the theory behind the radial distribution function, and provides a wide review of radial distribution functions measured in many systems performed prior to the date of the review.

From the foregoing discussion and examples, it should be clear that the crystallographic structure of thin films, whether epitaxial, polycrystalline or amorphous, depends heavily on the deposition conditions. In turn the film structure strongly influences the mechanical, thermal, optical and electronic properties.

29.2.3 Morphology

Perhaps the most striking morphological characteristic observed in thin films grown by the physical vapour deposition techniques is that of the columnar grain structure. Electron microscopy frequently reveals that high-density regions tend to grow in parallel columns. They tend to develop when the surface mobility of adsorbed atoms is relatively low, so that the area of the base of the columns is small. Further growth of the grains is away from the substrate. The direction of column growth is not necessarily perpendicular to the substrate surface, although it frequently is for depositing material arriving at normal incidence. For oblique incidence it has been observed that the columns are oriented away from the substrate normal and towards the vapour source, although not necessarily pointing directly towards it. If α is the angle between the substrate normal and the vapour stream, and β is the angle between the substrate normal and the direction of the columns, it has been shown experimentally that β is usually less than α. The tangent rule [29.124] relates these two angles and is given by

$$\tan\alpha = 2\tan\beta \qquad (29.47)$$

and was found to be followed for the case of directly evaporated aluminium. In experiments on CdS films, *Hussain* [29.125] deposited films of thickness up to 100 μm with angle of incidence α varying between 0 and 60°. Up to a certain film thickness the c-axis of the grains was perpendicular to the direction of vapour incidence, but above this thickness the c-axis started to shift towards the substrate normal, eventually becoming aligned with it. Etching of a thin film deposited at oblique incidence confirmed that the top layers of the film had their c-axis normal to the substrate, whereas in the bottom layers it was parallel to the direction of vapour incidence. In other work on this material [29.126] similar thickness-dependent orientation effects have been reported. In this case, films of thickness a few tens of nanometres had randomly orientated crystallites, while those of thickness of about 20 μm were aligned with the c-axis normal to the substrate.

Evaporated CdS films also show an interesting morphology on a somewhat larger scale, which is illustrated in Fig. 29.16 [29.123]. This illustrates features which are evident in many different types of thin films. The top surface of as-deposited films of thickness 25–30 μm were observed by scanning electron microscopy to consist of

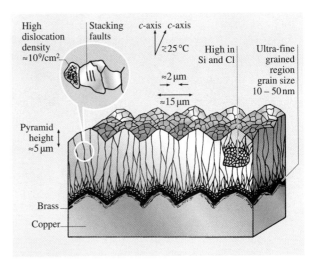

Fig. 29.16 Schematic representation of evaporated CdS film microstructure inferred from scanning electron microscopy studies. After [29.123] with permission from Elsevier Science

partially faceted hills with a height of 5 μm and a width of about 15 μm at the base. These workers also determined the grain size and crystal defect structure using transmission electron microscopy. The grain size was about 2 μm at the top and at the centre of the films, but was only of the order of 10–50 nm at the back surface adjacent to the substrate. Typical dislocation densities were of the order of 10^{13} m^{-2} (10^9 cm^{-2}). Ultrafine-grain regions of 2–15 μm in diameter were observed. These were thought to be spits of material rapidly ejected from the source. Similar features have also been observed in this laboratory [29.110]. The partially faceted protrusions were reported, and cracks in the films were also present, as a result of differential thermal contraction between the CdS films and the glass substrates during cooling. Larger regions of CdS were seen, which were considered to have been ejected from the evaporation boat during deposition. This phenomenon has been referred to as *spattering* or *splattering* by *Stanley* [29.127], and its elimination is an important goal for deposition processes operating at higher temperatures.

In early work on CdTe films [29.128] photovoltages were measured when there was an angle between the vapour stream and the normal to the substrate. This led to the suggestion that in this case the film growth was directionally anisotropic where the nucleating material depositing on a heated substrate has sufficient surface mobility to form islands rather than a uniform thin film. Additional material arrives at an angle, shadowing the

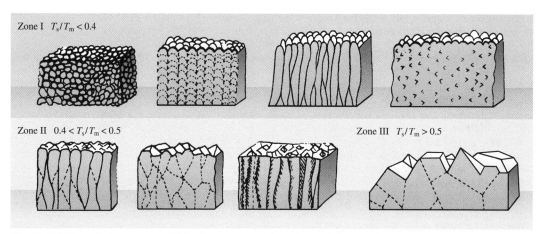

Fig. 29.17 Schematic three-zone model for the temperature dependence of the growth morphology in evaporated Cd_3As_2 films. T_s/T_m represents the ratio of the substrate temperature to the melting point of Cd_3As_2. After [29.131] with permission from Elsevier Science

far side of the island from the deposition source, and the shadowed side grows relatively slowly and is well ordered, while the exposed face grows towards the vapour source. When islands begin to touch each other there is expected to be a mismatch, leading to a disordered state in the region between the crystallites. Each interface will therefore have an ordered and a disordered side and at the interface between grains a photovoltage develops. Photomicrographs of epitaxial CdTe layers grown by the close-space technique [29.129] revealed smooth epitaxial layers when the deposition was on to (100), (110) and (111) Cd oriented substrates, but large hillocks grew on (111) Te substrate faces. When grown on mica substrates [29.130] it was observed that the morphology was significantly affected by the composition of the incident vapour phase. This was characterised by the ratios γ_{Cd} and γ_{Te}, which depended on the ratio of the actual concentration to the stoichiometric concentration of the Cd atoms or the Te_2 molecules in the gas phase. When $\gamma_{Cd} = 53$ when CdTe and Cd were co-evaporated the films were of high quality consisting of large crystallites with degenerated (111) faces. When CdTe and Te were co-evaporated and $\gamma_{Te} = 12$ formations resembling dendrites were observed.

CdSe film morphologies have been observed by several workers. *Tanaka* [29.132] prepared films using Ar and H_2 sputtering gases; some films were subsequently annealed in Ar. Microscopic examination showed that sputtering in H_2 gas led to uniform grain growth of the films over the whole substrate surface, while films sputtered and annealed using Ar gave rise to geometric pillars and a textured structure. An investigation of the optimum conditions for obtaining CdSe epitaxial layers on (0001) oriented sapphire substrates was performed by *Ratcheva-Stambolieva* et al. [29.133]. For substrate temperatures up to 460 °C the layers were polycrystalline, from 460 °C to 530 °C a small degree of preferred orientation was observed, with a more significant textured structure for temperatures of 530–570 °C. At higher temperatures of 570–620 °C a monocrystalline structure was obtained, varying in its degree of perfection depending on the actual temperature and the growth rate. Hexagonal pyramids with peaks, regular or irregular facets and flat tops were also identified. RF-sputtered CdSe films [29.134] were porous when deposited at a substrate temperature of 550 °C onto vitreous silica. When deposited onto substrates of single-crystal sapphire the surface appeared shiny, and showed features of hexagonal flat tops and pyramids with edge lengths in the range 1–5 μm.

We finally look at the morphological characteristics of another cadmium compound, Cd_3As_2. Evaporated films of this material have been investigated by *Jurusik* and *Żdanowicz* [29.131] using electron microscopy. It was found that the microstructure of the films was strongly dependent on the substrate temperature T_s. This type of behaviour is frequently addressed in terms of a zone model, and examples of some of these for other materials are discussed by *Ohring* [29.19]. In Cd_3As_2 the microstructures were interpreted in terms of a simple three-zone model in which the structure was correlated with the ratio T_s/T_m, where $T_m = 994$ K

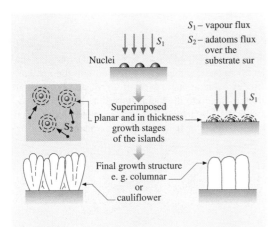

Fig. 29.18 Post-nucleation growth in evaporated Cd_3As_2 films considered as a consequence of two superimposed (planar and thickness) growth stages. After [29.135] with permission from Elsevier Science

is the melting point of Cd_3As_2 [29.131]. They found that for $T_s/T_m \leq 0.4$ the films were amorphous, consisting of spherical or elongated fibrous-like clusters, or were uniform with a large number of pores or cavities. For $0.4 < T_s/T_m < 0.5$ columnar growth was identified, while for $T_s/T_m \geq 0.5$ films with large crystallites and clear grain boundaries were observed. These structures are illustrated in Fig. 29.17 [29.131]. More recently a very detailed study was performed on amorphous Cd_3As_2 films evaporated onto substrates maintained at 293 K [29.135, 136]. It was found that the size of post-nucleation islands was a function of the substrate temperature, type of substrate and the deposition rate. Furthermore the shape and type of growth of the islands varied with these parameters. At a low deposition rate (1 nm/s) the initial nuclei grew until a certain size was reached, and new nuclei started to grow around the original islands. For a moderate deposition rate (3.5 nm/s) the initial nuclei grew into islands which in turn increased in size until the islands had only narrow interfaces between them. Some islands formed polygons. Films deposited at a high rate (50 nm/s) consisted of small islands connected by relatively large low-density areas. Films deposited at low rates were of approximately stoichiometric proportions (i.e. 40 at. % As) while those deposited at higher rates had excess arsenic (typically 51.8 at. % As). It was suggested that the growth process could be considered as a superposition of two growth stages, as shown in Fig. 29.18 [29.135]. The vapour flux onto the substrate is denoted by S_1 and the adatom flux over the substrate surface by S_2. In the first stage of growth the initial nuclei enlarge owing to the adatom flux S_2. The second stage of growth incorporates atoms immediately from the incident flux S_1. This is usually very slow, but increases as the island surface area increases. At this stage branched complex islands or only compact islands grow as the result of a shadowing mechanism. When the islands touch each other the molecular flux S_2 over the substrate surface is eliminated, and the films grow in thickness as the result of the incident flux S_1 and the structure initiated in the earlier stage (i.e. large branched or smaller compact islands). The final growth structure gives rise to cauliflower-like or columnar structures. In later work [29.136] Cd_3As_2 films were evaporated from bulk crystallites of Cd_3As_2 or $CdAs_2$. For films grown from the Cd_3As_2 source a pronounced columnar structure was observed, whereas those evaporated from the $CdAs_2$ source grew uniformly, and the columnar structure was absent. In the latter case the initial island structure was not observed in the early stages of film growth. It was concluded that prerequisites for columnar growth were a limited surface mobility and an oblique component in the deposited flux, particularly where the films are deposited normally to the substrate surface.

In this section some of the morphological characteristics of some cadmium compound films prepared by PVD methods have been described. The emphasis in this discussion of the films was not primarily because of their usefulness as semiconductors in various applications, bur because they exhibit many features typical of a wide range of other film materials. Surface mobility of atomic species on the substrate during deposition is clearly of great importance, as this tends to control the grain size at the substrate interface, while the angle of vapour incidence relative to the substrate normal is an important factor in determining the major direction of crystallite growth and whether a columnar structure develops. Shadowing of the vapour stream by islands during film growth can lead to differential growth rates at different points on the film surface. Bulk diffusion and desorption also contribute to the film morphology variations [29.19]. The morphological characteristics of many materials may be explained using zone models in which the substrate temperature and other variables, such as sputtering gas pressure, conspire to determine the overall morphology via the mechanisms detailed above. The morphology of thin films is particularly rich, and an important factor in influencing film properties.

29.3 Properties

As mentioned in the introduction, the properties covered in this section are restricted to the optical and electrical properties. Moreover, only the very fundamental properties, which are a direct consequence of the thin-film nature of the material, are covered. Thus the more advanced features, such as those exhibited by specific classes of materials, individual elements or compounds, and electronic, optical and optoelectronic devices are not normally discussed, although many of these are considered elsewhere in this Handbook. The basic thin-film properties, which have been exploited in many of these developments are stressed. Examples include the use of thin films in layers or stacks in anti-reflection coatings and mirrors [29.30], the contribution of grain boundaries in conducting thin films to the resistivity of tracks and connections in integrated circuits, the importance of high-field conduction in some electronic devices [29.17], and the possible influence of high-field conduction in nanostructures [29.11].

29.3.1 Optical Properties

This section comprises an introduction to the main optical properties of thin films, starting with a discussion of the properties of nonabsorbing single films at normal and non-normal incidence, and then considering modifications due to absorption. Finally a brief discussion of the use of multilayer film stacks is given. Thin-film optics is a particularly wide field, and readers interested in further details and a more in-depth analysis are referred to one of the standard texts [29.30, 137, 138], or indeed to most general textbooks on optics, nearly all of which cover thin films in some detail. As mentioned earlier, optical effects in thin films were first recorded in the 17th century, which predates by two centuries the emergence of Maxwell's equations, which encapsulate the behaviour of all electromagnetic radiation, including visible light. The boundary conditions at the interface between two different optical media are derivable directly from Maxwell's equations, and allow us to calculate the reflection and transmission coefficients for the amplitude of the electromagnetic wave at the interface. These are termed the Fresnel coefficients.

Figure 29.19a shows the simplest situation, where a beam of light in a medium having refractive index n_0 is incident at an angle φ_0 to the normal on a film of refractive index n_1. The following expressions for the Fresnel coefficients apply to all individual interfaces in a structure, and for all values of the refractive index. At this stage in the discussion we assume that the materials are isotropic and nonabsorbing. At the interface the electromagnetic radiation is refracted into the medium of index n_1 at an angle φ_1, with a reflected component directed back into the incident medium at an angle φ_0. Conventionally, the coefficients are quoted for components of the light beam resolved parallel (denoted by

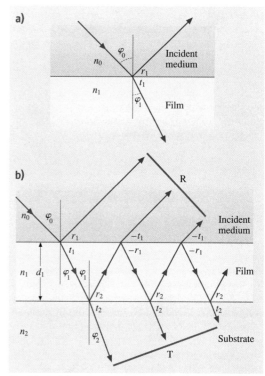

Fig. 29.19 (a) Reflection and transmission at a single interface between two media with refractive indices n_0 and n_1. r_1 and t_1 represent the amplitude reflection and transmission coefficients at the interface (the Fresnel coefficients). (b) Total reflection and transmission as the result of multiple reflection and transmission through a nonabsorbing thin film of thickness d_1 and refractive index n_1 deposited on a substrate of refractive index n_2. The Fresnel coefficients are indicated at each transit of a boundary, and change sign if the direction of incidence is reversed. R and T represent the total amplitude of reflectance and transmittance, respectively, each obtained by summing the amplitudes of an infinite number of components, although only three are shown

p) and perpendicular (denoted by s) to the plane of incidence. Therefore there are two reflection coefficients $r_{1\text{p}}$ and $r_{1\text{s}}$, and two transmission coefficients, $t_{1\text{p}}$ and $t_{1\text{s}}$. The 1 in the subscripts in this notation refers to the first interface, i.e. between the materials with refractive indices n_0 and n_1 in this case. More complex systems will have more than one interface, where 2, 3, etc., signify the second or third interfaces. The Fresnel coefficients for reflection are [29.1, 15]

$$r_{1\text{p}} = \frac{n_1 \cos\varphi_0 - n_0 \cos\varphi_1}{n_1 \cos\varphi_0 + n_0 \cos\varphi_1},$$

$$r_{1\text{s}} = \frac{n_1 \cos\varphi_1 - n_0 \cos\varphi_0}{n_1 \cos\varphi_1 + n_0 \cos\varphi_0}, \tag{29.48}$$

and those for transmission are

$$t_{1\text{p}} = \frac{2n_0 \cos\varphi_0}{n_1 \cos\varphi_0 + n_0 \cos\varphi_1},$$

$$t_{1\text{s}} = \frac{2n_0 \cos\varphi_0}{n_1 \cos\varphi_1 + n_0 \cos\varphi_0}. \tag{29.49}$$

Clearly these coefficients reduce to

$$r = \frac{n_1 - n_0}{n_1 + n_0}, \quad t = \frac{2n_0}{n_1 + n_0} \tag{29.50}$$

for the case of normal incidence where all the cosine terms become unity. There is no difference in this case between the coefficients for the p and s components. An expression for the energy reflectance at the interface R can be obtained by evaluating rr^*, where r^* is the complex conjugate of r. For nonabsorbing films r is real and this reduces to r^2, and gives for normal incidence

$$R = \left(\frac{n_1 - n_0}{n_1 + n_0}\right)^2. \tag{29.51}$$

A similar expression for the energy transmittance T at the boundary can also be derived from t, but in this case the incident and transmitted beams are in different media, and the expression is not simply r^2, but includes the ratio n_1/n_0. Full details are given elsewhere [29.18, 138].

We consider now the slightly more complex situation of a thin film of thickness d_1 and refractive index n_1 deposited on a substrate of refractive index n_2. In this case the value of n_0 for the incident medium will usually be the free-space value of unity. There will be multiple reflections as shown in Fig. 29.19b. It is clear that the total amplitude of the reflected beam is given by summing the amplitudes of the individual reflections into the incident medium, while the total amplitude of the transmitted beam is obtained by summing the amplitudes of the transmitted components into the substrate.

In traversing the film of thickness d_1, the radiation undergoes a change in phase, which is accounted for by a *phase thickness* δ_1, given by

$$\delta_1 = \frac{2\pi}{\lambda} n_1 d_1 \cos\varphi_1, \tag{29.52}$$

where λ is the wavelength in vacuum. In each case the summations lead to infinite series. These involve $\exp(-\text{i}2\delta_1)$ for reflection, where the beam traverses the film an even number of times, and $\exp(-\text{i}\delta_1)$ for transmission, where the beam traverses an odd number of times. The results of these summations give the amplitude reflectance R and transmittance T, respectively, as

$$R = \frac{r_1 + r_2 \exp(-2\text{i}\delta_1)}{1 + r_1 r_2 \exp(-2\text{i}\delta_1)} \tag{29.53}$$

and

$$T = \frac{t_1 t_2 \exp(-\text{i}\delta_1)}{1 + r_1 r_2 \exp(-2\text{i}\delta_1)} \tag{29.54}$$

In this case r_1 and t_1 represent the Fresnel coefficients between n_0 and n_1, and r_2 and t_2 the coefficients between the media n_1 and n_2. Depending on whether we are considering parallel or perpendicular polarisation (p or s) the relevant Fresnel coefficients from (29.48) and (29.49) should be used in these expressions. Equation (29.53) and (29.54) refer to the ratios of the *amplitudes* of the reflected and transmitted beams to that of the incident beam. In terms of the beam *energies*, the reflectance R and the transmittance T may be defined as the ratios of the reflected and transmitted beam energies to that of the incident energy. The energies of the reflected and transmitted beams are given by $n_0 RR^*$ and $n_2 TT^*$ respectively, where R^* and T^* represent the complex conjugates of R and T. We then obtain

$$R = \frac{n_0}{n_0} RR^* = \frac{r_1^2 + r_2^2 + 2r_1 r_2 \cos 2\delta_1}{1 + r_1^2 r_2^2 + 2r_1 r_2 \cos 2\delta_1} \tag{29.55}$$

for the reflectance, and

$$T = \frac{n_2}{n_0} TT^* = \frac{n_2}{n_0} \frac{t_1^2 t_2^2}{(1 + 2r_1 r_2 \cos 2\delta_1 + r_1^2 r_2^2)} \tag{29.56}$$

for the transmittance. Although these latter two expressions are reasonably compact, when the four Fresnel coefficients are substituted they become somewhat unwieldy. However, for the important case of normal incidence, where (29.50) applies, and where (29.52) simplifies to $\delta_1 = (2\pi/\lambda) n_1 d_1$, they reduce to still fairly long expressions, but involving only the three refractive indices and δ_1, i.e.

$$R = \Big((n_0^2+n_1^2)(n_1^2+n_2^2) - 4n_0n_1^2n_2$$
$$+ (n_0^2-n_1^2)(n_1^2-n_2^2)\cos 2\delta_1\Big) \Big/$$
$$\Big((n_0^2+n_1^2)(n_1^2+n_2^2)$$
$$+ 4n_0n_1^2n_2 + (n_0^2-n_1^2)(n_1^2-n_2^2)\cos 2\delta_1\Big)$$
(29.57)

and

$$T = 8n_0n_1^2n_2 \Big/ \Big[(n_0^2+n_1^2)(n_1^2+n_2^2) + 4n_0n_1^2n_2$$
$$+ (n_0^2-n_1^2)(n_1^2-n_2^2)\cos 2\delta_1\Big] \,.$$
(29.58)

The variations of R and T with δ_1 are shown in Fig. 29.20 [29.139] for normal incidence, where the refractive index of the incident medium $n_0 = 1$ and that of the substrate $n_2 = 1.5$. These correspond to air and glass respectively. Similar plots of these variations as a function of optical thickness n_1d_1 expressed as multiples of the wavelength λ are also common in the literature [29.1, 16, 138], where an optical thickness of unity corresponds to $\delta_1 = 2\pi$. The figure shows curves for three values of the film refractive index of 2.00, 1.70 and 1.23. For the first two of these $n_0 < n_1 > n_2$ and maxima in reflectance occur at $\pi/2$ and $3\pi/2$ with minima at 0, π and 2π, while for the third $n_0 < n_1 < n_2$ and the positions of the maxima and minima are reversed. In general the maximum and minimum reflectance values R_{max} and R_{min} depend on the relative values of n_0, n_1 and n_2. For the film index greater or less than those at both its boundaries, i.e. $n_0 < n_1 > n_2$ or $n_0 > n_1 < n_2$, we obtain

$$R_{max} = \left(\frac{n_1^2 - n_0n_2}{n_1^2 + n_0n_2}\right)^2 , \quad R_{min} = \left(\frac{n_2 - n_0}{n_2 + n_0}\right)^2$$
(29.59)

with maxima occurring when δ_1 is an odd multiple of $\pi/2$ (optical thickness n_1d_1 an odd number of quarter-wavelengths) and minima when δ_1 is an even multiple of $\pi/2$ (n_1d_1 an even number of quarter wavelengths). For the film index intermediate between those at its boundaries, i.e. $n_0 < n_1 < n_2$ or $n_0 > n_1 > n_2$ these results are reversed, and we obtain

$$R_{max} = \left(\frac{n_2 - n_0}{n_2 + n_0}\right)^2 , \quad R_{min} = \left(\frac{n_1^2 - n_0n_2}{n_1^2 + n_0n_2}\right)^2$$
(29.60)

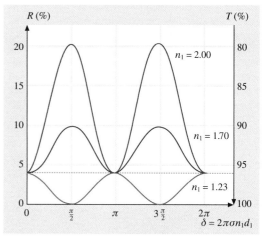

Fig. 29.20 Variation of the energy reflectance R and transmittance T with phase thickness δ_1 for a nonabsorbing film at normal incidence for various film indices n_1, with $n_0 = 1$ and $n_2 = 1.5$. The wavenumber σ is the reciprocal of the wavelength in vacuum λ. After [29.139] with permission from Elsevier Science

with maxima occurring when δ_1 is an even multiple of $\pi/2$, and minima at odd multiples (or maxima for n_1d_1 an even number of quarter wavelengths and minima an odd number). Equation (29.60) predicts that zero reflectance occurs when $n_1 = \sqrt{(n_0n_2)}$, and for the refractive index values of Fig. 29.20 give $n_1 = 1.23$. Thus a simple *anti-reflection coating* can be made by using a film of this index with optical thickness $\lambda/4$. This is clearly evident from the figure at the points where δ_1 has values of $\pi/2$ and $3\pi/2$. Another point which emerges from the figure is that if the refractive index of the film is the same as that of the glass, the system behaves as if the glass is uncoated. This situation is shown by the dotted line in the figure. Here the reflectance is 4%, exactly as predicted by (29.51) for an uncoated substrate (i.e. $n_0 = 1$ and $n_1 = 1.5$).

If the electromagnetic light wave is attenuated by a film layer, that layer is said to be *absorbing*, in contrast to the *transparent* layers considered previously. Providing the incident light beam is incident *normally* on the surface of the absorbing medium, the theory described above may be applied, with the proviso that the refractive index n is replaced by a *complex refractive index* $\mathbf{n} = n - \mathrm{i}k$. The quantity k is called the *extinction coefficient*, and characterises absorption of energy within the film. Substitution of the complex refractive index in the appropriate expressions for the Fresnel coefficient of (29.50), and in the expression for the phase thickness of (29.52) at normal incidence allows the amplitude reflectance R to be obtained. The energy reflectance R can be obtained by substituting the complex expressions for r and δ directly into (29.53) and evaluating RR^*. For more than two or three layers the results become intensely unwieldy, and matrix methods as mentioned below are considerably more useful. A simple example of the use of the complex refractive index will suffice. If we take the case of an incident beam in a nonabsorbing medium such as air with refractive index n_0, incident upon an absorbing medium with complex refractive index $\mathbf{n}_1 = n_1 - ik_1$, the amplitude reflectance can be determined by replacing n_1 by $n_1 - ik_1$ in (29.50) and evaluating rr^*. In this case the energy reflectance at the boundary becomes

$$R = \frac{(n_0 - n_1)^2 + k_1^2}{(n_0 + n_1)^2 + k_1^2} \quad (29.61)$$

in contrast to the result of (29.51) for a nonabsorbing medium.

In general, in absorbing films, the planes of constant phase are not parallel to the planes of constant amplitude for the electromagnetic wave disturbances. Only for the special case of normal incidence is this true, and thus in absorbing films it is usual to consider normal incidence only. The general case of oblique incidence is discussed in the standard texts on the optics of thin films [29.137, 138]. The behaviour of the energy reflectance and transmittance depend on the optical constants n and k. We have seen that in transparent films, R and T oscillate with the film thickness due to interference effects. In absorbing films with small k values, the oscillatory behaviour remains, but the amplitudes of the maxima decrease with increasing films thickness. For higher absorption values, particularly in the case of metals, intensity maxima are not observed, and an exponential decrease in transmittance with film thickness is seen.

Of particular use in the case of thin film stacks containing several layers is the matrix formalism, which allows the reflection and transmission properties to be determined. A 2×2 matrix characterises each layer of the stack, and allows the electric and magnetic field strengths to be expressed in terms of those of the adjacent layer. For transparent films this is given for the r-th layer in the form [29.140]

$$M_r = \begin{pmatrix} \cos \delta_r & i \sin \delta_r / n_r \\ i n_r \sin \delta_r & \cos \delta_r \end{pmatrix}, \quad (29.62)$$

where δ_r is the phase thickness and n_r is the refractive index of the r-th layer. The electric and magnetic field intensities in the adjacent $(r-1)$-th layer, E_{r-1} and H_{r-1}, are related to those in the r-th layer E_r and H_r by

$$\begin{pmatrix} E_{r-1} \\ H_{r-1} \end{pmatrix} = M_r \begin{pmatrix} E_r \\ H_r \end{pmatrix}. \quad (29.63)$$

Repeated application of (29.63) to a stack of layers enables E_0 and H_0 in the incident medium (n_0) to be expressed in terms of the field intensities in any other layer of the stack. Knowing the n values for each of the various layers of the stack, a *characteristic matrix* (also 2×2) for the entire stack is obtained by repeated matrix multiplication of the individual layer matrices, and allows E_0 and H_0 to be obtained in terms of the transmitted field intensities at the opposite side of the stack. The energy reflectance is then easily obtained. Similar matrices to those in (29.62) may also be defined for the case of absorbing films, although it is then usual to work in terms of the layer admittance η, defined as the ratio of the transverse components of the magnetic to the electric vectors [29.138].

A relatively simple example of an interesting system is a stack of layers of equal optical thickness, but with alternating high (H) and low (L) refractive indices [29.140]. If the optical thicknesses correspond to $\lambda/4$ (a quarter-wave stack), there is a highly reflective region at the wavelength λ, with other highly reflecting regions at 2λ, 3λ, etc. The reflectance value becomes higher the more layers there are in the stack, and approaches unity for a few layers. If the refractive index of the incident medium is n_0, that of the substrate n_S, and those of the high- and low-index regions n_H and n_L, respectively, the energy reflectance for an even number of layers $2s$ (n_L next to substrate) is given by

$$R_{2s} = \left(\frac{n_S f - n_0}{n_S f + n_0} \right)^2 \quad (29.64)$$

where $f = (n_H/n_L)^{2s}$. For an odd number of layers $(2s+1)$, with n_H next to the substrate, the energy reflectance becomes

$$R_{2s+1} = \left(\frac{n_H^2 f - n_0 n_S}{n_H^2 f + n_0 n_S} \right)^2. \quad (29.65)$$

The bandwidth of the highly reflecting region depends on $\sin^{-1}[(n_L - n_H)/(n_L + n_H)]$, and thus is limited by the refractive indices available. Bandwidth can be increased by using a stack with layer thicknesses increasing in geometrical progression.

In addition to antireflection and highly reflecting coatings, various types of optical filters can be designed. Perhaps the simplest is the Fabry–Perot interference filter, which consists of two highly reflecting stacks, separated by a layer of optical thickness $\lambda/2$. This results in a very narrow transmission band at wavelength λ, with a rejection region on either side. Filter designs are usually based on computer evaluations of the optical matrices. These include band-pass, band-stop and edge filters. These are well beyond the scope of the present article and the reader is referred to the literature [29.30, 138, 141, 142].

In this section some of the basic optical properties have been described, with the primary emphasis on non-absorbing films at normal incidence. Several interesting film arrangements have been described, again for non-absorbing films. Implicit to any system there is some absorption, but this can be accounted for if both the optical constants are known. Usually the design process involves the manipulation of several layer matrices, but this is handled effectively by modern computer technology. The optical properties of highly absorbing films are described by *Abelès* [29.143], who also considers inhomogeneous films. Other useful references on the theory and detailed optical properties of various aspects of thin films may be found in the literature: theory and calculations [29.144], antireflection coatings [29.145], filters [29.142, 146, 147] and mirrors [29.148].

29.3.2 Electrical Properties

The electrical properties of thin films are one of their most important physical features, and differ from those of the corresponding bulk material for various reasons. Electrical properties cover a particularly wide field, and in this section only two important examples are covered. The first of these is the lateral conductivity (or corresponding resistivity) of thin metallic films. In metals we are usually concerned only with one type of charge carrier, electrons, and thus variations in conductivity σ due to the presence of both carrier types (such as compensation effects) are not considered. The main cause of deviations from the bulk material conductivity values is curtailment of the bulk electron mean free path λ_0 by the dimensions of the film. Clearly, if the film thickness $d < \lambda_0$, many instances of electron scattering occur at the film surfaces, thus reducing the effective electron mean free path and decreasing the conductivity. Modern theories also take into account electron scattering by individual grain boundaries as well as by the film surfaces. The second example of electrical properties in thin films is that of high-field conduction in insulators and semiconductors. Electric fields in thin films may approach the dielectric breakdown strength, and under these conditions the usual ohmic behaviour is not followed, but a range of different limiting conduction processes may take place. Conduction in semiconductors at somewhat lower electric fields is not discussed here, but is covered in detail elsewhere in the literature [29.1, 15, 18, 149]. Semiconductor conductivity is a broad subject, underlying the operation of many different electronic devices. We might note that the conductivity of semiconductors in their surface regions (and therefore also in semiconductor thin films) differs from that of the bulk material owing to depletion or accumulation regions at the surfaces, and that the application of a surface electric field forms the basis of several semiconductor devices, such as metal–oxide–semiconductor (MOS) transistors. Furthermore, the mobility of carriers in semiconductors frequently follows similar variations to those of the conductivity in metallic films, for essentially the same regions, in that the surface and/or grain boundary scattering significantly influence the mobility and thus the conductivity.

Electrical conductivity in metallic films

The electrical conductivity in bulk metals σ_0 is given generally by an expression of the form

$$\sigma_0 = \frac{ne^2 \lambda_0}{mv}, \tag{29.66}$$

where n is the concentration of free electrons, e is the electronic charge, λ_0 the bulk electron mean free path, m the electron effective mass and v the electron Fermi velocity. This expression results from the free-electron gas theory of metals, derived from the work of Drude, Lorentz and Sommerfeld. The main feature of this result is that the electrical conductivity is proportional to the mean free path, and that if the mean free path is curtailed by any means, then the conductivity will be reduced, all other considerations remaining unaltered. A further consideration is the well-known result termed Matthiessen's rule,

$$\rho_0 = \rho_d + \rho_{ph}(T), \tag{29.67}$$

where the bulk resistivity $\rho_0\ (= 1/\sigma_0)$ is given by the sum of contributions from static defects and impurities ρ_d and from phonons $\rho_{ph}(T)$. The latter contribution effectively isolates the main temperature-dependent effect from other scattering mechanisms, unless they too are temperature-dependent. Temperature-dependent phonon interactions increase with increasing temperature, giving rise to the positive temperature coefficient of resistivity (TCR) in metals. At very low temperatures $\rho_{ph}(T) \to 0$, and only the residual resistivity ρ_d remains. For the case of a thin metal film, scattering at the surfaces also comes into play, and Matthiesen's rule is modified by the addition of an extra surface-scattering resistivity term ρ_s to give

$$\rho_f = \rho_d + \rho_{ph}(T) + \rho_s \tag{29.68}$$

for the thin-film resistivity ρ_f. For very thick films $\rho_s \approx 0$, since the mean free path is not then limited by the surfaces, and the film resistivity is effectively the same as the bulk value. When ρ_s is significant for thinner films, $\rho_f > \rho_0$, and the film conductivity $\sigma_f < \sigma_0$. The above discussions follow naturally from the treatise in Chapt. 2.

Over a century ago, *Thomson* [29.151] was the first to suggest that observed decreases in the conductivity of thin films as compared to the bulk values are the result of the limitation of the electron mean free path by the specimen size. He made the following assumption in obtaining an expression for the film conductivity: the bulk mean free path in the metal is a constant λ_0, greater than the film thickness d, and that when electrons collide with the internal film surfaces the scattering is independent of the angle at which they strike, i.e. the scattering is diffuse. Collisions occurring in the film were split into two regions, in the first the mean free path is determined by the surfaces (top and bottom) and in the second the mean free path is not limited by scattering at the surfaces. In terms of the parameter

$$k = \frac{d}{\lambda_0}, \tag{29.69}$$

the ratio of the film thickness to the bulk mean free path, the ratio of the film to bulk conductivity was determined as

$$\frac{\sigma_f}{\sigma_0} = \frac{\rho_0}{\rho_f} = \frac{k}{2}\left[\ln\left(\frac{1}{k}\right) + \frac{3}{2}\right]. \tag{29.70}$$

Fuchs [29.150] pointed out that this derivation neglected mean free paths starting at the film surfaces, and also the statistical distribution of the mean free paths about λ_0 in the bulk metal. A corrected version of Thomson's model for very thin films, i.e. $d \ll \lambda_0$ ($k \ll 1$), where effectively all the mean free paths start from one of the surfaces, yields an alternative result [29.152]

$$\frac{\sigma_f}{\sigma_0} = \frac{\rho_0}{\rho_f} = k\left[\ln\left(\frac{1}{k}\right) + 1\right]. \tag{29.71}$$

Although both expressions predict a decrease in conductivity below the bulk value for films, neither gives a sensible prediction for very thick films, i.e. $d \gg \lambda_0$ ($k \gg 1$).

A more useful general model to account for variations of conductivity with film thickness was introduced by *Fuchs* [29.150], who formulated the problem in terms of the Boltzmann transport equation, which describes the way in which the electron energy changes due to applied fields and collisions. Readers are referred to the original paper, or to one of the reviews mentioned at the end of this section, for details of the mathematical arguments. The final result of this model is

$$\frac{\sigma_f}{\sigma_0} = \frac{\rho_0}{\rho_f} = 1 + \frac{3}{4}\left(k - \frac{k^3}{12}\right)B(k)$$
$$- \frac{3}{8k}\left(1 - e^{-k}\right) - \left(\frac{5}{8} + \frac{k}{16} + \frac{k^2}{16}\right)e^{-k}, \tag{29.72}$$

where $B(k)$ is a representative of the set of exponential integral functions, given by

$$B(x) = -\text{Ei}(-x) = \int_x^\infty \frac{e^{-y}}{y}\,dy. \tag{29.73}$$

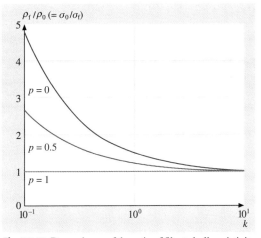

Fig. 29.21 Dependence of the ratio of film to bulk resistivity ρ_f/ρ_0 on the ratio of film thickness to bulk mean free path $k = d/\lambda_0$ predicted by the model of *Fuchs* [29.150]; p is the specularity parameter, with $p = 0$ representing totally diffuse scattering and $p = 1$ representing totally specular scattering, where for the latter $\rho_f/\rho_0 = 1$. The curve labelled $p = 0.5$ is intermediate between these two values

Figure 29.21 shows the dependence of ρ_f/ρ_0 ($= \sigma_0/\sigma_f$) on k over the range 0.1–10 for the curve labelled $p = 0$ (diffuse scattering). Here it is clear that the resistivity drops to the bulk value for thick films. Various approximations for (29.72) are frequently quoted to aid computation in specific thickness ranges. For thick films ($k \gg 1$)

$$\frac{\sigma_f}{\sigma_0} = \frac{\rho_0}{\rho_f} \approx 1 - \frac{3}{8k}. \tag{29.74}$$

This clearly shows that as $k \to \infty$, $\sigma_f \to \sigma_0$. For thinner films ($k \ll 1$), Campbell [29.153] quotes

$$\frac{\sigma_f}{\sigma_0} = \frac{\rho_0}{\rho_f} \approx \frac{3k}{4}\left[\ln\left(\frac{1}{k}\right) + 0.4228\right] \tag{29.75}$$

and for *very* thin films Sondheimer [29.154] gives

$$\frac{\sigma_f}{\sigma_0} = \frac{\rho_0}{\rho_f} \approx \frac{3k}{4}\left[\ln\left(\frac{1}{k}\right)\right]. \tag{29.76}$$

Equations (29.70), (29.71), (29.75) and (29.76) give some confidence in the general approach, since they are all of comparable form, with the right-hand sides each consisting of a constant $\times k \times [\ln(1/k) + $ another constant$]$.

Fuchs also considered the case where the electron scattering at the surfaces was not totally randomised, and that a constant proportion of electrons p are scattered specularly. p is termed the *specularity parameter*, where $p = 0$ for the case of totally diffuse scattering and $p = 1$ for totally specular scattering. Clearly in the latter case the film surfaces do not affect the conductivity, and σ_f reduces to the bulk value. Figure 29.21 also shows the variation of ρ_f/ρ_0 for $p = 0.5$ (intermediate between diffuse and specular scattering) and $p = 1$ (totally specular scattering). Again for the case of some specular scattering there are limiting expressions for thick films ($k \gg 1$) and thin films ($k \ll 1$), given respectively by

$$\frac{\sigma_f}{\sigma_0} = \frac{\rho_0}{\rho_f} \approx 1 - \frac{3(1-p)}{8k} \tag{29.77}$$

and

$$\frac{\sigma_f}{\sigma_0} = \frac{\rho_0}{\rho_f} \approx \frac{3k}{4}(1+2p)\left[\ln\left(\frac{1}{k}\right) + 0.4228\right], \tag{29.78}$$

where p is small in (29.78). For totally diffuse scattering, where $p = 0$, these two expressions reduce to the previous results of (29.74) and (29.75). An alternative expression for thin films is also given by Sondheimer [29.154],

$$\frac{\sigma_f}{\sigma_0} = \frac{\rho_0}{\rho_f} \approx \frac{3k}{4}\left(\frac{1+p}{1-p}\right)\ln\left(\frac{1}{k}\right). \tag{29.79}$$

This reduces to (29.76) for $p = 0$, and for small k and p is consistent with (29.78).

Both Lucas [29.155] and Juretschke [29.156, 157] have extended the specularity analysis by assuming a different specularity parameter for each side of the film, i.e. p and q. This is of course quite reasonable, in that the specularity parameter of the lower (substrate) surface is likely to be heavily influenced by the substrate surface. The thick- ($k \gg 1$) and thin-film ($k \ll 1$) approximations in this case are then

$$\frac{\sigma_f}{\sigma_0} = \frac{\rho_0}{\rho_f} \approx 1 - \frac{3}{8k}\left[1 - \left(\frac{p+q}{2}\right)\right] \tag{29.80}$$

and

$$\frac{\sigma_f}{\sigma_0} = \frac{\rho_0}{\rho_f} \approx \frac{3k}{4}\frac{(1+p)(1+q)}{(1-pq)}\ln\left(\frac{1}{k}\right), \tag{29.81}$$

respectively. For identical surfaces with $p = q$, these expressions reduce to (29.77) and (29.79) respectively. Tabulated values of ρ_f/ρ_0 are given by Chopra [29.18] as functions of k for various combinations of p and q. The whole topic of the size-dependent electrical conductivity in thin metallic films and wires is thoroughly reviewed by Larson [29.158], but omitting the important work on grain-boundary scattering of Mayadas and Shatzkes [29.159]. This is, however, covered in a more recent book reviewing size effects in thin films generally [29.160], which also includes useful tabulated conductivity data.

As mentioned above, an important development in size-effect models came about with the inclusion of grain-boundary scattering [29.159]. In a bulk sample, the individual grains are normally considerably larger than the bulk mean free path λ_0, and do not therefore affect the conductivity. However, in thin films, the grain sizes and orientations depend on the preparation method and conditions. Typical grain sizes are of the same order as λ_0 and therefore scattering at their boundaries will influence the conductivity. Mathematically, the model takes account of the fact that grains tend to grow in a columnar structure, with the axes normal to the film plane. Single grains usually extend between both film surfaces, and thus their upper and lower boundaries do not affect the conductivity. Mayadas and Shatzkes argue that only grain boundaries normal to the film plane influence the conductivity. The grain boundaries are represented by δ-function potentials, and are used in the model in the solution of the Boltzmann transport equation, subject to suitable boundary conditions. The model is developed in terms of a dimensionless parameter α, given by

$$\alpha = \frac{\lambda_0}{D}\frac{R}{1-R}, \tag{29.82}$$

where, as before λ_0 is the bulk mean free path, D is the mean grain dimension, and R is the grain-boundary electron reflection coefficient. The ratio of the grain-boundary conductivity σ_g (and the corresponding resistivity ρ_g) to the bulk values are then given by

$$\frac{\sigma_g}{\sigma_0} = \frac{\rho_0}{\rho_g} = 3\left[\frac{1}{3} - \frac{1}{2}\alpha + \alpha^2 - \alpha^3 \ln\left(1 + \frac{1}{\alpha}\right)\right]. \quad (29.83)$$

This is easy to evaluate, and the dependence of ρ_g/ρ_0 ($=\sigma_0/\sigma_g$) on α is shown in Fig. 29.22. For small values of α (i.e. large grain size D, small bulk mean free path λ_0, or small reflection coefficient R) the grain boundaries do not affect the overall conductivity, which remains at the bulk value. However, for large α the grain-boundary resistivity increases (conductivity decreases). For very large ($\alpha \gg 1$) and very small ($\alpha \ll 1$) values of α, (29.83) has the limiting approximations

$$\frac{\sigma_0}{\sigma_g} = \frac{\rho_g}{\rho_0} \approx \frac{4}{3}\alpha \quad (29.84)$$

and

$$\frac{\sigma_0}{\sigma_g} = \frac{\rho_g}{\rho_0} \approx 1 + \frac{3}{2}\alpha, \quad (29.85)$$

respectively. Note that following the original work [29.159] the latter two expressions are quoted as the ratio ρ_g/ρ_0 ($=\sigma_0/\sigma_g$), i.e. the reciprocal of the ratio in (29.83).

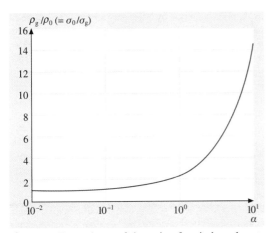

Fig. 29.22 Dependence of the ratio of grain-boundary to bulk resistivity ρ_g/ρ_0 on the parameter α from the model of *Mayadas* and *Shatzkes* [29.159]. The data are calculated from (29.83)

Mayadas and *Shatzkes* [29.159] also considered the case of grain-boundary scattering in conjunction with scattering at the film surfaces, and followed the *Fuchs* [29.150] formulation to arrive at an expression for the overall film resistivity ρ_f. Unfortunately, the resulting expression could not be evaluated analytically, requiring numerical solution. Detailed comparisons between the basic Fuchs model with $p=0$ and $p=0.5$, and their model using the same values of p and including grain-boundary scattering with $\alpha = 1$, led these workers to interpret their results as showing an enhancement of the intrinsic resistivity ρ_g in polycrystalline samples in which grain size is constant and independent of film thickness. However, discrepancies between the Fuchs and Mayadas–Shatzkes models become most apparent for very large α (very small grain size, D). For this situation the deviation corresponds to the additional resistivity arising from grain boundaries rather than from surface scattering [29.161, 162]. Useful tabulated data of resistivity ratios in samples with both grain-boundary and surface scattering are given elsewhere in the literature [29.160, 163].

Finally, we briefly discuss the thickness variations of the temperature coefficient of resistivity. This is a measure of the response of the resistivity to changes in temperature, and is usually denoted by the symbol α. To avoid confusion with the Mayadas–Shatzkes parameter, here we denote it by α_T. The temperature coefficient of resistivity is given for the bulk (α_{T0}) and for thin films (α_{Tf}) by

$$\alpha_{T0} = \frac{1}{\rho_0}\frac{d\rho_0}{dT}; \quad \alpha_{Tf} = \frac{1}{\rho_f}\frac{d\rho_f}{dT} \quad (29.86)$$

where ρ_0 and ρ_f are the bulk and thin-film resistivities, as given in (29.67) and (29.68). From the above relationships, and Matthiesen's rule, it is possible to derive expressions for α_{Tf}/α_{T0} in a similar manner as ρ_f/ρ_0. Referring to (29.68), ρ_d is essentially temperature-independent, while the temperature dependence of ρ_s depends on the film thickness. For thicker films ($k \gg 1$), ρ_s is also temperature-independent, and thus

$$\frac{d\rho_f}{dT} = \frac{d\rho_{ph}}{dT} \quad (29.87)$$

Similarly, from (29.67) for the bulk material

$$\frac{d\rho_0}{dT} = \frac{d\rho_{ph}}{dT} \quad (29.88)$$

and therefore both the differentials in (29.86) are equal, and we obtain

$$\alpha_{T0}\rho_0 = \alpha_{Tf}\rho_f. \quad (29.89)$$

This useful result implies that

$$\frac{\sigma_f}{\sigma_0} = \frac{\rho_0}{\rho_f} = \frac{\alpha_{Tf}}{\alpha_{T0}} \quad (29.90)$$

and that for the case of thicker films the ratio of the film-to-bulk temperature coefficient of resistivity is equal to the ratio of film-to-bulk conductivity. For the case of thinner films ($k \ll 1$), the situation is complicated by the fact that ρ_s is also temperature-dependent, because the mean free path decreases with increasing temperature and becomes significant for small k. Various calculated data for α_{Tf} are given elsewhere in the literature [29.6, 18, 153, 158, 160], to which the reader is referred for a fuller discussion. Discussions concerning the applicability of the various models described above to several types of metallic film are also given in some of these references, and a wider coverage including additional contributions to the size effect models, which cannot be included in the present work, is given in the book by *Tellier* and *Tosser* [29.160].

High-field conduction in insulating and semiconducting thin films

Here electrical conductivity is considered for the case where a voltage is applied across an insulating or semiconducting thin film prepared in a sandwich configuration between metallic electrodes. This situation is in direct contrast to that described in the previous section, in that the intrinsic film conductivity is significantly lower, and the electric field applied can be extremely high, even when only low voltages are applied. The latter point is clearly illustrated by the fact that in a film of thickness 1 μm subjected to a voltage of only 10 V, the mean electric field established across the film is 10^7 V/m, approaching the dielectric breakdown strength of many materials. Several high-field conduction mechanisms that may in principle occur in thin dielectric films have been thoroughly reviewed in the literature [29.164]. These have also been observed in relatively wide-band-gap semiconducting films. These processes are discussed later in this section, together with the appropriate current density–voltage expressions which aid in their identification and analysis. First, however, it is useful to discuss the types of electrical contact which may be applied to dielectric films, since the dominant conduction process is frequently determined by the type of electrical contact and its interfacial properties.

Simmons [29.164] has demonstrated that the type of electrode used significantly influences the conduction processes which may be observed; the various conduction mechanisms were classified as either *bulk-limited* or *electrode-limited*. For the former the charge carriers are generated in the bulk of the material and the electrodes merely serve to apply a potential which generates a drift current. In the latter, at the interface between the electrode and the insulator or semiconductor, a potential barrier to the flow of charge is established, which limits the current. Contacts themselves are said to be either *ohmic* or *blocking*, with an intermediate *neutral* contact. Blocking contacts are conventionally known as Schottky barriers. In principle, the type of contact formed depends on the relative work functions of the insulator or semiconductor ψ_i and that of the metal contact ψ_m. Excellent energy-band diagrams of the various types of contact are given in the literature [29.164], and are reproduced elsewhere in descriptions of contacts to cadmium compound films [29.38] and to nanostructures [29.11]. In equilibrium the vacuum and the Fermi levels are continuous across the interface, and with no applied voltage the Fermi level also must be flat, in order that no current flows. Away from the interface in the bulk of the material, the energy difference between the vacuum and Fermi levels will be equal to the work function ψ_i, and it can easily be seen that the potential barrier height at the interface φ_0 is given by

$$\varphi_0 = \psi_m - \chi, \quad (29.91)$$

where χ is the film electron affinity and ψ_m is the metal work function. However, it is the *type* of contact rather than its precise barrier height which determines the type of conductivity observed. We consider primarily electrons as the charge carriers, as it has been argued that for insulators conduction by holes may normally be ignored, owing to relatively low mobilities and to immobilisation by trapping effects. Similar arguments are also applicable in the case of wide-band-gap semiconductors, and thus for both insulators and some semiconductors n-type conductivity is usual. For p-type conductivity, the conditions concerning the relative values of the work functions are reversed [29.165].

Ohmic contacts occur for electron injection when $\psi_m < \psi_i$. Electrons are injected from the electrode into the insulator or semiconductor in order to comply with thermal equilibrium requirements. They are subsequently located within the insulator or semiconductor conduction band and penetrate a distance λ_0 below the interface, forming a region of negative space charge which is termed an *accumulation region*. Upwards band bending of the conduction band occurs in the accumulation region, beyond which the dielectric is shielded from the electrode. The accumulation region acts as a charge reservoir, and can supply electrons to the material as required by the bias conditions. It has been pointed out by *Simmons* [29.164] that the conductivity is limited by the rate at which electrons can flow into the bulk of the material, and is therefore bulk-limited. Conversely, blocking contacts or Schottky barriers occur when $\psi_m > \psi_i$. In this case the initial flow of charge is reversed, with electrons flowing from the insulator or semiconductor to the metal electrode. In contrast to the case for ohmic contacts, this results in a positive space-charge region near the interface in the insulator or semiconductor, which, owing to the absence of electrons, is termed a *depletion region*. An equal and opposite negative charge is induced on the electrode, and the interaction between the two charges establishes a local electric field near the interface. In this case the conduction band in the insulator or semiconductor bends downwards, and the free-electron concentration in the interfacial region is much lower than in the bulk of the film material. The rate of flow of electrons is thus limited by the rate at which they can flow over the interfacial barrier, and the conductivity is electrode-limited.

Although, in principle, the type of contact established depends on the relative values of ψ_m and ψ_i, in reality such considerations are usually inadequate in predicting the type of contact formed, because of the existence of *surface states*. These are due either to the effects of unsaturated bonds and impurities at the interface [29.166] or to the departure from periodicity in the structure at the interface. It is very difficult to accurately predict a priori the presence and concentration of surface states. If a significant concentration of surface states does exist, (29.91) is inapplicable and the type of contact and the conduction process are determined almost entirely by their presence.

In the remainder of this section the bulk-limited conduction processes are first considered, where the conductivity does not depend on the barrier height at the interface. Following this the electrode-limited processes are considered, where the barrier height becomes a significant feature in the current density–voltage (J–V) equations.

A particularly important conduction process in insulating and semiconducting thin films is that of space-charge-limited conductivity (SCLC). For SCLC to occur there is a requirement that the injecting electrode is an ohmic contact. Thus there is a charge reservoir available in the accumulation region, and electrons do not need to be excited over a potential barrier. At low voltages the thermally generated carrier concentration exceeds the injected concentration, and the current density J is proportional to the applied voltage V, following Ohm's law,

$$J = n_0 e \mu \frac{V}{d}, \quad (29.92)$$

where n_0 is the thermally generated carrier concentration (electrons), e is the electronic charge, μ the mobility and d the film thickness. When the injected electron concentration exceeds that of the thermally generated concentration the SCLC current becomes dominant. SCLC current density–voltage relationships are derived by taking into account both the drift and diffusion currents, and solving Poisson's equation. This results in an expression for the electric field as a function of distance from the interface. Charge-carrier traps normally exist within the insulating or semiconducting film, and these are effective in mopping up or immobilising most of the injected electrons. For shallow traps located at a discrete energy E_t below the conduction band edge the SCLC current density is given by [29.164]

$$J = \frac{9}{8} \varepsilon_r \varepsilon_0 \theta \mu \frac{V^2}{d^3}, \quad (29.93)$$

where ε_r is the relative permittivity of the material, ε_0 the permittivity of free space and θ is the ratio of free to trapped charge. This expression predicts that $J \propto V^2$ and $J \propto d^{-3}$, and the search for these relationships is an important tool in identifying SCLC dominated by a discrete trap level. For perfect materials which do not contain traps, θ is replaced by unity in (29.93). The ratio of free to trapped charge is given by [29.167]

$$\theta = \frac{N_c}{N_{t(s)}} \exp\left(-\frac{E_t}{k_B T}\right), \quad (29.94)$$

where N_c is the effective density of states at the conduction band edge, $N_{t(s)}$ is the shallow trap concentration located at the discrete energy level, k_B is Boltzmann's constant and T is the absolute temperature. A transition between ohmic conductivity and SCLC takes place when the injected carrier concentration first exceeds the thermally generated carrier concentration. An expression for the transition voltage V_t at which this occurs is obtained by simultaneously solving (29.92) and (29.93) to yield

$$V_t = \frac{8 e n_0 d^2}{9 \theta \varepsilon_r \varepsilon_0}. \quad (29.95)$$

SCLC of this type, showing square-law J–V behaviour, has been observed in many types of thin films. In SbS$_3$ dielectric films, with bismuth oxide ohmic contacts, the current was at first ohmic, before a transition to SCLC [29.168]. Current density was also found to be proportional to $1/d^3$ at constant voltage, and the transition voltage V_t showed the characteristic linear variation with d^2. It has also been observed in CdS films [29.40, 169, 170], particularly with In injecting electrodes, and a discrete-trap concentration of 2.5×10^{20} m^{-3} was deduced [29.40]. In CdTe, a square-law SCLC behaviour was recorded in electro-deposited Ni−CdTe−Au structures [29.171]. The work also showed a systematic increase in current density with temperature over the range 273–338 K, together with the expected $J \propto 1/d^3$ behaviour. *Dharmadhikari* [29.172] observed similar behaviour in evaporated CdTe films in the lower voltage range, although power-law behaviour became apparent at higher voltages. CdSe films have also been reported as showing square-law SCLC [29.173, 174]. In organic films, the phthalocyanines are well know for exhibiting SCLC behaviour [29.111]. Square-law behaviour with discrete trap levels has been seen in several of these materials, with typical activation energies of 0.77 eV [29.175] and 0.62 eV [29.176] in copper phthalocyanine, 0.42 eV in lead phthalocyanine [29.177], 0.58 eV in nickel phthalocyanine [29.178] and 0.45 eV and 0.69 eV in cobalt phthalocyanine [29.179]. These values appear to vary according to the central metal atomic species in the molecule, and also with the oxygen content. In general these materials are p-type, having a trap concentration for discrete levels in the range 5×10^{16}–7.1×10^{25} m^{-3}.

For amorphous and polycrystalline materials, where a single shallow trap level is unlikely to be present, several poorly defined trap levels can lead to an overall exponential distribution of traps $N(E)$ [29.167] given by

$$N(E) = N_0 \exp\left(-\frac{E}{k_B T_t}\right), \quad (29.96)$$

where $N(E)$ is the trap concentration per unit energy range at an energy E below the conduction band edge, N_0 is the value of $N(E)$ at the conduction band edge and $T_t > T$ is a temperature parameter which characterises the distribution. Furthermore, it can be shown that the total concentration of traps comprising the distribution $N_{t(e)}$ is [29.181]

$$N_{t(e)} = N_0 k_B T_t. \quad (29.97)$$

Lampert [29.182] obtained an expression for the current density by assuming an exponential trap distribution as described by (29.96) to yield

$$J = e\mu N_c \left(\frac{\varepsilon_r \varepsilon_0}{e N_0 k_B T_t}\right)^l \frac{V^{l+1}}{d^{2l+1}}. \quad (29.98)$$

This expression predicts a power-law dependence of J on V, with the exponent $n = l+1$, where l represents the ratio T_t/T. J is also expected to be proportional to $d^{-(2l+1)}$. Similarly to the case for a single shallow trap level, there is also a transition voltage between ohmic conduction and SCLC dominated by an exponential trap distribution. The transition voltage in this case is given by [29.183]

$$V_t = \left(\frac{n_0}{N_c}\right)^{1/l} \frac{e N_0 k_B T_t d^2}{\varepsilon_r \varepsilon_0}. \quad (29.99)$$

Many instances of power-law SCLC are also reported in the literature. *Zuleeg* [29.169] observed a cubic power-law dependence at higher voltages in his CdS films, which suggested that an exponential trap distribution was present. *Dharmadhikari* [29.172] appears to have been the first to observe a power-law J–V dependence in CdTe films at higher voltage levels, in addition to the square-law dependence at lower voltages. Very high exponents were recorded, although at the time these was not analysed using the exponential trap distribution expressions given above. This connection was made in a later series of measurements [29.180], where the intermediate square-law behaviour was absent. This

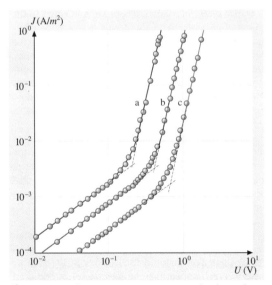

Fig. 29.23a–c Room-temperature current density–voltage characteristics for evaporated Al−CdTe−Al sandwich structures of thickness 0.31 μm (**a**), 0.45 μm (**b**) and 0.56 μm (**c**), showing ohmic conduction and SCLC. U is the applied voltage. After [29.180] with permission from Wiley–VCH

J–V dependence is shown in Fig. 29.23 [29.180] for three different film thicknesses, and is typical of many results on various different materials showing an exponential trap distribution. Similar exponential trap distributions have also been identified in some CdSe films [29.184–186], notwithstanding the earlier results which indicated discrete trap levels [29.173, 174]. The differences between these two classes of results appear to depend on the different deposition conditions and processing used. Exponential trap distributions are extremely prevalent in phthalocyanine films, and were originally investigated by *Sussman* [29.187] for copper phthalocyanine films with ohmic Au electrodes. The material was shown to be p-type, and exhibited exponent values n in the J–V characteristics in the range 2.6–4.0. The predicted thickness dependence for this type of conductivity was also observed. Many other phthalocyanine films have subsequently shown this type of conductivity, and data on these are tabulated and compared in the literature [29.111]. Total trap concentrations for the exponential distributions are in the range 6×10^{20}–9.3×10^{26} m^{-3}.

Using the above Eqs. (29.96–29.98) as a starting point, it was shown by the present author [29.188] that for SCLC dominated by an exponential trap distribution, measurements of J as a function of temperature at constant applied voltage in the SCLC region are sufficient to determine both the mobility and the trap concentration. If the data are plotted in the form $\log_{10} J$ against $1/T$ the curves should be linear, and when extrapolated to negative values of $1/T$, should all intersect at a common point irrespective of the applied voltage. The coordinates of this point are given by

$$\log_{10} J = \log_{10}\left(\frac{e^2 \mu d N_c N_{t(e)}}{\varepsilon_r \varepsilon_0}\right); \quad \frac{1}{T} = -\frac{1}{T_t}. \tag{29.100}$$

Other useful results enabling data to be extracted from these plots are that for the gradient, given by

$$\frac{\mathrm{d}(\log_{10} J)}{\mathrm{d}(1/T)} = T_t \log_{10}\left(\frac{\varepsilon_r \varepsilon_0 V}{e d^2 N_{t(e)}}\right) \tag{29.101}$$

and for the intercept $\log_{10} J_0$ on the $\log_{10} J$ axis, which is

$$\log_{10} J_0 = \log_{10}\left(\frac{e \mu N_c V}{d}\right) \tag{29.102}$$

This set of equations has been used to determine mobility and trap concentration in several different materials showing this form of SCLC, including copper phthalocyanine, where these quantities were measured as functions of the ratio of evaporation background pressure to deposition rate [29.189]. In another example the method was used to determine these quantities in semiconducting CdTe films [29.190].

We might mention that *Rose* [29.167] also explored the possibility of a uniform trap distribution, in addition to discrete energy levels and the exponential distribution. In this case $J \propto V \exp(tV)$, where t is a temperature-dependent constant, but this is observed far less frequently than the square-law and power-law behaviour described above. An example is in the case of electrodeposited p-type CdTe films [29.191], where Rose's expression was followed and the trap concentration per unit energy range was 10^{21} eV^{-1} m^{-3}. It was also observed in early work on copper phthalocyanine films with ohmic Au electrodes [29.192]. The discussion above for all types of SCLC considers carrier injection of a single type only. Double injection, when one electrode is ohmic for electrons and the other ohmic for holes, is covered in detail in the text by *Lampert* and *Mark* [29.193] on current injection in solids, and is also discussed by *Lamb* [29.166].

The second bulk-limited conduction process to be discussed is the Poole–Frenkel effect, which is essentially the field-assisted lowering of the Coulombic potential barrier φ between electrons located at impurity sites and the edge of the conduction band. Electrons in such centres are unable to contribute to the conductivity until they overcome the potential barrier φ and are promoted into the conduction band. For an applied electric field F the potential is reduced by an amount eFx where e is the electronic charge and x is the distance from the centre. For high electric fields the lowering of the potential can be significant. Such a variation is illustrated in the literature, usually for the case of a donor level located an energy E_d below the bottom of the conduction band [29.164]. The potential energy of the electron in the Coulombic field is $-e^2/4\pi\varepsilon_r\varepsilon_0 x$, and there is also another contribution to the potential energy, $-eFx$, resulting from the applied field F. The combined effect of both these contributions to the potential energy is that there is a maximum in the potential energy for emission at a distance x_m from the centre. The effective potential barrier for Poole–Frenkel emission is then lowered by an amount $\Delta\varphi_{\mathrm{PF}}$, which depends on the electric field according to the relationship

$$\Delta\varphi_{\mathrm{PF}} = \beta_{\mathrm{PF}} F^{1/2}, \tag{29.103}$$

where

$$\beta_{\mathrm{PF}} = \left(\frac{e^3}{\pi\varepsilon_r\varepsilon_0}\right)^{1/2} \tag{29.104}$$

is referred to as the Poole–Frenkel field-lowering coefficient. The conductivity, and consequently the current density, depend exponentially on the potential barrier for excitation of electrons into the conduction band, and since this is lowered by $\Delta\varphi_{PF} = \beta_{PF} F^{1/2}$ the current density is increased by an exponential factor. The current density is given by

$$J = J_0 \exp\left(\frac{\beta_{PF} F^{1/2}}{kT}\right), \quad (29.105)$$

where $J_0 = \sigma_0 F$ is the low-field current density. Although J_0 in this expression depends linearly on F, it is frequently ignored in comparison with the exponential term, which increases far more rapidly with F. Equation (29.105) is usually applicable for thin-film insulators and semiconductors in the presence of traps. If, rarely, traps are absent and the low-field conductivity σ_0 shows an intrinsic dependence proportional to the factor $\exp(-E_g/2kT)$, where E_g is the energy gap, then a factor of 2 appears in the denominator of the exponential term in (29.105) [29.164]. Since $F = V/d$, the above expression can be written

$$J = J_0 \exp\left(\frac{\beta_{PF} V^{1/2}}{kT d^{1/2}}\right). \quad (29.106)$$

Therefore a plot of $\log J$ against $V^{1/2}$ should show a linear relationship in the case of Poole–Frenkel conductivity, from which a value of the field-lowering coefficient can be deduced and compared with the theoretical value of (29.104). If it is necessary for the variation of the pre-exponential factor to also be taken into account, then a linear plot of $\log(J/V)$ against $V^{1/2}$ should be obtained.

Modifications to (29.105) have been proposed to account for the somewhat different current density dependence in the presence of various combinations of donors and traps. Simmons [29.164] mentioned the case where the material contains donor levels below the Fermi level as well as shallow neutral traps, in which the coefficient of $F^{1/2}/kT$ is one half of the theoretical value given by (29.104). For the case of a nonuniform electric field, having a maximum value of $\alpha^2 F$, where F is the mean field, the following expression has been proposed [29.195]

$$J = \frac{2J_0 kT}{\alpha \beta_{PF} F^{1/2}} \exp\left(\frac{\alpha \beta_{PF} F^{1/2}}{k_B T}\right). \quad (29.107)$$

Although this expression was originally proposed for the case of CdTe films, it has also been adopted for the interpretation of results in other materials.

The Poole–Frenkel effect has been observed in many types of thin film, including insulating oxides, semiconductors and organic materials. The main indicator of Poole–Frenkel conductivity is a linear dependence of $\log J$ on $V^{-1/2}$, although it is also seen in the case of the Schottky effect, as will be discussed later. Care must be taken in distinguishing between the two processes. In SiO films, good agreement with (29.106) was obtained [29.196], but the value of β derived exceeded the theoretical Poole–Frenkel value. This type of discrepancy had previously been discussed in the literature [29.197], and it was later proved that in this model the J–V characteristics were symmetric over five orders of magnitude of applied voltage, irrespective of electrode material, which ruled out the Schottky mechanism [29.198]. In another important insulator, silicon nitride, Sze [29.199] observed Poole–Frenkel behaviour in metal–insulator–semiconductor (MIS) diodes fabricated on a degenerate Si substrate, over a range of nearly four decades in current. It was found that the current density versus electric field characteristics were independent of film thickness, sample area, electrode materials and polarity, suggesting that the

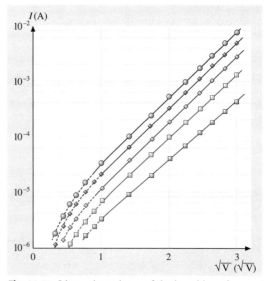

Fig. 29.24 Linear dependence of the logarithm of current on the square root of the applied voltage in an evaporated Al–CdTe–Al sandwich structure of thickness 722.5 nm at 0 °C (*bottom curve*), 30 °C, 55 °C, 80 °C and 108 °C (*top curve*). The behaviour was identified with a form of the Poole–Frenkel effect. After [29.194] with permission from Elsevier Science

conductivity was bulk-limited rather than electrode-limited, and therefore eliminating the Schottky effect. Poole–Frenkel conductivity has also been observed in several semiconducting cadmium compounds, such as RF-sputtered CdS films [29.200, 201] and more widely in CdTe films [29.194, 195, 202]. In Al–CdTe–Al samples, Poole–Frenkel-type conductivity was consistently observed over the temperature range 0–108 °C. These results are typical of Poole–Frenkel behaviour, and are illustrated in Fig. 29.24 [29.194]. However, the value of the experimental field-lowering coefficient increased from 4.95×10^{-5} eV m$^{1/2}$ V$^{-1/2}$ at 0 °C to 7.7×10^{-5} eV m$^{1/2}$ V$^{-1/2}$ at 108 °C, and was about twice the theoretical β_{PF} for CdTe. The Schottky effect was eliminated, as the results did not appear to depend on the electrode workfunction when the Al electrodes were replaced with either In or Ag. These workers adopted the suggestion of *Jonscher* and *Ansari* [29.198] that electrons can hop between sites as the result of thermal activation; on this assumption good agreement was found with the predicted β_{PF}, with an activation energy between hopping sites of 0.15 eV. Very similar results to these were also observed subsequently [29.195], and again the measured value of β exceeded the theoretical Poole–Frenkel value. This coefficient was also calculated from the gradients of the ln J–$V^{1/2}$ characteristics as a function of temperature, and consistent values were obtained. It was suggested that the similarity between the results of this work and the earlier work [29.194] was related to individual centres experiencing varying electric fields, i.e. the electric field was nonuniform. Equation (29.107) was proposed to account for this effect, and has been successful in accounting for the enhanced coefficients sometimes obtained. Poole–Frenkel conductivity was also identified in p-type CdTe, which also shows a change in conductivity type from SCLC to Poole–Frenkel conductivity [29.203]. This conduction mechanism has also been observed in other cadmium compounds, namely CdSe [29.186] and Cd$_3$As$_2$ [29.204]. In phthalocyanine films several instances of this effect have been reported. In oxygen-doped copper phthalocyanine films with a Au and an Al electrode, slightly enhanced values of β were obtained and attributed to a nonuniform electric field, whereas in annealed films the field appeared to be uniform [29.205]. Agreement with the basic model was also found for samples with a Au and a Pb electrode [29.175]. Lead phthalocyanine films with Au and Al electrodes exhibited two different values of β [29.177] consistent with Poole–Frenkel conductivity at higher fields and the Schottky effect at lower fields. However other workers had not considered that Poole–Frenkel conductivity was apparent in the same system [29.206], preferring an interpretation based on two different Schottky barrier widths. Both Schottky and Poole–Frenkel behaviour were observed in oxygen-containing lead phthalocyanine films with two Au electrodes [29.207].

Hopping is a third bulk-limited conduction process which is observed in thin films. This type of conductivity is observed particularly in noncrystalline materials, and is thoroughly discussed in various texts concerned with this topic [29.121, 208]. In this type of material the lack of long-range order results in a phenomenon known as localisation, where the energy levels do not merge into a continuum, particularly in the band-tail regions at the edge of the energy bands. Electrons are transported through the material in a series of jumps or *hops* from one localised energy level to another. The hopping process can occur between the localised energy levels even when only small amounts of thermal energy are available, because the localised levels are normally very closely spaced in energy. For this reason, the process can take place at very low temperatures when other processes are energetically impossible. *Mott* and *Davis* [29.121] argued that the conductivity σ exhibits different behaviour in different temperature regions. At higher temperatures thermal excitation of carriers to the band edges is possible and extended-state or free-band conductivity can take place, while at lower temperatures, where less thermal energy is available, hopping may occur. Different varieties of hopping may take place, and these are distinguished by the length of the hop. Nearest-neighbour hopping is self-explanatory, while in variable-range hopping the hops are on average further than to the nearest neighbour, but the energy difference between the states involved is lower. In the case of variable-range hopping the conductivity has been shown to follow a relationship of the form [29.208]

$$\sigma = \sigma_0 \exp\left(-\frac{A}{T}\right)^{1/4} \qquad (29.108)$$

where σ_0 and A are constants. This relationship is known as the Mott $T^{1/4}$ law, since a plot of log σ against $T^{-1/4}$ shows a linear characteristic with negative slope.

Hopping conduction has been observed in many materials, and is extensively reviewed in the literature [29.121, 208, 209]. The $T^{-1/4}$ law is observed in amorphous Ge [29.210], with the results of *Walley* and *Jonscher* [29.211] also showing this behaviour [29.121]. There is also an example of its appearance in amorphous carbon films [29.212]. AC hopping effects have been observed in several cadmium chalcogenides, and this work is reviewed elsewhere [29.38]. DC hopping was identified in CdTe films with Al electrodes [29.213] following earlier work on AC conductivity [29.172]. The results suggested a hopping process at temperatures below about 175 K, with a hopping centre concentration of about 9×10^{12} m^{-3} and a hopping mobility of 2.16×10^{-5} m^2/V s, the latter considerably lower than the free-band mobility applicable at higher temperatures. The extent of the electron wavefunction was estimated to be 0.69 nm, corresponding to the lattice parameter in CdTe, and supporting the existence of a nearest-neighbour hopping process. Hopping at lower fields was also identified in CdSe films [29.214] and associated with an impurity conduction process proposed previously [29.215]. Variable-range hopping following the $T^{-1/4}$ law has been observed in iron phthalocyanine and in triclinic lead phthalocyanine films [29.216, 217], although the temperature range was restricted. In the work on iron phthalocyanine the law was followed over the range 140–220 K, with particularly good agreement in the range 156–175 K. In the lead phthalocyanine films similar very good linear behaviour was observed in the temperature range of approximately 220–260 K. Hopping at low temperatures has also been indicated from AC measurements in various phthalocyanine systems [29.218–221].

The first electrode-limited conduction process to be considered is that of tunnelling, which is a quantum-mechanical effect, in which the electron wavefunction is attenuated only moderately by a *thin* barrier, resulting in there being a finite probability of its existence on the opposite side of the barrier. Tunnelling directly from the Fermi level of one electrode to the conduction band of the other is normally possible only for very thin films of thickness less than about 10 nm when subjected to a high electric field. *Simmons* [29.222] investigated the effects of tunnelling between similar electrodes separated by a thin insulating film of thickness s, where the barrier presented by the film was of arbitrary shape. The Wentzel–Kramers–Brillouin (WKB) approximation was used to predict the probability of an electron penetrating the barrier. Different approximations for the current density were found, depending on the relative values of the applied voltage V and φ_0/e, where φ_0 is the barrier height at the electrode interface and e is the electronic charge. These approximations are too unwieldy to be included here, however for the case of $V > \varphi_0/e$ it is only necessary for electrons tunnelling from the Fermi level of one electrode to penetrate a distance $\Delta s < s$ to reach unoccupied levels in the second electrode. This worker also suggested [29.223] that for $V > \varphi_0/e$ a modified

Fowler–Nordheim [29.224] expression may be applicable for tunnelling through an interfacial region. In this case, the electric field at the barrier is sufficiently high to reduce the barrier width, measured at the Fermi level, to about 5 nm. Under these circumstances the current density is related to the voltage and barrier thickness according to the expression

$$J = \frac{e^3 V^2}{8\pi h \varphi_0 d_t^2} \exp\left(-\frac{8\pi (2m)^{1/2} \varphi_0^{3/2} d_t}{3ehV}\right),$$

(29.109)

where h is Planck's constant, φ_0 and d_t are the barrier height and effective thickness, respectively, of the tunnelling barrier and m is the free electron mass. The tunnelling current is largely controlled by the barrier height φ_0, which appears both in the pre-exponential and in the exponential terms. Fowler–Nordheim tunnelling is indicated by a linear dependence of $\log J/V^2$ on $1/V$. A further interesting point for tunnelling in general is that there is very little temperature dependence in the current density, a feature allowing it to be distinguished from other temperature-dependent mechanisms, such as the Schottky effect. There is in fact a very slight quadratic dependence on temperature [29.164], but this is normally negligible within the limitations of experimental measurement.

Early measurements on tunnelling were made by *Fisher* and *Giaever* [29.225] on Al$_2$O$_3$ produced by oxidation of Al. There was good agreement with an early tunnelling model [29.226], providing the electron effective mass in the insulator was about 1/9 of the free electron mass. Very thin films of Al$_2$O$_3$ and BeO showed ohmic and faster-than-exponential dependencies of J on V at lower and higher voltages respectively [29.227]. These results were correlated with direct metal-to-metal electrode tunnelling at low voltages, and from the Al electrode to the insulator conduction band at higher voltages. The model of *Simmons* [29.164] was tested by measurements on thermally grown Al$_2$O$_3$ films on Al, which used several different types of evaporated counter electrodes [29.228]. An extended tunnelling model [29.229], which is applicable to samples with dissimilar electrodes, was followed over nine decades of current. Measurements on thermally grown SiO$_2$ films, of thickness 65–500 nm with Al or Au electrodes, showed strong evidence of Fowler–Nordheim tunnelling [29.230]. Since these films were relatively thick, high voltages could be applied, which reduced the Fermi level barrier thickness to that suitable for tunnelling. A linear dependence of log J/V^2 on $1/V$ was obtained, but the measured currents were somewhat lower than those predicted by the theory, probably as the result of trapping effects. Effective masses of $0.48m$ for Ag, $0.39m$ for Al and $0.42m$ for Si were estimated, where m is the free electron mass. Since these early tunnelling measurements, the mechanism has been observed in many different insulating films.

Tunnelling is semiconductors forms the basis of the operation of the tunnel diode. However, in this case a doubly degenerate p–n junction is used, and tunnelling occurs across the depletion region. The operation of this device is described in a standard text on semiconductor devices [29.231]. In sandwich samples of nondegenerate materials, tunnelling has been observed in CdTe [29.202]. In this case there was a linear dependence of the logarithm of the drift mobility on reciprocal electric field for fields in excess of 1.2×10^6 V/m, which is consistent with (29.109). Tunnelling has also been observed in sandwich structures consisting of a copper phthalocyanine film with In electrodes [29.232]. A modified Fowler–Nordheim expression for tunnelling through an interfacial barrier, when the electric field is sufficient to reduce the barrier width to about 5 nm, has been proposed [29.223], and was applied in this case to the In/Cu phthalocyanine interfacial region. Depending on the sample structure, barrier heights of 0.27 eV and 0.36 eV were derived. Current densities were somewhat lower than predicted by theory, but were accounted for by the fact that tunnelling areas are often considerably lower than the geometric area [29.18] and that in the presence of traps and space charge the current density is also reduced [29.233].

The second electrode-limited conduction process to be considered may occur when the insulating or semiconducting film is too thick for tunnelling to take place and the concentration of allowed states in the forbidden gap is too low for the hopping process to occur. It is known as the Schottky effect, and is the field-assisted lowering of a potential barrier at the injecting electrode. It is similar in origin to the Poole–Frenkel effect, the latter effect having previously been termed the bulk analogue of the Schottky effect [29.234]. The potential barrier at the injecting electrode interface is reduced by an amount $\Delta \varphi_S$ which is given by

$$\Delta \varphi_S = \beta_S F^{1/2}, \quad (29.110)$$

where

$$\beta_S = \left(\frac{e^3}{4\pi \varepsilon_r \varepsilon_0} \right)^{1/2} \quad (29.111)$$

is the Schottky field-lowering coefficient. The barrier lowering process is similar to that for the Poole–Frenkel effect, and is also illustrated in various literature reviews [29.164–166]. Differences between the standard Poole–Frenkel and Schottky coefficients given by (29.104) and (29.111) are related to the dissimilar symmetry of the potential barriers in the two cases. Clearly the two field-lowering coefficients are related, with $\beta_{PF} = 2\beta_S$. The fundamental thermionic emission equation of Richardson

$$J = AT^2 \exp\left(-\frac{\varphi}{kT}\right) \quad (29.112)$$

gives the current density flowing by a process of electron emission over a potential barrier of height φ at a temperature T. In this expression, the constant $A = 1.2 \times 10^6$ A/m^2, and is known as the Richardson constant. The current density depends only on the potential barrier height and the temperature, and does not require an applied voltage to flow, the electrons acquiring thermal energy only. For the case of Schottky emission, if φ_0, given by (29.91), is the zero-voltage barrier height then the reduced barrier height φ is given by $(\varphi_0 - \Delta\varphi_S)$, so that (29.112) becomes

$$J = AT^2 \exp\left(-\frac{\varphi_0}{kT}\right) \exp\left(\frac{\beta_S F^{1/2}}{kT}\right) \quad (29.113)$$

or

$$J = AT^2 \exp\left(-\frac{\varphi_0}{kT}\right) \exp\left(\frac{\beta_S V^{1/2}}{kTd^{1/2}}\right). \quad (29.114)$$

Thus, the Schottky effect should show a linear dependence of log J on $V^{1/2}$, as for the Poole–Frenkel effect described by (29.106). In principle the two effects can be distinguished by the measured value of the field-lowering coefficient, which should be twice as high for the Poole–Frenkel effect as for the Schottky effect.

The Schottky effect has been reported in many inorganic materials, both insulators and semiconductors. *Emtage* and *Tantraporn* [29.235] observed the effect in both Al_2O_3 and GeO_2 sandwich structures having Al electrodes, while *Pollack* [29.236] made similar observations in Al_2O_3 films with Pb electrodes. In this case, not only the constant voltage J–V characteristics were as predicted by (29.114), but the gradients of log J–$V^{1/2}$ curves increased linearly with reciprocal temperature for temperatures above 235 K, as also predicted by this expression. Below this temperature the variation with temperature was very small, and therefore tunnelling was identified in this region. The dependence of the Schottky effect on the species of electrode was determined in Ta–Ta_2O_5–Au samples prepared by electron beam evaporation and plasma oxidation [29.237]. In this system the Schottky effect was observed when the Ta electrode was biassed negatively and tunnelling when it was biassed positively. Similar results were obtained in CdS films [29.169] using In (ohmic) and Au (blocking) contacts. For electron injection from the Au electrode Schottky-type behaviour was observed, with a Schottky barrier height of 0.88 eV obtained from differential capacitance measurements, close to the value expected for Au contacts to CdS single crystals. Field-lowering behaviour has also been observed in Cd_3As_2 films with a bottom Ag electrode and a top electrode of Ag, Al or Au [29.204]. Although the Schottky barrier height is unlikely to be consistent with the simple model of (29.91) owing to the very narrow band gap and the presence of surface states, the experimental value of β was consistent with the value of β_S predicted by (29.111) for a relative permittivity value of 12.

The Schottky effect has also been observed in several organic thin-film materials. Examples are given here only for the phthalocyanines, which are invariably p-type semiconductors. The Schottky effect occurs for hole injection at the blocking contact. Both metal-free [29.238] and copper phthalocyanine films [29.239] having Al electrodes have shown this effect. Behaviour consistent with the Schottky effect has also been observed in planar copper phthalocyanine films with Al electrodes [29.240], with a depletion region width of thickness 96 nm. In sandwich structures having a copper phthalocyanine film and a positively biased Al electrode, values of β several times higher than the theoretical $\beta_S = 2 \times 10^{-5}$ eVm$^{1/2}$V$^{-1/2}$ were obtained [29.205]. However, if the theoretical value of β_S were assumed and also that most of the voltage was dropped over a depletion region of thickness d_S as suggested previously [29.240], values of $d_S = 120$ nm and $\varphi_0 = 0.88$ eV were derived. This type of assumption has also been subsequently applied in the case of hole injection from a Pb electrode into copper phthalocyanine films [29.175] and for hole injection in triclinic lead phthalocyanine films from an Al electrode [29.177, 206]. In both cases the Schottky barrier width was 50 nm with a barrier height of either 1.11 eV or 1.0 eV.

In this section the main high-field conduction processes that might be observed in insulating and semiconducting films have been reviewed, and illustrative examples quoted from the literature. It should be emphasised that identification of a particular mechanism is nontrivial, particularly in distinguishing between the Poole–Frenkel and Schottky effects which have very similar J–V dependencies. Furthermore J–V data obtained over a specified voltage range may also show a good correlation with both SCLC and the Poole–Frenkel effect. It is essential to measure temperature dependencies and also to determine the effects of electrode species on the conductivity, making a full consideration of the type of contacts (ohmic or blocking). As devices and structures become ever smaller with the advance of nanotechnology, some of these conduction processes will inevitably become important in nanoelectronic devices. A discussion concerning this aspect is given elsewhere in the literature [29.11].

29.4 Concluding Remarks

In this chapter, an overview of the deposition methods, structures and major properties of thin films has been given. In this section these are briefly summarised and some possible future developments outlined.

Films may be prepared by many different methods. However, these may usually be classified into chemical or physical methods. In the physical methods the film is deposited from a vapour, maintained at a pressure considerably below that of the atmosphere. Vacuum evaporation is the first of these, where the material is transformed into the vapour phase by heating at reduced pressure. Various types of process have been developed to ensure stoichiometry of the deposited material, such as reactive evaporation and co-evaporation. Taken to its limit of sophistication, the method may be operated under UHV conditions using several sources, each of which may be independently controlled. In this case epitaxial films may be deposited, and the method is known as molecular-beam epitaxy (MBE). The second widely used physical method is that of sputtering, which involves the removal of particles (atoms, molecules or larger fragments) from the surface of a target. This is normally achieved using a gas discharge, which delivers high-energy ions to the target surface. Sputtering has several advantages over evaporation, including deposition of stoichiometric materials, better adhesion to the substrate and the capability of depositing most materials. There are several varieties of sputtering, including DC (diode) sputtering, reactive sputtering and RF sputtering, the latter enabling the deposition of insulators as well as conducting materials. Reactive and co-sputtering give greater flexibility, and higher deposition rates are possible using magnetron sputtering. Various chemical methods may also be utilised for the deposition of films from materials having specific chemical properties. Electrodeposition and electroplating involve passing a high electric current through a solution containing the material, while chemical vapour deposition (CVD) entails the use of various chemical reactions to generate a vapour. The latter method is frequently used in the deposition of electronic materials, and substantial work has been performed in developing processes for several of these. Again there are several different subdivisions of this method, operating at different pressures and with different types of precursor materials. A third chemical method that has found favour in recent years is Langmuir–Blodgett deposition, where molecular layers are applied individually using a dipping process. It has the advantage that high vacuum and elevated temperatures are not required, and film thicknesses are easily derived from the number of layers deposited. It is, however, limited to certain defined materials, and does not have the flexibility of the other methods.

In common with bulk materials, the crystal structure and morphology of thin films may be investigated using diffraction methods and various microscopic techniques (electron microscopy, scanning tunnelling microscopy, etc.). Diffraction methods allow the type of crystal structure to be determined by measurement of the unit cell dimensions in the case of epitaxial and polycrystalline films. Frequently a material may exist in two or more crystalline modifications, and generally these may be distinguished by comparison of diffraction data from the film with tabulated standard data. The type of crystalline modification that is deposited can depend sensitively on the deposition conditions such as the substrate temperature and deposition rate, and this dependence may be established by systematically depositing films under a wide combination of conditions. Many films have a preferred crystalline orientation, and this too may sometimes be correlated with deposition conditions and the substrate morphology. In polycrystalline films, the crystallites are generally preferentially oriented and show a fibre texture. Estimates of the mean microcrystallite size may be obtained from the width of the diffraction peaks using the Scherrer method. The morphological characteristics of thin films commonly have a columnar grain structure when physical deposition methods are used, and the orientation of the grains may sometimes be related to the incident direction of the depositing atoms or molecules. Larger-scale features of the morphology may often be accounted for in terms of a zone model, where the observed features are identified with the ratio of the substrate temperature to the melting temperature of the film material. Examples of some of these effects in different thin-film materials have been given, and although these are representative of behaviour shown in many thin-film systems, it is clear that both crystal structure and morphology require determination in any specific process used to deposit a particular film material. These features are likely to be particularly important in influencing the optical and electrical properties.

The optical and electrical properties of thin films are those most relevant to this Handbook, and examples of some of the basic properties have been considered as a foundation for some of the more advanced applications described in other chapters. The interference properties of light are fundamental to most of the phenomena observed. In principle the properties of light propagating through media are described by Maxwell's equations, which allow a full analysis for both nonabsorbing and absorbing films for various states of polarisation and optical frequencies. The Fresnel coefficients for reflection and transmission of light at the boundaries between different optical media allow expressions for the energy reflectance and transmittance to be derived. Some examples of these have been given, together with an introduction to matrix methods which may be used to calculate the electric and magnetic field vectors for optical designs of considerable complexity, involving both nonabsorbing and absorbing films. The operation of optical filters, other optical components and fibre-optic waveguides may be predicted by advanced application of these methods. Examples of the basic electrical properties of thin films, which are directly related to the film thickness, are lateral electrical conductivity and high-field conduction processes. Lateral electrical conduction in thin films is a very important feature, which is determined primarily by electron scattering within the film. For metallic films, conductivity–film thickness relationships have been derived in terms of a parameter k, the ratio of the film thickness to the bulk electron mean free path. One would intuitively expect the conductivity to decrease (or the resistivity to increase) with decreasing k, since this implies a curtailment of the mean free path. This problem was first tackled over a century ago by Thomson, and more realistic expressions have been obtained subsequently by solving the Boltzmann transport equation. Unfortunately, the derived general expressions are somewhat complex, but a wide variety of approximations have been obtained for various ranges of k, and the conductivity is predicted to reduce to that of the bulk material for thick films. Further refinements in the analysis were made by introducing a specularity parameter, p, which takes into account the proportion of electrons which are specularly scattered at the surfaces. Better agreement with experimental results is generally obtained if the value of p is intermediate between that for totally diffuse scattering ($p = 0$) and totally specular scattering ($p = 1$). Another development of the theory includes assigning different specularity parameters to the two film surfaces, since scattering at the substrate surface is clearly different from scattering at the top surface. In general the various expressions derived from this type of model are consistent with each other, in that those derived from the more complex assumptions normally reduce to those obtained using simpler assumptions if suitable limits are chosen. Scattering by internal grain boundaries has also been investigated by solving the Boltzmann transport equation, and allows contributions to the conductivity resulting from internal features, as well as the surfaces, to be included in the analysis. The ratio of the thin-film to bulk temperature coefficient of resistivity has been shown to be equal to the ratio of thin-film to bulk conductivity for the case of thicker films, although no such convenient relationship applies for thinner films owing to the dependence of the bulk mean free path on temperature, which is particularly significant for very thin films. Considerations of these types of effect need to be fully assessed when designing interconnects to integrated circuits operating in particular temperature ranges. In insulating or semiconducting thin films, very high electric fields may be applied across the thickness of the film. Under these circumstances a variety of different conduction processes may occur, none of which are observed at lower fields. Crucial to the type of conduction process which develops is the type of contact at the interface. Ohmic contacts allow currents to pass unimpeded by the contact, and the conduction process is bulk-limited, while blocking contacts effectively limit the conductivity, which is then electrode-limited. Bulk-limited conduction processes include SCLC, Poole–Frenkel conductivity and hopping. Varies types of SCLC have been observed, depending on the type of trap distribution present, if any, and may be distinguished by their characteristic voltage and temperature dependencies. Poole–Frenkel conductivity results from field-dependent lowering of the potential barrier for emission of electrons into the conduction band. Again, various varieties of this process have been observed depending on the appearance of donors and/or traps, and also for the case of nonuniform fields. Hopping is observed particularly in noncrystalline solids, where electrons hop from one localised energy level to another. It is characterised by low mobility, but requires little thermal energy, and is therefore often observed at low temperatures. Different types of hopping are characterised by the mean distance of the electron hopping process, and show different characteristic behaviours. Electrode-limited conduction processes include tunnelling and the Schottky effect. Tunnelling is a quantum-mechanical effect, where for a thin barrier the electron wavefunction has a finite probability of existence on the opposite side of the barrier. Tunnelling may be either directly from one electrode to the other, or may be from the Fermi level of the injecting electrode into the conduction band of the insulator or semiconductor. The type of tunnelling depends on the relative values of the applied voltage and the barrier height at the interface. The tunnelling current is almost independent of temperature, a feature which allows it to be identified relatively easily. For structures where the barrier is too thick for tunnelling to occur, the Schottky effect may take place instead. This is the field-assisted lowering of a potential barrier at the interface, and has several similarities with the bulk Poole–Frenkel effect. Similar $J-V$ characteristics are predicted for both, but may usually be distinguished from the value of a measured field-lowering coefficient, or from considerations of the current level drawn with varying electrode species.

In conclusion, it should be noted that in this chapter only the main established deposition methods and some of the structural, optical and electrical properties of thin films have been reviewed. Although it is hoped that the references included herein are sufficient to allow the reader to explore the subject in some considerable depth, it is clear that these can only provide a snapshot of knowledge at the present time. During the past half century, there have been tremendous developments in all areas of thin-film knowledge, although the basic principles and properties stressed in this chapter underpin most of them. The main physical methods of deposition have been known for a century or so, but improvements have been contingent on advances in vacuum technology, gas-handling capabilities and theoretical work concerning atomic interactions in solids. MBE deposition systems of considerable complexity and sophistication now exist, enabling the deposition of a considerable range of materials in a variety of structures and designs, although only some of the most important processes have been developed to the level where they are employed in automated manufacture. Control of many factors are vital in successful deposition processes, and progress in this area has gone hand in hand with the development of computerised monitoring and data-handling systems. Progress in the manipulation of individual atoms and molecules has been made by the adoption of techniques originally developed for very high-resolution surface imaging, such as scanning tunnelling and atomic force microscopy. Chemical methods of film deposition have developed into many dedicated CVD systems, exploiting reactions which are specific to a particular material or class of materials. Additionally many types of films may be deposited using the Langmuir–Blodgett technique, although this relies on the ingenuity of chemists to develop suitable substituted organic compounds. Further development of many of these deposition methods will undoubtedly progress in future years, particularly as the power of computer systems increases. The most significant advances are likely to be in the area of nanotechnology, where MBE systems, equipped with facilities to build nanostructures atom by atom, are currently under development. The structure and morphology adopted by thin films are likely to have a significant influence on all their properties; this is not restricted to just the optical and electrical properties described in this chapter. Diffraction methods are becoming more precise, and can now be used for the investigation of real-time changes when synchrotron radiation is used. Scanning tunnelling and atomic force microscopic techniques have enabled the acquisition of surface images with resolution of atomic dimensions, and the increased use and development of these and associated techniques will be necessary for future investigation of nanostructures. Optical properties described in this chapter will no doubt be exploited in more complex and sophisticated filter and mirror designs, as well as their employment in fibre-optic cables and applications in photonics. Recent developments in the area of negative-refractive-index materials will no doubt also be exploited in due course. The electrical properties of thin films under both low- and high-field conditions are reasonably well understood, and it is anticipated that applications are likely to be made in the area of nanoelectronics. Some possible applications have been proposed elsewhere [29.11]. The provision of electrical contacts will remain of particular importance, although it may not be possible to consider the contact region separately from the remaining nanostructure. The efficient injection of charge into nanostructures using tunnelling and Schottky barrier (blocking) contacts may be achieved, and high current densities may be obtained with trap-free SCLC contacts. It is nevertheless clear, however, that in all areas of future development in thin-film technology, the control of deposition processes and materials processing aspects are of paramount importance. This is likely to remain a considerable challenge in the future.

References

29.1   L. Eckertová: *Physics of Thin Films*, 2nd edn. (Plenum, New York 1986)

29.2   J. Thewlis: *Concise Dictionary of Physics and Related Subjects*, 2nd edn. (Pergamon, Oxford 1979) p. 336

29.3   R. W. Hoffman: The mechanical properties of thin condensed films. In: *Physics of Thin Films*, Vol. 3, ed. by G. Hass, R. E. Thun (Academic, San Diego 1966) p. 211

29.4   D. S. Campbell: Mechanical properties of thin films. In: *Handbook of Thin Film Technology*, ed. by L. I. Maissel, R. Glang (McGraw–Hill, New York 1970) Chap. 12

29.5   C. A. Neugebauer, M. B. Webb: J. Appl. Phys. **33**, 74 (1962)

29.6   L. I. Maissel: Electronic properties of metallic thin films. In: *Handbook of Thin Film Technology*, Vol. 13, ed. by L. I. Maissel, R. Glang (McGraw–Hill, New York 1970)

29.7   K. H. Behrndt: Film-thickness and deposition-rate monitoring devices and techniques for producing films of uniform thickness. In: *Physics of Thin Films*, Vol. 3, ed. by G. Hass, R. E. Thun (Academic, San Diego 1966.) p. 1

29.8   W. A. Pliskin, S. J. Zanin: Film thickness and composition. In: *Handbook of Thin Film Technology*, Vol. 11, ed. by L. I. Maissel, R. Glang (McGraw–Hill, New York 1970)

29.9   J. P. Hirth, K. L. Moazed: Nucleation processes in thin film formation. In: *Physics of Thin Films*, Vol. 4, ed. by G. Hass, R. E. Thun (Academic, San Diego 1967) p. 97

29.10 C. A. Neugebauer: *Condensation, Nucleation and Growth of Thin Films*, Vol. 8, ed. by L. I. Maissel, R. Glang (McGraw–Hill, New York 1970)

29.11 R. D. Gould: High field conduction in nanostructures. In: *Encyclopedia of Nanoscience and Nanotechnology*, Vol. 3, ed. by H. S. Nalwa (American Scientific, Stevenson Ranch 2004) pp. 891–915

29.12 *Handbook of Thin Film Technology*, ed. by L. I. Maissel, R. Glang (McGraw–Hill, New York 1970)

29.13 *Thin Films*, Vol. 25, ed. by S. M. Rossnagel, A. Ulman, M. H. Francombe (Academic, San Diego 1998.)

29.14 *Handbook of Thin Film Materials*, ed. by H. S. Nalwa (Academic, San Diego 2001)
29.15 O. S. Heavens: *Thin Film Physics* (Methuen, London 1970)
29.16 L. I. Maissel, M. H. Francombe: *An Introduction to Thin Films* (Gordon and Breach, New York 1973)
29.17 *Active and Passive Thin Film Devices*, ed. by T. J. Coutts (Academic, New York 1978)
29.18 K. I. Chopra: *Thin Film Phenomena* (Krieger, New York 1979)(first published by McGraw–Hill, New York, 1969)
29.19 M. Ohring: *The Materials Science of Thin Films* (Academic, San Diego 1992)
29.20 R. E. Honig, D. A. Kramer: RCA Rev. **30**, 285 (1969)
29.21 *Scientific Foundation of Vacuum Technique*, 2nd edn., ed. by S. Dushman, J. M. Lafferty (Wiley, New York 1962)
29.22 R. Glang: Vacuum evaporation. In: *Handbook of Thin Film Technology*, ed. by L. I. Maissel, R. Glang (McGraw–Hill, New York 1970) Chap. 1
29.23 O. Kubaschewski, S. L. Evans, C. B. Alcock: *Metallurgical Thermochemistry*, 4th revised edn. (Pergamon, Oxford 1967)
29.24 R. E. Honig: RCA Rev. **23**, 567 (1962)
29.25 H. Hertz: Ann. Phys. Chem. **17**, 177 (1882)
29.26 M. Knudsen: Ann. Phys. **47**, 697 (1915)
29.27 J. P. Hirth, G. M. Pound: *Condensation and Evaporation* (Pergamon, Oxford 1963)
29.28 L. Holland, W. Steckelmacher: Vacuum **2**, 346 (1952)
29.29 A. von Hippel: Ann. Phys. **81**, 1043 (1926)
29.30 H. A. Macleod: *Thin-Film Optical Filters* (Adam Hilger, London 1969)
29.31 T. C. Tisone, J. B. Bindell: J. Vac. Sci. Technol. **11**, 72 (1974)
29.32 R. Brown: Thin film substrates. In: *Handbook of Thin Film Technology*, ed. by L. I. Maissel, R. Glang (McGraw–Hill, New York 1970) Chap. 6
29.33 G. Sauerbrey: Z. Phys. **155**, 206 (1959)
29.34 C. D. Stockbridge: Resonance frequency versus mass added to quartz crystals. In: *Vacuum Microbalance Techniques*, Vol. 5, ed. by K. H. Behrndt (Plenum, New York 1966) p. 193
29.35 C. J. Bowler, R. D. Gould: J. Vac. Sci. Technol. A **5**, 114 (1987)
29.36 R. Hultgren, R. L. Orr, P. D. Anderson, K. K. Kelley: *Selected Values of Thermodynamic Properties of Metals and Alloys* (Wiley, New York 1963)
29.37 G. Zinsmeister: Vakuum-Tech. **8**, 223 (1964)
29.38 R. D. Gould: Electrical conduction properties of thin films of cadmium compounds. In: *Handbook of Thin Film Materials*, ed. by H. S. Nalwa (Academic, San Diego 2001) Chap. 4, pp. 187–245
29.39 D. R. Stull: *JANAF Thermochemical Tables* (Dow Chemical Co., U.S. Clearinghouse, Springfield, Virginia 1970)
29.40 F. A. Pizzarello: J. Appl. Phys. **35**, 2730 (1964)
29.41 P. A. Timson, C. A. Hogarth: Thin Solid Films **8**, 237 (1971)
29.42 A. Y. Cho: Thin Solid Films **100**, 291 (1983)
29.43 B. A. Joyce: Rep. Prog. Phys. **48**, 1637 (1985)
29.44 Y. Ota: Thin Solid Films **106**, 3 (1983)
29.45 G. K. Wehner, G. S. Anderson: The nature of physical sputtering. In: *Handbook of Thin Film Technology*, ed. by L. I. Maissel, R. Glang (McGraw–Hill, New York 1970) Chap. 3
29.46 O. Almén, G. Bruce: Nucl. Instrum. Methods **11**, 279 (1961)
29.47 F. Keywell: Phys. Rev. **97**, 1611 (1955)
29.48 P. Sigmund: Sputtering by ion bombardment: Theoretical concepts. In: *Sputtering by Particle Bombardment I*, ed. by R. Behrisch (Springer, Berlin, Heidelberg 1981) Chap. 2, p. 9
29.49 R. S. Pease: *Rendiconti della Scuola Internatzionale di Fisica "Enrico Fermi"*, Corso XIII (Società Italiana di Fisica, Bologna 1959) p. 158
29.50 P. Sigmund: Phys. Rev. **184**, 383 (1969)
29.51 L. I. Maissel: The deposition of thin films by cathode sputtering. In: *Physics of Thin Films*, Vol. 3, ed. by G. Hass, R. E. Thun (Academic, San Diego 1966) p. 61
29.52 G. H. Kinchin, R. S. Pease: Rep. Prog. Phys. **18**, 1 (1955)
29.53 H. H. Andersen, H. L. Bay: Sputtering yield measurements. In: *Sputtering by Particle Bombardment I*, ed. by R. Behrisch (Springer, Berlin, Heidelberg 1981) Chap. 4, p. 145
29.54 J. Lindhard, V. Nielsen, M. Scharff: Kgl. Danske Videnskal. Selskab, Mat.-Fys. Medd. **36**(10), 1 (1968)
29.55 A. von Engel: *Ionized Gases*, 2nd edn. (Clarendon, Oxford 1965)
29.56 F. Llewellyn-Jones: *The Glow Discharge* (Methuen, London 1966)
29.57 L. I. Maissel: Applications of sputtering to the deposition of films. In: *Handbook of Thin Film Technology*, ed. by L. I. Maissel, R. Glang (McGraw–Hill, New York 1970) Chap. 4
29.58 G. Perny, M. Samirant, B. Laville Saint Martin: Compt. Rend. Acad. Sci. Ser. C **262**, 265 (1966)
29.59 G. Perny, B. Laville Saint Martin: Proceedings of the International Symposium on Basic Problems in Thin Film Physics, Clausthal 1965, ed. by R. Niedermayer, H. Mayer (Vandenhoeck and Ruprecht, Göttingen 1966) 709
29.60 S. A. Awan, R. D. Gould: Thin Solid Films **423**, 267 (2003)
29.61 W. D. Westwood: Reactive sputtering. In: *Physics of Thin Films*, Vol. 14, ed. by M. H. Francombe, J. C. Vossen (Academic, San Diego 1989) p. 1
29.62 W. R. Sinclair, F. G. Peters: Rev. Sci. Instrum. **33**, 744 (1992)
29.63 L. I. Maissel, P. M. Schaible: J. Appl. Phys. **36**, 237 (1965)
29.64 R. Frerichs: J. Appl. Phys. **33**, 1898 (1962)
29.65 H. C. Theuerer, J. J. Hauser: J. Appl. Phys. **35**, 554 (1964)
29.66 E. W. Williams: *The CD-ROM and Optical Disc Recording Systems* (Oxford Univ. Press, Oxford 1996)

29.67 H. E. Winters, E. Kay: J. Appl. Phys. **38**, 3928 (1967)
29.68 G. S. Anderson, W. N. Mayer, G. K. Wehner: J. Appl. Phys. **33**, 2991 (1962)
29.69 P. D. Davidse, L. I. Maissel: J. Appl. Phys. **37**, 574 (1966)
29.70 R. A. Powell, S. M. Rossnagel: *PVD for Microelectronics: Sputter Deposition Applied to Semiconductor Manufacturing*, Thin Films, Vol. 26 (Academic, San Diego 1999) p. 51
29.71 J. S. Logan, N. M. Mazza, P. D. Davidse: J. Vac. Sci. Technol. **6**, 120 (1969)
29.72 S. Rohde: Unbalanced magnetron sputtering. In: *Physics of Thin Films*, Vol. 18, ed. by M. H. Francombe, J. L. Vossen (Academic, San Diego 1994) p. 235
29.73 B. Window, N. Savvides: J. Vac. Sci. Technol. A **4**, 196 (1986)
29.74 Y. Ochiai, K. Aso, M. Hayakawa, H. Matsuda, K. Hayashi, W. Ishikawa, Y. Iwasaki: J. Vac. Sci. Technol. A **4**, 19 (1986)
29.75 D. S. Campbell: The deposition of thin films by chemical methods. In: *Handbook of Thin Film Technology*, ed. by L. I. Maissel, R. Glang (McGraw-Hill, New York 1970) Chap. 5
29.76 D. S. Campbell: Preparation methods for thin film devices. In: *Active and Passive Thin Film Devices*, ed. by T. J. Coutts (Academic, New York 1978) p. 23
29.77 A. Brenner: *Electrodeposition of Alloys*, Vol. 1, 2 (Academic, New York 1963)
29.78 K. R. Lawless: The growth and structure of electrodeposits. In: *Physics of Thin Films*, Vol. 4, ed. by G. Hass, R. E. Thun (Academic, San Diego 1967) p. 191
29.79 L. Young: *Anodic Oxide Films* (Academic, New York 1961)
29.80 C. J. Dell'Oca, D. L. Pulfrey, L. Young: Anodic oxide films. In: *Physics of Thin Films*, Vol. 6, ed. by M. H. Francombe, R. W. Hoffman (Academic, San Diego 1971) p. 1
29.81 W. M. Feist, S. R. Steele, D. W. Readey: The preparation of films by chemical vapor deposition. In: *Physics of Thin Films*, Vol. 5, ed. by G. Hass, R. E. Thun (Academic, San Diego 1969) p. 237
29.82 J. R. Knight, D. Effer, P. R. Evans: Solid State Electron. **8**, 178 (1965)
29.83 M. H. Francombe, J. E. Johnson: The preparation and properties of semiconductor films. In: *Physics of Thin Films*, Vol. 5, ed. by G. Hass, R. E. Thun (Academic, San Diego 1969) p. 143
29.84 S. M. Ojha: Plasma-enhanced chemical vapor deposition of thin films. In: *Physics of Thin Films*, Vol. 12, ed. by G. Hass, M. H. Francombe, J. L. Vossen (Academic, San Diego 1982) p. 237
29.85 A. C. Adams: Dielectric and polysilicon film deposition. In: *VLSI Technology*, 2nd edn., ed. by S. M. Sze (McGraw-Hill, New York 1988) Chap. 6, p. 233
29.86 O. A. Popov: Electron cyclotron resonance plasma sources and their use in plasma-assisted chemical vapor deposition of thin films. In: *Physics of Thin Films*, Vol. 18, ed. by M. H. Francombe, J. L. Vossen (Academic, San Diego 1994) p. 121
29.87 C. W. Pitt, L. M. Walpitta: Thin Solid Films **68**, 101 (1980)
29.88 N. G. Dhere: High-T_c superconducting thin films. In: *Physics of Thin Films*, Vol. 16, ed. by M. H. Francombe, J. L. Vossen (Academic, San Diego 1992) p. 1
29.89 I. Langmuir: Trans. Faraday Soc. **15**, 62 (1920)
29.90 K. B. Blodgett: J. Am. Chem. Soc. **56**, 495 (1934)
29.91 G. L. Gaines: Thin Solid Films **99**, ix (1983)
29.92 S. Baker, M. C. Petty, G. G. Roberts, M. V. Twigg: Thin Solid Films **99**, 53 (1983)
29.93 M. C. Petty: *Langmuir–Blodgett Films: An Introduction* (Cambridge Univ. Press, Cambridge 1996)
29.94 L. S. Miller, P. J. W. Stone: Thin Solid Films **210/211**, 19 (1992)
29.95 L. S. Miller, A. L. Rhoden: Thin Solid Films **243**, 339 (1994)
29.96 P. S. Vincett, G. G. Roberts: Thin Solid Films **68**, 135 (1980)
29.97 A. N. Broers, M. Pomerantz: Thin Solid Films **99**, 323 (1983)
29.98 R. E. Thun: Structure of thin films. In: *Physics of Thin Films*, Vol. 1, ed. by G. Hass (Academic, San Diego 1963) p. 187
29.99 C. A. Neugebauer: Structural disorder phenomena in thin metal films. In: *Physics of Thin Films*, Vol. 2, ed. by G. Hass, R. E. Thun (Academic, San Diego 1964) p. 1
29.100 I. H. Khan: The growth and structure of single-crystal films. In: *Handbook of Thin Film Technology*, ed. by L. I. Maissel, R. Glang (McGraw-Hill, New York 1970) Chap. 10
29.101 G. B. Stringfellow: Rep. Prog. Phys. **43**, 469 (1982)
29.102 D. Walton: J. Chem. Phys. **37**, 2182 (1962)
29.103 D. Walton, T. N. Rhodin, R. Rollins: J. Chem. Phys. **38**, 2695 (1963)
29.104 K. K. Muravjeva, I. P. Kalinkin, V. B. Aleskowsky, N. S. Bogomolov: Thin Solid Films **5**, 7 (1970)
29.105 V. N. E. Robinson, J. L. Robins: Thin Solid Films **20**, 155 (1974)
29.106 J. A. Venables, G. D. T. Spiller: Rep. Prog. Phys. **47**, 399 (1984)
29.107 I. P. Kalinkin, K. K. Muravyeva, L. A. Sergeyewa, V. B. Aleskowsky, N. S. Bogomolov: Krist. Tech. **5**, 51 (1970)
29.108 D. B. Holt: Thin Solid Films **24**, 1 (1974)
29.109 K. Zanio: Cadmium telluride. In: *Semiconductors and Semimetals*, Vol. 13, ed. by R. K. Willardson, A. C. Beer (Academic, New York 1978)
29.110 A. Ashour, R. D. Gould, A. A. Ramadan: Phys. Status Solidi A **125**, 541 (1991)
29.111 R. D. Gould: Coord. Chem. Rev. **156**, 237 (1996)
29.112 A. K. Hassan, R. D. Gould: Phys. Status Solidi A **132**, 91 (1992)
29.113 S. I. Shihub, R. D. Gould: Phys. Status Solidi A **139**, 129 (1993)
29.114 P. Scherrer: Gott. Nachr. **2**, 98 (1918)

29.115 M. Ashida, N. Uyeda, E. Suito: Bull. Chem. Soc. Jpn. **39**, 2616 (1966)
29.116 L. Żdanowicz, S. Miotkowska: Thin Solid Films **29**, 177 (1975)
29.117 G. A. Steigmann, J. Goodyear: Acta Crystallogr. B **24**, 1062 (1968)
29.118 A. Bienenstock, F. Betts, S. R. Ovshinsky: J. Non-Cryst. Solids **2**, 347 (1970)
29.119 S. R. Elliott: *Physics of Amorphous Materials*, 2nd edn. (Longman, London 1990)
29.120 J. R. Bosnell: Amorphous semiconducting films. In: *Active and Passive Thin Film Devices*, ed. by T. J. Coutts (Academic, New York 1978) p. 245
29.121 N. F. Mott, E. A. Davis: *Electronic Processes in Non-Crystalline Materials*, 2nd edn. (Oxford Univ. Press, Oxford 1979)
29.122 D. B. Dove: Electron diffraction analysis of the local atomic order in amorphous films. In: *Physics of Thin Films*, Vol. 7, ed. by G. Hass, M. H. Francombe, R. W. Hoffman (Academic, San Diego 1973) p. 1
29.123 K. H. Norian, J. W. Edington: Thin Solid Films **75**, 53 (1981)
29.124 J. M. Nieuwenhuizen, H. B. Haanstra: Philips Tech. Rev. **27**, 87 (1966)
29.125 S. B. Hussain: Thin Solid Films **22**, S5 (1974)
29.126 J. I. B. Wilson, J. Woods: J. Phys. Chem. Solids **34**, 171 (1973)
29.127 A. G. Stanley: Cadmium sulphide solar cells. In: *Applied Solid State Science*, Vol. 5, ed. by R. Wolfe (Academic, New York 1975) p. 251
29.128 B. Goldstein, L. Pensak: J. Appl. Phys. **30**, 155 (1959)
29.129 J. Saraie, M. Akiyama, T. Tanaka: Jpn. J. Appl. Phys. **11**, 1758 (1972)
29.130 Yu. K. Yezhovsky, I. P. Kalinkin: Thin Solid Films **18**, 127 (1973)
29.131 J. Jurusik, L. Żdanowicz: Thin Solid Films **67**, 285 (1980)
29.132 K. Tanaka: Jpn. J. Appl. Phys. **9**, 1070 (1970)
29.133 T. M. Ratcheva-Stambolieva, Yu. D. Tchistyakov, G. A. Krasulin, A. V. Vanyukov, D. H. Djoglev: Phys. Status Solidi A **16**, 315 (1973)
29.134 H. W. Lehmann, R. Widmer: Thin Solid Films **33**, 301 (1976)
29.135 J. Jurusik: Thin Solid Films **214**, 117 (1992)
29.136 J. Jurusik: Thin Solid Films **248**, 178 (1994)
29.137 Z. Knittl: *Optics of Thin Films* (Wiley, New York 1976)
29.138 O. S. Heavens: *Optical Properties of Thin Solid Films* (Dover, New York 1965)(first published by Butterworths, London, 1955)
29.139 O. S. Heavens: Measurement of optical constants of thin films. In: *Physics of Thin Films*, Vol. 2, ed. by G. Hass, R. E. Thun (Academic, San Diego 1964) p. 193
29.140 O. S. Heavens: Optical properties of thin films. In: *Encyclopaedic Dictionary of Physics*, Supplementary, Vol. 3, ed. by J. Thewlis (Pergamon, Oxford 1969) p. 412
29.141 H. A. Macleod: Thin film optical devices. In: *Passive and Active Thin Film Devices*, ed. by T. J. Coutts (Academic, New York 1978) p. 321
29.142 A. Thelen: Design of multilayer interference filters. In: *Physics of Thin Films*, Vol. 5, ed. by G. Hass, R. E. Thun (Academic, San Diego 1969) p. 47
29.143 F. Abelès: Optical properties of metallic thin films. In: *Physics of Thin Films*, Vol. 6, ed. by M. H. Francombe, R. W. Hoffman (Academic, San Diego 1971) p. 151
29.144 P. H. Berning: Theory and calculations of optical thin films. In: *Physics of Thin Films*, Vol. 1, ed. by G. Hass (Academic, San Diego 1963) p. 69
29.145 J. T. Cox, G. Hass: Antireflection coatings for optical and infrared optical materials. In: *Physics of Thin Films*, Vol. 2, ed. by G. Hass, R. E. Thun (Academic, San Diego 1964) p. 239
29.146 G. Hass, M. H. Francombe, R. W. Hoffman: Metal-dielectric interference filters. In: *Physics of Thin Films*, Vol. 9, ed. by G. Hass, M. H. Francombe, R. W. Hoffman (Academic, San Diego 1977) p. 74
29.147 J. M. Eastman: Scattering in all-dielectric multilayer bandpass filters and mirrors for lasers. In: *Physics of Thin Films*, Vol. 10, ed. by G. Hass, M. H. Francombe (Academic, San Diego 1978) p. 167
29.148 G. Hass, J. B. Heaney, W. R. Hunter: Reflectance and preparation of front surface mirrors for use at various angles of incidence from the ultraviolet to the far infrared. In: *Physics of Thin Films*, Vol. 12, ed. by G. Hass, M. H. Francombe, J. L. Vossen (Academic, San Diego 1982) p. 2
29.149 J. N. Zemel: Transport phenomena in hetero-epitaxial semiconductor films. In: *The Use of Thin Films in Physical Investigations*, ed. by J. C. Anderson (Academic, New York 1966) p. 319
29.150 K. Fuchs: Proc. Cambridge Phil. Soc. **34**, 100 (1938)
29.151 J. J. Thomson: Proc. Cambridge Phil. Soc. **11**, 120 (1901)
29.152 A. C. B. Lovell: Proc. R. Soc. A **157**, 311 (1936)
29.153 D. S. Campbell: The electrical properties of single-crystal metal films. In: *The Use of Thin Films in Physical Investigations*, ed. by J. C. Anderson (Academic, New York 1966) p. 299
29.154 E. H. Sondheimer: Adv. Phys. **1**, 1 (1952)
29.155 M. S. P. Lucas: J. Appl. Phys. **36**, 1632 (1965)
29.156 H. J. Juretschke: Surface Sci. **2**, 40 (1964)
29.157 H. J. Juretschke: J. Appl. Phys. **37**, 435 (1966)
29.158 D. C. Larson: Size-dependent electrical conductivity in thin metal films and wires. In: *Physics of Thin Films*, Vol. 6, ed. by M. H. Francombe, R. W. Hoffman (Academic, San Diego 1971) p. 81
29.159 A. F. Mayadas, M. Shatzkes: Phys. Rev. B **1**, 1382 (1970)
29.160 C. R. Tellier, A. J. Tosser: *Size Effects in Thin Films* (Elsevier Science, Amsterdam 1982)
29.161 E. E. Mola, J. M. Heras: Thin Solid Films **18**, 137 (1973)
29.162 T. J. Coutts: Electrical properties and applications of thin metallic and alloy films. In: *Active and Passive*

Thin Film Devices, ed. by T. J. Coutts (Academic, New York 1978) p. 57
29.163 E. E. Mola, J. M. Heras: Electrocomp. Sci. Technol. **1**, 77 (1974)
29.164 J. G. Simmons: J. Phys. D: Appl. Phys. **4**, 613 (1971)
29.165 E. H. Rhoderick: *Metal–Semiconductor Contacts* (Clarendon, Oxford 1978)
29.166 D. R. Lamb: *Electrical Conduction Mechanisms in Thin Insulating Films* (Metheun, London 1967)
29.167 A. Rose: Phys. Rev. **97**, 1538 (1955)
29.168 T. Budinas, P. Mackus, A. Smilga, J. Vivvakas: Phys. Stat. Sol. **31**, 375 (1969)
29.169 R. Zuleeg: Solid State Electron. **6**, 645 (1963)
29.170 R. Zuleeg, R. S. Muller: Solid State Electron. **7**, 575 (1964)
29.171 B. M. Basol, O. M. Stafsudd: Solid State Electron. **24**, 121 (1981)
29.172 V. S. Dharmadhikari: Int. J. Electron. **54**, 787 (1983)
29.173 R. Glew: Thin Solid Films **46**, 59 (1977)
29.174 R. K. Pandey, R. B. Gore, A. J. N. Rooz: J. Phys. D: Appl. Phys. **20**, 1059 (1987)
29.175 R. D. Gould, A. K. Hassan: Thin Solid Films **193/194**, 895 (1990)
29.176 A. K. Hassan, R. D. Gould: J. Phys: Condens. Matter **1**, 6679 (1989)
29.177 T. S. Shafai, R. D. Gould: Int. J. Electron. **73**, 307 (1992)
29.178 A. K. Hassan, R. D. Gould: Int. J. Electron. **74**, 59 (1993)
29.179 S. Gravano, A. K. Hassan, R. D. Gould: Int. J. Electron. **70**, 477 (1991)
29.180 B. B. Ismail, R. D. Gould: Phys. Status Solidi A **115**, 237 (1989)
29.181 R. D. Gould, M. S. Rahman: J. Phys. D: Appl. Phys. **14**, 79 (1981)
29.182 M. A. Lampert: Rep. Prog. Phys. **27**, 329 (1964)
29.183 R. D. Gould, B. A. Carter: J. Phys. D: Appl. Phys. **16**, L201 (1983)
29.184 K. N. Sharma, K. Barua: J. Phys. D: Appl. Phys. **12**, 1729 (1979)
29.185 A. O. Oduor, R. D. Gould: Thin Solid Films **270**, 387 (1995)
29.186 A. O. Oduor, R. D. Gould: Thin Solid Films **317**, 409 (1998)
29.187 A. Sussman: J. Appl. Phys. **38**, 2738 (1967)
29.188 R. D. Gould: J. Appl. Phys. **53**, 3353 (1982)
29.189 R. D. Gould: J. Phys. D: Appl. Phys. **9**, 1785 (1986)
29.190 R. D. Gould, B. B. Ismail: Int. J. Electron. **69**, 19 (1990)
29.191 S. S. Ou, O. M. Stafsudd, B. M. Basol: Thin Solid Films **112**, 301 (1984)
29.192 G. M. Delacote, J. P. Fillard, F. J. Marco: Solid State Commun. **2**, 373 (1964)
29.193 M. A. Lampert, P. Mark: *Current Injection in Solids* (Academic, New York 1970)
29.194 S. Gogoi, K. Barua: Thin Solid Films **92**, 227 (1982)
29.195 R. D. Gould, C. J. Bowler: Thin Solid Films **164**, 281 (1988)
29.196 A. Servini, A. K. Jonscher: Thin Solid Films **3**, 341 (1969)
29.197 A. K. Jonscher: Thin Solid Films **1**, 213 (1967)
29.198 A. K. Jonscher, A. A. Ansari: Phil. Mag. **23**, 205 (1971)
29.199 S. M. Sze: J. Appl. Phys. **38**, 2951 (1967)
29.200 H. Murray, A. Tosser: Thin Solid Films **36**, 247 (1976)
29.201 A. Piel, H. Murray: Thin Solid Films **44**, 65 (1977)
29.202 C. Canali, F. Nava, G. Ottaviani, K. Zanio: Solid State Commun. **13**, 1255 (1973)
29.203 R. D. Gould, B. B. Ismail: Vacuum **50**, 99 (1998)
29.204 M. Din, R. D. Gould: Thin Solid Films **340**, 28 (1999)
29.205 A. K. Hassan, R. D. Gould: Int. J. Electron. **69**, 11 (1990)
29.206 A. Ahmad, R. A. Collins: Phys. Status Solidi A **126**, 411 (1991)
29.207 J. Kašpar, I. Emmer, R. A. Collins: Int. J. Electron. **76**, 793 (1994)
29.208 N. F. Mott: *Metal–Insulator Transitions* (Taylor and Francis, London 1974)
29.209 H. Böttger, V. V. Bryksin: *Hopping Conduction in Solids* (VCH, Weinheim 1985)
29.210 A. H. Clark: Phys. Rev. **154**, 750 (1967)
29.211 P. A. Walley, A. K. Jonscher: Thin Solid Films **1**, 367 (1968)
29.212 C. J. Adkins, S. M. Freake, E. M. Hamilton: Phil. Mag. **22**, 183 (1970)
29.213 R. D. Gould, B. B. Ismail: Phys. Status Solidi A **134**, K65 (1992)
29.214 D. S. H. Chan, A. E. Hill: Thin Solid Films **35**, 337 (1976)
29.215 J. Mycielski: Phys. Rev. **123**, 99 (1961)
29.216 J. Le Moigne, R. Even: J. Chem. Phys. **83**, 6472 (1985)
29.217 A. Ahmad, R. A. Collins: Thin Solid Films **217**, 75 (1992)
29.218 S. A. James, A. K. Ray, J. Silver: Phys. Status Solidi A **129**, 435 (1992)
29.219 A. M. Saleh, R. D. Gould, A. K. Hassan: Phys. Status Solidi A **139**, 379 (1993)
29.220 R. D. Gould, A. K. Hassan: Thin Solid Films **223**, 334 (1993)
29.221 S. I. Shihub, R. D. Gould: Thin Solid Films **254**, 187 (1995)
29.222 J. G. Simmons: J. Appl. Phys. **34**, 1793 (1963)
29.223 J. G. Simmons: Phys. Rev. **166**, 912 (1968)
29.224 R. H. Fowler, L. W. Nordheim: Proc. R. Soc. A **119**, 173 (1928)
29.225 J. C. Fisher, I. Giaever: J. Appl. Phys. **32**, 172 (1961)
29.226 R. Holm: J. Appl. Phys. **22**, 569 (1951)
29.227 D. Meyerhofer, S. A. Ochs: J. Appl. Phys. **34**, 2535 (1963)
29.228 S. R. Pollack, C. E. Morris: J. Appl. Phys. **35**, 1503 (1964)
29.229 J. G. Simmons: J. Appl. Phys. **34**, 2581 (1964)
29.230 M. Lenzlinger, E. H. Snow: J. Appl. Phys. **40**, 278 (1969)
29.231 S. M. Sze: *Physics of Semiconductor Devices*, 2nd edn. (Wiley, New York 1981)

29.232 A. K. Hassan, R. D. Gould: J. Phys. D: Appl. Phys. **22**, 1162 (1989)
29.233 D. V. Geppert: J. Appl. Phys. **33**, 2993 (1962)
29.234 J. G. Simmons: Electronic conduction through thin insulating films. In: *Handbook of Thin Film Technology*, ed. by L. I. Maissel, R. Glang (McGraw–Hill, New York 1970) Chap. 14
29.235 P. R. Emtage, W. Tantraporn: Phys. Rev. Lett. **8**, 267 (1962)
29.236 S. R. Pollack: J. Appl. Phys. **34**, 877 (1963)
29.237 W. E. Flannery, S. R. Pollack: J. Appl. Phys. **37**, 4417 (1966)
29.238 M. Fusstöss-Wegner: Thin Solid Films **36**, 89 (1976)
29.239 B. Sh. Barkhalov, Yu. A. Vidadi: Thin Solid Films **40**, L5 (1977)
29.240 A. Wilson, R. A. Collins: Sens. Actuators **12**, 389 (1987)

30. Thick Films

Thick film technology is an example of one of the earliest forms of microelectronics-enabling technologies and it has its origins in the 1950s. At that time it offered an alternative approach to printed circuit board technology and the ability to produce miniature, integrated, robust circuits. It has largely lived in the shadow of silicon technology since the 1960s. The films are deposited by screen printing (stenciling), a graphic reproduction technique that can be dated back to the great Chinese dynasties of around a thousand years ago. Indeed, there is evidence that even early Palaeolithic cave paintings from circa 15 000 BC may have been created using primitive stenciling techniques. With the advent of surface-mounted electronic devices in the 1980s, thick film technology again became popular because it allowed the fabrication of circuits without through-hole components.

This chapter will review the main stages of the thick film fabrication process and discuss some of the commonly used materials and substrates. It will highlight the way in which the technology can be used to manufacture hybrid microelectronic circuits. The latter stages of the chapter will demonstrate how the technology has evolved over the past twenty years or so to become an important method in the production of solid state sensors.

30.1	**Thick Film Processing**	718
	30.1.1 Screen Printing	718
	30.1.2 The Drying and Firing Process	719
30.2	**Substrates**	720
	30.2.1 Alumina	720
	30.2.2 Stainless Steel	720
	30.2.3 Polymer Substrates	720
30.3	**Thick Film Materials**	721
	30.3.1 Conductors	721
	30.3.2 Resistors	722
	30.3.3 Dielectrics	723
	30.3.4 Polymer Thick Films	723
30.4	**Components and Assembly**	724
	30.4.1 Passive Components	724
	30.4.2 Active Components	725
	30.4.3 Trimming	725
	30.4.4 Wire Bonding	726
	30.4.5 Soldering of Surface-Mounted Components	727
	30.4.6 Packaging and Testing	727
30.5	**Sensors**	728
	30.5.1 Mechanical	728
	30.5.2 Thermal	729
	30.5.3 Optical	730
	30.5.4 Chemical	730
	30.5.5 Magnetic	730
	30.5.6 Actuators	731
References		731

The term "thick film" is often misinterpreted, and so it is worth noting from the outset that it has little to do with the actual thickness of the film itself. The preferred definition encompasses the fabrication process, namely screen printing, used to deposit the films. The typical range of thicknesses for thick film layers is 0.1 μm to 100 μm. Screen printing is one of the oldest forms of graphic art reproduction and involves the deposition of an ink (or paste) onto a base material (or substrate) through the use of a finely-woven screen with an etched pattern of the desired geometry. The process is commonly used for the production of graphics and text onto items such as T-shirts, mugs, pencils, textiles and so on, and is very similar to that used for microelectronic thick films. The degree of sophistication for the latter is, however, significantly greater, resulting in high-quality, reproducible films for use in a variety of electronic systems.

The technology used to manufacture thick film hybrid microelectronic circuits was introduced in the 1950s. Such circuits typically comprised semiconductor devices, monolithic integrated circuits, discrete passive components and the thick films themselves [30.1, 2]. In the early days of the technology, the thick films were mainly resistors, conductors or dielectric layers. Evidence of thick film circuits can still be found in many of today's commercial devices such as televisions, calculators and telephones. The use of thick film technology was overshadowed in the 1960s by the impact of silicon technology. It found popularity again in the 1980s as a result of the advent of surface-mounted devices, which can be attached to circuit boards using screen-printed solder layers. Nowadays, thick films are not only used in hybrid circuits but also in advanced solid state sensors and actuators, as we shall see later. In broad terms, thick films can be classified as either *cermets* (ceramic/metallics), requiring high-temperature processing, or *polymer* thick films, which are cured at significantly lower temperatures. Polymer thick films will be covered in more detail in Sect. 30.3.4.

30.1 Thick Film Processing

Thick film technology is sometimes referred to as an *additive* technology in that the layers are built up in sequence without the need to remove (or subtract) parts of the film by techniques such as etching. Compare this with, say, a standard printed circuit board (PCB), where the conductive tracks are formed by selectively etching away the undesired areas (gaps) from a continuous copper layer. In the early days of the technology, the ability to fabricate together components made with different enabling technologies to produce a *hybrid* circuit was seen as a way of opening up a whole new field of electronics. This is still one of the most endearing features of the technology today.

30.1.1 Screen Printing

This is the distinguishing feature of thick film technology and is the method of depositing the desired films. The fundamental aspects of the process, traditionally recognized as an art, have been modernized and updated to provide a scientific tool for microelectronic technology. Figure 30.1 illustrates the main aspects of the screen printing process [30.3].

The screen fabric is permanently attached to the screen frame, which is firmly held within the screen printing machine. The distance between the screen and the substrate is typically around 0.5 mm, although the precise gap size is dependent on the overall screen dimensions. The substrate is held in place in the holder by either a vacuum or a mechanical clamp. The position of the screen can be finely adjusted to ensure good registration between consecutive layers. The paste is applied to the upper surface of the screen and the flexible *squeegee* is pulled across the screen over the ink, which is forced through the open areas of the screen mesh. At a point immediately behind the squeegee, the screen peels away from the substrate and, due to the surface tension between the ink and the substrate, leaves a deposit of the paste in the desired pattern on the substrate.

The squeegee is a flexible blade whose function is to transfer the ink through the screen and onto the substrate. It is usually made from materials such as polyurethane or neoprene. The squeegee pressure is adjustable and facilitates accurate and repeatable print thicknesses. The pastes generally exhibit pseudoplastic behavior in that

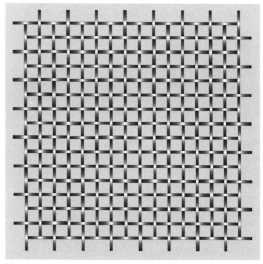

Fig. 30.2 A plain weave screen mesh

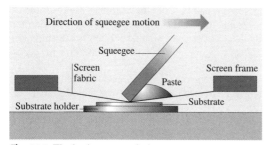

Fig. 30.1 The basic screen printing process

the viscosity varies with the applied shear force. This is a necessary property because the ink must have minimum viscosity to ensure good transfer though the screen, but it also must become more viscous after printing to provide a good definition of the film.

The function of the screen is to define the pattern of the printed film and also to control the amount of paste being deposited. The screens used in graphic artwork are generally made of silk and the process is often termed silk screen printing. For microelectronic circuits, however, the screens tend to be made from polyester, nylon or stainless steel. The resolution of the printed line widths is largely determined by the mesh count (in lines per inch or centimeter) and the mesh filament diameter. Nylon has very good elastic properties but relatively poor resilience. Stainless steel screens produce excellent line definition and durability and are particularly well suited to flat substrates. Polyester screens have the highest resilience and offer a long lifetime and low squeegee wear. Figure 30.2 shows a plain weave pattern on a typical thick film screen. An ultraviolet (UV)-sensitive emulsion covers the underside of the screen. The desired pattern is formed onto screen emulsion using a photographic process. The screen printing process can be used to deposit both cermet and polymer thick films; the processing steps that follow are, however, different.

30.1.2 The Drying and Firing Process

The final form of a cermet thick film is a fired, composite layer that is firmly attached to the substrate. Essentially, there are three main stages in the production process: screen printing, drying and firing. At each of these stages, the film is in a slightly different state. Commercial thick film pastes are purchased from the manufacturer in plastic jars and are similar to standard printing inks in many respects. The three main components in a thick film paste are:

- The active material;
- A glass frit;
- An organic vehicle.

The active material is a finely ground powder with a typical particle size of a few microns. A conductor paste, for example, will contain a precious metal or metal alloy. The glass frit serves as a binder that holds the active particles together and also bonds the film to the substrate. The organic vehicle is necessary to give the paste the correct viscosity for screen printing. It usually contains a resin dissolved in a solvent together with a surfactant that ensures good dispersion of the solid particles.

After screen printing, it is usual to allow the film to stand in air for a few minutes to let the paste level off. The film is then dried in an infrared belt drier or a conventional box oven at temperatures up to 150 °C. The purpose of drying is to remove the organic solvents so that the film and substrate can be handled during further processing steps. After drying, the films then proceed to the firing stage. In some circumstances, however, they may be overprinted with another thick film layer if the nature of the film allows this.

The high-temperature firing cycle performs three main functions: the remaining organic carrier is removed from the film, the electrical properties are developed, and the film is bonded to the substrate. In order to achieve these, it is necessary to subject the films to temperatures of up to 1000 °C in a moving belt furnace. Inside the furnace there are several heating zones, which can be independently controlled to a profile specified by the paste manufacturer. Figure 30.3 shows a typical firing profile that is used with commercial thick film resistor pastes. The substrates enter the furnace on a metal chain mesh belt at ambient temperature and slowly (at a rate of a few centimeters per minute) travel through the heating zones, which have their temperatures set to form the desired profile. The substrates remain at the peak temperature for about ten minutes and the electrical properties of the film are formed during this phase. Most thick film materials are fired in a clean, filtered air atmosphere, as this produces high-quality films with repeatable characteristics. Occasionally, however, it is necessary to provide an inert atmosphere such as nitrogen, so that materials such as copper can be processed.

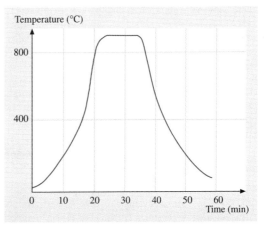

Fig. 30.3 A typical thick film firing profile with a peak temperature of 850 °C

30.2 Substrates

The main functions of the substrate are to provide mechanical support and electrical insulation for the thick films and hybrid circuits. Some of the main considerations for selecting substrates are listed below:

Dielectric constant: This determines the capacitance associated with different elements fabricated onto the substrate. The dielectric strength will also determine the breakdown properties of the substrate.

Thermal conductivity: Substrates with a high thermal conductivity can be used in applications where the circuit generates significant amounts of heat.

Thermal coefficient of expansion (TCE): In general terms, the TCE of the substrate should be closely matched to the thick film materials and other components mounted on it. In some cases this cannot be assured, and in this case the consequences (in terms of thermal strains) must be fully considered.

The main substrate materials used in thick film technology are the ceramic materials alumina (Al_2O_3), beryllia (BeO) and aluminium nitride (AlN). Enamelled or insulated stainless steel substrates are sometimes used in some applications. Silicon has also been used in specialist transducer applications. Alumina, however, is the most common substrate material and it possesses desirable physical and chemical properties in addition to providing an economical solution. Alumina of 96% purity is used in the vast majority of worldwide commercial circuits. The remaining 4% weight fraction of the content is made up of magnesia and silica, which improve the densification and electrical properties. Beryllia has a high thermal conductivity and is used in applications where rapid heat removal is required. It is, however, a very toxic material and is therefore only used in limited application areas. Aluminium nitride is, essentially, an alternative for beryllia, with a high thermal conductivity and also improved mechanical properties such as higher flexural strength. Insulated stainless steel substrates are sometimes used in applications where a high thermal dissipation and mechanical ruggedness are required. They are particularly well suited to mechanical sensor applications.

30.2.1 Alumina

Alumina substrates are manufactured by blending alumina powder, with an average particle size of around 1 μm, together with small amounts of silica, magnesia and calcia. These are either ball- or roll-milled for about 10 h with lubricants, binders and solvents that ensure thorough mixing. Most thick film substrates are less than one millimeter thick, and the preferred method of fabrication is sheet casting. A slurry is allowed to flow out onto a smooth belt, and it passes under a metal doctor blade which controls the resultant thickness. The material is then dried in air to remove the solvent and, at this stage, it is sometimes referred to as the *green state* because of its color. The substrates are then fired in a kiln for at least 12 h. A peak temperature of around 1500 °C ensures that the materials are properly sintered. During firing, the substrates can shrink by up to 20%, and this needs to be taken into consideration for the formation of the final substrate. The surface finish can be improved by coating the surface with a thin, glassy layer (glazing), which is done as an additional step at a lower temperature.

30.2.2 Stainless Steel

Stainless steel is a strong, elastic material with a relatively high thermal conductivity. Being a good electrical conductor, the steel must be coated with an insulating layer before it can be used as a substrate for thick film circuits. Porcelain enamelled steel substrates are made by coating a stainless steel plate with a glassy layer between 100 and 200 μm in thickness. The steel is enamelled either by dipping, electrostatic spraying or electrophoretic deposition of a low-alkali glass and subsequent firing at several hundred degrees Celsius. Some commercial paste manufacturers produce an insulating dielectric thick film ink that can be screen printed directly onto various types of stainless steel. The substrates are fired at a temperature of around 900 °C. The paste contains a devitrifying glass that does not recrystallize on further firings and therefore provides compatibility with other standard thick film materials. Insulated stainless steel substrates also offer the advantages of having a built-in ground plane (the steel itself) and excellent electromagnetic and electrostatic shielding properties. It is also possible to machine the substrate, using conventional workshop facilities, prior to the circuit fabrication.

30.2.3 Polymer Substrates

In some applications it is desirable to fabricate a circuit onto a flexible substrate; evidence of these can be found in mobile telephones, calculators and notebook computers. Cermet thick film materials are not compatible with flexible substrates and hence special polymer thick film materials are used (Sect. 30.3.4). Polymer substrate materials are mainly based on polyesters, polycarbonates and polyimide plastics. The maximum processing temperature is usually limited to around 200 °C.

30.3 Thick Film Materials

30.3.1 Conductors

Thick film conductors are the most widely used material in thick film hybrid circuits. Their main function is to provide interconnection between the components in the circuit. For a multilayered circuit, the conductor tracks are separated by dielectric layers and connection between each layer is achieved with metallized vias. Conductors are also used to form attachment pads for surface-mounted components such as integrated circuits or discrete passive components (resistors, capacitors and inductors). They can also be used as bond pads for naked dice, which may be attached directly to the thick film circuit. Another function of conductors is to provide the terminations for thick film resistors. With such a diverse range of applications, it is no surprise that a wide range of conductor materials is available.

The characteristics of thick film conductors are dependent upon the composition of the functional phase of the paste. Typically, these comprise finely divided particles of precious metals such as silver, gold, platinum or palladium. Base metals such as aluminium, copper, nickel, chromium, tungsten or molybdenum are also used. The particle size, distribution and shape also have an effect on the electrical and physical properties of the fired film.

The resistance of a conductor film is given by

$$R = \frac{\rho l}{wt},$$

where ρ is the bulk resistivity (Ω cm), l and w are the length and width of the conductor respectively, and t is the fired thickness of the film. The term *sheet resistivity* is often used for thick films and is defined as the bulk resistivity at a given thickness (ρ/t), expressed in ohms per square ($\Omega/\square$). This is convenient because the resistance of the track can then be calculated by simply multiplying the sheet resistivity by the aspect ratio (l/w) of the film. Table 30.1 summarizes some of the most common metals and metal alloys used in thick film conductors together with their sheet resistivities.

Silver Conductors

Silver pastes were one of the earliest thick film conductors to be developed. They possess good bond strength and high conductivity. There are, however, several disadvantages to using pure silver which prevent it being widely used in many applications. These include:

- Poor leach resistance to solder;
- Oxidation in air over time;
- Susceptible to electromigration in the presence of moisture, elevated temperature and bias voltage.

Silver/palladium conductors

The alloy of silver and palladium is the most common type of thick film conductor and is the alloy most widely used in the hybrid circuit industry. It can be used for interconnecting tracks, attachment pads and resistor terminations, but is generally not recommended for wiring bonding pads. Silver/palladium conductors overcome many of the problems associated with pure silver, and low-migration formulations are available from several commercial suppliers.

Gold Conductors

Gold pastes have a high conductivity and are mainly used in applications where high reliability is required. Gold is a particularly good material for wire bonding pads, although it has relatively poor solderability. Gold is a precious material and hence very expensive; it is therefore not used for general purpose applications and is limited to those areas that can justify the higher costs.

Table 30.1 Most of the common metals and metal alloys used in thick film conductors and their sheet resistivities

Metallurgy	Sheet resistivity (m$\Omega/\square$)
Silver (Ag)	1–3
Gold (Au)	3–5
Copper (Cu)	2–3
Silver/palladium (Ag/Pd)	10–50
Gold/palladium (Au/Pd)	10–80
Gold/platinum (Au/Pt)	50–100

Copper Conductors

Copper thick films have to be processed in an inert atmosphere; nitrogen is typically used in this case. Copper is widely used as the conductor material for printed circuit boards owing to its good electrical conductivity and solderability. These features are also applicable to thick film copper conductors, which also offer ability to handle larger currents than most other thick film conductors.

Platinum Conductors

Not too surprisingly, perhaps, these are the most expensive of all the commercial thick film conductor materials. Platinum films have a very high resistance to solder leaching and exhibit similar electrical properties to those of the bulk material: a linear, well-defined temperature coefficient of resistance (TCR). Platinum films are therefore used in specialist applications such as heaters, temperature sensors and screen-printed chemical sensors.

Gold Alloy Conductors (Au/Pd, Au/Pt)

These have good bond strength, good solder leach resistance and are relatively easy to solder. The conductivity of the gold alloys tends to be inferior to that of the other type of thick film conductor, but they can be used with both ultrasonic (aluminium) and thermosonic (gold) wire bonds.

Once the conductor has been fired, a composite structure is formed. Figure 30.4 shows a simplified view of the cross-section of a fired thick film conductor on a substrate. Many of the metallic particles have joined together to make a continuous chain. The glass is mainly evident at the interface between the bulk of the film and the substrate. The presence of voids (gaps), both within the film and on its surface, can also be noted.

30.3.2 Resistors

Resistor inks consist of a mixture of the three main phases listed earlier. The relative proportions of the active material to glass frit can have a dramatic effect on the electrical properties of the fired film. The earliest resistors were made from materials such as carbon, silver and iron oxide and were found to suffer from poor long-term stability, unpredictability of fired resistivity, and unacceptably high temperature coefficients. Modern thick film resistor systems are mainly based on ruthenium dioxide (RuO_2) and have much-improved characteristics. This material has a high conductivity and is extremely stable at high temperatures. Within the resistive paste, the conductive phase comprises submicron

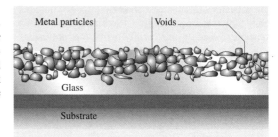

Fig. 30.4 Idealized cross-sectional view of a cermet thick film conductor

particles, and these are mixed with larger glass particles (several microns in diameter). Various additives are added to the formulation to improve the stability and electrical properties of the fired film. The nature of these additives is proprietary knowledge of the paste manufacturer.

Figure 30.5 shows an idealized cross-section of a thick film resistor fabricated onto a substrate. The end terminations are first printed and fired and then the resistor is formed in the same manner, allowing for a slight overlap on the terminations so that small misalignments during the printing of the resistor can be accounted for. The actual surface profile of a real resistor would show a nonuniform thickness across its length.

Commercial thick film resistor pastes are available in a range of sheet resistivities from 1 to 10^9 $\Omega/\square$, and the value of the fired resistor is determined by the selected sheet resistivity and the ratio of the length to the width, as previously described for conductors. For example, a $10\,k\Omega$ resistor could be made by printing a square ($l = w$) of $10\,k\Omega/\square$ resistor paste. Similarly, a $5\,k\Omega$ resistor requires an aspect ratio of $1 : 2$ ($l : w$) if printed with the same paste. Note that the absolute value of a thick film resistor is not limited to a preferred value, as is the case for discrete resistors.

Figure 30.6 shows the temperature dependence of resistance for a typical thick film resistor. The shape of

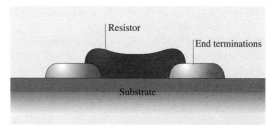

Fig. 30.5 An idealized cross-section of a thick film resistor

the plot is unusual because there is a point at which the resistance is a minimum (T_{min}). In general terms, the sensitivity of a resistor to temperature is denoted by the temperature coefficient of resistance (TCR). Mathematically, this can be expressed by

$$\text{TCR} = \frac{\Delta R/R}{\Delta T},$$

where $\Delta R/R$ is the relative change in resistance and ΔT is a small change in temperature. Metals generally exhibit an increase in resistance with increasing temperature, so the TCR is positive. A thick film resistor appears to have both a positive and negative TCR at different regions of the resistance versus temperature curve [30.4]. The point of minimum resistance occurs around room temperature, which means that the resistance is stable under normal operating conditions. It is common practice to refer to the 'hot' and 'cold' TCR for the regions above and below T_{min}, respectively.

The long-term stability of thick film resistors in air at room temperature is excellent, with a typical change of less than 0.2% over 10^5 h (about 100 years!). At elevated temperatures the stability is still impressive, with most resistance values changing by only 0.5% over 1000 h at 150 °C. The tolerance of fired resistors is around ±20% due to a variety of reasons, including variations in resistor thickness, the tolerance on the quoted sheet resistivity and the effects of the firing process. If accurate values of resistance are required, it is therefore necessary to trim the resistor. This is achieved by removing areas of the resistor with either an air-abrasive jet or a laser. Resistor values can only be increased by trimming, so the printed value is always designed to be less than that of the post-trimmed value.

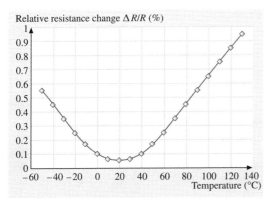

Fig. 30.6 A plot of relative resistance change against temperature for a thick film resistor

30.3.3 Dielectrics

Dielectric pastes have four main uses in thick film hybrid circuit applications:

- Cross-over insulators for multilayer circuits;
- Thick film capacitor dielectrics;
- Passivation layers;
- Insulation layers for stainless steel substrates.

Cross-over dielectrics comprise a ceramic material such as alumina, together with the usual glass frit and organic vehicle. These films are required to have a low dielectric constant in order to minimize capacitive coupling between the conducting tracks. They must also have a good insulation resistance and a smooth, pinhole-free surface finish.

There is an occasional requirement to fabricate thick film capacitors, although the wide availability of surface-mounted capacitors has largely precluded their need. It is also difficult to ensure that post-fired values of thick film capacitors closely match those for which they were designed. An expensive trimming process is needed to ensure that the correct valves are obtained.

Overglazes are mainly used to protect resistors from overspray during trimming and also the circuit from environmental attack during operation. The passivation layer is usually the last one to be processed and it is therefore fired at a reduced temperature in order to minimize any adverse refiring effects on other layers. Overglaze materials are therefore almost exclusively made from low-temperature glasses that fire at temperatures of 450 to 500 °C.

The use of insulated stainless steel substrates was mentioned earlier. The dielectric materials used for this purpose must provide a high insulation resistance (in excess of $10^{12}\,\Omega$) and also possess a high breakdown voltage (greater than 2 kV/mm). There is an inherent mismatch between the thermal coefficient of expansion of the metal substrate and that of the insulation layer, which limits the type of steels that can be used for this purpose.

30.3.4 Polymer Thick Films

The processing temperatures for polymer thick films are significantly less than those needed for cermet materials. Rather than being 'fired', polymer materials are cured at temperatures below 200 °C. One advantage of polymer thick films over cermets is a reduction in processing and material costs. As with cermet pastes, the formulation of polymer thick films comprises the active material, a polymer matrix and various solvents. The polymer matrix acts as 'glue' for the active component. Three types of polymer organic composition are used in polymer thick films:

- Thermoplastic;
- Thermosetting;
- Ultraviolet (UV)-curable.

With thermoplastics, the required viscosity for screen printing is achieved via solvents. The polymer material is typically acrylic, polyester, urethane or vinyl. After printing, the paste is hardened by drying in a belt or box oven. These types of film have relatively poor resistance to environmental conditions and are not resistant to elevated temperatures and solvents. Thermosetting pastes have polymers that are partially cured and are typically epoxy, silicone or phenolic resin. After printing, the polymer is fully cured, providing a strong and stable matrix. Solvents are still needed to provide the correct rheology for printing. The UV curable pastes are generally used for dielectric inks and can be cured at room temperature under an ultraviolet light source.

For conductors, the most commonly used active phases are silver, copper and nickel. Owing to their poor stability at high temperatures, polymer thick film conductors cannot be soldered and alternative forms of attachment must be adopted. Carbon is typically used as the active material in polymer thick film resistors. The performance of these resistors is inferior to that of their cermet counterparts and they are therefore seldom used in critical applications. The dielectric pastes are similar in nature to conductors and resistors except that the conducting phase is omitted from the formulation. Some manufacturers add minerals to improve the electrical and mechanical properties of the dielectric films. Polymer thick films are a popular choice of material for disposable biosensors such as those used in

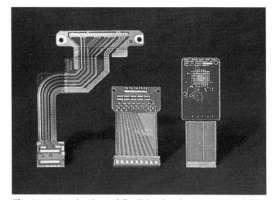

Fig. 30.7 A selection of flexible circuits (courtesy of Flex Interconnect Technologies, Milpitas, USA)

the home testing of levels of glucose in human blood samples.

The process for fabricating polymer thick films is similar to that used with cermet materials. Once they have been screen printed, the layers are left to stand in air for a few minutes to ensure that the surface is level and contains no residual mesh patterns. The curing process is achieved in a box oven or an infrared belt dryer at temperatures in the range 150–200 °C. For a thermoset polymer, the higher the temperature and longer the curing time, the greater the cross-linking of the polymer chains in the matrix. This can lead to improved film stability and increased shrinkage.

An early and successful application of polymer thick film technology was the fabrication of membrane switches for keyboards. Today, examples of polymer thick film circuits can be found in many consumer products, such as mobile phones, portable computers, personal digital assistants and calculators. Figure 30.7 shows some typical flexible polymer circuits.

30.4 Components and Assembly

30.4.1 Passive Components

Passive electronic components are those that do not require an external energy source to function. Examples are resistors, capacitors and inductors. As we have seen earlier, thick film technology allows the fabrication of high-quality, stable resistors. It is also possible to add resistors to a thick film circuit in the form of an additional chip component. Such devices are available as surface-mounted devices, which do not require holes to be drilled into the circuit board. Interestingly, chip resistors are often manufactured as multiple parts using thick film techniques on ceramic substrates. These are then diced and the terminations are added. An example of a typical chip component is shown in Fig. 30.8. Resistor values can range $1\,\Omega$ to $10\,M\Omega$, with typical tolerances of between $\pm 1\%$ to $\pm 20\%$.

Planar screen-printed capacitors are rarely used in thick film hybrid circuits owing to their poor stability and high production costs. They typically comprise at least three layers (two electrodes and one dielectric layer), and trimming is often required to achieve the target value of capacitance. Chip capacitors offer improved performance at a lower cost. Capacitances in the range 1 pF to 100 μF are readily available in a range of sizes from the 0201 series (0.6 mm × 0.3 mm × 0.3 mm) to the 2220 series (5.7 mm × 5.0 mm × 3.2 mm). Chip inductors are also available (typical range: 0.1 μH to 1000 μH) in a variety of package sizes. There is also a wide choice of variable passive components such as potentiometers and variable capacitors/inductors that are currently available from many major component suppliers.

30.4.2 Active Components

Active components are those that require an external energy source to function. Examples are transistors, diodes and semiconductor integrated circuits. A wide variety of semiconductor components are available to the thick film circuit designer. Transistors and diodes, being relatively small devices, are obtained in a small plastic package with three terminals. This is known as a SOT-23 package and is shown in Fig. 30.9a. Many standard integrated circuits that are available in dual-in-line (DIL) packages for through-hole printed circuit boards are also available in small outline (SO) packages for hybrid circuits. An example of a small outline device is shown in Fig. 30.9b, and these usually have between 8 and 40 pins. For devices with higher pin counts such as microprocessor, gate arrays and so on, it is usual to place the leads on all four sides of the package, as with the flatpack device shown in Fig. 30.9c. Occasionally, ICs may be obtained in the form of a naked die, without any packaging. In such cases, it is necessary to glue the chip to the board and to bond very fine wires from the chip to the board. Care has to be taken to ensure that the naked device is suitably encapsulated for use afterwards.

Other forms of IC include flip chips, which are essentially naked chips with raised connection contacts (bumps) made of solder, gold or aluminium. These are mounted by turning over the chip (flipping) and bonding directly to the substrate. Beam lead chips also examples of naked dice with either gold or aluminium leads protruding from the edge. The leads (beams) are an integral part of the chip metallization process. Such devices are usually passivated with a layer of silicon nitride during processing. Tab automated bonding (TAB) refers to a technique by which the naked chip is attached to metallized fingers on a continuous strip of film. Large quantities of devices can be produced on a single roll and the leads of the devices are welded onto the boards before the carrier film is removed. In this manner, it is possible to achieve good yields on high-density circuit populations.

30.4.3 Trimming

The tolerance on the printed value of thick film components such as resistors is around ±20% of the desired value. Many applications require a much tighter tolerance, and so the components need to be trimmed. The two most popular techniques used are trimming by laser or by an air-abrasive jet. Both of these are capable of producing resistors with a tolerance of ±0.1%. Laser trimming, however, is more amenable to large-scale component adjustment.

Air-abrasive trimming uses a pressurized jet of air containing a fine abrasive powder to remove a small

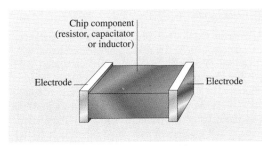

Fig. 30.8 A typical surface-mounted chip component

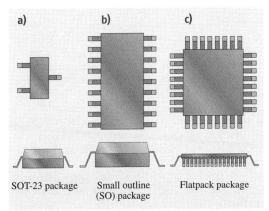

Fig. 30.9a–c Examples of surface-mounted packages (**a**) SOT-23 package (**b**) Small outline (SO) package (**c**) Flatpack package

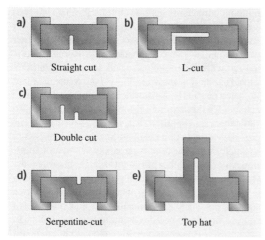

Fig. 30.10a–e Examples of trim cuts for resistors (**a**) Straight cut (**b**) L-cut (**c**) Double cut (**d**) Serpentine cut (**e**) Top hat

area of the fired thick film. The diameter of the jet nozzle is around 0.5 to 1 mm. Alumina particles of average diameter 25 μm are often used as the abrasive medium. The substrate containing the component to be trimmed is held underneath the nozzle at a distance of around 0.6 mm and electrical probes are attached to the device. During trimming, the particles in the jet stream remove material from the component. The debris is removed from the substrate by a vacuum exhaust system. In the case of resistor trimming, the value can only increase because the material is being removed. Resistor values cannot be reduced by the trimming process. For this reason, resistors requiring trimming are designed to be between 25 and 30% lower than the post-trimmed value.

Laser trimming has the advantage of offering a fully automated, high-speed way of adjusting component values. The thick film is vaporized with high-energy laser beam pulses. The laser is typically a Q-switched neodymium-doped YAG (yttrium aluminium garnet) type. A single pulse removes a hole of material and a line is achieved by overlapping consecutive pulses.

Resistors are the most common component requiring trimming. Sometimes conductor tracks need to be adjusted in special circumstances. For example, a platinum conductor being used as a classic Pt 100 resistance thermometer must have a resistance of 100 Ω at 0 °C. In rare circumstances, thick film capacitors can be trimmed by removing an area of one of the plates, although the post-trimmed stability is poor.

Figure 30.10 shows some examples of different types of cut that can be used to trim resistors. The straight cut is the fastest way to trim, but it does not provide a reliable way of achieving a high accuracy. The L-cut overcomes this problem; the resistor is first trimmed straight and then the cut runs parallel to its length, providing a finer adjustment of the value. Both the serpentine and double cut require additional cuts perpendicular to an initial straight cut. The top hat structure is used in situations where a large change in resistance is needed.

30.4.4 Wire Bonding

When a naked integrated circuit is needed as part of a hybrid circuit, connection must be made directly to the bond pads on the die. The chip can be attached to the substrate using an epoxy or by a gold/silicon eutectic bond. Once the chip is firmly held in position, wire bonding can commence. Three main methods are:

- Thermocompression;
- Ultrasonic;
- Thermosonic.

Thermocompression bonding relies on a combination of heat and pressure. The wire is usually made of gold with a diameter of around 25 μm. Gold or palladium/gold pads are deposited onto the hybrid substrate prior to bonding. The wire is fed through a ceramic capillary and a ball is formed at the end of the wire by a flame or spark discharge. A temperature of 350 °C is required for the bond and is achieved by heating either the substrate or the capillary. An epoxy chip bond cannot be used with this technique as it will soften during the bonding process. The first part of the bond is made on the aluminium bond pad on the chip; the capillary is lowered onto the pad and a force is applied to form a ball shape on the pad. The wire feeds out of the capillary and is then positioned over the desired pad on the substrate. As the capillary is lowered, a combination of heat and pressure forms the bond and the wire is then broken so that the process can be repeated for further bonds.

Ultrasonic bonding uses either a gold or aluminium wire and does not require external heat. The ultrasonic energy is supplied from a 40 kHz transducer. The combination of pressure and ultrasonic vibration causes the materials to bond together at the interface of the wire and the bond pad. This is generally a faster technique than the thermocompression method.

The final category, thermosonic bonding, is a combination of the other techniques. The substrate is heated to around 150 °C and the bond is made using the ultrasonic vibrations. This method is amenable to multilevel and multidirectional bonding and is therefore the preferred method, allowing bonding of up to 100 wires per minute.

30.4.5 Soldering of Surface-Mounted Components

Several techniques exist for the attachment of surface-mounted components to a thick film hybrid circuit. Surface-mounted components are generally much smaller than their through-hole counterparts. Of course, soldering by hand is also possible, although this is a tricky task requiring good operator skill and is often impractical because of the relatively long length of time needed.

Solder dipping requires the components to be placed on the board, either by hand or by a special pick-and-place machine. The components are fixed in position on the substrate by adding a small dot of glue and elevating the temperature to between 120 °C and 180 °C, which is sufficient to cure the adhesive. The board can then be dipped into a bath of molten solder at a temperature of 200 °C and then withdrawn at a sufficient rate to ensure that an adequate solder coating is obtained.

Wave soldering also requires the components to be fixed in position prior to the soldering process. Wave soldering machines were originally used for soldering through-hole components onto printed circuit boards, but they can also be used effectively with surface-mounted devices. The substrates are placed on a moving belt component side-down and initially pass through a flux bath before entering a solder bath. A wave of molten solder then flows over the substrate and creates a good joint at the desired location. This process can expose the components to a great thermal shock and it is therefore common to have a preheating phase which minimizes such effects. With both of these techniques, it is also necessary to ensure that a solder resist layer is applied to the substrate to cover all the areas that are not required to be soldered. With very densely populated circuits, there can also be a masking effect where some areas are not sufficiently coated with solder.

Reflow soldering is the preferred method of attaching surface-mounted devices. A solder cream is deposited onto the component pads either by screen printing or by a solder dispenser. The flux within the cream is sufficiently tacky to hold the component in place so that handling of the substrate is possible. After all of the components have been positioned on the circuit, the solder cream is dried and then reflowed. This process takes place by belt reflow, vapor phase or infrared belt system. A typical belt reflow system comprises a thermally conducting belt upon which the substrates are placed. The belt then travels through a number of heating stages, which causes the solder cream to melt (reflow). Vapor-phase soldering requires the substrates to be lowered into a vessel containing a boiling, inert fluorocarbon. The vapor condenses onto the substrate and raises the temperature uniformly to that of the liquid below. Infrared belt reflow systems are similar to those used for drying thick film materials. The substrate is placed on a wire-mesh belt, which travels through several infrared radiator zones.

30.4.6 Packaging and Testing

Thick film hybrid circuits are very versatile and offer advantages over other forms of enabling technologies. Owing to this flexibility, the circuits have a wide range of shapes and sizes and hence there is no "standard" package type. Selection of a particular form of packaging must therefore involve the consideration of issues such as:

- Protection of the circuit from harsh environmental conditions;
- Protection from mechanical damage;
- Avoidance of water ingress;
- Electrical or mechanical connections to other parts of the system;
- Thermal mismatches of different materials.

A simple way of protecting the circuit is to screen print an overglaze layer over the substrate, covering all areas of the substrate except those where components are to be added. A lead frame can then be added to the substrate to allow external connections to be made. An example of a thick film hybrid circuit (without overglaze) is depicted

Fig. 30.11 A thick film hybrid circuit

in Fig. 30.11. The two resistors on the left of the circuit have been trimmed and the straight cuts are visible.

Conformal coatings are often used to protect the circuit from environmental attack. These are applied in the form of either a powder or fluid. In the former case, the substrate is heated and immersed into the powder. The temperature is then increased so that the coating dries. For fluids, the substrate is dipped into the coating material and subsequently dried at a temperature of around 70 °C. Typically, the thickness of a conformal coating is between 300 and 1200 μm.

For circuits requiring operation in harsh environments, a special hermetic packaging is needed. The package can be made from ceramics, metals, ceramic/metal or glass/ceramic compositions. The hermetic seal is made by brazing, welding or glass sealing techniques. This form of packaging is often very expensive and is therefore only used in special application areas.

The final stage of the process is to test the circuit to see that its performance matches the design specification. Electrical testing can be difficult if the circuit has been coated or hermetically sealed, as physical access to components may be restricted. It is therefore usual to ensure that key test points are brought out to an external pin on the package. Environmental testing over a range of temperature and humidity may also be required in some circumstances. High-reliability circuits are often subject to a so-called burn-in phase, which involves holding the circuits at an elevated temperature for a given time to simulate the ageing process.

30.5 Sensors

Advances in the field of sensor development are greatly affected by the technologies that are used for their fabrication. The use of thick film processes as an enabling technology for modern-day sensors continues to expand. As we have already seen, the ability to produce miniaturized circuits is clearly one area in which thick film technology excels. The hybrid electronic circuitry can be integrated into the sensor housing to produce the basis of a smart (or intelligent) sensor [30.5]. Thick film technology also offers the advantage that it can provide a supporting structure onto which other materials can be deposited, possibly using other enabling technologies [30.6].

A major contribution of the technology to sensor development, however, results from the fact that the thick film itself can act as a primary sensing element. As an example, the thick film strain gauge, described below, is merely a conventional thick film resistor that is configured in such a way as to exploit one of its physical characteristics. Commercial thick film platinum conductors can be trimmed and used as calibrated temperature sensors. Most standard pastes, however, have not been specifically developed for sensor applications and do not necessarily have optimum sensing properties. The formulation of special-purpose thick film sensor pastes is the subject of intensive research activity [30.7].

For the purpose of this text, a *sensor* is considered as being a device that translates a signal from one of the common sensing domains (mechanical, thermal, optical, chemical or magnetic) into an electrical signal. An *actuator* is a device that converts an electrical signal into one of the other domains (mainly mechanical).

30.5.1 Mechanical

In broad terms, thick film mechanical sensors are mainly based on piezoresistive, piezoelectric or capacitive techniques. Materials that exhibit a change in bulk resistivity when subjected to deformation by an external force are termed *piezoresistive*. A more common term is the strain gauge, denoting the fact that such devices produce a change in resistance when strained. The effect can be observed in standard cermet thick film resistors [30.6,8]. The sensitivity of a strain gauge is called the gauge factor and is defined as:

$$\mathrm{GF} = \frac{\Delta R/R}{\varepsilon},$$

where $\Delta R/R$ is the relative change in resistance and ε is the applied strain (dimensionless). The gauge factors of metal foil strain gauges and thick film resistors are around 2 and 10 respectively. The former have typical resistance values of either 120 Ω or 350 Ω. As we have already seen, however, it is possible to produce thick film resistors with a wide range of resistance values, and this allows greater flexibility in strain gauge design. It is usual to place the strain gauges in a Wheatstone bridge configuration in order to produce a linear output analog voltage change that is proportional to the mechanical measurand. A wide range of thick film piezoresistive sensors exist, including accelerometers, pressure sensors and load cells.

Piezoelectric materials exhibit the property of producing an electric charge when subjected to an applied mechanical force. They also deform in response to an externally applied electric field. This is an unusual effect, as the material can act as both a sensor and actuator. Certain crystals such as quartz and Rochelle salt are naturally occurring piezoelectrics, whilst others, like the ceramic materials barium titanate, lead zirconate titanate (PZT) and the polymer material polyvinylidene fluoride (PVDF), are *ferroelectric*. Ferroelectric materials are those that exhibit spontaneous polarization upon the application of an applied electric field. This means that ferroelectrics must be poled (polarized) prior to use in order to obtain piezoelectric behavior.

Thick film piezoelectrics have been made by mixing together PZT powder, a glass binder and an organic carrier [30.9]. A conducting layer is first screen printed, dried and fired onto a substrate and then several layers of the PZT film are deposited onto this lower electrode. The piezoelectric layer can be processed in a similar manner to conventional thick films. An upper electrode layer is then deposited onto the PZT in order to make a sandwich structure similar to that shown in Fig. 30.12. This is essentially a planar capacitor and the as-fired film must undergo a poling process by applying a DC electric field of around 4 MV/m and elevating the temperature to about 120 °C.

Thick film piezoelectrics have been used in a variety of sensor and actuator applications, including accelerometers, pressure sensors [30.10], micromachined pumps [30.11], surface acoustic wave (SAW) [30.12] and resonant sensors [30.13].

The capacitance C of a parallel plate capacitor is given by

$$C = \frac{\varepsilon_0 \varepsilon_r A}{d}$$

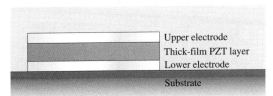

Fig. 30.12 Cross-section of a thick film piezoelectric sample

where ε_0 is the permittivity of free space, ε_r is the relative permittivity of the material between the electrodes, A is the area of overlap, and d is the separation of the electrodes. A mechanical sensor exhibiting a change in capacitance can be made by varying A, d or by displacing the dielectric (changing ε_r). The most popular configuration is to vary d in accordance with the desired measurand. This results in a nonlinear relationship between displacement and capacitance. If the variable plate is positioned between two fixed outer electrodes, then a differential structure with a linear response can be obtained. This is a common arrangement in many types of pressure sensor.

30.5.2 Thermal

Devices that exhibit a change in resistance in accordance with variations in temperature are termed *thermoresistive*. For metals, such devices have a linear response to temperature and are known as resistance thermometers. Thermally sensitive semiconductors, typically having a nonlinear response, are termed *thermistors*. Thick film platinum conductor layers can be used as resistance thermometers and exhibit a linear TCR of around 3800 ppm/°C, slightly lower than that of bulk platinum, which is around 4000 ppm/°C. Platinum resistance thermometers (PRTs) are often trimmed so that they have a resistance (R_0) of 100 Ω at 0 °C. Such sensors are sometimes referred to as Pt 100s and are often made of wound platinum wire. Thick film Pt 100s, fabricated onto alumina substrates, are available commercially and are considerably cheaper than the bulk versions. The following expression applies to a PRT over the linear part of the temperature characteristic (-200 °C to $+500$ °C)

$$R = R_0(1 + \alpha T),$$

where R is the resistance at a temperature T and α is the TCR (ppm/°C).

Thermistors are usually available in the form of discs, rods or beads comprising a sintered composite of a ceramic and a metallic oxide (typically manganese, copper or iron). Thermistor pastes are commercially available and screen-printed sensors are fabricated in a similar manner to conventional thick film resistors. The electrodes are first printed and fired onto an alumina substrate. The thermistor is then deposited across the electrodes and can be trimmed to a specific value if desired. Most thermistors have a negative temperature coefficient (NTC) of resistance; in other words their resistance decreases as the temperature increases. The resistance versus temperature relationship is of the form

$$R = R_0 \exp\left[\beta\left(\frac{1}{T} - \frac{1}{T_0}\right)\right]$$

where R_0 is the resistance at a reference temperature T_0 (usually 25 °C).

Positive temperature coefficient (PTC) thermistors are also available, although they are generally not as stable or repeatable as NTCs and are therefore used as simple thermal detectors rather than calibrated devices.

Another type of temperature sensor can be made by joining together two dissimilar metals (or semiconductors). If a temperature difference exists between the joined and open ends, then an open-circuit voltage can be measured between the open ends. This is known as the Seebeck effect and is the basis of a *thermocouple*. A thick film version can be made by overlapping different conductor materials on a substrate, although the thermal sensitivity is much less than traditional wire-based devices.

30.5.3 Optical

Screen-printed photosensors are probably one of the earliest examples of thick film sensors, and their use dates back to the mid 1950s. Materials that exhibit a change in electrical conductivity due to absorbed electromagnetic radiation are known as *photoconductors*. Cadmium sulfide (CdS) is an example of such a material and is notable because of its highly sensitive response in the visible range (450–700 nm). The resistances of such devices can drop from several tens of MΩ in the dark to a few tens of ohms in bright sunlight.

Thick film photoconductor pastes are not widely available commercially, but have been the subject of some research activity [30.14]. Such pastes, based on cadmium sulfide and selenide, are prepared by sintering powdered CdS or CdSe with a small amount of cadmium chloride (which acts as a flux) at a temperature of around 600 °C. This is then ground into a powder and mixed with an organic carrier to make a screen-printable paste. This can then be printed over metal electrodes and fired at a temperature of around 600 °C in air.

30.5.4 Chemical

Thick film materials have been used in a variety of chemical sensor applications for the measurement of gas and liquid composition, acidity and humidity [30.15]. The two main techniques are impedance-based sensors and electrochemical sensors. With the former method, the measurand causes a variation in resistance or capacitance, whilst the latter relies on the sensed quantity changing an electrochemical potential or current.

Impedance-based gas sensor pastes usually comprise a semiconducting metal oxide powder, inorganic additives and organic binders [30.16]. The paste is printed over metal electrodes and a back-heated resistor on an alumina substrate. The heating element is necessary to promote the reaction between the gas being measured and the sensing layer. Figure 30.13 shows an example of a thick film sensor, without a heating element, that can be used to measure humidity. The interlocking 'finger' electrodes are often referred to as interdigitated electrodes and are screen printed and fired onto an alumina substrate. A porous dielectric layer is then screen printed onto the electrodes. As the humidity increases, moisture will penetrate the surface of the dielectric layer causing a change in dielectric constant within the sensitive layer. This results in a change in capacitance between the electrodes.

Electrochemical techniques can be used to realise pH sensors. These are often used in biomedical, fermentation, process control and environmental applications. These devices often make use of a solid electrolyte which generates an electrochemical potential between two electrodes in response to the measurand.

Perhaps the most common example of a thick film chemical sensor is the disposable, polymer-based glucose sensor used in many home testing kits for diabetic patients. This illustrates how thick film sensors can offer robust, compact and cost-effective solutions to many modern-day requirements.

30.5.5 Magnetic

Some screen-printable conductors, particularly those containing nickel, exhibit a change in resistivity in response to an applied magnetic field. Such devices are referred to as *magnetoresistive* sensors. Air-fireable

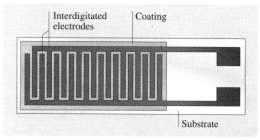

Fig. 30.13 A thick film humidity sensor

nickel-based conductors have been shown to exhibit a nonlinear change in resistance for a linear increase in applied magnetic field [30.17]. A peak change in resistance of around 1% can occur at an applied field of 0.1 T. Researchers have made linear and rotary displacement sensors based on thick film nickel pastes, although it should be noted that such devices are also thermoresistive and therefore the magnetic measurements need to be taken in a temperature-controlled environment.

30.5.6 Actuators

We have previously defined an actuator as a device that converts a signal from the electrical domain into one of the other signal domains. It was noted earlier that piezoelectric materials produce a mechanical stress in response to an electrical charge. Such materials can therefore be used as actuators. Thick film piezoelectric layers have been screen printed onto thin silicon diaphragms in order to form the basis of a micropump [30.11].

Photovoltaic devices convert incident optical radiation into electric current and are often termed *solar cells*. They are used to power devices such as calculators, clocks, pumps and lighting. In general terms, the output power level is proportional to the physical size of the photovoltaic cell. The device is essentially a heterojunction between n-type and p-type semiconductors. Thick film solar cells have been made comprising CdS (n-type) and CdTe (p-type) as the junction materials [30.18]. Such thick film actuators have been shown to have relatively low conversion efficiencies (between 1% and 10%).

References

30.1 R. A. Rikoski: *Hybrid Microelectronic Circuits: The Thick-Film* (Wiley, New York 1973)
30.2 M. A. Topfer: *Thick-Film Microelectronics: Fabrication, Design and Fabrication* (Van Nostrand-Reinhold, New York 1971)
30.3 P. J. Holmes, R. G. Loasby: *Handbook of Thick Film Technology* (Electrochemical Publ., Ayr 1976)
30.4 M. Prudenziati, A. Rizzi, P. Davioli, A. Mattei: Nuovo Cim. **3**, 697–710 (1983)
30.5 J. E. Brignell: *Thick-Film Sensors*, ed. by M. Prudenziati (Elsevier, Amsterdam 1994)
30.6 J. E. Brignell, N. M. White, A. W. J. Cranny: Sensor applications of thick-film technology, IEE Proceedings Part I, Solid State and Electron Devices **135**(4), 77–84 (1988)
30.7 N. M. White, J. D. Turner: Meas. Sci. Technol. **8**, 1–20 (1997)
30.8 C. Canali, D. Malavisi, B. Morten, M. Prudenziati: J. Appl. Phys. **51**, 3282–3286 (1980)
30.9 H. Baudry: Screen printing piezoelectric devices, 6th European Microelectronics, 456–463 (Bournemouth, UK, 1987)
30.10 M. Prudenziati, B. Morten, G. De Cicco: Microelectron. Int. **38**, 5–11 (1995)
30.11 M. Koch, N. Harris, A. G. R. Evans, N. M. White, A. Brunnschweiler: Sensors Actuat. A **70**(1–2), 98–103 (1998)
30.12 N. M. White, V. T. K. Ko: Electron. Lett. **29**, 1807–1808 (1993)
30.13 S. P. Beeby, N. M. White: Sensors Actuat. A **88**, 189–197 (2001)
30.14 J. N. Ross: Meas. Sci. Technol. **6**, 405–409 (1995)
30.15 M. Prudenziati, B. Morten: Microelectron. J. **23**, 133–141 (1992)
30.16 G. Martinelli, M. C. Carotta: Sensors Actuat. B **23**, 157–161 (1995)
30.17 B. Morten, M. Prudenziati, F. Sirotti, G. De Cicco, A. Alberigi-Quaranta, L. Olumekor: J. Mater. Sci. Mater. El. **1**, 118–122 (1990)
30.18 N. Nakayama, H. Matsumoto, A. Nakano, S. Ikegami, H. Uda, T. Yamashita: Jpn. J. Appl. Phys. **19**, 703–712 (1980)